ASTROBIOLOGY
An Evolutionary Approach

T0303869

ASTROBIOLOGY
An Evolutionary Approach

Edited by

Vera M. Kolb

University of Wisconsin-Parkside
Kenosha, Wisconsin, USA

CRC Press
Taylor & Francis Group
Boca Raton London New York

CRC Press is an imprint of the
Taylor & Francis Group, an **informa** business

Cover Image credits: Front cover main: courtesy of NASA/JPL-Caltech/MSSS/JHU-APL/Brown Univ.

Top front cover images, from left to right: 1. modified from Wacey, D., In situ morphologic, elemental and isotopic analysis of Archean life, in: Dilek, Y. and Furnes, H., eds., Evolution of Archean Crust and Early Life, Modern Approaches in Solid Earth Sciences, Vol. 7, Springer, Dordrecht, the Netherlands, pp. 351–365, 2013. 2. modified from Wacey, D. et al., Proc. Natl. Acad. Sci. USA, 110, 8020, 2013b. 3. modified from Wacey, D., Astrobiology, 10, 381, 2010. 4. Author supplied.

CRC Press
Taylor & Francis Group
6000 Broken Sound Parkway NW, Suite 300
Boca Raton, FL 33487-2742

First issued in paperback 2019

© 2015 by Taylor & Francis Group, LLC
CRC Press is an imprint of Taylor & Francis Group, an Informa business

No claim to original U.S. Government works

ISBN-13: 978-1-4665-8461-7 (pbk)

Visit the Taylor & Francis Web site at
http://www.taylorandfrancis.com

and the CRC Press Web site at
http://www.crcpress.com

This book is dedicated to all my teachers who have quenched my thirst for knowledge, taught me how to ask questions, and inspired me to learn more.

Contents

Preface

THIS BOOK, *ASTROBIOLOGY: AN EVOLUTIONARY APPROACH*, provides a slice through time, an angle by which we may look at the past and imagine the future of this exciting field. The contributors tell the stories of different aspects of astrobiology, reflecting the exciting journeys of their own research. Some authors are established scientists; some are at the beginning of their careers. Yet all of them breathe the urgency of this field, the excitement and drive to formulate and begin to answer the big questions. As the field evolves more rapidly than ever, the boundaries between the past, the present, and the future are blurred.

We hope to inspire students at all levels to learn about and explore the endless possibilities in astrobiology.

Vera M. Kolb

Acknowledgments

I thank the authors for contributing their chapters and the publisher for providing guidance and help, which made this book possible.

Vera M. Kolb

Editor

Vera M. Kolb earned her BS in chemical engineering and MS in organic chemistry from Belgrade University. In 1976, she earned her PhD in organic chemistry from Southern Illinois University at Carbondale. She currently serves as a chemistry professor at the University of Wisconsin–Parkside, where she has been working since 1985. She received her training in exobiology during her first sabbatical from 1992 to 1994 at the NSCORT (NASA Specialized Center of Research and Training) in San Diego, where she has worked with Leslie Orgel from the Salk Institute and Stanley Miller from the University of California–San Diego. She has worked in the field of astrobiology ever since. She has also studied sugar organosilicates and their astrobiological importance with Joseph B. Lambert at Northwestern University from 2002 to 2003 during her second sabbatical. In 2002, she was inducted into the Southeastern Wisconsin Educators' Hall of Fame. She was also named Wisconsin Teaching Scholar, 2007–2008. In 2013, she received the Phi Delta Kappa Outstanding Educator Award. She is also a recipient of numerous grants from Wisconsin Space Consortium/NASA. Currently, she has over 150 publications in organic chemistry, medicinal chemistry, and astrobiology.

Contributors

Jaana K.H. Bamford
Centre of Excellence in Biological
   Interactions
Department of Biological and
   Environmental Science
University of Jyväskylä
Jyväskylä, Finland

Sonny Clary
Nevada-Meteorites & Science
Las Vegas, Nevada

Henderson James (Jim) Cleaves
Earth-Life Science Institute
Tokyo Institute of Technology
Tokyo, Japan

and

Institute for Advanced Study
Princeton, New Jersey

Alfonso F. Davila
Space Science Division
Ames Research Center
National Aeronautics and Space
   Administration
Moffett Field, California

and

Search for Extraterrestrial Intelligence
   Institute
Mountain View, California

Daniel C. Dewey
Department of Chemistry
Pennsylvania State University
University Park, Pennsylvania

Alberto G. Fairén
Department of Astronomy
Cornell University
Ithaca, New York

Erica A. Frankel
Department of Chemistry
Pennsylvania State University
University Park, Pennsylvania

Zelimir Gabelica
ENSCMu
Université de Haute Alsace
Mulhouse, France

Istvan Gebefügi
Analytical BioGeoChemistry
Helmholtz-Zentrum Muenchen
Neuherberg, Germany

Aaron David Goldman
Department of Biology
Oberlin College
Oberlin, Ohio

Régis Gougeon
Institut Universitaire de la Vigne et du Vin,
   Jules Guyot
Université de Bourgogne
Dijon, France

Senthil Andavan Gurusamy-Thangavelu
Polymer Division
CSIR–CLRI
Chennai, India

Mourad Harir
Analytical BioGeoChemistry
Helmholtz-Zentrum Muenchen
Neuherberg, Germany

Norbert Hertkorn
Analytical BioGeoChemistry
Helmholtz-Zentrum Muenchen
Neuherberg, Germany

Gerda Horneck
Radiation Biology
Institute of Aerospace Medicine
DLR German Aerospace Center
Koeln, Germany

Louis N. Irwin
University of Texas at El Paso
El Paso, Texas

Matti Jalasvuori
Centre of Excellence in Biological
   Interactions
Department of Biological and
   Environmental Science
University of Jyväskylä
Jyväskylä, Finland

Basem Kanawati
Analytical BioGeoChemistry
Helmholtz-Zentrum Muenchen
Neuherberg, Germany

Christine D. Keating
Department of Chemistry
Pennsylvania State University
University Park, Pennsylvania

Vera M. Kolb
Department of Chemistry
University of Wisconsin-Parkside
Kenosha, Wisconsin

Joseph B. Lambert
Department of Chemistry
Trinity University
San Antonio, Texas

Jesús Martínez-Frías
Geosciences Institute IGEO (CSIC-UCM)
Faculty of Geological Sciences
Madrid, Spain

Gene D. McDonald
Biology Instructional Office
University of Texas at Austin
Austin, Texas

Christopher P. McKay
Space Science Division
Ames Research Center
National Aeronautics and Space
   Administration
Moffett Field, California

Ralf Moeller
Radiation Biology
Institute of Aerospace Medicine
DLR German Aerospace Center
Koeln, Germany

Franco Moritz
Analytical BioGeoChemistry
Helmholtz-Zentrum Muenchen
Neuherberg, Germany

Matthew Pasek
School of Geosciences
University of South Florida
Tampa, Florida

Radu Popa
Department of Biological Sciences
University of Southern California
Los Angeles, California

Ken Rice
Institute for Astronomy
School of Physics and Astronomy
The University of Edinburgh
Edinburgh, United Kingdom

Philippe Schmitt-Kopplin
Analytical BioGeoChemistry
Helmholtz-Zentrum Muenchen
Neuherberg, Germany

and

Analytical Food Chemistry
Technische Universität München
Munich, Germany

Dirk Schulze-Makuch
Washington State University
Pullman, Washington

and

Technical University Berlin
Berlin, Germany

Timothy F. Slater
CAPER Center for Astronomy & Physics
    Education Research
and
University of Wyoming
Laramie, Wyoming

Henry J. Sun
Desert Research Institute
Las Vegas, Nevada

Luis P. Villarreal
Department of Molecular Biology and
    Biochemistry
University of California
Irvine, California

David Wacey
Centre for Microscopy, Characterisation
    and Analysis
Australian Research Council Centre of
    Excellence for Core to Crust Fluid
    Systems
The University of Western Australia
Perth, Australia

Sara Imari Walker
School of Earth and Space Exploration
and
Beyond Center for Fundamental Concepts
    in Science
Arizona State University
Tempe, Arizona

and

Blue Marble Space Institute of Science
Seattle, Washington

Guenther Witzany
Telos-Philosophische Praxis
Buermoos, Austria

Introduction

THIS BOOK IS AN ASTROBIOLOGY TEXTBOOK for fourth-year students in sciences, beginning graduate students, and astrobiologists in general at any level. It has 21 chapters authored by 37 contributors from 9 different countries. The contributors have a broad range of specialties within the interdisciplinary field of astrobiology. These specialties are in astrobiology education, astronomy, biological sciences, chemistry (analytical chemistry, biochemistry, biogeochemistry, cosmochemistry, organic chemistry, silicon chemistry, phosphorus chemistry), desert research, earth sciences, environmental sciences, fossils, fundamental concepts of science, geology, life sciences, meteorites, microbiology, microscopy, molecular biology, philosophy, physics, space sciences, and virology among others.

The title of the book, *Astrobiology: An Evolutionary Approach*, reflects the strong evolutionary component in astrobiology. Evolution of matter, stars, chemical evolution, prebiotic evolution, and biotic evolution at all levels are all critical parts of astrobiology. The origins of life and the possibility of life elsewhere are also a subject of philosophical examination. These also evolve with time as our understanding of life itself and the laws of chemical and biological evolution evolve.

Astrobiology is not a mature science. While there are defined laws and satisfying explanations for many aspects, there is still a lot of ground to be covered. Astrobiology is a new field where the future lies. It is our hope that students will be inspired by this book to explore and advance many different fields of astrobiology in the future.

The chapters typically have a glossary of terms, review questions, and recommended references. Some chapters have primers either as separate units or incorporated into the text. This was done as needed for select topics. We hope that students and other readers will find these useful.

Chapter 1 by Aaron David Goldman provides an overview of astrobiology. It provides a good background for the further reading of the book. The chapter is pedagogically well-suited to be the introductory chapter.

In Chapter 2, Ken Rice covers origins of elements and the formation of the solar system, planets, and exoplanets. The chapter explains difficult concepts in a clear and engaging manner.

Chapter 3 by Timothy F. Slater addresses astrobiology education and public outreach. It reflects significant research in this area and provides guidance to astrobiologists on ways to successfully share their enthusiasm and knowledge with students and the general public.

In Chapter 4, Philippe Schmitt-Kopplin and coworkers update us on the analysis of organics on the Murchison meteorite. The importance of knowing which chemicals are found on meteorites cannot be overstated. Meteorites bring chemicals to Earth from the extraterrestrial media. These chemicals might have been involved in the original prebiotic chemistry and the evolution of life on Earth and elsewhere.

In Chapter 5, Jim Cleaves provides a critical analysis of the prebiotic syntheses of biochemical precursors. The chapter clearly describes important prebiotic reactions on Earth in an environmental context, such as atmospheric syntheses and syntheses in hydrothermal vents. It covers the synthesis of organic compounds by classes, such as lipids, amino acids, nucleic acids, etc. It also addresses chemicals on meteorites and much more.

In Chapter 6, Gene D. McDonald covers biochemical pathways as models for prebiotic syntheses. No reader will get lost in this exciting journey, since an excellent primer is offered first. Among many critically important topics that are covered, we select here LUCA, last universal common ancestor, and the minimum requirement for a metabolic system.

Much has been said and discussed about silicon-based life. Is it possible? The answer comes from Joseph B. Lambert and Senthil Andavan Gurusamy-Thangavelu in Chapter 7, which discusses the role of silicon in life on Earth and elsewhere. The focus is on the potential of silicon to make chemical bonds and its capacity to produce chemical diversity as compared to carbon.

Chapter 8 by David Wacey covers the fossil records of early life on Earth and describes the latest developments in the field. It does help that the author is from Australia, which has wonderful sites full of such early fossils.

Chapter 9 by Vera M. Kolb (the editor) provides insights in the prebiotic organic reactions in water, which occur even when organic materials do not dissolve in water. She also covers prebiotic organic chemistry in the solid state, which is applicable to the chemistry on asteroids, for example.

Chapter 10 by Christine D. Keating and coworkers addresses the new and exciting work on the encapsulation of organic materials in protocells. Various aspects of early compartmentalization are considered, both from a theoretical and practical point of view. Vesicles, micelles, membranes, coacervates, and other systems are examples. This chapter has a built-in primer.

Matthew Pasek addresses the role of phosphorus in prebiotic chemistry in Chapter 11. Phosphorus is critical for life, but its availability in the proper chemical form for incorporation into biotic chemical systems is still a puzzle. Pasek updates us on the latest discoveries in this important field.

The cold and dry limits of life are covered by Christopher P. McKay and coworkers in Chapter 12. It is a wonder how life survives and thrives under the harsh conditions found on Earth. Studies by McKay and coworkers open the door for the evaluation of the possibility of life on Mars and other extraterrestrial harsh environments. The chapter also outlines the questions that need to be addressed in the future.

Gerda Horneck and Ralf Moeller address microorganisms in space in Chapter 13. Some microorganisms survive in space, which has important implications for transport of life through space. In this comprehensive chapter, many critical issues are covered, such as the likelihood of panspermia and much more.

In Chapter 14, Jesús Martínez-Frías describes search for life on Mars through an astro-geological approach. The chapter is comprehensive and covers planetary geology, the relevant sections of NASA's Astrobiology Roadmap, Mars meteorites, Mars missions, and Mars analogs among other important topics.

Radu Popa provides a detailed survey of the main ideas on the elusive definition of life in Chapter 15. In a surprising twist, he delves deep into the origins of life and the RNA world and makes an important contribution in this area.

Chapter 16 is written by a philosopher, Guenther Witzany. He addresses language and communication as universal requirements for life. Witzany covers the ground about communication and life for various life forms at all levels of complexity. Much is to be learned from this chapter, since these topics are not sufficiently addressed in traditional astrobiology.

The key question of astrobiology is transition from abiotic to biotic, and, naturally, we would like to know if there is an algorithm for it. In Chapter 17, Sara Imari Walker handles this question expertly. Her chapter goes above and beyond verbal descriptions. She has provided mathematical explanations, but does not leave the readers in the dark. For those who are somewhat rusty in math, she has incorporated a math primer at strategic places across the chapter.

In Chapter 18, Dirk Schulze-Makuch and coworkers address the search for extraterrestrial life and what we are looking for in such a search. They describe the challenges in finding extraterrestrial life and examine the possibility of life in various extraterrestrial environments (e.g., Venus, Mars, Titan). They also discuss geoindicators for life and various biosignatures among other topics.

Matti Jalasvuori and Jaana K.H. Bamford cover the evolution of viruses and their astrobiological significance in Chapter 19. This is a clearly written chapter, which is also a primer on viruses. The question that astrobiologists ask most often about viruses is whether or not they are alive and whether they fit into the definition of life, which requires self-replication. Scientists are only recently becoming aware that viruses are ancient. Viruses are well connected to the origins of life and to the evolution of life in general, via participation in host evolution.

Chapters 20 and 21 have been written by Luis P. Villarreal, who is a virologist and a molecular biologist. These chapters combine advanced material with complex ideas. Chapter 20 explores how viruses are connected to the origins of life, especially the cooperation element in the so-called quasispecies consortia. Viruses interact with the host in a manner that has evolutionary consequences. In Chapter 21, Villarreal explores how *virolution* can help us understand recent human evolution. He also provides two primers in Chapter 21 on the topics that require more extensive knowledge in molecular biology and virology.

Finally, the reader might notice some repetition of topics in different chapters (Drake's equation, quasispecies consortia, Eigen's quasispecies, etc.). This has been done on purpose, as revisiting topics is helpful to students. Moreover, these topics have been covered in different astrobiological contexts. It is therefore important for students to experience these repetitions so as to make the concepts clear, which is critical for the interdisciplinary field of astrobiology.

Vera M. Kolb

Astrobiology
An Overview

Aaron David Goldman

1.1 EXTRATERRESTRIALS FROM THE SCIENTIFIC REVOLUTION TO THE SPACE AGE

Medieval European scholars spent little time thinking about whether and where extraterrestrial life may exist. The Aristotelian cosmology adopted by most medieval scholars held that the Sun, Moon, planets, and stars all comprised perfect spheres orbiting the Earth in concentric circles. Because the celestial bodies were thought to be perfect spheres without geologically complex terrains, their habitation was rarely considered. In the mid-sixteenth century, Nicolaus Copernicus's mathematical model of the Earth and the five other known planets orbiting the Sun led to an enormous shift in the broader understanding of the Earth's place in the cosmos.

In addition to this scientific discovery, an idea later called the Copernican principle swept across the scholarly world, which maintained that the Earth is not a special part of the universe. A consequence of this perspective is that earthlike environments should also exist on other planets and that they should be inhabited by extraterrestrial life. This opinion was later captured by Bernard le Bovier de Fontenelle's 1686 publication, *Conversations on the Plurality of Worlds*, which explained the work of Copernicus in clear language intended for a broad public audience and included discussions on the nature of extraterrestrials.

By the mid-eighteenth century, nearly half of scholars associated with the enlightenment (the defining scholarly movement of the time) had written about extraterrestrial life. In the latter half of the century, astronomers began to expand the Copernican principle by seriously considering other stars to be suns with planets inhabited by organisms orbiting around them. The principal stellar astronomers of the time, Thomas Wright, Immanuel Kant, Johann Lambert, and William Herschel, assumed that planets, moons, and even comets were likely to be inhabited.

The nineteenth century began with an overwhelming majority of scientists, philosophers, and academic theologians sharing a common belief in extraterrestrial life. What can now be seen as wild speculation was at the time considered to be a sound assumption based on a growing scientific understanding of the cosmos. The overwhelming differences in gravity, light, and heat on the different celestial bodies of the solar system were often disregarded. However, as the power of telescopes increased and the advent of spectroscopy allowed astronomers to measure the chemistry of nearby planetary atmospheres, the idea that these bodies were inhabited, at least by animal-like organisms, steadily decreased in popularity.

Mars was the last holdout for this kind of extraterrestrial life beyond Earth. Between the late nineteenth century and the early twentieth century, some astronomers argued that markings on the Martian surface that would later be identified as optical illusions or streaks of dust entrained by wind were in fact canals manufactured by a population of intelligent organisms (Figure 1.1). Though this interpretation was out of favor by the 1920s, a seasonal darkening pattern around the Martian polar region was still being attributed to simple plantlike life. This interpretation was ultimately dispelled in 1964 by the Mariner 4 spacecraft, which provided the first images of the Martian surface taken from close range.

The Mariner spacecrafts are one example of how the space age beginning in the late 1950s brought with it the ability to directly image and examine planetary and lunar surfaces within the solar system, thereby discrediting claims of easily recognizable, multicellular, extraterrestrial life forms. But, at the same time, subsequent missions to Mars and the outer planets produced a detailed picture of those bodies and their moons and, especially in the case of Mars, a partial understanding of its early history. In several extraterrestrial locales that will be discussed later on in this chapter, planetary scientists found

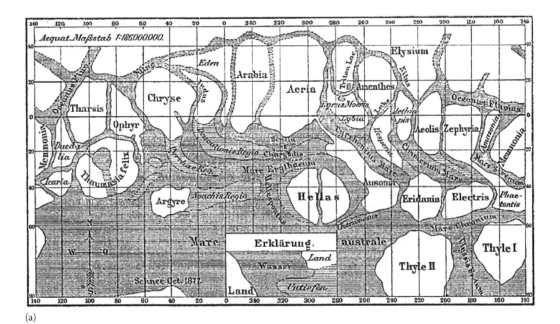

(a)

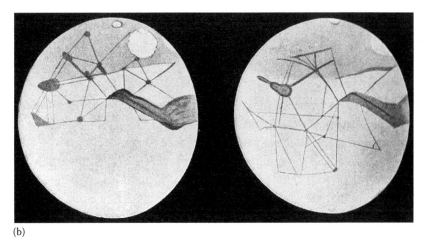

(b)

FIGURE 1.1   Martian maps of the late nineteenth century. A map by Giovanni Schiaparelli (a) was published in 1888 and first described a series of interconnected lines on the Martian surface. He referred to these lines as *canali*, which can be translated into English as *channels* or *canals*, the latter implying that they resulted from engineering by intelligent Martians. Many astronomers at the time believed that these lines were in fact a sign of intelligent life. Most notable was Percival Lowell, who first published his own maps of Martian canals in 1895 (b) and argued that the canals were dug by intelligent Martians in order to bring water from the melting polar ice caps to the arid equatorial region. Later comparisons of these maps to images of the same region demonstrate that the canals were optical illusions. (These images are from Wikimedia commons and are in public domain.)

environments not too dissimilar from inhabited environments found on Earth. Where once the existence of extraterrestrials was based on speculative assumptions, it could now be studied through careful consideration of chemical and biological evidence.

1.2 DRAKE EQUATION AND THE ANTHROPIC PRINCIPLE

Though hypotheses of intelligent extraterrestrials in our solar system were effectively the advancement of technology, the same technology made contact with intelligent extraterrestrials from other, distant solar systems possible. This was the thinking of Giuseppe Cocconi and Philip Morrison, who in 1959 published an article entitled "Searching for Interstellar Communications." In it, they wrote that interstellar communication is possible by means of electromagnetic waves and went on to suggest certain frequency ranges and targets for detection. One year later, Frank Drake, who had independently arrived at the same idea, conducted the first search for extraterrestrial communications. This test is seen as the birth of the search for extraterrestrial intelligence (SETI).

One year later, Drake, in conjunction with the National Academy of Sciences, organized a meeting to discuss the new search for intelligent life. In preparation for the meeting, he wrote down all of the factors that needed to be known or at least estimated in order to conduct a careful and informed search. Drake realized that writing the list as a series of probabilities multiplied together would equal the number of civilizations with which communication is possible. This formulation became known as the *Drake equation*:

$$N = R_* \cdot f_p \cdot n_e \cdot f_l \cdot f_i \cdot f_c \cdot L \tag{1.1}$$

where

N is the number of civilizations in our galaxy with which communication is possible
R_* is the average rate of star formation in the galaxy
f_p is the fraction of stars that have planets
n_e is (if a star has planets) the number of those planets that can potentially support life
f_l is the fraction of planets able to support life that actually go on to develop it
f_i is the fraction of planets where life exists that have at least one intelligent species
f_c is the fraction of planets with an intelligent species where that species has developed detectable technology
L is the length of time for which such civilizations produce detectable signals

Though some SETI researchers use the Drake equation to hone their search, the broad utility of the Drake equation is that it breaks the problem of extraterrestrial life into specific questions that can be addressed within and between real scientific disciplines. The majority of astrobiologists, today, are concerned with finding any extraterrestrial, be it microbial, multicellular, or intelligent. In this case, the Drake equation without the last three terms still provides a useful guide for thinking about extraterrestrial life through questions that can be addressed by science.

R_*, the rate of star formation in the galaxy, and f_p, the fraction of stars that have planets, ask clear astronomical questions. When the Drake equation was first proposed, there was very little evidence one could use to estimate R_* and f_p. Today, strong evidence suggests that R_*, the rate of star formation in the galaxy, is about seven stars per year. More amazing is the recent revelation that f_p is very high, averaging one planet per star in the Milky Way galaxy. This recent discovery will be discussed later in the chapter.

As the equation moves on, its terms increasingly require input from disciplines other than astronomy. Estimating n_e requires astronomical observation, but knowing how to interpret those observations depends on our understanding of the limits of life. The term f_l requires an understanding of how life can emerge from nonliving chemistry, a question that, itself, requires contributions from chemistry, geology, and molecular biology. The term f_i is dependent on the evolution of multicellularity and the subsequent evolution of at least one intelligent species. The final two terms, f_c and L, fall within the purview of sociology, if they can be answered at all.

At first glance, it is tempting to give f_l and f_i and f_c values of 100% because we find that life on Earth was able to develop and did so, life was able to evolve an intelligent species and did so, and that this intelligent species was able to develop communication technology and did so. This argument about the seeming inevitability of intelligent life can be explained by the anthropic principle, which states that we see a world perfectly arranged for our own existence because we would not be able to observe any other. If the world were not perfectly arranged for our own existence, we would not be present to observe it. In other words, the values of f_l, f_i, and f_c have to be 100% for Earth because otherwise, there would be no one around to discuss the Drake equation.

This does not mean that the values of f_l, f_i, and f_c should be 100% throughout the galaxy. In fact, many evolutionary biologists and psychologists now think that the value of f_i is probably just above 0%. This conclusion is based not on the mere fact of our existence, but a detailed understanding of the evolutionary steps leading to animal life and the subsequent evolution of one species intelligent enough to develop science and technology. Thus, any consideration of life beyond Earth should be based on our best understanding of how life emerges and evolves.

1.3 ESSENTIAL FEATURES OF LIFE

It is unclear exactly how different an extraterrestrial organism would be from life on Earth. But our understanding of what constitutes a habitable environment, what primitive organisms on Earth may have been like, and ultimately how we can detect extraterrestrial life depends on what life actually is and what properties of life on Earth we should expect to see in extraterrestrial life. Does life have to evolve by natural selection? Does life have to be based on carbon chemistry? Does life have to be made of cells? Does life have to be made of matter? The problem of defining life is one that will be addressed in greater detail in the following chapters. For now, we will simply survey general properties of life on Earth with an eye toward those features that seem likely to be universal properties of life.

1.3.1 Reproduction and Heredity

Organisms on Earth are able to reproduce themselves. Most organisms do this by direct replication through cell division. The components of the cell are copied and separated into two identical daughter cells. Other organisms reproduce sexually, dividing their genetic material in half and combining it with a mate to produce a new organism that represents a hybrid of the two. Either way, reproduction appears to be a fundamental feature of life, given that all life on Earth does it. Furthermore, it is difficult to imagine what nonreproductive life would even be.

One requirement of reproduction is that the information that produces and directs the organism must be stably passed on to subsequent generations. For life on Earth, the organisms, their chemistry, and their behavior are mostly produced by the functions of proteins, the primary functional molecule of cells. Proteins are composed of a set of 20 amino acids that fold into 3D structures. These structures can interact with other proteins or nonprotein components of the cell, catalyze chemical reactions including synthesis of cellular components, receive and transmit signals, give their cells shape and internal organization, and facilitate a cell's interaction with the environment.

The chief heritable material of life is deoxyribonucleic acid (DNA), which contains the instructions to synthesize proteins, along with the timing and magnitude of their synthesis. Individual genes encode proteins as sequences of four nucleotides that are read in sets of three and converted into a sequence of amino acids corresponding to the protein. This conversion from DNA to protein is performed by first transcribing the sequence to a related molecule, ribonucleic acid (RNA), and then converting the RNA to protein through a genetic code.

Many features of this genetic system are probably a result of historical chance during the process of evolution and are unlikely to be shared by all life throughout the universe. As we will see later, some ancient forms of life most likely did not use this three-component genetic system of DNA, RNA, and proteins. Still, the storage of genetic material in genomes and the translation of a specialized information molecule like DNA into a specialized functional molecule like protein are useful strategies for optimal reproduction that we might expect to find in an extraterrestrial life form.

1.3.2 Evolution

Evolution is used in other sciences to refer to different sorts of long-term change. Astronomers use *stellar evolution* to describe the predictable changes that a star undergoes between the onset and end of the nuclear fusion that powers it. Geologists use the term evolution, as in *crustal evolution* or *isotopic evolution*, to signify a slow process of change in composition or character of rocks, landforms, or continental plates. Biological evolution is very different from these concepts. It is not merely change over long periods of time, but change under some selection pressure that promotes traits that are best adapted to an organism's ability to reproduce.

In order for biological evolution to occur, (1) a population of organisms must reproduce, (2) the form of reproduction must produce a variety of traits that can be inherited by a subsequent generation, and (3) a mixture of selection pressures must cause some organisms to contribute genetically to the next generation more than others due to advantages from their specific combination of traits. Over long time spans, these small changes produced by selection pressures can accumulate to yield drastically different species of organisms with dramatically different traits.

Charles Darwin discovered the most important mechanism of evolution, natural selection, in which some members of a species are more successful than others because they are, for example, more likely to survive or gain access to greater resources. Darwin later discovered a second form of selection, sexual selection, in which sexually reproducing organisms may contribute more or less to the subsequent generation because they are better at attracting a mate. The genetic and genomic mechanisms of evolution are now understood in great detail, and evolution, itself, can be seen not only at the level of organisms but at smaller levels like individual genes and, some believe, at larger levels like groups of organisms.

Indeed, anything will evolve as long as it replicates, produces variation, and is subject to some form of selection. Cultural ideas, or *memes*, evolve in a manner similar to biological evolution, and many computer programs now take advantage of the evolutionary process in order to optimize their parameters. It is hard to see how a process of evolution like that observed in life on Earth would not occur in other forms of life, even if the form of reproduction and genetic transmission was very different.

1.3.3 Cellularity

Cells are the basic unit of organismal life. Every organism is made of one or more cells, which are bounded by a membrane. Cell membranes are composed of *amphipathic* molecules, meaning that they have one end that is water soluble and another that is fat soluble. In water, these amphipathic compounds can form layers that are two molecules thick in which the fat-soluble ends of the compounds face each other in the interior of the membrane and the water-soluble ends of the compounds face either the cell interior or the external environment.

Cellular membranes are stable enough to act as a barrier between the interior of the cell and the extracellular environment, but are fluid enough to allow the organism to change shape, absorb other membrane material, and fill the membrane with channels, transporters, signal receptors, and other macromolecules that impart the sorts of functions associated with the barrier between an organism and the external environment. It is possible that other forms of life could exist without cellular membranes. Most models of the origin of life start with replicating systems of genetic macromolecules or chemical networks that are not organized into individual cells. But the cellular structure of organisms on Earth is ubiquitous and was possibly a prerequisite for life to evolve a diversity of species and spread across the biosphere.

1.3.4 Metabolism

Organisms maintain themselves through a network of controlled chemical reactions referred to collectively as metabolism. Metabolic reactions are linked so that the products of one reaction are the reactants of the next. The majority of these reactions are catalyzed by protein enzymes. Metabolism is responsible for extracting energy from the environment, often in the form of light, or extracting chemical potential energy stored in organic compounds produced by other organisms. Metabolism is also responsible for using that stored energy to produce the building blocks of the cell, such as sugars, starches, nucleic acids, proteins, lipids, and a host of other compounds.

It is not clear whether an organism could exist without some kind of metabolism under its own control. Much of what organisms do in the first place falls under the category of metabolism. Metabolism allows cells to grow and reproduce subsequent generations. Metabolism may also require the cellular structure of organisms to better concentrate precursor compounds, remove waste products from the organism's interior, and keep other organisms from appropriating the organism's metabolic products. As we shall see later, many of the detection methods for extraterrestrial life rely on the presence of metabolic waste products released into the environment.

1.4 ORIGIN(S) OF LIFE

By the time of the Drake equation, study of the origin of life was already an established scientific field. Like extraterrestrial life, early ideas about origins were overly optimistic in hindsight. The theory of spontaneous generation held that organisms regularly emerged from nonliving material given the right conditions, for example, worms emerging from mud or flies emerging from putrefying earth. The slow decline of the belief in spontaneous generation theories culminated with the development of cell theory in the mid-nineteenth century, which stated that all life is made of one or more cells and that cells only come from other cells.

This new understanding of life's most basic structural unit implied that origins of life were not regular occurrences. The scientific study of the origin of life was initiated by Alexander Oparin in 1936, who proposed that the methane- and ammonia-rich atmosphere of the early Earth could have produced larger organic compounds like those present in life, today. This hypothesis was experimentally tested and validated 17 years later by Stanley Miller and Harold Urey, who electrically shocked clouds of ammonia and methane gas and were able to synthesize amino acids and other organic compounds essential for life. The Miller–Urey study marked the beginning of experimental research on the origin of life, a field that has now grown to include a broad range of methodologies and disciplines.

1.4.1 Prebiotic Chemistry

The Miller–Urey experiment also represents the first study in a branch of research that endeavors to understand how the chemistry of life can be produced through nonbiological processes. Soon after Miller and Urey demonstrated the synthesis of amino acids, the building blocks of proteins, Juan Oró demonstrated that nucleobases, the genetic component of DNA and RNA, could be generated by methods of synthetic organic chemistry. Other laboratory conditions meant to mimic natural environments are now known to produce polymers of amino acids and nucleotides mimicking the proteins and nucleic acids that are responsible for molecular function and genetic inheritance in life, today.

This sort of experimental prebiotic chemistry is bolstered by studying the organic chemistry of carbon-rich meteors. Amino acids and nucleobases are found in such meteors, indicating that these compounds not only are generated without biological enzymes in a laboratory but are generated by natural processes in the solar system. The lipids found in these same meteorites can spontaneously form bilayer membranes like those of cells. These two parallel lines of research have demonstrated the ability of natural processes to yield large organic compounds and the importance of identifying real geochemical scenarios that would facilitate many of these disparate reactions at once.

A number of possible locations for the origin of life have been proposed, each with its own set of favorable traits. Iron–sulfur mineral surfaces are a popular potential setting for origins because of their ability to catalyze oxidation/reduction reactions. These minerals are also found in hydrothermal systems that could provide an energy source while their porous structure could serve to concentrate compounds. The presence of iron–sulfur clusters in the active sites of many enzymes involved in oxidation and reduction reactions supports the potential of an iron–sulfur setting for prebiotic chemistry.

Other surface environments have also been proposed as settings for the origin of life. For example, certain types of clay have a number of properties beneficial to prebiotic chemistry. Clay layers can serve to concentrate organic compounds and their surfaces can catalyze the formation of protein and nucleic acid polymers. Water ice has also been proposed as a setting for the origin of life. The freezing of ice has the ability to concentrate organic compounds, and the surface of ice crystals can stabilize RNA and promote its catalytic functions. The same organic compounds that formed through electrical discharges in the Miller–Urey experiment can also form in cold environments over longer periods of time.

Hydrothermal vents rich in iron–sulfur minerals, ordered layers of clay, and cold liquid inclusions within ice crystals represent only a fraction of proposed settings for the origin of life. It is not clear that there is any way to distinguish which of these scenarios, if any, is the most similar to the true setting of the origin of life on Earth. But for the purposes of astrobiology, it is more important that we now know that it is possible to generate complex organic compounds without enzymes and that there is a sense of which compounds are more likely to have been present prior to life and which were likely not present. As we look for habitable planets beyond Earth, we may also ask the question of whether settings capable of prebiotic chemistry are present on these planets.

1.4.2 Ancient Life

Evolutionary conservation and fossilization have preserved evidence of the character of early life and the major transitions that took place in life's early development. As we look deeper into history, however, the evidence grows thinner. Evolutionary biology will probably never give us a detailed picture of life's origin, but the historical evidence provided by the preservation of ancient gene families can suggest a context for prebiotic chemistry and geological settings for the origin of life and helps us understand how the fundamental features of life, such as the genome, the genetic code, and metabolism, first took shape.

The earliest broadly accepted fossil evidence for life on Earth dates to around 3.5 billion years ago, whereas controversial evidence of life reaches back to 3.8 billion years ago. Some of the 3.5 billion-year-old fossils resemble structures similar to those of modern bacteria. These fossils often contain carbon with an isotopic signature, the ratio of heavy atoms to light atoms that is similar to life. The biological origin of these fossils has not, yet, been confirmed, but it is generally accepted that life was present on Earth at least as late as 3.5 billion years ago.

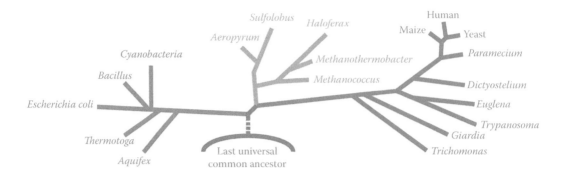

FIGURE 1.2   Evolutionary tree of life on Earth. The tree coalesces into a single root commonly referred to as the LUCA and subsequently branches early into three major taxonomic domains, Bacteria, Archaea, and Eukarya. Modern evolutionary trees are usually generated by comparing one or more shared gene or protein sequences from each organism. (Adapted from Alberts, B. et al., *Essential Cell Biology*, 3rd edn., Garland Science, Taylor & Francis, New York, 2010.)

To go beyond the limits of geological preservation, ancient life can be examined from the perspective of gene and genome evolution. Comparing the sequences of related genes found in every species can be used to infer the evolutionary history of life. At the base of this tree is a single point of coalescence typically referred to as the last universal common ancestor (LUCA), which subsequently splits into two branches to produce the domain Bacteria and the ancestor of the domains Archaea and Eukarya (Figure 1.2). By surveying the gene families present in genomes across the tree of life, studies have deduced that hundreds of gene families were likely present in the genomes of these ancestors. Among these are genes related to protein synthesis and the genetic code, metabolism, and the maintenance of a cellular membrane. Not present on this list are the majority of genes required to synthesize DNA and its nucleotide components, suggesting that the DNA genome arose either around the time of the LUCA or soon after the split between Bacteria and Archaea/Eukarya.

In contrast to the machinery involved in synthesizing the DNA genome, the genes required for protein synthesis were numerous in the LUCA and remain highly conserved across the tree of life. Today, the genetic code that is used to direct the synthesis of proteins almost never differs between species, suggesting that it was well established in the LUCA. The presence of many gene families encoding large proteins in the LUCA further indicates that a capable system of protein synthesis by nucleic acid translation was in place.

Proteins play the chief functional role in almost every metabolic pathway. But protein synthesis, itself, relies on a number of functional RNA molecules that are central to the process. The genetic code is translated by transfer RNAs and the catalyst of protein elongation is located in one of the RNA components of the ribosome. The genes, themselves, are read off of messenger RNA rather than DNA. The discovery in the mid-1980s that RNA has catalytic capabilities as well as genetic functions led to the idea that the original genetic system may have been composed of heritable RNA molecules that could be copied like DNA, but could also perform catalytic functions like proteins. While the details and accuracy of this hypothesis are still being scrutinized, the so-called RNA world hypothesis remains the most widely accepted explanation for how our complex system composed of DNA, RNA, and proteins arose from a simpler system composed of only one type of macromolecule.

1.4.3 Unanswered Questions about the Origins of Life

A number of subjects related to origins and early evolution of life have been omitted from this account for the sake of brevity. How chemical systems self-organize into replicating lifelike entities has been under investigation for several decades. The role of membranes in the origin and early evolution of life is also an active area of research. We do not know how many times life may have originated and whether the LUCA truly represents a single lineage or a mixture of lineages that shared genes across organismal boundaries at a rate higher than that of today. We do not even know whether life originated on Earth or whether it arose elsewhere and was subsequently transported.

The volume of unanswered questions regarding the origin of life appears daunting. But the understanding of these various problems and the sophistication with which they are studied has steadily increased since the research was initiated theoretically by Alexander Oparin and experimentally by Stanley Miller. For the purpose of astrobiology, it is less important to know exactly how life originated on Earth and much more important to know the mechanisms underlying potential origins of life on any planet. Research into prebiotic chemistry, origin of life settings, and early evolution from simpler to more complex forms of life all inform the search for extraterrestrial forms of life that may have originated on other planets or moons.

1.5 LOOKING FOR LIFE BEYOND EARTH

The primary goal of astrobiology is finding life beyond Earth. So far, the exploration of our solar system has not presented any signs of extraterrestrial life. If there is life to be found elsewhere in the solar system or the galaxy, it will only be discovered with a sophisticated search that takes seriously the questions of which environments are capable of supporting life, where beyond Earth might we find those environments, and how we will detect the inhabitants of those environments with the limitations of robotic probes and remote observations. This effort will require a survey of the most extreme and unexpected conditions in which we find life on Earth, an intimate knowledge of the local conditions present on planets and moons within our solar system, a better understanding of planetary systems beyond our solar system, and well-designed methods for detecting the presence of life.

1.5.1 Environmental Limits of Life on Earth

The Earth has many environments that are too hot, too cold, too acidic, too basic, too salty, too high pressure, or otherwise too poisonous or too mutagenic for human survival. But the majority of such environments are now known to host ecosystems of microbial organisms and sometimes even animals and plants. Exploring the communities of the so-called extremophiles that live in these environments can further our understanding of the limits of earthlike life and aid in the search for extraterrestrials.

For example, one might assume that the temperature range of life would at most be limited by the temperature range of liquid water, 0°C–100°C. But liquid water can exist below 0°C if it contains dissolved salts and can exist above 100°C if the water is under high pressure. Organisms that live at extremely low-temperature environments, *psychrophilic*, may also require the high-salt conditions of the environment, *halophilic*, and organisms that live at extremely high temperatures, *thermophilic*, are also often tolerant of high pressures, *barophilic* or *piezophilic*.

Currently, the most heat-tolerant organism known is an archaean called *Methanopyrus kandleri* strain 116, which was isolated from hydrothermal vents and was shown to reproduce at 122°C under high pressure. The lowest growth temperature, −20°C, can be tolerated by a number of bacterial and archaeal species. Small amounts of unfrozen water in permafrosts or brine channels in ice create liquid water environments at these low temperatures. Similar environmental limits for life have been described for acidity (pH 0–13), pressure (102 MPa), radiation (60 Gray h⁻¹), and other parameters.

These specific environmental limits are due to features of biochemistry such as the temperature and pressure range of liquid water and the stability of the major macromolecules of life, especially membranes, proteins, and DNA. We have no way of knowing what the environmental limitations of extraterrestrial life would be if that life had an entirely different biochemistry. But given that any search for life has limitations on time and resources, it is sensible to limit these searches to extraterrestrial environments that mimic inhabited environments on Earth.

1.5.2 Extraterrestrial Habitable Environments

As mentioned earlier in the chapter, space exploration has identified several bodies within our solar system that are likely to host environments that are currently or formerly habitable. Landforms that resemble dry river beds and flood-carved channels are prevalent on the surface of Mars, and subsequent chemical analysis by the Mars Curiosity rover has established that at least some of these features were created by surface liquid water in the early history of Mars. Orbiting spacecrafts have provided further chemical evidence for bodies of water that once existed on the planet's surface.

The time span over which Mars was able to maintain a habitable surface environment in its early history is still being investigated. The Curiosity rover has also identified the presence of elements required for life (but no organic matter) as well as evidence of a chemical gradient on ancient Mars that could have provided energy for life. Today, the thin atmosphere on Mars is not capable of warming much beyond the freezing point of water and the average surface temperature is −55°C. But, it is still possible that life can inhabit subsurface aquifers that remain heated through geothermal activity.

Other locations that may host extraterrestrial life in our solar system are the icy moons known to orbit the large gaseous planets beyond the asteroid belt. Europa is the sixth closest moon of Jupiter and is slightly smaller than the Earth's Moon. It is covered by ice that is cracked and relatively free of craters, which is taken to mean that beneath this layer of ice is a vast liquid water ocean assumed to be much deeper than Earth's oceans. Further evidence for a liquid water ocean comes from Europa's magnetic field, which is affected by currents of salt-rich liquid (presumably water). Tidal forces created by the gravitational pull of Jupiter are thought to provide an internal heat source that may melt large pockets of surface ice above

FIGURE 1.3   Rich ecosystem surrounding a hydrothermal vent in the Juan de Fuca ridge in the northeast Pacific Ocean. The fluid emitted by the vent is 360°C and contains a high proportion of metal sulfide compounds. The exterior of the vent is covered by tube worms that contain symbiotic microorganisms that live off of the metal sulfides emitted by the vent. Other multicellular organisms found in these environments include clams and spider crabs. (http://oceanexplorer.noaa.gov/explorations/05lostcity/background/overview/media/fig2strawberry.html; Image courtesy of University of Washington.)

and may result in deep-sea vents like those seen on Earth. Such vents on Earth host ecosystems that are nearly energetically independent from the rest of the biosphere (Figure 1.3).

Enceladus, an icy moon that orbits Saturn, is too small to have any internal heating like the Earth, but it does appear to have a heated polar region that produces jets of water vapor, implying a liquid water ocean beneath. The water vapor plume also contains organic compounds, making a viable case for habitability or perhaps even habitation. Another moon of Saturn, Titan, presents a more peculiar case for life. Like the icy moons, Titan has a cold surface (−180°C). The water on that surface is frozen, but organic compounds like methane and ethane, which exist as gases on the surface of the Earth, are liquid on Titan and likely form rivers and lakes. There is a possibility that subsurface liquid water exists on Titan, but a case has also been made for life based in liquid methane/ethane rather than water. Such life would have a radically different biochemistry and would also have to contend with a very slow rate of chemical reactions due to the extremely low surface temperatures. Still, some have argued that if life is a consequence of chemistry, any chemistry, then we should find it on the surface of Titan.

1.5.3 Habitable Planets around Other Stars

Potentially habitable environments within our solar system are convenient in that they are mostly within reach of robotic landers and orbiters and perhaps, someday, human scientists. Mars has been visited by a steady stream of landers and orbiters, and most of the technology required for a probe to fly through the plume of Enceladus already exists. However, recent advances in astronomy have pushed the search for life beyond our solar system in ways that very few scientists imagined several decades ago. At the time of the Drake equation's formulation, astronomers had very little evidence that planets, and by association, moons, existed outside of the solar system. Now, strong evidence suggests that there are at least as many planets in the galaxy as there are stars and that some of these planets are similar to Earth in their size and the amount of energy they receive from their host star.

The first confirmed extrasolar planets, or *exoplanets*, were discovered in the mid-1990s using a technique that depends on the planet's tug on its star, which causes the star to move toward or away from the observer. Just like the pitch of a passing ambulance's siren goes from high to low as it approaches, passes, then recedes, so starlight goes from blue to red as an orbiting planet pulls its star toward and away from the observer. This technique is called the *radial velocity* method because it uses the change in velocity of the star to indirectly detect the planet. Though it has been successful in finding exoplanets, the technique suffers from a selection bias favoring the discovery of large planets close to their stars, because their tug will be the strongest. Other exoplanet detection methods like astrometry, which measures the movement of the star directly, and transit, which measures the dimming of a star by a passing planet, are similar in their preference for large planets close to their host star. Refinements of these techniques over the past two decades have removed much of this bias, allowing a more accurate understanding of the distribution of planets in our galaxy.

NASA's Kepler mission, which ran for 3.5 years from 2009 to 2013, was the most concerted effort to understand the distribution of exoplanets in the galaxy. The Kepler mission consisted primarily of an orbiting space telescope that stared at a small region constituting 0.25% of the total sky. The Kepler telescope measured light from stars in this field of view to detect exoplanets by transit. As of this writing, the Kepler mission has detected 3538 candidate exoplanets, over 1000 of which are likely to be rocky planets. Some of these planets are orbiting within their star's *habitable zone*, the orbital distance from a star in which liquid water is expected to be present on the surface (Figure 1.4).

The success of these exoplanet-detection methods has provided a sense of the abundance of planets that are in the galaxy in addition to the overall distribution of planet sizes, densities, and orbital distances. In some cases, if the planet is of similar size to the Earth and it orbits within the habitable zone of its star, its habitability can be asserted with greater confidence. The next generation of space telescopes may be capable of directly imaging putatively habitable exoplanets and measuring the chemistry of their atmospheres. As we will see, this sort of observation may provide evidence not only for habitability but also for habitation.

1.5.4 Detecting Extraterrestrial Life

Up until this point, we have discussed what life is, how it emerges from nonliving chemistry, and where else it might be in the solar system and the rest of the galaxy. But the question of how to accurately identify the presence of extraterrestrial life is just as formidable and integral to astrobiology as these other topics. Important considerations include how similar the extraterrestrial life is expected to be to life on Earth as well as whether the detection will be done by a probe that can perform experiments on a sample (which is usually the case for planets and moons in our solar system) or whether the detection must be based on remote observation (which is unavoidably the case for exoplanets). Forms of evidence called *biosignatures* have been developed and debated for just this purpose.

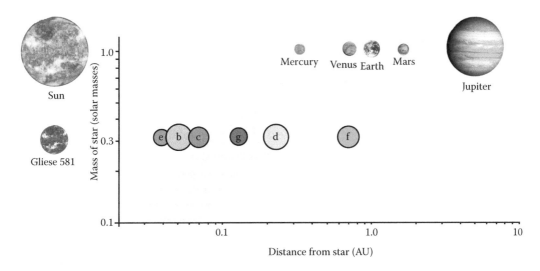

FIGURE 1.4   Circumstellar habitable zone (in blue) of our own solar system compared to the puta-tive planetary system of the star, Gliese 581. Gliese 581 is an M-class star meaning that it is smaller and cooler than our Sun. The habitable zone of such a star, the orbital radius in which liquid water is likely to be present on a rocky planet's surface, is much closer to Gliese 581 than that of our Sun. As of writing, three of the exoplanets orbiting Gliese 581, b, c, and e, are confirmed; d is likely, and f and g are controversial. Also as of writing, the current general estimates for stars hosting an earth-sized planet in its habitable zone range from 5% to 50%, based on the type of star and the definition of habitable zone. (Adapted from an original image produced by the European Southern Observatory; http://www.eso.org/public/images/eso0915b/.)

The first attempt by scientists to find microbial extraterrestrial life was a set of experiments on board the Viking 1 and Viking 2 spacecrafts, which landed on Mars in 1976 and 1978, respectively. One of these experiments radiolabeled nutrients similar to those produced by the Miller–Urey experiments and subjected them to Martian soil. The gas released from the soil was then analyzed for labeled CO_2, a sign that the nutrients were metabolized. Surprisingly, the experiment did detect labeled CO_2, but this conflicted with the results of the other experiments conducted by Viking and was later explained as a geochemical process rather than a biological one. This result demonstrates an important principle of biosignatures: that they not only have to be produced by life, but that they also should not be produced by anything else.

On the other hand, some biosignatures might be too specific to life on Earth, for example, complex organic compounds that are made by life on Earth but are probably not universal to all life. These are more likely to fail to yield a positive result, even if the sample contains extraterrestrial life forms. One solution to this problem relies on the hypothesis that life will probably always use only a small subset of available compounds within a given class of molecules. For example, out of all the possible amino acids that can exist, the majority of proteins contain only 20. Another seemingly universal biosignature is purposeful motion, that is, movement that cannot be explained by the flow of a solution. Features of life that do not rely on specific biomolecules, such as chemical selectivity and purposeful motion, are both detectable and likely to be shared more broadly by potential extraterrestrials.

Biomarkers such as these require that samples be taken and often tested on location, limiting their utility to the exploration of our solar system. But we now know that the galaxy is teeming with planets orbiting distant stars. Computer simulations of atmospheric chemical properties have shown that the presence of biological processes can be detected by their production of chemical constituents that are far out of equilibrium with regard to their oxidation states. For example, detecting the dual presence of methane (a reduced gas) and oxygen (an oxidized gas) in a planetary atmosphere would build a strong case for life. Indeed, keeping a chemical system out of thermodynamic equilibrium is a property of metabolism that we expect would be true for life throughout the universe. The ability to expand the search for life from the solar system to the galaxy (or at least our galactic neighborhood) will dramatically increase the chances of success.

1.6 FUTURE OF ASTROBIOLOGY

The scientific search to find extraterrestrial life has so far been unsuccessful. But as we have discussed, that search has become more and more sophisticated as astrobiological research produces new insights into the nature and setting of life's origin, the habitability of environments on Earth and on other planets and moons, and the techniques required to find and confirm extraterrestrial life. The next decade will likely see greater efforts to explore Mars and the icy moons of Jupiter and Saturn, an unprecedented ability to observe and analyze extrasolar planets, the potential to engineer and study simpler forms of life in the laboratory, and breakthroughs that have not yet been imagined.

All of these advances will undoubtedly increase our likelihood of finding life on Earth-like planets. But perhaps, extraterrestrial life does not exist or is so sparsely distributed throughout the galaxy that we have almost no chance of finding it. Even if this pessimistic scenario is true, astrobiology research increasingly broadens our perspective of life on Earth and its place in the universe.

ACKNOWLEDGMENTS

Thanks to Sanjoy Som and Shawn Domagal-Goldman for the scientific review of this chapter and Ren Wisons, Eint Kyi, Sarah Page, and Thomas Molyneaux for review of its clarity.

GLOSSARY

Anthropic principle: A set of explanations for the observation that the universe seems ideal for the evolution of intelligent life. One variant, the weak anthropic principle, states that only a universe capable of supporting intelligent life can be observed because only this sort of universe would produce the intelligent observer.

Biosignature: One of a set of forms of evidence that can be used to recognize the presence of life.

Copernican principle: A perspective based loosely on Nicolaus Copernicus's refutation of an Earth-centered universe in the mid-sixteenth century. The broader principle expands on this initial observation to refute any claim that our place in the universe is special.

Exoplanet: Short for *extrasolar planet*; refers to any planet orbiting a star other than the Sun.

Extremophile: A population of organisms that live in an environment in which one or more parameters would be lethal to most other organisms on Earth.

Habitability: The ability of an environment on Earth or elsewhere to support life.

Habitable zone: Also sometimes called the *Goldilocks zone*, the orbital distance around a star in which liquid water likely exists on the surface. The original concept has been expanded to include the effects of climate and the gradual changes in a star's energy output over time.

Hydrothermal vent: An ocean vent that emanates hot water produced through either geological or chemical activity. They are considered a possible setting for the origin of life and also likely habitable environments on icy moons.

Icy moon: A type of moon covered in surface ice and often likely to have liquid water beneath. Examples from our solar system include Europa, a moon of Jupiter, and Enceladus, a moon of Saturn.

LUCA: Last universal common ancestor; the population or populations of organisms represented by the root of the tree of life.

Prebiotic chemistry: The study of increasing chemical complexity that preceded and was necessary for the origin of life.

RNA world: A hypothesis about an early stage of evolution in which RNAs were both the chief genetic molecules and the chief executive molecules; functions that are now mostly performed by DNA and proteins, respectively.

SETI: Search for extraterrestrial intelligence; an astronomical survey for radio signals that likely originated from an intelligent extraterrestrial species.

REVIEW QUESTIONS

1. Describe two pieces of evidence suggesting that present-day life was preceded by an *RNA world* in which RNAs served as both the chief genetic molecule and the chief functional molecule.

2. In what ways does research in biology contribute to the study of astrobiology? How about research in chemistry?

3. The term *evolution* is used by biologists, geologists, and astronomers in different contexts. How does biological evolution differ in meaning from these other uses of the word?

4. What are some key features of life that we might expect to find on other planets?

5. A decade from now, NASA decides to send a mission to Europa looking for life. Where on Europa would you suggest they look and what signs of life should they look for?

ADDITIONAL READING

Alberts, B., D. Bray, J. Lewis, M. Raff, P. Walter, K. Hopkin, A. Johnson, K. Roberts. 2010. *Essential Cell Biology*, 3rd edn. Garland Science, Taylor & Francis, New York and Abingdon, UK.

Barlow, C. (Ed.). 1995. *Evolution Extended: Biological Debates on the Meaning of Life*. The MIT Press, Cambridge, MA.

Becerra, A., L. Delaye, S. Islas, A. Lazcano. 2007. The very early stages of biological evolution and the nature of the last common ancestor of the three major cell domains. *Annual Review of Ecology, Evolution, and Systematics*, 38:361–379.

Bell, G. 1997. *Selection: The Mechanism of Evolution*. ITP, New York.

Bennett, J., S. Shostak, B. Jakosky. 2003. *Life in the Universe*. Pearson Ed., Addison Wesley, San Francisco, CA.

Chela-Flores, J. 2001. *The New Science of Astrobiology: From Genesis of the Living Cell to Evolution of Intelligent Behaviour in the Universe*. Kluwer Academic Publishers, Dordrecht, the Netherlands.

Committee on the Limits of Organic Life in Planetary Systems, Committee on the Origins and Evolution of Life, National Research Council. 2007. *The Limits of Organic Life in Planetary Systems*. National Academies Press, Washington, DC.

Crowe, M.J., M.F. Dowd. 2013. The extraterrestrial life debate from antiquity to 1900. In: *Astrobiology, History, and Society: Life beyond Earth and the Impact of Discovery* (Vakoch, D.A., ed.). Springer, Heidelberg, Germany.

Cziko, G. 1995. *Without Miracles: Universal Selection Theory and the Second Darwinian Revolution*. The MIT Press, Cambridge, MA.

Dennet, D.C. 1996. *Darwin's Dangerous Idea, Evolution and the Meanings of Life*. Simon & Schuster, New York.

Des Marais, D.J., J.A. Nuth III, L.J. Allamandola, A.P. Boss, J.D. Farmer, T.M. Hoehler, B.M. Jakosky et al. 2008. The NASA astrobiology roadmap. *Astrobiology*, 8:715–730.

Dick, S.J. 2013. The twentieth century history of the extraterrestrial life debate: Major themes and lessons learned. In: *Astrobiology, History, and Society: Life beyond Earth and the Impact of Discovery* (Vakoch, D.A., ed.). Springer, Heidelberg, Germany.

Gesteland, R.F., T. Cech, J.F. Atkins (Eds.). 2005. *The RNA World*, 3rd edn. Cold Spring Harbor Laboratory Press, Cold Spring Harbor, NY.

Gilmour, I., M.A. Sephton (Eds.). 2003. *An Introduction to Astrobiology*. Cambridge University Press, Cambridge, U.K.

Jones, B.W. 2004. *Life in the Solar System and Beyond*. Springer-Praxis, Chichester, U.K.

Margulis, L., D. Sagan. 1995. *What is Life?* Simon & Schuster, New York.

Mason, S.F. 1991. *Chemical Evolution: Origins of the Elements, Molecules, and Living Systems*. Clarendon Press, Oxford, U.K.

Miller, S.L., L.E. Orgel. 1974. *The Origins of Life on the Earth*. Prentice-Hall, Inc., Englewood Cliffs, NJ.

Mix, L.J. (Ed.). 2006. The astrobiology primer: An outline of general knowledge. *Astrobiology*, 6:735–813.

Morowitz, H.J. 1992. *Beginnings of Cellular Life, Metabolism Recapitulates Biogenesis*. Yale University Press, New Haven, CT.

Olomucki, M. 1993. *The Chemistry of Life*. McGraw-Hill, Inc., New York.

Oparin, A.I. 1953. *Origins of Life*. Dover Publications, Inc., New York (republication of the 1938 book).

Oparin, A.I. 1968. *Genesis and Evolutionary Development of Life*. Academic Press, New York.

Pinker, S. 1997. *How the Mind Works*. Norton & Company, New York.

Shapiro, R. 1987. *Origins: A Skeptic's Guide to the Creation of Life on Earth*. Bantam Books, New York.

Shaw, A. 2006. *Astrochemistry: From Astronomy to Astrobiology*. Wiley, Chichester, U.K.

Sullivan, W.T., J. Baross (Eds.). 2007. *Planets and Life: The Emerging Science of Astrobiology*. Cambridge University Press, Cambridge, U.K.

Zubay, G. 2000. *Origins of Life on the Earth and in the Cosmos*, 2nd edn. Academic Press, San Diego, CA.

CHAPTER 2

Origin of Elements and Formation of Solar System, Planets, and Exoplanets

Ken Rice

CONTENTS

2.1 INTRODUCTION

The standard picture today is that the universe is formed 13.8 billion years ago (Bennett et al. 2013) in what is typically called the Big Bang. There are two aspects of the period immediately after the Big Bang that are of interest with regard to planet formation and the existence of life. One is Big Bang nucleosynthesis that created the initial elements in the universe (Peebles 1966). The other is the formation of the initial structure that later collapsed to form the galaxies, stars, and planets that we see today.

Immediately after the Big Bang, the universe was extremely hot. At this stage, nuclear reactions were converting protons into neutrons and neutrons back into protons. At these high temperatures ($\sim 10^{11}$ K), there were an equal number of protons and neutrons. These protons and neutrons were also continually combining to form deuterium (one proton and one neutron) and helium (two protons, two neutrons). However, at this stage, it was too hot for these elements to survive for very long.

Neutrons are very slightly more massive than protons and hence it requires energy for a proton to become a neutron. As the universe expanded and cooled, the rate at which protons changed into neutrons decreased, and the number of protons started to exceed the number of neutrons. After about 3 min, the temperature dropped sufficiently so that the heavier elements (deuterium and helium) could also survive. At this point in time, there was about one neutron for every seven protons, and all of these neutrons were incorporated into helium nuclei. A helium nucleus has two neutrons and two protons (and hence a mass almost four times that of a proton). Therefore, if there are 2 neutrons for every 14 protons, creating a helium atom leaves 12 protons and 1 helium nucleus containing 2 protons and 2 neutrons. Hence, at the end of Big Bang nucleosynthesis, helium made up (by mass) 25% of the baryonic matter (normal matter composed of protons and neutrons) in the universe, with hydrogen (protons) making up the other 75%. A small amount of deuterium (one proton, one neutron), helium-3 (two protons, one neutron), and lithium (which forms when helium-3 combines with helium-4) also formed. Essentially, at the end of Big Bang nucleosynthesis, the baryonic matter in the universe consisted of, by mass, 75% hydrogen, 25% helium, and a tiny amount of lithium, helium-3, and deuterium (Walker et al. 1991). All the other elements that exist in the universe today—the building blocks of life—were formed through nuclear reactions in stars.

Another aspect of the early universe that is of relevance to star formation, planet formation, and life is the formation of structure. Initially, the universe is thought to have been full of tiny quantum fluctuations. It is theorized that a period of rapid inflation (Guth 1981) occurred immediately after the Big Bang. The reason this is important is that it is thought that this rapid inflation (in which the universe expanded by a factor of 10^{30} in only about 10^{-36} s) enhanced these quantum fluctuations so that the resulting density

FIGURE 2.1   Tiny temperature fluctuations in the CMB radiation, observed by ESA's Planck satellite. These fluctuations are the seeds for all the structure in the universe today. (Courtesy of ESA and the Planck Collaboration, Paris, France.)

enhancements could then collapse gravitationally to form the galaxies, stars, planets, and all other structures that we see in the universe today.

We can, actually, observe the fluctuations in the early universe in the cosmic microwave background (CMB) radiation (Penzias and Wilson 1965). This is the thermal radiation left over after the Big Bang that has now cooled, as the universe expanded, to a temperature of only 2.73 K. Figure 2.1 shows the temperature anisotropies in the CMB radiation, as observed by the European Space Agency's (ESA) Planck satellite. These tiny temperature fluctuations were imprinted on the sky when the universe was only 380,000 years old and show the structure in the universe at the instant when it cooled sufficiently for the electrons and protons to combine to form neutral atoms. These tiny fluctuations represent the seeds of all the structure in the universe today (Davis et al. 1985).

2.2   STELLAR NUCLEOSYNTHESIS

Big Bang nucleosynthesis led to the universe consisting, initially, of hydrogen, helium, and a small amount of lithium, deuterium, and helium-3. All other elements in the universe today formed in stars or are associated with stellar evolution (Hoyle et al. 1956). In a simple sense, we can divide stars into three basic categories, low-mass stars, intermediate-mass stars, and high-mass stars. Low-mass stars have masses less than 2 solar masses, stars with masses between 2 and 8 solar masses are regarded as intermediate-mass stars, and those with masses above 8 solar masses are high-mass stars.

2.2.1   Low-Mass Stars

Stars form in the densest, coldest parts of the interstellar medium, the region between the stars. The process starts when a clump of gas and dust is sufficiently dense so as to collapse under its own gravity. Initially, young stars are powered by the release of gravitational potential energy, but once their cores reach a temperature of 10 million K, they are able to start nuclear fusion, which then provides the energy for the rest of their lives.

In 1905, Einstein published his theory of special relativity. This theory included a relationship between mass and energy, the now famous equation $E = mc^2$. In low-mass stars, the process is quite simple. Four hydrogen nuclei (protons) combine to form a helium nucleus (two protons and two neutrons). The mass of a helium nucleus is less than that of four protons and so the difference in mass is released as energy. Each reaction, therefore, *uses up* four hydrogen nuclei. This is known as the proton–proton chain. We can calculate how much energy is released in every reaction. We know how much energy is released by the Sun every second. Consequently, we can determine how much mass is *used up* every second in powering the Sun. If we assume that, during its life, the Sun will *use up* 10% of its hydrogen in core nuclear burning, we can estimate the lifetime of the Sun as being about 10–12 billion years.

When a star like the Sun has *used up* the hydrogen in its core, the core will be composed entirely of helium and will start to contract because the core is no longer generating energy. A shell of compressed hydrogen around the core starts to undergo hydrogen fusion and adds helium to the inert helium core. Eventually, the helium core becomes hot enough for helium fusion to occur. Thus, a second phase of core burning will occur in which helium is converted into carbon with, occasionally, a helium nucleus fusing with a carbon nucleus to form oxygen.

Interestingly, helium burning is possibly the only example of a successful anthropic prediction (a prediction that is based on the existence of humans). Helium burning involves two helium nuclei colliding to form beryllium, followed by a third helium nucleus colliding to then form carbon. Known as the triple-alpha reaction, initial calculations suggested that the chance of a third helium nucleus colliding before the beryllium decayed was very small. Fred Hoyle suggested that our very existence indicated that there needed to be a resonance (in this case an energy level in beryllium) so that such a collision was much more likely than expected. Such a resonance was indeed discovered, in 1952, by Willy Fowler and hence explained the generation of carbon through stellar nucleosynthesis (Salpeter 1952).

Core helium burning in a low-mass star only lasts for a very short time, and the helium is typically exhausted within a few hundred million years. The core, now composed of carbon and maybe some oxygen, will shrink and there will be a short phase when helium and hydrogen are burnt in shells around the now inert carbon core. The carbon core of a low-mass star will, however, never become hot enough for carbon burning, so the core will contract until it is supported by what is called electron degeneracy pressure. The outer layers of the star will be ejected into space and will form, for a very brief period, what is known as a planetary nebula. This has nothing to do with planets but is simply the name given to these objects when they were first observed (and looked like planets) through small telescopes. Figure 2.2 shows an image, taken by the Hubble Space Telescope, of such a planetary nebula called the Ring Nebula.

In some low-mass stars, the atmospheres are enriched, through convection, in carbon generated by helium fusion. This carbon is ejected out into space, during the planetary

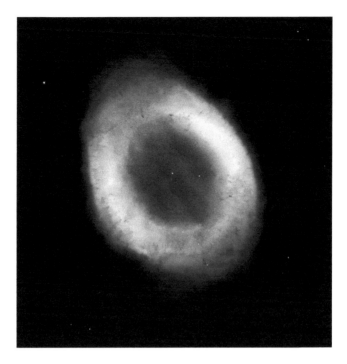

FIGURE 2.2 Image, taken by the Hubble Space Telescope, of a planetary nebula called the Ring Nebula. The progenitor star was probably a star slightly more massive than the Sun that has now exhausted its core hydrogen and helium and is shedding its outer layers to produce this relatively short-lived nebula. (Courtesy of The Hubble Heritage Team—AURA/STScI/NASA, Baltimore, MD.)

nebula phase, and it is thought that most of the carbon in the universe, and hence most of the carbon used for life, was created in such stars. The carbon core will remain as what is called a white dwarf, which will have a radius similar to that of the Earth and a mass about one-third that of the original low-mass star.

2.2.2 High-Mass Stars

Low-mass stars play two basic roles in the origin of life. One is that they have lifetimes of billions of years, giving sufficient time for life to form and evolve on a planetary companion. The other is that they are the origin of most of the carbon in the universe today. High-mass stars (stars with masses greater than 8 solar masses), however, play a very different role. All stars spend most of their lives burning hydrogen in their cores. Although the process in the core of a high-mass star also essentially converts four protons into a helium nucleus and releases the same amount of energy per reaction as in the core of a low-mass star, high-mass stars use carbon as a catalyst. During hydrogen burning in the core of a high-mass star, carbon is converted into nitrogen and then oxygen, and at the end of this process, four protons have been converted into helium and the carbon is recovered to be used again.

Since carbon (C), nitrogen (N), and oxygen (O) are involved in hydrogen burning in the core of a high-mass star, the process is known as the CNO cycle. It is a much faster process than the proton–proton chain that takes place in the cores of low-mass stars, and consequently, high-mass stars are much more luminous than low-mass stars. A star with a mass 10 times that of the Sun is about 10,000 times as luminous. This means that it converts hydrogen to helium in its core 10,000 times faster than the Sun. A 10 solar mass star will therefore run out of fuel 1000 times faster than a solar mass star and will therefore have a lifetime of only about 10 million years. High-mass stars, therefore, do not have lives that are long enough for any form of life to form or evolve on any orbiting planets, were any to exist.

When the core of a high-mass star runs out of hydrogen, the star also undergoes a phase of hydrogen burning in a shell around the inert helium core. As with low-mass stars, this adds helium to the core so that it eventually becomes hot enough for carbon to form via helium fusion. The helium is exhausted in a very short time, after which there is a phase of double shell burning (helium in a shell around the inert carbon core and hydrogen in a shell around the helium shell). Unlike low-mass stars, however, the core of a high-mass star does become hot enough (600 million K) for carbon burning. Carbon burning forms neon. This stage is very short, lasting only hundreds of years.

After carbon burning ceases, shell burning resumes. Now there will be a carbon shell, surrounded by a helium shell, surrounded by a hydrogen shell. The temperature in the core, however, continues to increase and becomes hot enough for advanced nuclear burning. This typically happens through helium-capture reactions, adding two protons and two neutrons to the various elements that have already formed in the core. For example, carbon and helium produce oxygen, oxygen and helium produce neon, and neon and helium produce magnesium. If the temperature is high enough, then heavy elements can combine directly, such as carbon and oxygen to form silicon, and oxygen and oxygen to form sulfur. Again, each stage is very short lived and is followed by another stage of shell burning. Eventually, the core becomes composed entirely of iron, and this whole process stops. The reason it stops at this stage is because iron has the lowest mass per nuclear particle. A fusion reaction with iron would produce an element with a total mass greater than the total initial mass and hence requires energy. This is sometimes referred to as the *iron catastrophe*. What this means is that iron fusion is not possible in the core of a high-mass star, and so the core can no longer generate energy and starts to contract. The core contracts until the protons in the iron nuclei combine with the free electrons to produce neutrons, and the core becomes composed entirely of neutrons. This reaction produces high-energy neutrinos and the outer layers of the star are ejected in what is known as a supernova explosion. The core remains as a neutron star, which typically has a radius of only a few kilometers and a mass similar to that of the Sun. If the core mass exceeds about 2.5 solar masses, it is unable to prevent further collapse and contracts further to form a black hole.

So far, stellar nucleosynthesis has produced carbon from fusion in low-mass stars and elements up to iron through fusion in high-mass stars. Elements more massive than iron form during the supernova explosion at the end of a high-mass star's life. When the core of a high-mass star collapses to form a neutron star, the outer layers of the star are ejected in an extremely energetic event known as a supernova explosion. During this process, high-energy neutrons bombard the elements in the supernova ejecta, increasing the number of neutrons in the nuclei of these elements. These neutrons then decay to protons, producing newer, and heavier, elements.

2.2.3 Intermediate-Mass Stars

We have discussed low-mass (<2 solar masses) and high-mass (>8 solar masses) stars, but have not discussed intermediate-mass stars (between 2 and 8 solar masses). This is because they behave partly like low-mass stars and partly like high-mass stars. The initial stage of an intermediate-mass star's life is similar to that of a high-mass star: hydrogen fusion in the core of an intermediate-mass star occurs through the CNO cycle, and the process proceeds until the core is composed of carbon. The core of an intermediate-mass star, however, does not get hot enough to create elements heavier than carbon and so it ends its life like a low-mass star. The outer layers are ejected to form a planetary nebula and the core remains as a white dwarf.

2.2.4 Hertzsprung–Russell Diagram

A typical way to present the different types of stars and to illustrate their evolution is to consider what is called the Hertzsprung–Russell diagram. This is really just a plot of temperature/spectral type (on the x-axis) and luminosity/magnitude (on the y-axis) and is illustrated in Figure 2.3. The top x-axis is temperature in Kelvin, increasing to the left, while the bottom is the corresponding spectral class. The right-hand y-axis is luminosity relative to the Sun, while the left-hand y-axis is the corresponding magnitude. Stars spend most of their lives on the main sequence (the diagonal band in Figure 2.3), which is the stage at which they are burning hydrogen in their cores. Massive, hot (blue) stars are at the top left of the main sequence, while low-mass, cool (red) stars are at the bottom right of the main sequence. The most massive stars on the main sequence in Figure 2.3 will be a few tens of solar masses, while the lowest-mass stars will be about one-tenth of a solar mass. This figure illustrates how luminosity increases dramatically with mass and, hence, that massive stars have very short lives compared to that of the Sun. The Sun has a surface temperature of about 5700 K and so is in the yellow region of the main sequence. Figure 2.3 also includes some stars visible in the night sky.

When stars with masses similar to that of the Sun evolve off the main sequence, they move into the region labeled *giants* before expelling their atmospheres to produce a short-lived planetary nebulae and leaving their cores as a white dwarf (which ends up in the bottom left region of the Hertzsprung–Russell diagram). Massive stars move off the main sequence to become *supergiants* before undergoing a supernova explosion and leaving their cores as a neutron star or, possibly, a black hole.

2.2.5 Evidence for Nucleosynthesis

Is there any evidence to support the picture presented earlier for the origin of the elements? Yes, because we can consider the abundances of elements in our own solar system (Lodders 2003). Figure 2.4 illustrates the observed abundances of elements in our solar system. The y-axis is a log scale, so each division is a change of a factor of 10. What the figure shows is that hydrogen (H) and helium (He) are the most abundant elements in the solar system and that there is a small amount of lithium (Li). This is consistent with Big Bang nucleosynthesis. We then notice that from carbon (C) to iron (Fe), elements with an even number of protons are more common than those with an odd number of protons.

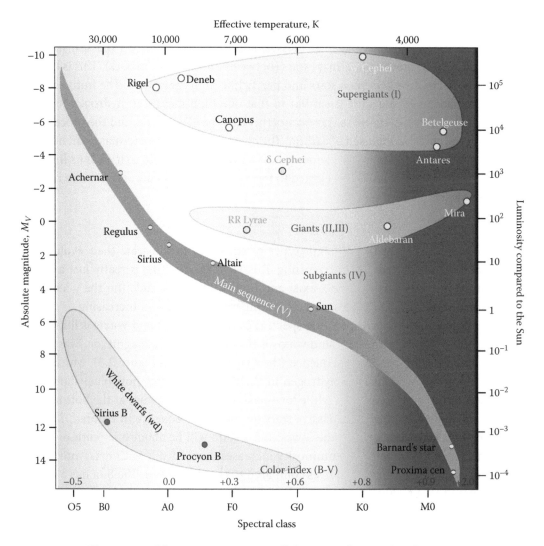

FIGURE 2.3   Illustration of the Hertzsprung–Russell diagram. This is a plot of temperature/spectral type against luminosity/magnitude, for different types of stars, and also illustrates something of how stars evolve. Stars spend most of their lives on the main sequence with massive, hot stars at the top left and low-mass, cool stars at the bottom right. Sun-like stars evolve off the main sequence to become *giants* and eventually white dwarfs. Massive stars evolve to become *supergiants* and end their lives in a supernova explosion. The figure also shows the positions of some stars visible in the night sky. (Courtesy of CSIRO, Epping, New South Wales, Australia.)

This is because stellar nucleosynthesis normally proceeds via reactions that involve helium. Helium has two protons and hence even-numbered elements are more common than odd-numbered elements. There are other processes that can produce elements with an odd number of protons, but these processes occur less often than helium-capture reactions.

Beyond iron (Fe), there is a substantial drop in abundance. This is because these elements are produced in supernova explosions (which are themselves rare) via neutron capture

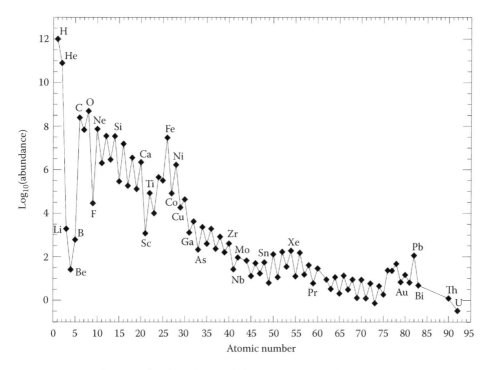

FIGURE 2.4 Figure showing the abundance of elements in our solar system with data taken from Lodders (2003). The figure illustrates that the abundances observed are consistent with what we would expect from the Big Bang nucleosynthesis producing the hydrogen (H), helium (He), and a small amount of lithium (Li). Heavier elements are then formed through stellar nucleosynthesis either in the star itself [up to iron (Fe)] or in neutron capture reactions in the supernova explosions associated with the death of high-mass stars (elements beyond iron). (Data taken from Lodders, K., *Astrophys. J.*, 591, 1220, 2003.)

reactions (also rare). Ultimately, the abundances of elements in our solar system (and by inference in our galaxy) are consistent with the basic picture of the Big Bang nucleosynthesis followed by stellar nucleosynthesis.

2.3 STAR AND PLANET FORMATION

We have considered how almost all elements heavier than hydrogen and helium in the universe today were formed primarily through stellar nucleosynthesis, but we have not considered how these stars actually form. The initial small-scale structure in the universe collapsed gravitationally to form the stars and galaxies that we observe today. Star formation occurs in cold and dense regions of the interstellar medium, typically called giant molecular clouds. These are clouds primarily composed of molecular hydrogen and dust. For star formation to occur, the gravity in the cloud must be sufficient so as to overcome the pressure forces that are acting to prevent collapse.

In our own galaxy, a typical giant molecular cloud will need to have a mass of a few hundred solar masses in order for gravitational collapse to occur. The cloud is, however, initially optically thin and so the temperature remains almost constant as the density increases.

This means that rather than forming a single massive star, the cloud fragments into many low-mass cores (Bonnell et al. 2001). These initial protostellar cores typically have masses only a few times that of Jupiter, but continue to accrete mass from the surrounding cloud, with their final mass determined by the density of the region in which they are growing. Most stars will grow to masses similar to that of the Sun (between 0.3 and 2 solar masses). Some will be ejected or grow slowly and end up with a very low mass (called brown dwarfs if their mass ends up less than 0.08 solar masses), while a few may grow to become very massive.

Initially, stars are powered by the release of gravitational potential energy. This process also heats the core of the young protostar, which continues to contract until it reaches a temperature of 10 million K. At this point, nuclear fusion commences and the star will be in an approximate equilibrium, which it will maintain for most of its life. As discussed before, the lifetime depends on the mass of the star and decreases rapidly with increasing stellar mass.

2.3.1 Protostellar Discs

The giant molecular clouds, from which stars form, are typically very large. As the material collapses to form a young star, it needs to conserve angular momentum. As the individual protostellar cores contract, they spin faster and faster. This means that most of the material cannot fall directly onto the young protostar. If it did, the young protostar would end up spinning so fast that it would simply break apart, rather than ultimately contracting to form a star. Instead of this material falling directly onto the central protostar, it falls parallel to the rotation axis and forms a thin disc around the young protostar (Terebey et al. 1984). The radial extent of the disc depends on the angular momentum of the material from which it forms, but these discs are typically a hundred to a few hundred astronomical units (AUs—distance from the Sun to the Earth) in size. Figure 2.5 shows images of discs around young stars in the Orion star-forming nebula. The bright spot in the center of each image is the young protostar, and the dark-surrounding region is the dusty circumstellar disc that is acting to block light from the brighter nebula in the background.

The disc then plays two important roles in star and planet formation. One is that instabilities in the disc provide a mechanism for transporting angular momentum outwards, allowing mass to accrete onto the young protostar and hence allowing the star to grow. Observations and modeling suggest that typical disc lifetimes are about 5 million years (Haisch et al. 2001). The other role that these discs play is that they are also the sites of planet formation, and so the lifetime of such discs places certain constraints on the planet formation process.

2.3.2 Planet Formation

Although most of the material in the giant molecular clouds is molecular hydrogen and helium, 2%—by mass—are elements heavier than hydrogen and helium (often called metals by astronomers), and half of this is in the form of micron-sized dust grains. Consequently, about 1% of the mass in the disc around young protostars is solid, planet-building material. Additionally, the disc temperature decreases with increasing distance

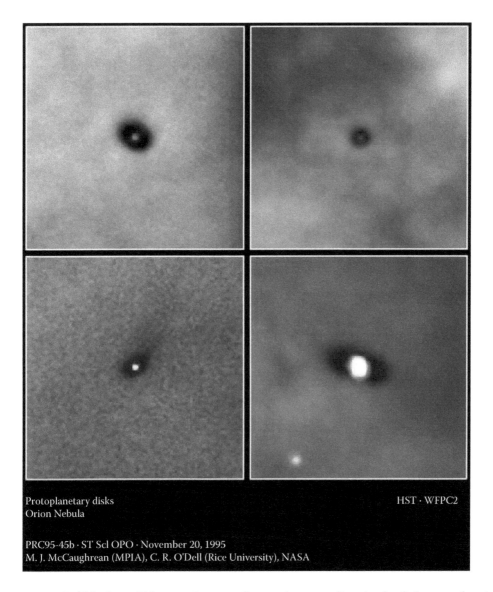

Protoplanetary disks
Orion Nebula

HST · WFPC2

PRC95-45b · ST ScI OPO · November 20, 1995
M. J. McCaughrean (MPIA), C. R. O'Dell (Rice University), NASA

FIGURE 2.5 Hubble Space Telescope images of protoplanetary discs in the Orion star forming nebula. The bright spot in the center of each image is the young protostar, and the dark region around each protostar is a protoplanetary disc that is blocking the brighter background nebula. (Courtesy of M.J. McCaughrean and C.R. O'Dell, STSci, Baltimore, MD.)

from the central protostar. At some distance from the star, the disc becomes cold enough for ices (water, ammonia, methane) to form on the solid dust grains. This means that the solids in the inner part of the disc are dust grains only, while in the outer part of the disc, these are dust grains with substantial ice mantles. The radius beyond which the ices form on the dust grains is known as the ice line or snow line. In a disc around a young star that will evolve to become like the Sun, it occurs at about 2.7 AU. Since the dust grains beyond the snow line are coated in ice mantles, there is actually more solid, planet-building material just beyond the snow line than just inside the snow line.

The basic planet formation scenario is that the solids grow via collisions to ultimately form kilometer-sized planetesimals. These then coagulate to form planetary mass bodies. Due to the enhancement of solids (dust plus ice) beyond the snow line, planetary mass bodies just beyond the snow line grow faster than those inside the snow line. Beyond the snow line is where basic planet formation theory would suggest most massive gas/ice giant planets should form (planets similar to Jupiter, Saturn, Uranus, and Neptune). The enhancement in the solids beyond the snow line means that the solid/icy cores of gas giant planets can form before the gas in the disc disappears (within 5 million years) and can become massive enough (>10 Earth masses) so as to gravitationally attract a massive, gaseous envelope (Pollack et al. 1996).

Terrestrial (earthlike) planets are thought to form inside the snow line where the disc is too warm for ices to form on the solid dust grains. Since terrestrial planets do not have substantial gaseous atmospheres, they do not need to form before the gas disc has dissipated, and, indeed, there is evidence that the formation of terrestrial planets can take tens of millions of years.

2.4 SOLAR SYSTEM

The solar system comprises all the bodies in our own planetary system. This includes the Sun, eight planets, three dwarf planets, asteroids, comets, a region known as the Kuiper belt, and a region known as the Oort cloud. The Sun is a fairly typical low-mass star. It is currently on the main sequence, which means that its energy comes from the fusion of hydrogen to helium in its core.

2.4.1 Planets

The solar system has eight planets, four inner rocky terrestrial planets and four outer gas/ice giant planets. The four inner planets are Mercury, Venus, Earth, and Mars. The four outer planets are Jupiter, Saturn, Uranus, and Neptune. The properties of the solar system planets are consistent with the basic planet formation picture we developed in the previous section. All of the solar system planets orbit in the same plane and, with the exception of Mercury, their orbits are close to being circular. This is consistent with the idea that the planets formed in a thin disc around the young Sun. We also find the rocky, terrestrial planets in the inner regions—where the planet-forming disc would have been too hot for ices to form on the dust grains—and the gas/ice giants in the outer regions, where ices would have formed on the dust grains, enhancing the amount of planet-forming material and allowing these planets to form before the gas disc dissipated.

2.4.2 Asteroids, Comets, and Dwarf Planets

Asteroids and comets are both small bodies that orbit the Sun. They are both thought to be remnants of the planet formation process. Asteroids typically orbit in the region between Mars and Jupiter, although there are some that orbit within the inner solar system and some (called near-earth objects [NEOs]) that pass quite close to the Earth. The composition of asteroids can be quite varied, but most are thought to be a loose collection of rocks and metal, typically referred to as a *rubble pile*. The asteroid belt also contains Ceres, one of the three dwarf planets.

Comets originate in the outer parts of the solar system and, consequently, tend to be a mixture of ices and other solids, hence the nickname *dirty snowballs*. They often have very eccentric orbits, passing close to the Sun and then disappearing back into the outer parts of the solar system. When close to the Sun, the outer layers of the comet melt, producing a coma (small atmosphere) that is blown away to form, what can be, an impressive tail.

There are two types of comets, short-period comets and long-period comets. Short-period comets orbit in the same plane as the planets and tend to have periods of hundreds of years. Long-period comets come from all different directions and can have periods of thousands or tens of thousands of years. The existence of such comets tells us that there must be two reservoirs of bodies in the solar system: a belt of objects, called the Kuiper belt, beyond the orbits of the planets and a cloud of material surrounding the solar system, called the Oort cloud, and extending out to many thousands of AU.

The Kuiper belt is composed of hundreds or thousands of comet-like bodies that orbit beyond Neptune. This region also contains two of the three dwarf planets, Pluto and Eris. We have also discovered at least 10 other bodies with masses similar to that of Pluto and expect to find many more such bodies in the coming years. We have not definitively found any bodies in the Oort cloud, but have observed one object (Sedna) that may be an inner Oort cloud object.

2.4.3 Age of the Solar System

The Sun is a typical low-mass star and based on its mass and luminosity, we can estimate its lifetime as 10–12 billion years. We can also use radioactive dating to estimate the actual age of the solar system. Potassium-40 decays to argon-40 with a half-life of 1.25 billion years. When the solar system was formed, argon-40—being gaseous—was not incorporated into the solid bodies that formed at that time. Hence, all the argon-40 we find today must be from the radioactive decay of potassium-40. After 1.25 billion years, there would have been equal amounts of argon-40 and potassium-40. Today, we find that there are 11.5 argon-40 atoms for every potassium-40 atom, which tells us, remarkably precisely, that the solar system is 4.56 billion years old. Given that the Sun has a lifetime of 10–12 billion years, this suggests that we are about halfway through the Sun's life. The Sun, however, will continue to get more luminous as it ages, and so the Earth will become hotter and will likely be uninhabitable in about 1.6 billion years (Franck et al. 2005).

2.5 EXTRASOLAR PLANETS

In a sense, the properties of the solar system are consistent with our basic picture of planet formation. Solar system properties have, however, also informed how we expect planet formation to proceed. When we first started searching for planets around other stars, we expected to find systems similar to our own. We have now, however, detected numerous planets—known as extrasolar planets or exoplanets—around other stars, and what we have found has been quite surprising.

2.5.1 Detecting Exoplanets

One of the obvious ways in which we could detect exoplanets would be to simply look for them. This, however, is extremely difficult as stars are very bright compared to planets. In the visible, a star like the Sun is a billion times brighter than an earthlike planet. We can, however, directly detect massive planets (many Jupiter masses) at large distances from their host stars. We now have a couple of examples of such planets. One example is HR8799, shown in Figure 2.6, which has three Jupiter-like planets orbiting at distances of 24, 38, and 68 AU. Systems like this create a bit of a problem for theories of planet formation as these planets lie outside the region where we would expect the standard planet formation scenario to work.

To date, we have, however, confirmed the existence of over 1000 exoplanets and have an additional 3000 that are planetary candidates. Most of these have been detected via indirect methods. Rather than directly detecting the planet, we detect the planet's influence on its host star (Murray 2012). There are a number of different methods that we will discuss in more detail in the following text.

2.5.1.1 Radial Velocity Method

Although we think of the planets in our solar system as orbiting the Sun, in fact, all the objects in the solar system orbit the system's common center of mass. This means that even the Sun is technically orbiting within the solar system. The Sun does, however, dominate the mass of the solar system, and so the center of mass is very near the Sun and often inside it. The same is going to be true for any other planetary system of interest.

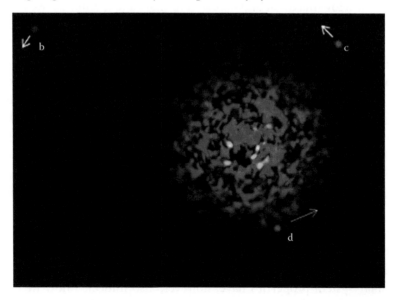

FIGURE 2.6   Image, taken with the 10 m Keck telescope, shows three Jupiter-like planets in the HR8799 system. The image was taken using a differential imaging technique that minimizes the light from the host star, but not from the orbiting planets. These planets, however, orbit at large distances (24, 38, and 68 AU) from their host star, and it is still not clear how such planets can form. (From Marois, C. et al., *Science*, 322, 1348, 2008. With permission.)

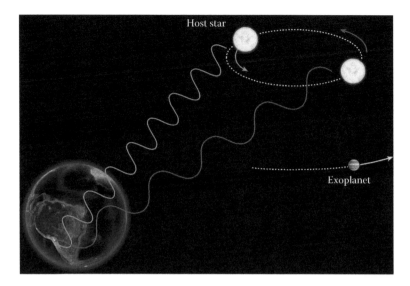

FIGURE 2.7   Illustration of the radial velocity or *Doppler wobble* method for detecting extrasolar planets. The planet and star both orbit the system's common center of mass. This means that the star's spectral lines are shifted through the Doppler effect. This can then be used to determine the line-of-sight (or radial) velocity of the star and hence determine the properties of the orbiting planet. (Courtesy of European Southern Observatory [ESO], Garching, Germany.)

The radial velocity, or Doppler wobble, method works by observing stellar spectra. If the star is moving, relative to us, the wavelength of the star's spectral lines will be different to what they would be if the star were stationary. This is known as the Doppler effect. We can use this shift in wavelength to determine the line-of-sight velocity of the star. If the star has a body orbiting it, then there will be periods when the star appears to be moving towards us and other times when it is moving away. This is illustrated in Figure 2.7. The spectrum of the star tells us what type of star it is and hence its mass, and so we can use this measurement to determine the mass of the orbiting body, the radius of the orbit, and the eccentricity of the orbit.

There are, however, a number of basic issues with the radial velocity technique. One is that it only measures the line-of-sight velocity. This means that the mass estimate is a minimum, because it could be a more massive object on an orbit inclined with respect to our line of sight. Fortunately, it is only weakly dependent on inclination and so the *error* is typically going to be small. Another issue is that, for a given planet mass, the star's radial velocity decreases the further the planet is from the star. Therefore, it is easier to detect lower-mass planets close to the star than further from the star. The orbital period also increases as the orbital separation increases, and so detecting planets further from their host star takes longer than detecting planets closer to their parent star.

Radial velocity spectrometers today are, however, sufficiently accurate so that they can measure the velocity of the star to an accuracy of better than 1 m/s (Cosentino et al. 2012). This means it is possible to detect Earth mass planets within a few tenths of an AU of a Sun-like star, planets with masses a few times that of Earth at a distance of 1 AU, and can easily detect a Jupiter analog as long as the observations cover a sufficiently long time interval.

The radial velocity method is the most successful planet-hunting method to date, with more than 500 exoplanet detections. This ranges from planets with masses half that of the Earth, but with orbital periods of less than a day, to planets with masses more than 10 times that of Jupiter orbiting 6 times further from their star than the Earth is from the Sun.

2.5.1.2 Transit Method

Another indirect method for detecting exoplanets works by looking for dips in the brightness of a star that might indicate that a body has passed between us and the host star, blocking some of the starlight. This is known as the transit method and is illustrated in Figure 2.8, which also shows actual data from the Wide Angle Search for Planets (WASP)-22 system (Johnson et al. 2009). If a transit repeats, then you can be reasonably confident that it is an orbiting body, rather than some transient event. The orbital period can then be used to determine orbital distance. The fraction of the starlight blocked can also be used to estimate the radius of the orbiting body and, hence, whether or not it might be a planet.

The transit method alone, however, cannot typically be used to establish if the dip is due to an orbiting planet. It could be a brown dwarf star, which is a very-low-mass star with a radius similar to that of Jupiter-like planets. It could be an unresolved, distant binary star system that makes the brightness of the nearby star appears to dip periodically (in this case, even the periodicity would not imply a companion to the nearby star). It could be a grazing eclipse of a stellar companion. There are some ways of eliminating very obvious

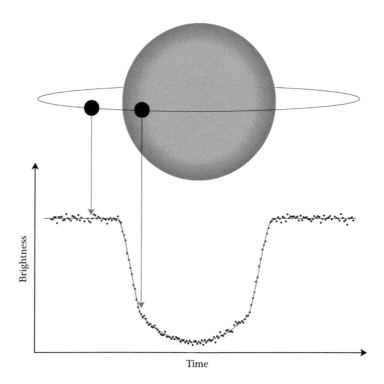

FIGURE 2.8　Illustration of the transit method showing how the measured brightness of a star dips as an orbiting planet passes between us and its host star. (Courtesy of John Johnson.)

false-positives, but even then, transiting systems would typically be followed up, using radial velocity spectrometers, to measure the radial velocity of the star and, hence, confirm the presence of a planetary companion.

In addition to radial velocity measurements being able to confirm if a transiting system is a planet or not, it also gives you a planet mass and hence a planet density. Additionally, in these systems, the orbit is almost exactly edge on, and so the mass estimate is accurate, rather than being a minimum. Hence, we can characterize these planets much more accurately than a single measurement would allow.

The transit method has become a very powerful method for detecting exoplanets, second only to the radial velocity method in terms of confirmed exoplanets. In addition to the almost 400 confirmed exoplanets, there are around 3000 planetary candidates detected by the National Aeronautics and Space Administration's (NASA) Kepler satellite. These have not yet been confirmed to be exoplanets but the Kepler data are so exquisite that they are possible to eliminate many of the false-positives and, hence, the expectation is that many of these will turn out to actually be planets, even if we do not yet know their masses or densities.

2.5.1.3 Gravitational Microlensing

Maybe the most interesting of the planet-hunting techniques is the one that uses Einstein's theory of general relativity. A premise of general relativity is that spacetime curves in the presence of mass (such as galaxies, stars, and planets). A consequence of this is that light is bent when it passes near massive bodies. In some cases, one can even observe distorted images of distant galaxies.

What we are interested in, however, is when a distant star in our galaxy passes behind a star that is nearer to us. The nearer star acts as a gravitational lens producing multiple images of the distant star. However, no telescope today is able to resolve these images and so what we observe is an apparent gradual brightening of the nearer star (as it focuses the light from the distant star) and then a gradual dimming as the distant star passes out from behind the nearer star. This whole process can last about 30–40 days.

If, however, the nearer star has an orbiting planet and if this planet happens to be in the right place at the right time, it can also act as an additional gravitational lens and can further amplify the signal for a short time (compared to the full duration of the event). This signal can then be analyzed to estimate the mass of the planet and its orbital radius. It is a powerful method in that it is sensitive to planets with orbital radii between 1 and 10 AU and can detect planets with masses similar to that of the Earth.

Figure 2.9 shows the microlensing light curve for one of the first exoplanets discovered using this method (Beaulieu et al. 2005). The figure shows the amplification of the light from the distant star by the nearer lens star. The x-axis is in Julian days (a dating scheme in which day 0 is January 1, 4713 BC), and so the entire event lasts slightly more than 40 days. The little blip on the right-hand side of the light curve (and shown blown up in the insert) is the signal from the planet orbiting the lens star. Analysis indicates that this is likely a 5.5 Earth mass planet orbiting at a distance of around 2.6 AU from a star with a mass about 0.2 times that of the Sun.

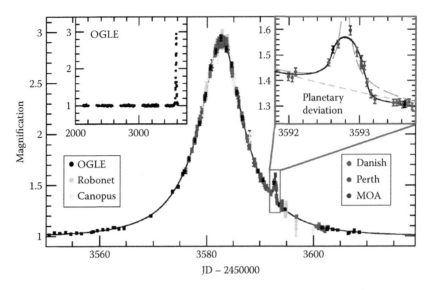

FIGURE 2.9   OGLE-2005-BLG-290Lb light curve showing an amplification of the light from a distant star through gravitational microlensing. The small extra amplification on the right-hand side of the curve (and shown in more detail in the insert) is due to an orbiting planet with a mass of about 5.5 Earth masses and an orbital radius of about 2.5 AU. (From Beaulieu, J.-P. et al., *Nature*, 439, 437, 2005. With permission.)

To date, however, we have only detected 24 planets using this method, although a robotic telescope network (ROBONET) hopes to detect many more over the coming years. This method also suffers from some issues. The process for a particular system cannot be repeated, as these chance alignments are extremely rare. Also, the stars being observed tend to be quite distant and hence performing any kind of follow-up is virtually impossible. However, since this method can probe parts of the galaxy inaccessible to other methods and can probe orbital regions that are difficult for other methods, it is likely to play a very important role in expanding our understanding of the general characteristics of exoplanets.

2.5.2 Exoplanet Characteristics

Earlier in this chapter, we considered the standard ideas for how planets form and how our solar system appears consistent with this basic picture. We find rocky, terrestrial planets and rocky remnants (asteroids) in the inner solar system, and we find big gas/ice giants and icy remnants (comets) in the outer solar system. The planets all orbit in the same plane and, with the exception of Mercury, have orbits that are very nearly circular. All are consistent with formation in a disc around the young Sun, which was hot on the inner parts and cooler in the outer parts.

We expected exoplanets to have properties similar to what we find in our own solar system. What we discovered was quite different. The first confirmed planet around a Sun-like star was discovered by a team from Geneva using the radial velocity method (Mayor and Queloz 1995). It was found to have a mass of 0.47 Jupiter masses, but to have an orbital period of only 4.2 days. Hence, this is a gas giant planet (similar to Jupiter and Saturn) that

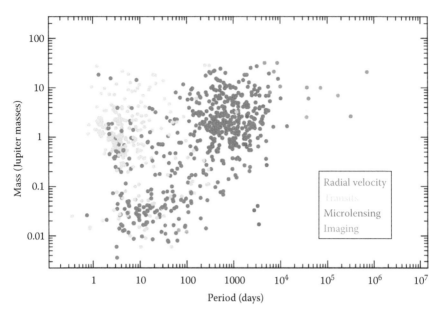

FIGURE 2.10   Graph showing exoplanet masses plotted against orbital period. The different colors represent the different detection techniques. The figure shows that we can find planets of all masses in the region where detection is possible. It also shows, surprisingly, the existence of *hot* Jupiters, Jupiter mass planets orbiting closer to their parent stars than Mercury is to the Sun. (From NASA Exoplanet Archive. This research has made use of the NASA Exoplanet Archive, which is operated by the California Institute of Technology, under contract with the National Aeronautics and Space Administration under the Exoplanet Exploration Program. With permission.)

is orbiting its parent star with an orbital radius smaller than that of Mercury. It is what is known as a *hot* Jupiter and many more have since been discovered.

Figure 2.10 is a plot showing exoplanet masses plotted against an orbital period (for reference, 1 Earth mass is 0.003 Jupiter masses). The different colors represent the different detection techniques. Green shows those detected via the transit method and illustrates how this method typically detects planets that are quite close to their parent star. Red is for exoplanets detected via the radial velocity method, which can detect exoplanets with a large range of orbital periods and hence a large range of orbital radii. The apparent diagonal boundary, however, illustrates that it is more sensitive to low-mass planets close to their parent star than further from the parent star. Blue is for exoplanets detected via microlensing and it is clear that this is typically sensitive to those at modest distances from their parent star (around a Sun-like star, periods between 1,000 and 10,000 days correspond to orbital distances of between 2 and 9 AU) and can detect even quite low-mass planets at these distances. The pink dots are for exoplanets detected via direct imaging, all of which are massive planets (>Jupiter mass) at large orbital radii.

What is intriguing about the results shown in Figure 2.10 is that we find planets of all masses in all regions of parameter space where we have the ability to detect planets. We have low-mass (<0.01 Jupiter masses) and high-mass (>10 Jupiter mass) planets, known as *hot* Jupiters, with periods of less than 10 days (corresponding to orbital separations of

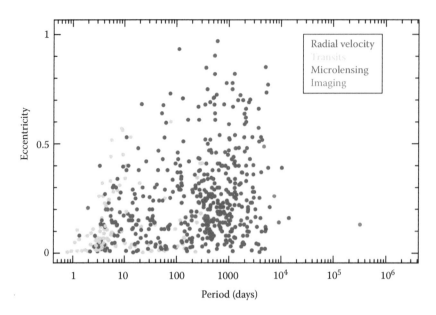

FIGURE 2.11   Graph showing exoplanet eccentricity plotted against orbital period. As in Figure 2.10, the different colors are for the different detection techniques. Unlike the solar system planets, exoplanets have a wide range of eccentricities. (From NASA Exoplanet Archive. This research has made use of the NASA Exoplanet Archive, which is operated by the California Institute of Technology, under contract with the National Aeronautics and Space Administration under the Exoplanet Exploration Program. With permission.)

less than 0.1 AU). There are also planets with similar masses but with periods in excess of 1000 days. Based on our own solar system, and on theories of planet formation, this is not what we would have expected. We would have expected massive gas giant planets in the outer parts of these systems (periods in excess of 1000 days) and small terrestrial, rocky planets in the inner regions.

We also expected that exoplanets would, typically, have quite circular orbits. As shown in Figure 2.11, what we have discovered is very different to what was expected. Many exoplanets have orbits that are remarkably eccentric ($e > 0.9$). What Figure 2.11 also shows is that as the orbital period decreases, the range of eccentricity also decreases. This is because as the planet gets closer to its parent star, it is able to tidally interact with the star that causes it to exchange angular momentum with the star, resulting in its orbit becoming more circular.

As already mentioned, combining transit observations and radial velocity measurements allows us to determine both the mass and radius of an exoplanet. Figure 2.12 shows a plot of the masses and radii of a number of recently detected exoplanets (Zeng and Sasselov 2013). Also shown are curves of the mass–radius relation for various different possible compositions. There are some exoplanets with compositions similar to that of the Earth (50% iron–50% $MgSiO_3$), but there are also some that appear to be more than 50% water, suggesting the possible existence of water worlds.

Essentially, it is now clear that the properties of the exoplanets discovered to date are quite different to what we would have expected only a few years ago and, although

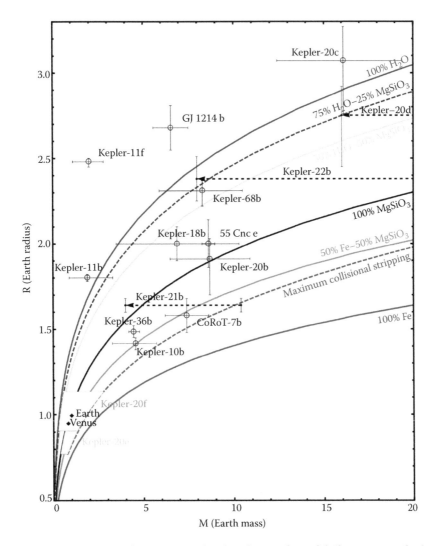

FIGURE 2.12   Figure showing the masses and radii of a number of different, recently discovered, exoplanets. The curves are mass–radius relations for different possible compositions. There are a few exoplanets with compositions similar to that of the Earth, but also a number that appear to be at least half water. This suggests the possible existence of water worlds. (From Zeng, L. and Sasselov, D., *Publications of the Astronomical Society of the Pacific*, 125, 227, 2013. With permission.)

fascinating, clearly means that understanding possible habitability is much more complicated than maybe we anticipated.

2.5.3 How Do We Explain the Properties of Exoplanetary Systems?

As discussed earlier, the basic properties of the exoplanet systems differ quite substantially from what was expected. How do we explain this? One of the main explanations for the existence of *hot* Jupiters was actually done before we had even discovered any exoplanets and was an attempt to explain the existence of close stellar companions. Planets form in discs and are able to exchange angular momentum with the surrounding disc material.

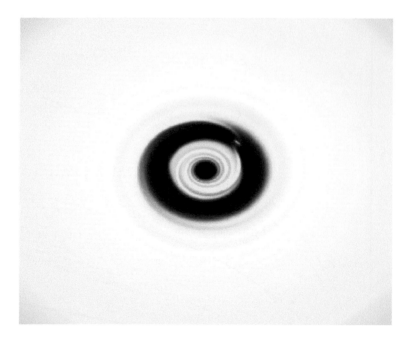

FIGURE 2.13   Figure, from a computer simulation, showing a planet opening a gap in a protostellar disc. In such a situation, the planet will exchange angular momentum with the surrounding disc and migrate in towards the central star, potentially explaining the existence of *hot* Jupiters. (Courtesy of Ken Rice.)

As illustrated in Figure 2.13, if a planet gets sufficiently massive, it can open a gap in the disc and then, typically, gets pushed towards the central star together with the disc material that is slowly flowing inwards. This means that gas giant planets can form in the outer parts of discs, beyond the snow line, but can end up much closer to their parent stars.

Explaining the range of eccentricities is a little less certain, and it is possible—although generally regarded as unlikely—that migration through the gaseous protostellar disc can also drive eccentricity growth. What seems more likely, though, is that dynamical interactions between bodies in multiplanet systems can eject some of the bodies, to produce free-floating planets, leaving the remaining planets on orbits that are more eccentric than they were initially.

Additionally, it is possible for stellar companions to also influence the orbital properties of exoplanets. What is particularly interesting are recent observations showing that some close-in exoplanets have inclined orbits, relative to the spin of the central star. Determining this is a fantastic illustration of how to combine basic measurements so as to determine something quite complex. As mentioned earlier, the radial velocity method involves using the Doppler shift of a star's spectral lines so as to determine the radial velocity of possible planetary host star. If the radial velocity shows a sinusoidal-like variation, then one can use that to determine the mass, orbital radius, and eccentricity of the companion.

If a planet detected via the radial velocity method also happens to transit the parent star, then these two methods can be combined to determine the alignment of the orbital plane with respect to the spin of the host star (Triaud et al. 2010). This is known as the

Rossiter–McLaughlin method and was first developed in the 1920s to analyze binary star systems (McLaughlin 1924; Rossiter 1924). As the planet moves in front of the star, it first blocks light from one side of the star and then the other. Depending on the orientation of the spin of the star, the planet will first block some blue-shifted light and then red-shifted (or vice versa). This changes the Doppler shift of the star's spectral lines and hence appears to change the star's radial velocity. The small kink in the radial velocity curve, seen in Figure 2.14, can then be used to determine the inclination of the orbit with respect to the spin of the central star. In this case, the planet is orbiting at 140° to what would be expected.

There are now a number of exoplanets with orbits inclined with respect to the central star, including one that is almost completely retrograde. It is thought that these systems are a result of interactions with a stellar companion that is on a very inclined orbit with respect to the initial orbit of the planet (Fabrycky and Tremaine 2007). This stellar companion can excite large variations in the orbital properties of the planet, in some cases producing changes in orbital inclination and extremely large eccentricities (Kozai 1962). This can result in the planet passing very close to its parent star. Tidal interactions between the planet and star can then lead to the orbit shrinking and circularizing, and so the planet ends up very close to its parent star and on an orbit that is now inclined with respect to its initial orbital plane.

2.6 HABITABLE ZONES

The habitable zone, sometimes referred to as the Goldilocks zone, is the region around a star where a planet, with sufficient atmospheric pressure, could support liquid water on its surface. There are a number of things that can influence where the habitable zone lies, and so different calculations produce slightly different answers. One factor is the planet's albedo, which is how much of the incoming radiation is simply reflected back into space. Another is the influence of greenhouse gases in the atmosphere that can act to warm the planet. Other more complex factors (carbon cycle, orbital properties) can also influence the range of the habitable zone, but typically, around a Sun-like star, we would expect it to be between about 0.75 AU and 1.4 AU (Kasting et al. 1993).

2.6.1 Habitable Exoplanets

If we consider Figure 2.10, there are clearly a number of exoplanets that likely fall within their star's habitable zone. Most, however, are gas giants or super-Earths (planets with masses between two and eight times that of the Earth). If we ignore all those that are clearly gas/ice giants, there are only a few that are orbiting stars similar to our own Sun. Possibly the most interesting at the moment is Kepler-22b, which has a radius 2.4 times that of the Earth but does not yet have an accurate mass estimate (Borucki et al. 2012). With a radius 2.4 times that of the Earth, Kepler-22b is either quite massive (>10 Earth masses) or a planet with a liquid or gaseous outer shell. This means that even if it is in its star's habitable zone, it is likely not an Earth analog. That, however, does not mean that it cannot harbor life.

One possibility is that a massive planet in the habitable zone of its parent star could have an orbiting satellite (moon) that could be habitable (Forgan and Kipping 2013). To retain an atmosphere, such a moon would need a mass similar to that of the Earth. The most massive planetary satellite in our own solar system is Jupiter's moon, Ganymede, which has a

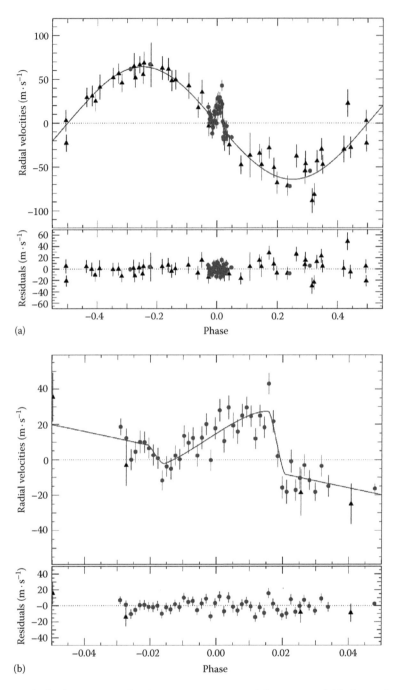

FIGURE 2.14   Radial velocity curve for the WASP-15 system. The top panel (a) shows the sinusoidal variation in the star's radial velocity and indicates the presence of an exoplanet. In this particular system, the planet also transits its host star. When it does so, it blocks some of the starlight and produces the kink in the radial velocity, which is shown in more detail in the lower panel (b). This can be analyzed to show that the plane of the planet's orbit is at 140° to what would be expected. (Reproduced from Triaud, A.H.M.J. et al., *Astronomy and Astrophysics*, 524, article id. A25, 22pp, 2010. With permission.)

mass of 0.025 Earth masses. Hence, we do not even know if sufficiently massive satellites can actually exist. It is, actually, possible to search for such exomoons (Kipping et al. 2013). As the planet and satellite orbit their common center of mass, the timing of the transits of the planet in front of the star changes very slightly. Light curves from satellite measurements, such as from NASA's Kepler satellite, have sufficient precision to detect such small timing variations. To date, however, no massive (>0.5 Earth masses) exomoons have been detected (Kipping et al. 2013).

2.6.1.1 Habitability around Cool Stars

Although there are only a few super-Earths in the habitable zones of Sun-like stars, there are about 10 known super-Earths in the habitable zones of stars cooler than the Sun. These stars are known as K-dwarfs and M-dwarfs and have masses less than 90% that of the Sun and surface temperatures below 5200 K. The habitable zone around such stars is closer than around a star like the Sun. Additionally, the star has a lower mass than the Sun and so the influence of the planet on the star is greater than for a similar planet in the habitable zone of a Sun-like star. Hence, it is easier to find planets in the habitable zones of K-dwarfs and M-dwarfs than around stars like the Sun.

Although it is easier to find low-mass planets in the habitable zones of K- and M-dwarfs than around stars like the Sun, it is not immediately clear that habitability is possible in such systems. The stars are cooler than the Sun and so emit most of their radiation at longer wavelengths than the Sun. This would probably require that processes such as photosynthesis would operate differently on such planets than on planets orbiting Sun-like stars. One suggestion (Wolstencroft and Raven 2002) is that around cooler stars, photosynthesis might operate via a three or four-photon mechanism, unlike the two-photon mechanism that operates on Earth.

An additional issue with habitability on planets orbiting cooler stars is that the habitable zone is close enough to the star that the planet becomes tidally locked. It is this process that has caused the same side of the Moon to always face the Earth, and so planets in the habitable zone of stars with masses less than that of the Sun are also likely to always have one side facing the star. This can make the atmospheres of these planets very unstable (Joshi et al. 1997) and can induce rapid climate shifts that could make the planet uninhabitable (Kite et al. 2011).

Figure 2.15 (Kasting et al. 2014) illustrates how the habitable zone changes with stellar type. For stars cooler than the Sun, it is closer to its parent star and narrower than it would be around a Sun-like star. Around stars with masses less than 90% the mass of the Sun, it also falls inside the tidal lock radius, meaning that one side of the planet will always face its parent star. Figure 2.15 also shows the positions of the known super-Earth exoplanets that fall inside their host stars' habitable zone and shows the positions of the planets in our own solar system.

2.6.2 Possible Exotic Systems

Another recent surprise has been the discovery of circumbinary planetary systems (Doyle et al. 2011). These are planetary systems in which the planet orbits both stars in a

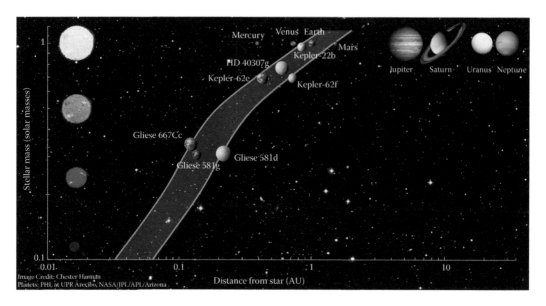

FIGURE 2.15   Figure illustrating how the habitable zone varies with stellar type. For stars cooler than the Sun, it is closer to the star and narrower than it is around a Sun-like star. Additionally, for stars with masses less than 90% that of the Sun, it falls inside the tidal lock radius, and so planets in the habitable zone of such stars would have one side that always faces the host star. The figure also shows the positions of the super-Earth exoplanets that happen to lie inside their star's habitable zone and shows the positions of all the solar system planets. (From Kasting, J.F. et al., *Proceedings of the National Academy of Sciences*, 2014. With permission.)

binary system. Although we had detected a number of planets orbiting one of the stars in a binary system, it had been assumed that the influence of both stars would make planet formation in circumbinary discs very difficult. That we now know of a number of such planets does mean that it must indeed be possible. It is also possible to determine the potential habitable zone in such systems (Forgan 2014), and in one case—Kepler-47c—the planet is in the system's habitable zone. Kepler-47c is, however, a Neptune-sized planet and is, hence, unlikely to be habitable. As mentioned earlier, however, exomoons around such a planet, if sufficiently massive, could potentially support life.

2.7 SOME FINAL COMMENTS

It is now clear that the properties of exoplanetary systems can be very different to what we would expect based on observations of our own solar system. There have been suggestions (Beer et al. 2004) that the solar system may be very rare. Figure 2.10, however, shows that although we have not yet detected an Earth mass exoplanet with an orbital radius of 1 AU around a Sun-like star, we have detected planets with masses similar to the Earth, but closer to their parent stars than the Earth is to the Sun, and we have detected planets with masses similar to the Earth, which are further from their parent star than the Earth is to the Sun. We have also detected Jupiter-like planets at similar distances to their parent stars as Jupiter is to the Sun. So, although we have not actually detected a true Earth analog or another system like the solar system, it seems likely that they must exist.

So, although we may not yet know how common systems like the solar system actually are, we are starting to detect systems that at least have similar properties. For example, we are aware of multiple-planet systems, such as Kepler-11 (Lissauer et al. 2011), where the planets all lie in the same plane and have orbits that are all close to being circular. So, although there are hints that we might be close to finding a solar system analog, we have yet to actually find one.

We also do not know, with respect to habitability at least, how important our solar system properties are. Does habitability require a Sun-like host star? We know that it cannot be a very massive star because their lives are too short, but stars less massive than the Sun have very long lives. However, they also emit most of their radiation in the infrared that may have implications for how processes like photosynthesis can operate. Does a habitable planet need a moon similar to our own? The moon acts to stabilize our rotation axis. Without the moon, the gravitational influence of the Sun and the other planets would have caused the Earth's rotation axis to change quite substantially on relatively short timescales (millions of years). Do we need a massive planet orbiting at a larger radius than the Earth? There is evidence, from craters on the Moon, that the solar system underwent a period of very heavy bombardment when only a few hundred million years old. It is therefore possible that Jupiter played a role in quickly clearing the inner system of dangerous asteroid-like bodies (Greaves 2006).

It does seem that the properties of our own solar system are almost ideally suited for our existence, but we do not know if this is coincidental or if it tells us something about what is needed for the existence of complex life. It does, at least, give us some idea of where we might consider searching for life. Currently, we probably do not quite have the ability to detect a true Earth analog, but the next few years will almost certainly see a significant increase in the number of known, potentially habitable exoplanets around stars less massive than the Sun. By the 2020s, we should be aware of the existence of true Earth analogs and can then start on the work of determining if such planets are actually habited, rather than only potentially habitable.

GLOSSARY

Astronomical unit (AU): The average distance from the Sun to the Earth.
Baryonic matter: Matter that is composed of protons and neutrons.
Big Bang: A term referring to the standard view of how the universe formed.
Coma: The atmosphere of a comet that is blown off to form the comet's tail.
CSIRO: Commonwealth Scientific and Industrial Research Organisation.
Deuterium: A form of hydrogen that has one proton and one neutron in its nucleus.
Eccentricity: The ellipticity of a planet's orbit.
Electron: An elementary particle with a negative charge. In atoms, it orbits the nucleus.
ESA: European Space Agency.
Exomoon: A moon (satellite) in orbit around an exoplanet.
Extrasolar planet (or exoplanet): A planet orbiting a star other than the Sun.
Fusion: A nuclear reaction in which low-mass elements combine to form a more massive element and which also releases energy.

Galaxy: Large collections of stars, gas, and dust. The solar system is in the Milky Way galaxy.
Giant molecular clouds: Dense regions of the interstellar medium where stars form.
Helium: An element with a nucleus containing two protons and two neutrons.
High-mass stars: Stars with masses greater than eight times that of the Sun.
Hydrogen: An element with a nucleus containing one proton.
Intermediate-mass stars: Stars with masses between two and eight times that of the Sun.
Interstellar medium: The material found in regions between the stars in a galaxy.
Julian day: A dating scheme in which day 0 is January 1, 4713 BC.
Lithium: An element with three protons and four neutrons formed via the Big Bang nucleosynthesis.
Low-mass stars: Stars with masses less than twice that of the Sun.
NASA: National Aeronautics and Space Administration.
Nebula: Originally any diffuse luminous object but, now, typically refers to clouds of gas and dust.
Neutrino: A very-low-mass, chargeless elementary particle.
Neutron: A chargeless nuclear particle with a mass similar to, but slightly smaller than, that of a proton.
Neutron star: The remnant core of a high-mass star that is composed entirely of neutrons.
Nuclear fusion: A process, associated with nucleosynthesis, in which low-mass elements combine to form a new, more massive, element.
Nucleosynthesis: Refers to the process in which elements combine to form a new element.
Nucleus: The central part of an atom containing protons and neutrons.
Proton: A nuclear particle with a positive charge.
Retrograde: Moving in the opposite direction to what would be expected.
Semimajor axis: The average orbital distance of a planet from its parent star.
Solar: Refers to the Sun.
Stellar: Refers to stars.
White dwarf: The remnant core of low-mass and intermediate-mass stars that is typically composed of carbon.

REVIEW QUESTIONS

1. What aspects of the Big Bang theory for the formation of the universe are relevant for the existence of life?

2. Explain how nucleosynthesis releases energy and hence powers stars.

3. Why do discs form around young protostars?

4. What is the snow line and how does it influence the formation of planets?

5. Explain the relevance of both low-mass and high-mass stars to the existence of life.

6. How can we use the Doppler effect to detect exoplanets?

7. What characteristics of exoplanets are quite different to what we might have expected?

8 What is the habitable zone?

9. Why would life on planets around very-low-mass stars (mass less than that of the Sun) probably be different to life on planets around Sun-like stars?

10. Discuss whether or not we might expect other habitable planetary systems to be similar to our own solar system.

REFERENCES

Beaulieu J.-P., Bennett D.P., Fouqué P. et al., Discovery of a cool planet of 5.5 Earth masses through gravitational microlensing, *Nature*, **439**, 437–440, 2005.

Beer M.E., King A.R., Livio M., Pringle J.E., How special is the Solar system? *Monthly Notices of the Royal Astronomical Society*, **354**, 763–768, 2004.

Bennett C.L., Larson D., Weiland J.L. et al., Nine-year Wilkinson microwave anisotropy probe (WMAP) observations: Final maps and results, *Astrophysical Journal Supplement*, **208**, article id. 20, 54pp, 2013.

Bonnell I.A., Bate M.R., Clarke C.J., Pringle J.E., Competitive accretion in embedded stellar clusters, *Monthly Notices of the Royal Astronomical Society*, **323**, 785–794, 2001.

Borucki W.J. et al., Kepler-22b: A 2.4 Earth-radius planet in the habitable zone of a Sun-like star, *Astrophysical Journal*, **745**, Article ID 120, 2012.

Cosentino R., Lovis C., Pepe F. et al., HARPS-N: The new planet hunter at TNG, In: *Ground-Based and Airborne Instrumentation for Astronomy IV. Proceedings of the SPIE*, **8446**, article id. 84461V, 20pp, 2012.

Davis M., Efstathiou G., Frenk C.S., White S.D.M., The evolution of large-scale structure in a universe dominated by cold dark matter, *Astrophysical Journal*, **292**, 371–394, 1985.

Doyle L.R., Carter J.A., Fabrycky D.C. et al., Kepler-16: A transiting circumbinary planet, *Science*, **333**, 1602–1606, 2011.

Fabrycky D., Tremaine S., Shrinking binary and planetary orbits by Kozai cycles with tidal friction, *Astrophysical Journal*, **669**, 1298–1315, 2007.

Forgan D., Assessing circumbinary habitable zones using latitudinal energy balance modeling, *Monthly Notices of the Royal Astronomical Society*, **437**, 1352–1361, 2014.

Forgan D., Kipping D., Dynamical effects on the habitable zone for Earth-like exomoons, *Monthly Notices of the Royal Astronomical Society*, **432**, 2994–3004, 2013.

Franck S., Bounama C., von Bloh W., Causes and timings of future biosphere extinction, *Biogeosciences Discussions*, **2**, 1665–1679, 2005.

Greaves J.S., Persistent hazardous environments around stars older than the Sun, *International Journal of Astrobiology*, **5**, 187–190, 2006.

Guth A.H., Inflationary universe: A possible solution to the horizon and flatness problem, *Physics Reviews D*, **23**, 347–356, 1981.

Haisch K.E., Lada E.A., Lada C.J., Disk frequencies and lifetimes in young clusters, *Astrophysical Journal*, **553**, L153–L156, 2001.

Hoyle F., Fowler W.A., Burbridge G.R., Burbridge E.M., Origin of the elements in stars, *Science*, **124**, 611–614, 1956.

Johnson J.A., Winn J.N., Cabrera N.E., Carter J.A., A smaller radius for the transiting exoplanet WASP-10b, *Astrophysical Journal*, **692**, L100–L104, 2009.

Joshi M.M., Haberle R.M., Reynolds R.T., Simulations of the atmospheres of synchronously rotating terrestrial planets orbiting M dwarfs: Conditions for atmospheric collapse and the implications for habitability, *Icarus*, **129**, 450–465, 1997.

Kasting J.F., Kopparapu R., Ramirez R.M., Harman C., Remote life detection criteria, habitable zone boundaries, and the frequency of Earth-like planets around M and Late-K stars, *Proceedings of the National Academy of Sciences*, in press, 2014.

Kasting J.F., Whitmire D.P., Reynolds R.T., Habitable zones around main sequence stars, *Icarus*, **101**, 108–128, 1993.

Kipping D.M., Forgan D., Hartman J. et al., The hunt for exomoons with Kepler (HEK): III. The first search for an exomoon around a habitable-zone planet, *Astrophysical Journal*, **777**, Article ID 134, 2013.

Kite E.S., Gaidos E., Manga M., Climate instability on tidally locked exoplanets, *Astrophysical Journal*, **743**, article id. 41, 12pp., 2011.

Kozai Y., Secular perturbations of asteroids with high inclinations and eccentricity, *Astronomical Journal*, **67**, 591–598, 1962.

Lissauer J.J., Fabrycky D.C., Ford E.B. et al., A closely packed system of low-mass, low-density planets transiting Kepler-11, *Nature*, **470**, 53–58, 2011.

Lodders K., Solar system abundances and condensation temperatures of the elements, *Astrophysical Journal*, **591**, 1220–1247, 2003.

Marois C., Macintosh B., Barman T. et al., Direct imaging of multiple planets orbiting the star HR 8799, *Science*, **322**, 1348–1352, 2008.

Mayor M., Queloz D., A Jupiter-mass companion to a solar-type star, *Nature*, **378**, 355–359, 1995.

McLaughlin D.B., Some results of a spectrographic study of the Algol system, *Astrophysical Journal*, **60**, 22–31, 1924.

Murray N., Evidence of things not seen, *Science*, **336**, 1121–1122, 2012.

Peebles P.J.E., Primordial helium abundance and the primordial fireball. II, *Astrophysical Journal*, **146**, 542–552, 1966.

Penzias A.A., Wilson R.W., A measurement of excess antenna temperature at 4080 Mc/s, *Astrophysical Journal*, **142**, 419–421, 1965.

Pollack J.C.B., Hubickyj O., Bodenheimer P. et al., Formation of the giant planets by concurrent accretion of solids and gas, *Astrophysical Journal*, **124**, 62–85, 1996.

Rossiter R.A., On the detection of an effect of rotation during eclipse in the velocity of the brighter component of beta Lyrae, and on the constancy of velocity of this system, *Astrophysical Journal*, **60**, 15–21, 1924.

Salpeter E.E., Nuclear reactions in stars without hydrogen, *Astrophysical Journal*, **115**, 326–328, 1952.

Terebey S., Shu F.H., Cassen P., The collapse of the cores of slowly rotating isothermal clouds, *Astrophysical Journal*, **286**, 529–551, 1984.

Triaud A.H.M.J., Cameron A.C., Queloz D. et al., Spin-orbit measurements for six southern transiting planets. New insights into the dynamical origins of hot Jupiters, *Astronomy and Astrophysics*, **524**, article id. A25, 22pp., 2010.

Walker T.P., Steigman G., Schramm D.N. et al., Primordial nucleosynthesis redux, *Astrophysical Journal*, **376**, 51–69, 1991.

Wolstencroft R.D., Raven J.A., Photosynthesis: Likelihood of occurrence and possibility of detection on Earth-like planets, *Icarus*, **157**, 535–548, 2002.

Zeng L., Sasselov D., A detailed model grid for solid planets from 0.1 through 100 Earth masses, *Publications of the Astronomical Society of the Pacific*, **125**, 227–239, 2013.

Astrobiology Education and Public Outreach

Timothy F. Slater

CONTENTS

3.1 INTRODUCTION

Over the centuries, the scientist's job description has changed significantly. In far earlier times, astrologers carefully monitored the predictable motion of the stars to advise clients about favorable times to wage war, plant crops, and engage in ceremony. Healers collected and administered specific herbs to restore health and fight off sickness. Alchemists mixed magical ores from the ground to create metals for machines and weapons. In more recent eras, astronomers now devise the structure of the universe by dissecting the faint light from distant objects, biochemists track the rapid evolution of living organisms by capturing their DNA, and geological engineers remotely search deep underground to recover fossil fuels that power our airplanes, computers, and home air-conditioning systems. Yet, unchanging across time is that all of these individuals are required to share what they have learned in order to bring their discoveries to profitable financial markets and to teach the next generation of scientists their trade to continue on the quest to understand the workings of the universe.

Astrobiologists, too, hold a deep responsibility to successfully and honestly share their science with other scientists (for peer review), with the younger generation (to keep the science career training pipeline full), and with the general public (to keep granted financial support through taxation flowing). Sometimes, this responsibility comes from an externally imposed requirement from scientists' funding entities to teach others about discoveries and methods of astrobiology: other times, this responsibility to teach seems to emerge unprompted from within an individual's intrinsic motivations. Regardless of the source of motivation, sharing astrobiology through education and public outreach (E/PO) is a necessary component of any successful scientific enterprise, just as are requirements for intellectual integrity in ethically reporting research methods and results as well as an unwavering commitment to avoiding unnecessary harm to humans. In other words, you might be surprised to know that successful E/PO efforts are just as important to doing science as any other aspect of science.

When most scientists think about education, they immediately think about their own schooling experiences. All scientists have gone to school—successfully—and this leads scientists to mistakenly think that they fully understand education already. Education, as it turns out, is a surprisingly complex entity all on its own. Just like there are interdisciplinary astrobiologists that specialize in microbial life and other interdisciplinary astrobiologists that specialize in comparative planetology, there are interdisciplinary astrobiologists who specialize in the scholarly pursuit of E/PO.

As a matter of analogy, consider that the most often failing small business started in the United States is restaurant ownership. This type of new business endeavor frequently fails because people mistakenly, albeit understandingly, think that because they have enjoyed eating at a variety of restaurants that they must have the knowledge necessary to open and successfully run a restaurant of their own design. The problem here is that simply because one has been to a restaurant, that does not mean that they know how to design or operate a restaurant.

In much the same way, when one observes a musical conductor wave their arms in front of a symphony orchestra, it seems plausible to assume that almost anyone could simply wave their arms in a similar way and just as effectively lead an orchestra or symphony. In truth, the job of a highly successful musical conductor sometimes deceptively looks easy to the outside world, but is, of course, far more complicated. This notion of assuming a task is easy because it looks easy when it really is quite complicated is a replicable phenomenon known as a mistaken *apprenticeship of observation* and occurs frequently in the realm of E/PO. Astrobiologists too are subject to this, all too often naturally assuming that they are already fully prepared to conduct E/PO simply because they have been a student of science for many years in the pursuit of becoming an astrobiologist. In this chapter, we take the first steps toward improving your understanding of E/PO in astrobiology.

3.2 SCIENCE EDUCATION AND PUBLIC OUTREACH LANDSCAPE

E/PO encompasses a surprisingly wide range of activities under the broader category of scientific communication. The most narrow view of E/PO is that of an astrobiologist giving talks during grade school classroom visits or teaching classes in astrobiology at a college or university. Fortunately, there is a much larger spectrum of ways astrobiologists can impact E/PO. To help make sense of this wide range of E/PO, we use four broad, but somewhat overlapping, categories of effort, based largely on the specific audience targeted.

3.2.1 Formal Education Settings

The most often targeted category of E/PO is known as formal education (see Table 3.1). Formal education is that learning that occurs in traditional school environments. This includes teaching in elementary, middle, and high school classrooms. It also includes teaching in formal courses in college and university lecture halls. Less often, but still important, the formal education category includes supervised research experiences for undergraduate or graduate students.

Within this category, astrobiologists frequently contribute to the E/PO enterprise by speaking to students during classroom visits. Although this is usually the first thing astrobiologists imagine when they consider engaging in E/PO, it turns out that this is the most time consuming and least effective E/PO effort. Considerable travel and preparation time, logistics coordination, and overall effort go into making a classroom visit, which rarely lasts more than 1 h. It is not that classroom visits are not fun for astrobiologists or the students they are talking to; rather, the lack of effectiveness stems from the fact that most classroom visits occur in complete isolation from the regular classroom curriculum and as a result fails to be integrated into the larger portfolio of a student's experience. The end result is that although everyone usually has a good time, its lasting impacts are at best negligible. More worrisome, many students hold tacit stereotypes that scientists can only be Caucasian males, and classroom visits, at their worst, can unintentionally confirm some students' negative stereotypes of who can become a scientist and remove themselves from the science career pipeline.

A more productive role for astrobiologists in formal education settings can be to assist professional educators in the development of high-quality and scientifically accurate curriculum materials. Again, astrobiologists can unintentionally fall subject to the aforementioned mistaken apprenticeship of observation, and most can best help professional curriculum developers by providing access to scientific databases and interpretation of recent discoveries.

Even the most well-meaning astrobiologists sometimes find it incredibly difficult to gain entry into schools to give lectures or to help create classroom curriculum materials. This is because most K-12 school teachers and schools feel tremendous pressure from national legislation that was not in place when most astrobiologists themselves were in school. For the

TABLE 3.1   E/PO Categories

Formal Education	Informal Education	Public Outreach
Participants are required to attend.	Participants choose to attend.	Information is delivered to participants wherever they currently are.
Fewest numbers impacted.	←--------→	Greatest numbers impacted.
Interactions are longest in duration.	←--------→	Interactions are shorter in duration.

past 200 years, schools in the United States have been coordinated, directed, and run by local school boards with only some state-level controls and essentially no national-level federal controls. In the past decade, this has changed dramatically with unprecedented federal-level legislation known as *No Child Left Behind* (NCLB). NCLB requires that all schools demonstrate improvement in student achievement each and every year through standardized testing and that every student is required to be taught by a highly qualified teacher. These are important goals that nearly everyone would agree are important in principle but have far-reaching implications when put into practice. For the amateur astronomer, the most important aspect of NCLB is to recognize that today's students are tested, tested, and tested and that these test scores really do matter. As such, teachers and schools feel considerable pressure to focus on helping students increase their achievement test scores above all else—even if that *else* includes unique opportunities to learn about astronomy and technology. In other words, teachers are pressured to teach what material is covered by the mandated achievement tests. The enthusiastic astrobiologist needs to determine the extent to which students learning about astrobiology will help improve achievement test scores before delving too far into a particular school's environment because this varies considerably school to school and state to state.

Perhaps the most helpful astrobiologists can be for informal classroom education settings is to serve as advocates for the importance and value of high-quality science education, rather than making classroom visits. A sizeable portion of school parents believe the concepts of biological evolution, human-induced global climate change, and cosmological big bang evolution to be in conflict with spiritual beliefs, and schools are constantly under pressure to remove these concepts from their teaching goals. Astrobiologists are perhaps the best positioned of all scientists to be of great assistance to schools in keeping these critically important scientific concepts as an integral part of school curriculum (see Table 3.1).

3.2.2 Informal Education Settings

A second category of E/PO is the realm of informal education (see Table 3.1). Informal education is generally considered to be the domain of museums, planetariums (the technically correct, but awkward, plural of planetarium is planetaria), nature centers, visitor centers, and comprehensive science centers. These informal education facilities and/or programs vary considerably from one to another in how many people they serve as well as how broad their subject areas are. As a few examples, New York's *American Museum of Natural History* covers nearly all of the science disciplines, whereas Chicago's *Adler Museum* or the Smithsonian *Air and Space Museum* focuses on the much narrower domain of exploring the universe of topics beyond Earth's atmosphere. For astrobiologists, these informal education venues can serve as important rallying points to serve as a meeting facility or source of advertising mailing lists for public lectures. In fact, astrobiologists can volunteer to play a critical functional role for museums working as docents, or volunteer tour guides, to enhance the experience of visitors.

What most easily distinguishes informal education from the previously described formal education is that informal education relies heavily on the concept of *free-choice learning*. Free-choice learning was widely popularized by John Falk, who wrote a book of the same title (Falk and Dierking 2002). Free-choice learning is the type of learning guided by an individual's needs and interests and is an essential component of lifelong learning. Indeed, it is likely that free-choice learning is the most common type of learning that people engage in. This sort of individually motivated learning is so common that most astrobiologists often simply taken for granted. Because people have control over what and how they learn, and because they can choose to learn in appropriate and supportive contexts, free-choice learning can be highly efficient and the ideas acquired this way can be highly durable and memorable.

What is important for astrobiologists to know about the ubiquitous dominance of free-choice learning in museums is that informal educators work diligently to create highly flexible learning environments where participants can devote either only a few seconds or up to tens of minutes for exploring a scientific concept. Astrobiologists working in museums sometimes find this to be rather disconcerting and somewhat difficult to adapt to if they are used to, and expect, people to stand around their telescope for several minutes politely listening to a mini lecture. Sometimes, people walking in and out when you are in midsentence takes some real getting used to, but, when mastered, informal education learning environments often make a fantastic environment to hook people into attending other astrobiology learning events.

3.2.3 Astrobiology Public Outreach

As a final category of E/PO, let us briefly consider the domain of public outreach (see Table 3.1). Public outreach is an activity where scientific information is delivered to the public in situ—that is, people's private homes through cable TV, in their cars over the radio, and in their doctor's waiting rooms through magazine subscriptions. In other words, this is where astrobiologists reach out to the public rather than asking learners to come to them.

Public outreach is typically delivered in relatively short bursts to incredibly large audiences in one-size-fits-all approach, with little or no direct interaction with the receiver. In this sense, public outreach delivered via video can take the form of 40–60 min video documentaries and IMAX films, such as those found on the *Discovery Channel* and the *Science Channel*. It can also be found on the radio in the form of 2 min audio shorts, such as those known as McDonald Observatory's *StarDate* or *Earth and Sky*. Of course, newspapers, news magazines, and astronomy-specific magazines share the excitement of astronomy on a regular basis. Even the quickly growing and evolving Internet participates in public outreach too, particularly in the form of websites such as *Space.com* and *UniverseToday.com* or through Internet-delivered podcasts. All of these examples, and many others, can be most easily found by conducting an Internet-based search using a common search engine, an example of which is *Google.com*. In the end, public outreach is generally considered to use a strategic E/PO approach of "take it to the people," whereas the approach of informal education relies on a "build it and they will come" strategy.

3.3 MOST IMPORTANT IDEAS ASTROBIOLOGY E/PO CAN SHARE

Providing learners with an academic, comprehensive, and technical definition of astrobiology is a guaranteed way to keep potentially interested learners at arm's length. Instead, framing astrobiology in such a way that audiences can connect with the science is the ultimate goal of astrobiology E/PO. One framework for defining astrobiology for the public is defining astrobiology as the study of the origin, evolution, distribution, and destiny of life in the universe.

There are an overwhelming number of scientific concepts described in this book, ranging from microbiology through evolution of planets. Even going beyond the scope of this book, ideas from sociology and philosophy are fundamentally important components of the astrobiology enterprise. Given such a wide variety of topics one could use in beginning an E/PO plan, we propose three here as departure points. These are as follows: (1) the search for life elsewhere in the universe is equivalent to the search for water; (2) microscopic bacteria are the most common form of the life on Earth and are what are most likely living out there; and (3) recent research results suggest that many of the components of the Drake equation's predictions for the frequency of intelligent life in the galaxy are much larger than earlier imagined, making extraterrestrial life even more likely.

3.3.1 Search for Water

It bears repeating what most people already know—unique among all known planets and moons discovered so far, Earth's surface is mostly covered by liquid water. In much the same way, most people readily understand that water is an essential ingredient for the existence of life. However, at the same time, what is perhaps quite surprising to many people is that other objects in our solar system have surprisingly large amounts of water.

If one were to collect up all of Earth's water, it would be a ball about 860 mi in diameter. If you were to burst this imaginary ball of water, the resulting flow would cover the contiguous United States (lower 48 states) to a depth of about 107 mi. This is similar in total to how much water is suspected to be hiding on Mars and Saturn's moon Enceladus and far less water than exists beneath the surfaces of Jupiter's moon Europa or Saturn's moon Titan.

Astrobiologists are exceptionally well positioned to help learners understand the unique properties of water that make it such a critical part of life, and why looking for life beyond Earth is nearly equivalent to looking for water beyond Earth. Water has all sorts of fascinating properties. One is that when water gets cold enough to freeze, it expands in size. Perhaps, this has unexpectedly happened to you before when you put a plastic bottle of water in the freezer. This is a unique property, because most other substances get smaller and take up less space when they get colder. This is important for fish living in lakes that freeze in the winter, because ice takes up more space than the same amount of water. Ice also floats to the top of a lake, rather than sinking to the bottom, leaving space for fish to swim all winter long. This also impacts the landscape because when water seeps into rocks and expands upon freezing, it breaks rocks apart.

Another important property of water is its great capacity for holding energy without changing temperature. You might already know that swimming pools and lakes are quite slow to heat up in the summer and equally slow to cool down in the autumn seasons. Water's great resistance to changing temperature compared to other substances makes it a stable substance that life can depend upon even when short-term weather changes the temperature quickly and dramatically. This same ability to hold tremendous amounts of energy allows water vapor to move energy around our planet quickly wherever the wind takes it.

Finally, different sides of an individual water molecule itself have different characteristics, giving water additional unique properties you might not expect. One is that water molecules tend to cling to each other making it flow from one place to another in continuous streams while sticking together. This property accounts for water collecting together and falling from the sky in the form of semiround drops. Water also tends to stubbornly stick to other surfaces. This same property allows it to spread out evenly across a surface, like a lake, or even within a substance, like inside a rock. Another property, compared to other liquids, is that water has a wide ability to break apart and dissolve a large variety of substances, including common substances such as salt, sugar, and ammonia. When mixed with acid, such as pollution-forming acid rain, water can even dissolve solid rock, like marble and limestone. As far as we know, life cannot exist without water. It is these unique properties taken together that make water a highly useful substance for life to thrive on our planet.

3.3.2 Bacteria Are Alive

A second important notion the astrobiology E/PO can focus on is the nature of life. It is mostly unknown to learners that microscopic bacteria constitute the most common form of the life on Earth today. Moreover, as modern humans have a history of about 100,000 years on Earth, bacteria have been on Earth much longer, perhaps as long as about 4 billion years. What this means is that over Earth's 4.5-billion-year history, if an alien spacecraft visited at a random time since the formation of Earth, it is highly likely that the visitor would have only found bacteria living here. This has tremendous implications for the search for life elsewhere in the solar system. If Earth is typical of planets, this means that most planets we explore and find to have life will have bacteria rather than farmers, musicians, and politicians living there.

The challenge here for astrobiology E/PO is twofold. The first is successfully explaining to the learner that much of searching for extraterrestrial life is about looking for strange-looking bacteria, rather than looking for the somewhat more familiar humanoid-looking creatures often portrayed in science fiction movies. The second, perhaps more challenging, is convincing the learner that bacteria themselves are a critically important part of *life* in the first place and that finding bacteria elsewhere in the universe is a huge triumph of a successful astrobiology research program.

3.3.3 How Long Do Civilizations Exist?

Whereas the framework of a professional scientist's view of astrobiology looks mostly like the table of contents of this book, the target audience member for astrobiology E/PO uses a dramatically different framework. The astrobiology E/PO magic hook to capture a listener's attention is the search for intelligent life. Only after these audience questions are acknowledged can astrobiology E/PO cover more academic ground.

Probably, the only mathematical equation an expert astrobiology E/PO presenter would use would be the Drake equation. When used delicately, the Drake equation can help provide an organizing framework to present some of the major ideas in astrobiology while, at the same time, address the public's questions about the search for intelligent life. The wise astrobiology E/PO specialist recognizes that the general public is easily discouraged when looking at mathematical equations and will slowly consider each characteristic, one at a time:

$$\text{Drake equation: } N = R^* f_p n_e f_l f_i f_c L$$

where

N is the number of technologically advanced civilizations in the galaxy whose messages we might be able to detect

R^* is the rate at which solar-type stars form in the galaxy

f_p is the fraction of stars that have planets

n_e is the number of planets per solar system that are earthlike (i.e., suitable for life)

f_l is the fraction of those earthlike planets on which life actually arises

f_i is the fraction of those life-forms that evolve into intelligent species

f_c is the fraction of those species that develop adequate technology and then choose to send messages out into space

L is the lifetime of a technologically advanced civilization

A wide range of values has been proposed for the many terms in the Drake equation, and these various guesses produce vastly different estimates of N. Some astrobiologists argue that there is exactly one advanced civilization in the galaxy and that we are it. Others speculate that there may be hundreds or thousands of planets inhabited by intelligent creatures, and extraterrestrial communication is just too difficult or too expensive in which to participate. At this point, we just do not know whether our galaxy is devoid of other intelligence, teeming with civilizations, or something in between—and this unknown is a captivating hook to engage audiences in astrobiology.

3.4 DIFFICULT ASTROBIOLOGY CONCEPTS

The astrobiology E/PO expert knows that some ideas are easy for learners to grasp, whereas others are more complicated. The most challenging of these have been identified as (1) the nature of life, (2) conceptions of temperature, and (3) the nature of stars.

Seminal work in astronomy and astrobiology education research by Offerdahl and her colleagues (2002) established the baseline for astrobiology E/PO. She found that the majority of students she studied, ranging in age from 12 to 22 years of age, correctly identified that liquid water is necessary for life and that life-forms can exist without sunlight. At the same time, most of these same students incorrectly stated that life cannot survive without oxygen. Furthermore, when students were asked to reason about life in extreme environments, they most often cited complex organisms—such as plants, animals, and humans—rather than microorganisms. This is summarized in Table 3.2.

It is generally known that people are often surprised to learn that our Sun is a star, only being larger because it is much closer. Astronomy education research on students'

TABLE 3.2   Findings by Offerdahl and Colleagues (2002) on Astronomy and Astrobiology
Education Research

Can life exist in places that never receive sunlight?	Citing animals that live in caves, 65% of students believe life can exist without sunlight.
Would you agree that for a life-form to exist, liquid water must be present in the local environment in at least small amounts or for short time periods?	75% of students believe that liquid water is necessary.
Can any forms of life exist in environments with temperatures much less than 0°C (32°F)?	Citing penguins and polar bears, 80% of middle school students and 90% of high school and college students responded that life can exist in this temperature range.
Can any forms of life exist in environments with temperatures much greater than 100°C (212°F)?	74% of middle school students, 54% of high school, and 42% of the college students responded that life cannot exist in this temperature range, while others disagreed and stated that life in deserts can flourish even at high temperatures.
Are there any environments on Earth that would *not allow any* form of life to exist?	• 72% of middle school students, 54% of high school students, and 51% of college students listed volcanoes. • Approximately 20% of students listed environments where no water exists. • 16% of middle school students, 25% of high school students, and 21% of college students described environments with no air or no oxygen.

understanding of the nature of stars by Bailey and her colleagues (2009) reports when giving college students open, student-supplied response questions about the nature of stars, nearly 80% of students responded that a star is made of gas or a gas/dust combination. Nearly half of all student responses included the phrase *ball of gas* to describe a star and that stars shine because they are *burning*. Less than half think of stars as forming a gravitational collapse of gas and dust to form a shining star, and only 16% even mentioned gravity as a mechanism.

Taken together, E/PO experts now recognize that there are many things that trained astrobiologists know that college-educated adults do not know. The imminent danger here is that if an astrobiologist makes faulty assumptions about their audiences' basic scientific literacy, then they will be unable to successfully communicate their ideas (Offerdahl et al. 2004, 2005). It is because of these existing misunderstandings about basic scientific principles that traditional lecture modes of teaching fail (Slater 2006, 2008). This requires successful astrobiology E/PO to use more contemporary teaching methods if they are to be successful.

3.5 MODERN ASTROBIOLOGY TEACHING METHODS

Quite different from the days gone by of when the average-aged astrobiologist was a student in school, the nature of what modern classroom instruction looks like has changed dramatically. Certainly more so in some schools and less so in others, in general, modern educators and public outreach specialists in the twenty-first century do not often lecture for hours on end and ask students to take copious notes for memorization. Instead, there exists a somewhat different perspective of how best to help students learn.

Astrobiology E/PO professionals who wish to be successful in schools need to understand this perspective and utilize it if they hope to be invited back for a return visit.

Contemporary E/PO is based upon a philosophical foundation known as *constructivism*. In a broad sense, classroom instruction built upon a constructivist viewpoint leans heavily on the idea that students already think that they know a considerable amount about how the world works—they believe they understand why the Moon appears to change shape over the month (the Earth's shadow blocks the Sun's illuminating light), why it is hotter in the summer time than the winter time (the Earth's closer to the Sun in the summer), and that gravity is caused by a planet's atmosphere (the Moon has no atmosphere, therefore cannot have any gravity), just to name a few of the most tenacious misconceptions. In other words, the teaching perspective is one that could be paraphrased as, "determine what each student knows and teach them accordingly." This is in direct contrast to the more traditional viewpoint of days gone by that students do not really know anything, they are essentially blank slates with which knowledge is to be precisely written, and it is the teacher's responsibility to clearly explain everything to them so they will then know the things we most want them to know.

This has important ramifications for the astrobiology E/PO provider in teaching in today's classrooms if they want to be effective. In days gone by, when a guest scientist came to visit a classroom, the students would most often respectfully listen to the speaker, no matter what was being said or how it was being said. However, students of today fully expect, if not demand, to understand *why* what they are being told has value to them in order for students to stay tuned-in to the presentation. The most important aspect of contemporary classroom-based instruction is that it is critical for scientists visiting the classroom to directly and intentionally ask students what they think about an idea and what preexisting knowledge they bring to the class about the topic before *teaching* or else students will quickly become bored and lose their attentions after the visitor's novelty has worn off. The key piece here for the effective astrobiology E/PO specialist teaching in a school is to ask students a lot of questions and give only few answers. This is summarized in Table 3.3.

TABLE 3.3   Suggested Astrobiology E/PO Toolkit

Astrobiology E/PO Toolkit to Actively Engage Students in Learning

- Ask students questions (not all questions are equal)
- Use demonstrations (interactive lecture demos)
- Surprise quizzes (graded/ungraded)
- In-class writing (with/without discussion)
  - Muddiest point
  - Summary of today's main points
  - 5 min free writing
- Think–pair–share (peer instruction–concept tests)
- Small group tasks (closed/open; in/out of class)
- Student debates (individual/group)
- Student presentations (oral, poster, role play)
- Jigsawing

As a matter of example, consider that when giving public talks on astrobiology and searching for extraterrestrial life, a frequently posed end of lecture question is, "Why are we looking for bacteria instead of for living things?" One approach to answering this question is to reflect the question back on the audience about what it means to be *alive*. An organism that is *alive* takes and recycles energy from their environment, altering their environment by their being there and reproducing—bacteria meet this definition. Another reason people ask is because they are concerned that NASA might not be wise by spending scarce tax dollars on looking for *lesser* forms of life, and describing why this is important is a critical aspect of successful astrobiology E/PO.

A more modern approach would be to start an E/PO event, perhaps a public lecture or a school visit, by specifically targeting this idea from the very beginning. Suppose instead of starting out by lecturing, one were to pose to listeners at the very beginning to "imagine they were an astronaut being sent to a newly discovered planet where life existed—and to creatively sketch what creatures they might find there and to provide three important facts about these new life-forms." You might imagine what your audience would sketch at the outset—nearly 100% draw anthropomorphic things with arms, mouths, eyes, and legs.

The next step for a modern teaching approach would be to elicit the listeners' ideas, thus engaging them on what they think is important. Delaying a lecture as long as possible, one could poll the audience and ask them to raise their hands and how many drew something with eyes. Then query, "Could several people give examples of living things on Earth that have eyes?" Next, query some specific examples of living things on Earth that do *not* have eyes—things like trees and worms (and of course, bats and sharks might be mentioned too as having lousy eyesight). The hook to start the lecture then is, "I wonder if there are more living things on Earth with eyes or without eyes?"

This teaching approach starts where the audience's knowledge actually is, instead of where one wishes they were. It is also a teaching approach that is focused on generating questions rather than on giving answers. Imagine that at this point, the speaker challenges the learners and poses, "Imagine you are an alien from another planet sent to land on Earth and look for life…would you look for life with eyes or life without eyes? If your alien exploration ship just landed at some random spot on Earth, would you land in the water or on land?"

One could simply tell the audience the answer. Or, if modern teaching approaches are best, here is an opportunity to get the audience kinesthetically involved. Research by Slater and her colleagues (2008) finds that learners are more engaged if they are physically engaged. Imagine using an inflatable beach ball globe of the Earth, randomly being tossed around the room. With each person's catch of the globe, record the responses to, "Imagine you landed where your right thumb is, would you likely look at the window of your spacecraft and see a person?" After 15 or so tosses, it becomes clear that most of the time, a randomly landing thumb (alien spacecraft) does not hit land, let alone a population center. Now, this is the culminating piece of evidence for the audience—aliens might not even know there were living things with eyes, mouths, arms, and legs if they visited Earth unless they are lucky. And, this, of course, supports the notion that when we go to other planets, we need to be open-minded about what life there might look like and it probably does not look much like us. Only after these experiences engaging the audience, are they ready to hear a short lecture.

But even then, active intellectual engagement is important. Asking learners questions to which they debate answers during a learning experience with other learners takes full advantage of what is known about how people learn. Questions posed by the presenter to the students can either be simple recall (*e.g., Which planet has the highest surface temperature?*) or can be questions of application (*e.g., At which phase of the moon will a solar eclipse occur?*). Less often, but still quite effective at enhancing understanding, presenters can pose questions encouraging learners to encounter a widely known misconception (*e.g., How often does the Moon's appearance change?*).

The underlying astronomy education research supporting these approaches is solid. Sadler (1992) and Bailey and Slater (2005), among others, found that college graduates, from even the US's top universities, harbor serious misconceptions about even the most basic concepts. Hake (1998) found that when surveying 6000 physics students, students that were taught using some sort of *interactive engagement techniques* scored statistically better than students that were taught using conventional lecture courses. Francis and his colleagues (1998) found that students from his interactive engagement physics course had no decrease in achievement 2, 3, and 4 years after the course. Taken together, astronomy education research as an entire body of scholarly literature points to the unquestionable need for contemporary teaching approaches as opposed to even the most cleverly delivered and entertaining lectures if the goal of astrobiology E/PO is enhanced understanding.

REVIEW QUESTIONS

1. What specific trainings do professional astrobiologists have in E/PO?

2. Describe in detail a learning experience you have had in one of the three realms of E/PO.

3. Justify which of the major ideas in astrobiology is the most important to share with students or the general public.

4. Explain why finding bacteria on another planet may not be judged as finding *life* by the media.

5. Which of the astrobiology misconceptions is most prevalent?

6. The anticipated cost of transporting a gallon of water from Earth to the Moon is $15,000. Estimate the cost of taking a single-day's supply of water for your class to the Moon by determining how much water each of your class members use in a single day.

REFERENCES

Bailey, J.M., Prather, E.E., Johnson, B., and Slater, T.F. (2009). College students' preinstructional ideas about stars and star formation. *Astronomy Education Review*, **8**(1).
Bailey, J.M. and Slater, T.F. (2005). Resource letter on astronomy education research. *American Journal of Physics*, **73**(8), 677–685.

Falk, J. and Dierking, L. (2002). *Lessons Without Limit: How Free-Choice Learning is Transforming Education*, AltaMira Press, Walnut Creek, CA.

Francis, G., Adams, J., and Noonan, E. (1998). Do they stay fixed? *The Physics Teacher*, **36**, 488–490.

Hake, R. (1998). Interactive engagement versus traditional methods: A six-thousand student survey of mechanics test data for introductory physics courses. *American Journal of Physics*, **66**(1), 64–74.

Offerdahl, E.G., Morrow, C.A., Prather, E.E., and Slater, T.F. (2005). Journey across the disciplines: A foundation for scientific communication in bioastronomy. *Astrobiology*, **5**(5), 651–657.

Offerdahl, E.G., Prather, E.E., and Slater, T.F. (2002). Student beliefs and reasoning difficulties in astrobiology. *Astronomy Education Review*, **2**(1), 5–27.

Offerdahl, E.G., Prather, E.E., and Slater, T.F. (2004). Emphasizing astrobiology: Highlighting communication in an elective course for science majors. *Journal of College Science Teaching*, **34**(3), 30–34.

Sadler, P. (1992). The initial knowledge state of high school astronomy students, dissertation. Harvard School of Education Dissertation Abstracts International, 53(05), 1470A (University Microfilms No. AAC-9228416).

Slater, S.J., Slater, T.F., and Morrow, C.A. (2008). The impact of a kinesthetic astronomy curriculum on the content knowledge of at-risk students. *National Association of Research in Science Teaching Conference*, Baltimore, MD, April 2008.

Slater, T.F. (2006). Teaching astrobiology with dilemmas and paradoxes. *Journal of College Science Teaching*, **35**(6), 42–45.

Slater, T.F. (2008). First steps toward increasing student engagement during lecture. *The Physics Teacher*, **46**(8), 554–555.

CHAPTER 4

Analysis of Extraterrestrial Organic Matter in Murchison Meteorite

A Progress Report

Philippe Schmitt-Kopplin, Mourad Harir, Basem Kanawati,
Régis Gougeon, Franco Moritz, Norbert Hertkorn,
Sonny Clary, Istvan Gebefügi, and Zelimir Gabelica

CONTENTS

4.1 INTRODUCTION

Natural organic matter (NOM) on Earth occurs in soils; in freshwater, marine, and hydrothermal environments; or in the atmosphere and represents an exceedingly complex mixture of organic compounds that collectively exhibits a nearly continuous range of properties (size–reactivity continuum). The composition and structure of NOM in the bio- and geosphere is established and governed according to the rather fundamental constraints of thermodynamics and kinetics. In these intricate materials, the *classical* signatures of the (geogenic or ultimately biogenic) precursor molecules, like lipids, glycans, proteins, and natural products, have been attenuated, often beyond recognition, during a succession of biotic and abiotic (e.g., photo and redox chemistry) reactions. NOM incorporates the hugely disparate characteristics of abiotic and biotic complexity.

Extraterrestrial organic matter (EOM) in particular can be considered as end members of pure chemical synthesis because its chemical composition is exclusively governed by abiotic chemical reactions in the absence of terrestrial contamination. Numerous descriptions of organic molecules present in ordinary chondrite-type meteorites and, in particular, in carbonaceous chondrites (chondritic organic matter [COM]) have improved our understanding of the early interstellar chemistry that operated at or just before the birth of our solar system (Pizzarello et al., 2013; Sephton, 2013). However, all molecular analyses were so far targeted towards selected classes of compounds with a particular emphasis on biologically active compounds in the context of prebiotic chemistry (Cleaves, 2013). Here, we demonstrate that a nontargeted molecular analysis of the solvent-accessible organic fraction extracted under mild conditions from Murchison CM2.5 and a few other related CM-type meteorites allows one to extend its indigenous chemical diversity to tens of thousands of different molecular compositions and likely millions of diverse structures.

4.2 ORIGIN OF CHEMICAL DIVERSITIES

Modern analytical tools like molecular *targeted* or *nontargeted* organic structural spectroscopy enable rapid and accurate detection and description of chemical diversities in relation with health and environmental sciences. *Metabolomics* is the comprehensive study of the *metabolome* (Figure 4.1). It describes metabolic reactions in biological systems and integrates the knowledge of earlier developed *omics* branches such as *genomics*, *proteomics*, or *transcriptomics*.

Following a traditional definition in the field of biology, metabolomics measures the concentrations of a large number of naturally occurring small molecules (called metabolites), which are produced as intermediates and end products of all metabolic processes. They are measured from biological samples and body fluids such as urine, saliva, blood plasma, or tissue samples. Even the simple breathing (exhaled breath condensates) can carry the information about the state of health. The total number of different metabolites in biological systems is still unknown. Some estimates extend from a few tens to almost hundreds of thousands of compounds, but even this latter estimate is probably conservative as far as in supersystems such as gut microbiome samples or food samples (wine, etc.) involving plant, bacterial, and secondary metabolites are concerned. The count of

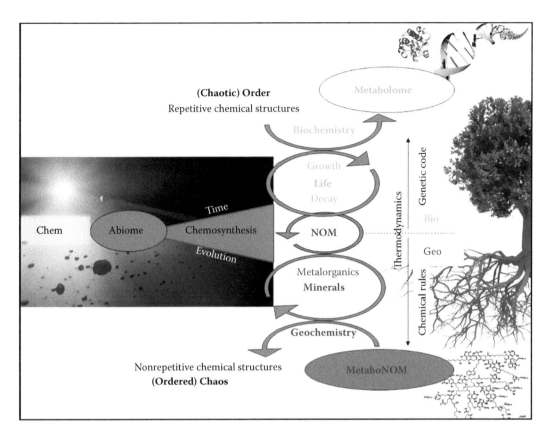

FIGURE 4.1 Terrestrial NOM is at the interface of life bioprocesses and geochemistry. The science describing the chemical diversity in biological systems is nowadays called *metabolomics*. The chemical diversity in biology is mainly a result and is limited to highly regulated bioprocesses. Geochemistry involves stronger interactions with mineral phases in the environment and leads to a higher molecular diversity. EOM results solely from abiotic chemosynthesis and therefore structural assignments are more difficult because of nonexisting database. Here, higher-resolving analytical tools are mandatory for a trustworthy depiction of abiotic and biogeochemical molecular complexity.

metabolites is also considerably larger than the number of corresponding genes, so the currently available databases cover at best 2% of the total number of existing metabolites (Meringer et al., 2013).

In environmental samples, an approach analogous to metabolomics can be followed by looking holistically to all small molecules (*NOM*) detectable in a given system in various time/space scales, integrating metabolite information from living organisms and all their biotic/abiotic transformation products (per analogy named *MetaboNOM*, Figure 4.1). These (organic) geochemistry studies in natural environments require understanding of the main processes in carbon, sulfur, nitrogen, and phosphorus cycling in surface or groundwater, marine water, atmosphere, and even extreme environments such as medium-temperature hydrothermal (geothermal) systems. In environmental systems, the molecular assemblages become more complex because both biotic and abiotic diagenetic reactions increase the chemical diversity to reach a steady state of hundreds of thousands of molecular structures.

The *Abiome* finally integrates all abiotic organic chemosynthesis representative of given extraterrestrial materials (Figure 4.1, inset), which are directly accessible when fresh-fallen meteorites are analyzed. Molecular complexity in extraterrestrial material was only recently approached and had rapidly found interest in the scientific community with some nontargeted high-resolving organic analysis of selected solvent extracts. The targeted analysis of known organic molecules in meteorites such as amino acids, fatty acids, and nucleic acids as building blocks of life was always of high interest (Pizzarello et al., 2006) and had some apogee at the time of the race to the Moon in the 1960s, thanks to the development of (gas) chromatographic analytical tools. In all cases, the combination of low carbon concentrations and extensive molecular complexity of the sample matrix had limited the scientific advances in this field to the rate of the development of ever more sensitive and resolving analytical tools (separation sciences, spectrometry, spectroscopy, etc.). In this new decade, we are facing a paradigm shift and a significant breakthrough in this field of research (Callahan et al., 2013; Danger et al., 2013).

High-resolution analytical technologies integrating separation sciences, spectroscopy and spectrometry, are needed for a molecular-level structural description of a system and the further evaluation of its spatial/temporal changes. Here, we describe the use of the high-field ion cyclotron resonance Fourier transform mass spectrometry (ICR-FT/MS) technique as an ultrahigh-molecular-resolution tool allowing the description of very complex materials on the level of the elemental composition distribution. In other words, ICR-FT/MS is an ideal tool for describing the chemical space of samples in various fields from biology to organic geochemistry. We illustrate here the particular use of this fantastic analytical tool in the case of ambient temperature solvent extracts of EOM entrapped in Murchison CM2.5 meteorite, with the emphasis of a few selected new results obtained after the publication of our first extended report (Schmitt-Kopplin et al., 2010a). These findings are critically compared to the recent ICR-FT/MS data we recorded for other chondrites, namely, the ordinary chondrite Sołtmany L6 (Schmitt-Kopplin et al., 2014); two other freshly fallen CM2 meteorites, Maribo (Haack et al., 2012) and Sutter's Mill (Jenniskens et al., 2012); and, by extension, for Moapa Valley, one of the only two CM1 petrological types found out of Antarctica (original data presented below).

4.3 ORGANIC MATERIAL IN CHONDRITES

The most common types of meteorites are *ordinary chondrites* that actually represent the most primitive solid matter in our solar system. Their classification is essentially based on the elemental and stable isotope composition as well as on their petrology (Weisberg et al., 2006). The presence of diverse mineralogical and rock phases in ordinary chondrites and especially the alteration of their composition and texture involving thermal metamorphism, water alteration, or shock metamorphism allow the scientists to reconstitute their history in space and time. *Carbonaceous chondrites* belong to a relatively scarce subcategory of ordinary chondrites. They contain a significantly large amount of carbon (0.5% to almost 3%), with sizable contribution of EOM on the top of carbon allotropes and carbides.

Much information on the organic matter and carbon content/speciation contained in carbonaceous chondrites remain to be obtained. COM is currently classified into three categories: (1) soluble organic matter (SOM), (2) insoluble organic matter (IOM), and (3) elemental carbon (graphite, diamond, fullerenes, etc.) and carbides (either present in metal grains or dispersed within the matrix). Generally, the total carbon content is lower than 0.5% in most of ordinary chondrites, decreasing with shock grade and thermal metamorphism, and is correlated to the chondritic water content (Figure 4.2).

It is thus conceivable that the composition and structure of the organic matter in fresh materials follows the classification of the chondrites with respect to various petrology types and subtypes.

One of the most commonly used techniques to follow the metamorphic grade of organic matter in case of, for example, unequilibrated ordinary chondrites is Raman spectroscopy

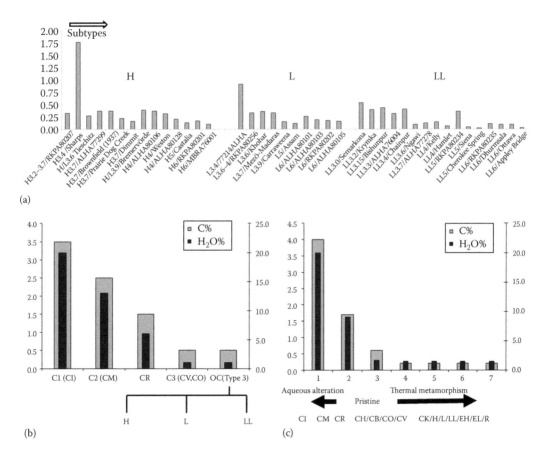

FIGURE 4.2 Petrologic groups and types of selected meteorites showing various water and carbon contents. Carbon is ubiquitous and not only relevant to carbonaceous chondrites. (a: Adapted from Grady, M.M. et al., *Meteoritics*, 24, 147, 1989; Makjanic, J. et al., *Meteoritics*, 28, 63, 1993; b: Adapted from Wood, J.A., The chondrite types and their origins, in: *Chondrites and the Protoplanetary Disk*, ASP Conference Series, eds. A.N. Krot, E.R.D. Scott, and B. Reipurth, Astronomical Society of the Pacific, San Francisco, CA, Vol. 341, pp. 953–971, 2005; c: Adapted from Weisberg, M.K. et al., *Meteorites and the Early Solar System II*, University of Arizona Press, Tucson, AZ, 2006.)

(Quirico et al., 2003) besides infrared spectroscopy (Kebukawa et al., 2011) and x-ray absorption near edge structure (XANES) at the carbon edge (Cody et al., 2008). Raman spectroscopy has reached its limits for higher petrological types 5 and 6 in which graphite was established as the main *surviving* carbon constituent although recent results suggest that sp^2-bonded carbon in presolar meteoritic matter could be amorphous and described as either glassy inorganic carbon or organic *kerogen-like* carbon (Wopenka et al., 2013). In a targeted way, the analysis of organic compounds has been often directed towards trace concentrations of selected classes of organic compounds such as polyaromatic hydrocarbons (PAHs) and amino acid derivatives (Zenobi et al., 1992; Gilmour, 2005; Pizzarello et al., 2006; Herd et al., 2011). These studies, based on petrologic characters, try to understand the early history of our solar system from the meteoritic perspective (Alexander et al., 2001).

High-resolution organic structural spectroscopy, a combination of ICR-FT/MS and NMR spectroscopy, revealed an astonishingly large chemical diversity of EOM in many types of chondrites with hundreds of thousands of distinct CHNOS elemental compositions (Popova et al., 2013; Schmitt-Kopplin et al., 2014). This opens the speculation of the role of such an unexpected chemical diversity in the early times of the solar system and its impact on the initial steps of life on Earth.

4.4 PRINCIPLES OF ICR-FT/MS

ICR-FT/MS refers to the measurement of the cyclotron frequency of ions trapped inside a confined cylindrical geometry located inside a magnet (Marshall et al., 1998). Figure 4.3 shows the ICR cell with an orbiting ion inside.

Ions can be detected by their energy absorption from an external waveform generator that causes expansion of their cyclotron radius so that their orbits range in proximity of two detector plates located in the central ring electrode of the cell. Each ion nominal mass (m/z) corresponds to a specific cyclotron frequency (ω_c). The cyclotron frequency ω_c is inversely proportional to the ion mass m (in amu) and is also directly proportional to the product of the ion charge q and the magnetic field strength B (in tesla). A higher magnetic field strength causes larger cyclotron frequencies of the confined ions, and this translates into a higher mass resolving power. A mass spectrum represents a histogram, which shows different ion abundances as a function of ion (m/z) ratios. Mass spectra result from the detection of the cyclotron frequencies of all confined ions. This is achieved by performing a mathematical Fourier transform of the detected induced charge transient, which is collected as a function of time (Figure 4.3, inset c). The mass accuracy reached in ICR-FT/MS is the highest possible in mass spectrometry with a routine full scan resolution higher than 500,000 at m/z 500. Two masses that differ by the mass of an electron can be separated with the precision of the mass of an electron. The result is a direct combinatorial assignment of elemental compositions (including fine structures of isotopologues) over the whole mass range analyzed (120–1000 amu). Each signal can be assigned to an elemental composition based on the combination of C, H, N, O, S, and P and their corresponding isotopes with an error lower than 100 ppb (Kim et al., 2006; Hertkorn et al., 2007; Schmitt-Kopplin et al., 2010b; Tziotis et al., 2011; Schmitt-Kopplin et al., 2012). Great attention is always paid

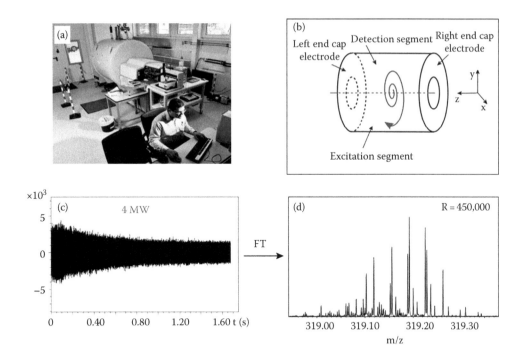

FIGURE 4.3 (a) 12 T ICR-FT/MS Instrument at Helmholtz Zentrum Munchen. (b) A cylindrical geometry of an ICR cell with two bored end-cap electrodes. The blue spiral indicates a trajectory of an ion during radial (x, y) ion excitation prior to detection. The central electrode is segmented fourfold with both counterpart segments representing either detector pairs or transmitter pairs. (c) Acquisition of a methanol extract collected from a Murchison fragment. The time domain detected transient is acquired with 4M data points within 1.6 s. (d) The mass spectrum is obtained as a result of a Fourier transform of the induced charge transient. Here, the mass resolving power is 450,000 at m/z = 319.

to work under extremely clean conditions including comparison with blank samples that were almost devoid from mass peaks. To further ascertain that the observed mass peaks were solely generated from the meteorite and not from solvent and/or from any other kind of contamination, three different meteoritic samples are currently handled under identical conditions for all measurements. Detailed sampling procedures and ICR-FT/MS spectra/ data acquisition are described elsewhere (Schmitt-Kopplin et al., 2010a).

4.5 CHEMICAL DIVERSITY IN BIOGEOSYSTEMS AS OBSERVED WITH ICR-FT/MS

Various solvents were tested for Murchison CM2.5 fragment organics extraction prior to analysis with negative electrospray ionization (ESI(−)) ICR-FT/MS. Among all protic, aprotic, polar, and apolar solvents, methanol has emerged as the one showing the largest number of signals corresponding to oxygenated polar compounds (Schmitt-Kopplin et al., 2010a). Apolar solvents would be more adapted for photoionization (APPI) or chemical ionization (APCI) types of ion sources allowing the analysis of the nonpolar organic fraction. The comparison of atmospheric pressure ion sources recently showed in the case of Suwannee

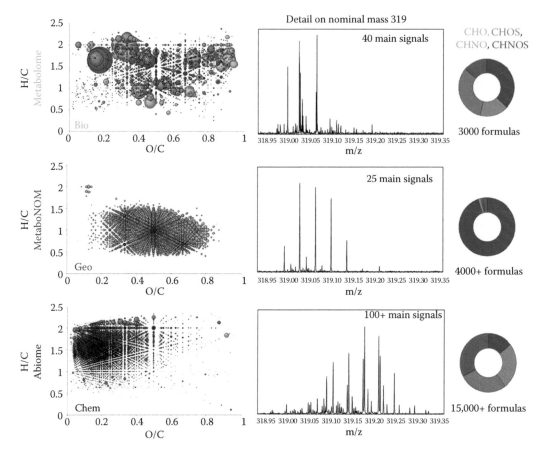

FIGURE 4.4 From top to down; left column: representation of all elemental compositions in van Krevelen diagrams (the color codes are blue, **CHO**; green, **CHOS**; orange, **CHNO**; and red, **CHNOS**; bubble sizes are proportional to signal intensity) showing the loose and irregular distribution of the mass peaks in the biological wine sample (metabolome), a rather homogeneous distribution of the signals for the Suwannee River surface water NOM (MetaboNOM) and for the meteoritic solvent extracts (Abiome). Middle column: signal regularity can be observed for each nominal mass as illustrated here for the nominal mass 319. Note that the wine sample reflecting biological signatures from grapes and yeast (fermentation) exhibits highly variable signal intensities. The last column provides the corresponding distribution (%) of CHO, CHOS, CHNO, and CHNOS in each extract. The abundance of CHOS type of molecules in wines is obviously due to the intentional addition of sulfites before/during fermentation.

River NOM solution (official NOM reference) that ESI(−) only accounts for half of the NOM compositional space relative to APCI and APPI (Hertkorn et al., 2008). Figure 4.4 illustrates the signal density in the chosen nominal mass 319 for three complex systems.

These are (1) a biological supersystem (wine, Gougeon et al., 2009) showing the random intensity distribution reflecting the presence of metabolite formulas based on C, H, N, O, and S found in that particular system and (2) a NOM (International Humic Substances Society [IHSS] NOM standard from Suwannee River) reflecting a significant enrichment in CHO-type molecules and a depletion in sulfur and nitrogen. The skewed near-Gaussian mass peak distribution is typical of terrestrial NOM and reflects the presence of many isomer compositions behind each mass peak (elemental composition), revealing the successive biotic and abiotic transformation, which this type of material had undergone in this N- and S-depleted environment; (3) Murchison methanol extract shows more than 100 signals per nominal mass and defining a new record in chemical diversity of complex organic mixtures.

The representation of all these compositions out of the list of more than 15,000 elemental compositions can be schematized in van Krevelen diagrams that show the variation of the atomic ratios H/C versus O/C in a color code for the **CHO, CHNO, CHOS,** and **CHNOS** type of molecules, with a dot size proportional to the intensity of the corresponding mass peaks in the mass spectra (Figure 4.4, left column). The isotopologues are not shown in this representation to avoid structural information redundancy. While the biological mixtures show molecular signatures with irregular intensity variations all over the H/C–O/C space, the Suwannee River NOM and the Murchison van Krevelen diagrams clearly show a high regularity in the compositional pattern, expressing the thermodynamically driven abiotic processes and the presence of many isomers behind each individual exact mass.

4.6 SOME CHARACTERISTICS OF CM-TYPE METEORITES INVESTIGATED

4.6.1 Murchison (CM2.5)

On September 28, 1969, near the town of Murchison, Victoria, in Australia, a bright fireball was observed leaving a cloud of smoke, before a deaf tremor was heard resulting in the fall of many meteoritic fragments dispersed over an area larger than 13 km^2. Most of the 100+ kg fragments of Murchison were collected shortly after it fell so that neither of these fresh samples suffered from intensive terrestrial weathering (Zolensky and Gooding, 1986). Murchison CM2.5, relatively rich in carbon, is one among the most chemically primitive (least-altered) chondrites (Browning et al., 1996). It therefore carries the signature of the solar system from around the time of the Sun's formation, roughly 4.6 billion years ago, freezing a record of some of the earliest chemistry taking place in the solar system that we have access to. Its aqueous alteration index (2.5 petrologic subtype), based on petrographic and mineralogical properties, indicates that its primary lithology had only experienced a relatively low and uniform degree of aqueous alteration by water-rich fluids at very low temperatures (20°C–50°C) on its parent body before falling on Earth, in contrast to the thermal metamorphism of ordinary chondrites that occurred in the range of 600°C–900°C under very dry conditions. More than 70% of the Murchison carbon content has been classified as (macromolecular) IOM of high aromaticity, whereas the soluble fraction contains extensive suites of organic molecules with more than 500 structures identified so far (Cronin, 1998). These structures basically resemble known biomolecules but are considered to result from abiotic synthesis because of peculiar occurrence patterns, racemic mixtures, and stable isotope contents and distributions.

4.6.2 Maribo (CM2)

Maribo is a new Danish CM2 chondrite that fell on January 17, 2009. The fall was recorded by a surveillance camera, an all-sky camera, a seismic station, and meteor radar observatories in Germany. Only a single fragment of Maribo with a dry weight of 25.8 g could be recovered 6 weeks after the fall. The oxygen isotopic composition of Maribo has an unusual low $\Delta^{17}O$ suggesting that Maribo is among the least aqueously altered CM chondrites (Haack et al., 2012).

4.6.3 Sutter's Mill (CM2-like)

Sutter's Mill is the first carbonaceous chondrite meteorite very rapidly recovered after it fell in April 2012, based on the detection of falling meteorites by Doppler weather radars. This is the fastest meteor known from which meteorites were recovered and also one having the highest disruption altitude on record. Sutter's Mill is a regolith breccia, with significant textural and mineralogical variations, containing from 1.3% to 1.6% carbon. Many characteristics of this meteorite resulting from the investigation of several fragments by an impressive series of techniques and teams (Jenniskens et al., 2012) are in line with a terrestrial alteration of the fallen fragments due to a series of parameters nonrelated to the SM parent body composition. Indeed, as SM rapidly broke at high altitude because of its brecciated texture, it could have been readily exposed to heating and/or to air oxidation while hot and probably further to a terrestrial postrain alteration, for samples recovered 3 days after the fall.

4.6.4 Moapa Valley (CM1)

A single dark gray, flattened stone (698.8 g), exhibiting subparallel contraction cracks and partially coated with black, vesicular fusion crust, was found by one of the coauthors (S.C.) in September 2004, in the Moapa Valley, southeast of Logandale, Nevada. So far, this is only the second CM1 found out of Antarctica. The petrological type 1 designates chondrites that have experienced a high degree of (low temperature) aqueous alteration in his parent body, where most primary minerals have been replaced by secondary phases and where chondrules are generally absent (Meteoritical Bulletin Database). Despite the long terrestrial age evaluated between 5 and 10 Ky (Irving et al., 2010), its degree of terrestrial weathering is low, probably because of the specific dry desert conditions.

4.7 STATE OF THE ART OF THE MURCHISON COMPOSITIONAL SPACE

In all analyses of the methanol extracts originating from various Murchison samples obtained from diverse sources or repositories, the spectra resulted in more than 25,000 resolved mass peaks in ESI(−) mass spectra and 20,000 resolved mass peaks in ESI(+) mass spectra. In ESI(−) mode, the peaks were converted into elemental compositions of which more than 15,000 could be confirmed considering the elements C, H, N, O, and S based on our NetCalc compositional approach presented in Tziotis et al. (2011). The key difficulty is to evaluate the entire adjunctive multidimensional space and to visualize simultaneously the extracted structure/process information of thousands of elemental compositions in two dimensions. Some of the possible visual and audio transformations carrying compositional and structural information were presented already in our first

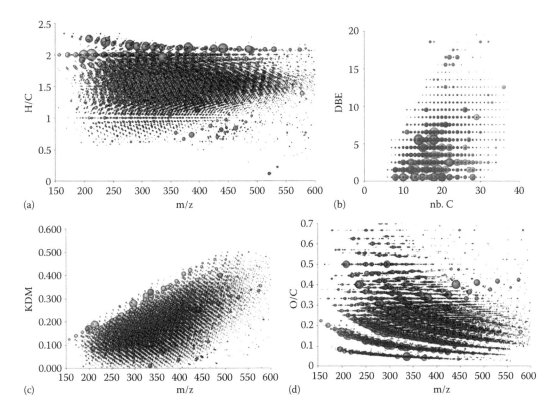

FIGURE 4.5 Various representations expending the multidimensional space providing structural information from ICR-FT/MS into complementary 2D space (property as a function of m/z). (a) H/C versus nominal mass, (b) DBE versus number of carbons, (c) KDM versus nominal mass, (d) O/C versus nominal mass (color code: blue, **CHO**; green, **CHOS**; orange, **CHNO**; and red, **CHNOS**; bubble size proportional to signal intensity).

paper on Murchison organic content (Schmitt-Kopplin et al., 2010a). In the meantime, the most effective representation complementary to the van Krevelen diagrams showing all data points proved to be the plot of the *H/C ratio versus nominal mass (m/z)* as shown on Figure 4.5, using the color codes as reported in Figure 4.4. One can discern the quantitative and qualitative compositional space continuum over the whole 150–550 mass range, only altered in harmony by randomly placed and oddly sized impurity signals (larger bubbles within the harmonious background meteoritic signature corresponding essentially to a very few fatty acids and some polyalkylsulfonates ubiquitous at low concentrations).

Whole homologous series with systematic repetitive structural units can be visualized that way. Alternatively, the *O/C versus nominal mass* diagrams are available (Figures 4.5 and 4.6).

Similarly, diagrams such as double-bond equivalents (DBEs) versus number of carbon atoms as used routinely in petroleomics (Marshall and Rodgers, 2008) or the Kendrick mass defect (KDM) versus the nominal mass express the abundance and diversity of the meteoritic SOM on the compositional scale (Danger et al., 2013).

Considering the distributions in the numbers and intensities of signals (relative positions and sizes in the diagrams), each particular compositional family (Figure 4.6) provides an integrative way to propose compositional similarities and transformation processes. From Figure 4.6, it becomes clear that the compositional space on the DBE level shares key characteristics between CHNO and CHNOS compounds as well as between CHO and CHOS compounds. This can be explained by considering a sulfurization process from

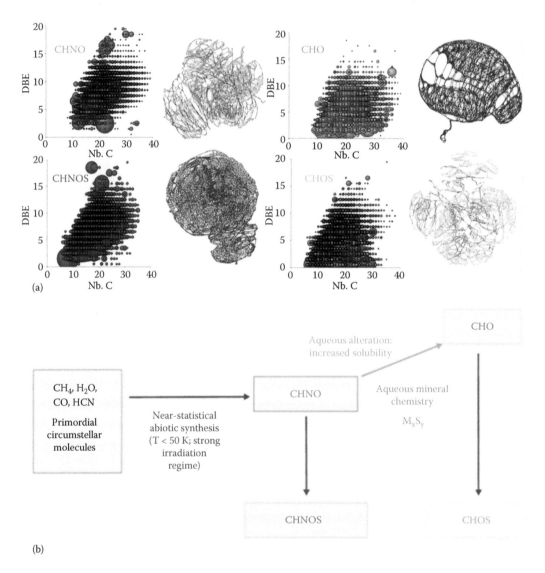

(a)

(b)

FIGURE 4.6 (a) Individual **CHO, CHNO, CHOS,** and **CHNOS** chemical footprints of Murchison methanolic extract in the diagrams *DBE versus number of carbons*. Corresponding similarity networks for each footprint showing extremely condensed features reflecting close structural relations within the SOM compositional space. (b) In the complex structure datasets, the sulfonation process affecting both CHO and CHNO types of compounds (potentially yielding the corresponding CHOS and CHNOS) is hidden.

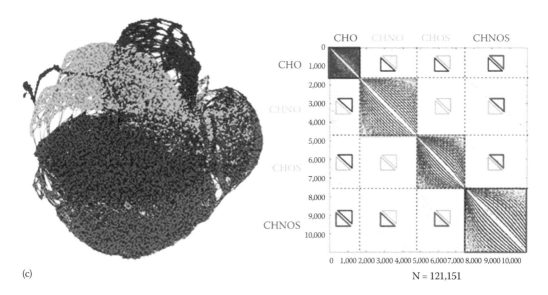

(c)

N = 121,151

FIGURE 4.6 (continued)   (c) This is verified when setting up the similarity networks involving all formula. In the corresponding correlation matrix, almost no structural similarity is found between CHNO and CHOS as well as between CHO and CHNOS.

S-sources likely involving metallic sulfides (M_xS_y) such as troilite or pentlandite, which are relatively abundant in chondrites, which affects two distinct populations of CHO and CHNO compounds in analogous ways as proposed previously (Schmitt-Kopplin et al., 2010a). Similarly, the visualization of all elementary compositions as similarity networks within the compositional classes shows condensed forms that reflect the close and systematic structural transitions and functional variations. Similarity networks among compositional classes show a strict separation between compositional classes. In analogy to the previously stated hypothesis, the compositional similarity matrix of all four compositional classes shows dissimilarity between CHO and CHNOS as well as between CHNO and CHOS. The similarity patterns match the proposed context displayed in Figure 4.6.

A limitation in this holistic approach is that the contemplation of elemental compositions will not provide detailed structural information because the various and numerous isomers cannot be distinguished. Isomeric differentiation needs an additional analytical technique, that is, chromatography hyphenated to mass spectrometry, separating physically the different isomers based on their polarity and interaction with the stationary phase.

Figure 4.7 shows a successful attempt separating the isobars highlighted in only m/z 319 in a 30 min time sequence using reverse phase liquid chromatography (LC) (ultraperformance liquid chromatography [UPLC]) separation with mass spectrometry detection (quadrupole time of flight mass spectrometry [qTOF/MS], MAXIS).

The 2D plot of this nominal mass 319 reveals over 300 individual signals only within the nominal mass 319, thereby providing another experimental proof of the record breaking of the high chemical diversity of meteoritic solvent-soluble fraction (Schmitt-Kopplin et al., 2010a). The plot reveals the most intense signals from the almost 100 observed mass peaks in ICR-FT/MS spectrum at the same nominal mass 319. Assuming an effective

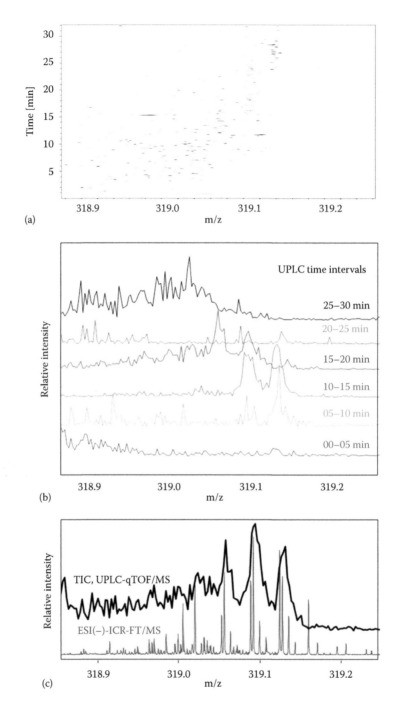

FIGURE 4.7 UPLC–qTOF/MS complements ICR-FT/MS data. (a) 2D LC separation in reverse phase (UPLC) versus qTOF/MS detection plot of Murchison methanol extract detailed on nominal mass 319, (b) corresponding extracted mass traces in mass 319 in the various time sections of UPLC, and (c) superimposed qTOF versus ICR-FT/MS mass information in the same nominal mass 319. The broad peaks in the TIC reflect the low resolution of 40,000 reached in UPLC-qTOF/MS relative to the ultrahigh resolution of 500,000 achieved in ICR-FT/MS.

molecular range from m/z 150 to 800, one can extrapolate the existence of almost 200,000 different compounds as detected experimentally with ESI(−) detection. Figure 4.7 also shows the extracted mass traces in the various elution time segments and the superimposition of the ESI(−)-ICR-FT/MS signal within the mass 319 with the total intensity current (TIC) reflecting the summed signals over all the separation time (note the lower resolution of near 40,000 of the qTOF/MS relative to the ultrahigh resolution of 500,000 of ICR-FT/MS).

The last example (Figure 4.8) considers one of the rare types of carbonaceous chondrites, the non–Antarctic Moapa Valley CM1, as compared to Maribo and Sutter's Mill CM2 and the fresh-fallen ordinary chondrite Sołtmany L6 described in detail elsewhere (Schmitt-Kopplin et al., 2014).

The chemical diversity in Moapa Valley is significantly lower than in Maribo CM2 that shows a very similar pattern as Murchison CM2.5 (Haack et al., 2012). This is easily explained if one considers that Maribo and Murchison were very rapidly collected after their falls (respectively, in 1969 and 2009), while Moapa Valley CM1, having a far longer terrestrial age, probably underwent progressive weathering in the desert open space for thousands of years. A direct consequence is the presence of a significant amount of

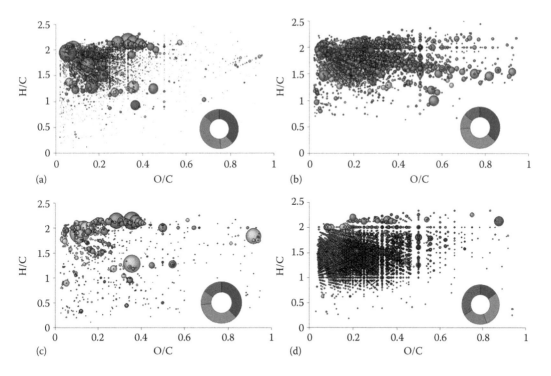

FIGURE 4.8 van Krevelen diagrams and mass-resolved H/C plot of different meteoritic methanol extracts. (a) Moapa Valley CM1. (b) Sołtmany L6. (c) Sutter's Mill CM2-like (sample SM2, collected before the rain fall). (d) Maribo CM2. In contrast to the exhaustively populated compositional space fullness of Murchison CM2.5 methanolic extract, the Moapa Valley CM1 extract reveals a significant signal scarcity, with mainly CHO and CHNO types of compounds and less sulfur-bearing compounds in general (restricted green and red areas in the A circle).

intense signals that can be assigned to degradation products or impurities (with respect to the meteoritic indigenous composition) accumulated during the meteorite terrestrial stay. The lack of sulfur chemistry is also very specific to Moapa Valley, in contrast to all the other fresh CM2 meteorites we have analyzed to date, suggesting a slow hydrolysis of accessible CHOS and/or CHNOS compounds in ambient terrestrial moisture. It may also have happened that water alteration already significantly affected the original composition of the meteorite parent body before it got shock-fragmented and that some of such fragments fell on Earth. Indeed, it was recently shown in the case of Chelyabinsk meteorite that a loss of sulfur-bearing compounds and an increase in N-bearing compounds can be the result of extraterrestrial thermal shock events (Popova et al., 2013). A loss in organic signatures was also detected in the case of Sutter's Mill CM2 fall that was significantly altered during the burst by pyrolysis. In that case, however, the signature in the polysulfur domain had increased with respect to the other freshly fallen CM2 analyzed (Jenniskens et al., 2012) arguing for a different degradation process of the genuine organics during the oxidative heating of the Sutter's Mill fragments in air during the last stages of the fall.

Finally, the Sołtmany L6 meteorite fallen in 2010 offered specific insights into the molecular diversity of a large number of organic compounds, with a very typical aliphatic oxygenized sulfur chemistry (Schmitt-Kopplin et al., 2014). Such a highly complex carbon chemistry detected for the first time in significant levels in an ordinary chondrite is surprising. This chemical signature integrates in time the chemical/collision history of the meteorites since their formation in the early times of the solar system.

4.8 CONCLUSION/OUTLOOK

We report the state of the art in the organic composition analysis of the solvent-soluble polar fractions extracted by grinding freshly exposed Murchison CM2.5 carbonaceous chondrite meteorite. The organic content of the extracts was analyzed with nontargeted ultrahigh-resolution mass spectrometry and discussed in the scope of the surprising but extremely important finding that such a high chemical diversity was also found in every fresh meteoritic material investigated so far by the same set of techniques, even in low carbon-containing ordinary chondrites such as Sołtmany L6. The first highlighted report on the high organic diversity in Murchison CM2.5 was soon followed by the investigation of the very fresh Maribo CM2 that was analyzed a few weeks after its fall in 2009 (pristine CM2 sample, Haack et al., 2012) and of Sutter's Mill CM2-like that was found to have been highly pyrolyzed during its burst in the atmosphere before it touched the Earth soil (extremely thermally altered samples, Jenniskens et al., 2012). Over the last years, many fresh-fallen noncarbonaceous meteorites also showed significant and specific organic signatures reflecting their time-integrated reactivity history of thermal alteration, water alteration, and shock events that occurred either within the specific parent body of each meteorite or during the oxidative thermal degradation of the meteoroid during the last stages of its fall. Carbonaceous chondrites represent a wide range of carbon saturation and oxygenation states, while ordinary chondrites, as recently shown for Chelyabinsk LL5

(Popova et al., 2013) or Sołtmany L6 (Schmitt-Kopplin et al., 2014, Figure 4.8b), have more restricted organic structures in a relatively oxygenated aliphatic range.

All the studied exemplary falls mentioned earlier show different ESI(–)-ICR-FT/MS spectral signatures that correspond to their very specific genesis within their parent body, their shock events in space, the extraterrestrial life of the so-generated fragments (meteoroids), and their eventual fall conditions as meteors or true meteorites. These few examples demonstrate how the diversity of the ultrahighly resolved MS peaks recorded, thanks to the unprecedented performances of the ICR-FT/MS technique, could be correlated to these various events that directly or indirectly affect the chemistry of each body in space and time. The total amount of carbon impacted on Earth, integrating not only simple forms of carbon but also highly complex and oxygenated carbonaceous compounds, can be significant in the mass accretion history of the Earth over the last few giga-years.

GLOSSARY

Carbonaceous chondrites: *Chondrites* that are rich in carbon. Thought to represent the earliest formed material in our solar system or even having a cometary origin.

COM: Chondritic organic matter.

EOM: Extraterrestrial organic matter.

ICR-FT/MS: Ion cyclotron resonance Fourier transform mass spectrometry.

IOM: Insoluble organic matter.

LC: Liquid chromatography.

Meteor: A bright trail or streak that appears in the sky when a meteoroid is heated to incandescence by friction with the Earth's atmosphere. Also called *falling star*, *meteor burst*, or *shooting star*.

Meteorite: A stony or metallic mass of matter that has survived entry through the atmosphere and reached the Earth's surface.

Meteoroid: A solid body, moving in space that is smaller than an asteroid and at least as large as a speck of dust.

MS: Mass spectrometry.

NMR: Nuclear magnetic resonance spectroscopy.

NOM: Natural organic matter.

Ordinary chondrite: The most common *meteorites* found. This term encompasses *meteorites* belonging to the *H chondrite* ("H" for high iron content), *L chondrite* ("L" for low iron content), or *LL chondrite* ("LL" for low total iron, low metallic iron content) group.

Parent body (of a meteorite): A larger body (located in the asteroidal belt or of cometary origin) from which a *meteorite* is thought to originate.

qTOF: Quadrupole time of flight.

SOM: Soluble organic matter.

UPLC: Ultraperformance liquid chromatography.

XANES: X-ray absorption near edge structure.

REVIEW QUESTIONS

1. What is Murchison meteorite and why is it important?

2. What kinds of chemicals are found in Murchison?

3. List other important meteorites that contain organic material, other than Murchison. Are the organic materials on these meteorites identical to those found in Murchison?

4. What instruments are used to analyze organic material in carbonaceous meteorites (meteorites that contain organic materials)? What are the advantages and disadvantages of each instrument? Which instrumental technique has identified the most organic compounds?

5. Are organic materials that are found in carbonaceous chondrites also found on Earth? Are there any organic materials in these meteorites that are unique to them and are not found on Earth?

6. What are the implications of such a diverse organic matter in meteorites in the light of the organics and life on Earth?

REFERENCES

Alexander C.M., Boss A.P., Carlson R.W., 2001. The early evolution of the inner solar system: A meteorite perspective. *Science*, 293, 64–68.

Browning L.B., McSween H.Y., Zolensky M.E., 1996. Correlated alteration effects in CM carbonaceous chondrites. *Geochim. Cosmochim. Acta*, 60, 2621–2633.

Callahan M.P., Gerakines P.A., Martin M.G., Peeters Z., Hudson R.L., 2013. Irradiated benzene ice provides clues to meteoritic organic chemistry. *Icarus*, 226, 1201–1209.

Cleaves, H.J., 2013. Prebiotic chemistry: Geochemical context and reaction screening. *Life*, 3, 331–345.

Cody G.D., Alexander C.M.O., Yabuta H., Kilcoyne A.L.D., Araki T., Ade H., Dera R., Fogel M., Militzer B., Mysen B.O., 2008. Organic thermometry for chondritic parent bodies. *Earth Planet. Sci. Lett.*, 272, 446–455.

Cronin J.R., 1998. In: *The Molecular Origins of Life: Assembling Pieces of the Puzzle*, ed. A. Brack, Cambridge University Press, Cambridge, U.K., pp. 119–146.

Danger G., Orthous-Daunay F.R., de Marcellus P., Modica P., Vuitton V., Duvernay F., Flandinet L., Le Sergeant d'Hendecourt L., Thissen R., Chiavassa T., 2013. Characterization of laboratory analogs of interstellar/cometary organic residues using very high resolution mass spectrometry. *Geochim. Cosmochim. Acta*, 118, 184–201.

Gilmour I., 2005. Structural isotopic analysis of organic matter in carbonaceous chondrites. In: *Meteorites, Comets and Planets*, ed. A.M. Davis in the Series *Treatise on Geochemistry*, eds. H.D. Holland and K.K. Turekian, Elsevier, Amsterdam, the Netherlands, pp. 269–290.

Gougeon R.D., Lucio M., Frommberger M., Peyron D., Chassagne D., Alexandre H., Feuillat F. et al., 2009. The chemodiversity of wines can reveal a metabologeography of cooperage oak wood. *Proc. Natl. Acad. Sci. (PNAS)*, 106, 9174–9179.

Grady M.M., Wright I.P., Pillinger C.T., 1989. A preliminary investigation into the nature of carbonaceous material in ordinary chondrites. *Meteoritics*, 24, 147–154.

Haack H., Grau Th., Bischoff A., Horstmann M., Wasson J., Norup Sørensen A., Laubenstein M. et al., 2012. Maribo—A new CM fall from Denmark. *Meteorit. Planet. Sci.*, 47, 30–50.

Herd C.D.K., Blinova A., Simkus D.N., Huang Y., Tarozo R., Alexander C.O.D., Gyngard F. et al., 2011. Origin and evolution of prebiotic organic matter as inferred from the Tagish Lake meteorite. *Science*, 332, 1304–1307.

Hertkorn N., Frommberger M., Schmitt-Kopplin Ph., Witt M., Koch B., Perdue E.M., 2008. Natural organic matter and the event horizon of mass spectrometry. *Anal. Chem.*, 80, 8908–8919.

Hertkorn N., Ruecker C., Meringer M., Gugisch R., Frommberger M., Perdue E.M., Witt M., Schmitt-Kopplin Ph., 2007. High-precision frequency measurements: Indispensible tools at the core of the molecular-level analysis of complex systems. *Anal. Bioanal. Chem.*, 389, 1311–1327.

Irving A.J., Kuehner S.M., Rumble III D., Korotev R.L., Clary S., 2010. Moapa Valley: A second non-Antarctic CM1 chondrite from Nevada, USA. *Meteorit. Planet. Sci.*, A96, 5372.

Jenniskens P., Fries M.D., Yin Q.Z., Zolensky M., Krot A.N., Sandford S.A., Sears D. et al., 2012. Radar enabled recovery of the Sutter's Mill meteorite, a carbonaceous chondrite regolith breccia. *Science*, 338, 1583–1587.

Kebukawa, Y., Conel, A., Cody, G.D., 2011. Compositional diversity in insoluble organic matter in type 1, 2 and 3 chondrites as detected by infrared spectroscopy. *Geochim. Cosmochim. Acta*, 75, 3530–3541.

Kim S., Rodgers R.P., Marshall A.G., 2006. Truly "exact" mass: Elemental composition can be determined uniquely from molecular mass measurement at similar to 0.1 mDa accuracy for molecules up to similar to 500 Da. *Int. J. Mass Spectrom.*, 251, 260–265.

Makjanic J., Vis R.D., Hovenier J.W., Heymann D., 1993. Carbon in the matrices of ordinary chondrites. *Meteoritics*, 28, 63–70.

Marshall A.G., Hendrickson C.L., Jackson G.S., 1998. Fourier transform ion cyclotron resonance mass spectrometry: A primer. *Mass Spectrom. Rev.*, 1, 1–35.

Marshall A.G., Rodgers R.P., 2008. Petroleomics: Chemistry of the underworld. *Proc. Natl. Acad. Sci. (PNAS)*, 105, 18090–18095.

Meringer M., Cleaves H.J., Freeland S., 2013. Beyond terrestrial biology: Charting the chemical universe of α-amino acid structures. *J. Chem. Inf. Model.*, 53, 2851–2862.

Pizzarello S., Cooper G.W., Flynn G.J., 2006. The nature and distribution of the organic material in carbonaceous chondrites and interplanetary dust particles. In: *Meteorites and the Early Solar System II*, eds. D.S. Lauretta and H.Y. McSween, University of Arizona Press, Tuscon, AZ, pp. 625–651.

Pizzarello S., Davidowski S.K., Holland G.P., Williams, L.B., 2013. Processing of meteoritic organic materials as a possible analog of early molecular evolution in planetary environments. *Proc. Natl. Acad. Sci. (PNAS)*, 110, 15614–15619.

Popova O., Jenniskens P., Emel'yanenko V., Kartashova A., Biryukov E., Khaibrakhmanov S., Shuvalov V. et al., 2013. Chelyabinsk airburst, damage, assessment, meteorite recovery and characterization. *Science*, 342(6162), 1069–1073.

Quirico E., Raynal P.I., Bourot-Denise M., 2003. Metamorphic grade of organic matter in six unequilibrated ordinary chondrites. *Meteorit. Planet. Sci.*, 38, 795–811.

Schmitt-Kopplin Ph., Gabelica Z., Gougeon R.D., Fekete A., Kanawati B., Harir M., Gebefügi I., Eckel G., Hertkorn N., 2010a. High molecular diversity of extraterrestrial organic matter in Murchison meteorite revealed 40 years after its fall. *Proc. Natl. Acad. Sci. (PNAS)*, 107, 2763–2768.

Schmitt-Kopplin Ph., Gelencsér A., Dabek-Zlotorzynska E., Kiss G., Hertkorn N., Harir M., Hong Y., Gebefügi I., 2010b. Analysis of the unresolved organic fraction in atmospheric aerosols with ultra-high resolution mass spectrometry and nuclear magnetic resonance spectroscopy: Organosulfates as photochemical smog constituents. *Anal. Chem.*, 82, 8017–8026.

Schmitt-Kopplin Ph., Harir M., Kanawati B., Tziotis D., Hertkorn N., Gabelica Z., 2014. Chemical footprint of the solvent soluble extraterrestrial organic matter occluded in Sołtmany ordinary chondrite, *Meteorites*, 1–2 (2), 79–92.

Schmitt-Kopplin Ph., Harir M., Tziotis D., Gabelica Z., Hertkorn N., 2012. Ultrahigh resolution Fourier Transform Ion Cyclotron Resonance Mass Spectrometry for the analysis of natural organic matter from various environmental systems. In: *Comprehensive Environmental Mass Spectrometry*, ed. A. Lebedev, ILM Publications, Glendale, AZ, pp. 443–459.

Sephton M.A., 2013. Aromatic units from the macromolecular material in meteorites: Molecular probes of cosmic environments. *Geochim. Cosmochim. Acta*, 107, 231–241.

Tziotis D., Hertkorn N., Schmitt-Kopplin Ph., 2011. Kendrick-analogous network visualisation of Ion Cyclotron Resonance Fourier Transform (FTICR) Mass Spectra: Improved options to assign elemental compositions and to classify organic molecular complexity. *Eur. J. Mass Spectrom.*, 17, 415–421.

Weisberg M.K., McCoy T., Krot A.N., 2006. Systematics and evaluation of meteorite classification. In: *Meteorites and the Early Solar System II*, eds. D.S. Lauretta and H.Y. McSween, University of Arizona Press, Tucson, AZ, pp. 19–52.

Wood J.A., 2005. The chondrite types and their origins. In: *Chondrites and the Protoplanetary Disk*, ASP Conference Series, eds. A.N. Krot, E.R.D. Scott, and B. Reipurth, Astronomical Society of the Pacific, San Francisco, CA, Vol. 341, pp. 953–971.

Wopenka B., Xu Y.C., Zinner E., Amari S., 2013. Murchison presolar carbon grains of different density fractions: A Raman spectroscopic perspective. *Geochim. Cosmochim. Acta*, 106, 463–489.

Zenobi R., Philippos J.M., Zare R.N., Wing M.R., Bada J.L., Marti K., 1992. Organic compounds in the Forest Vale, H4 ordinary chondrite. *Geochim. Cosmochim. Acta*, 56, 2899–2905.

Zolensky M.E., Gooding J.L., 1986. Aqueous alteration on carbonaceous chondrite parent bodies as inferred from weathering of meteorites in Antarctica. *Meteoritics*, 21, 548–549.

CHAPTER 5

Prebiotic Synthesis of Biochemical Compounds

An Overview

Henderson James (Jim) Cleaves

CONTENTS

5.1 BRIEF OVERVIEW OF ORGANIC CHEMICAL COMPLEXITY AND THE DEFINITION OF PREBIOTIC CHEMISTRY

The origin of life has a long and storied history as a scientific problem (see, e.g., Farley 1977). For long periods of time, the notion of spontaneous generation was generally accepted as a physical phenomenon among some classes of organisms (Fry 2000). With the invention of the microscope and the discovery of the microbial world, the idea of spontaneous generation was pressed into more definable terms. By the mid-nineteenth century while it was well accepted that higher organisms were not generated spontaneously, it remained unclear whether microbes were. Various experimental evidences eventually suggested that microbes were also not spontaneously generated (Geison 1997). At the same time, the idea that all life was related by a Darwinian system of descent with modification led to the notion of an *ur*-organism, which would logically need an origin. Concurrently, the development of geology as a science supported the notion that the Earth's environment had changed over time, and this led to the logical conclusion that perhaps the conditions that allowed for the generation of the first organism were no longer present on the surface of the Earth (Dalrymple 1994).

Shortly after the turn of the twentieth century, two European scientists, one in the United Kingdom and one in the Soviet Union, proposed that conditions no longer present on Earth, which allowed for the synthesis of organic compounds, might help explain the origin of life. John Burdon Sanderson Haldane and Alexander Ivanovich Oparin, in a pair of closely temporally published papers (Oparin 1924; Haldane 1929), suggested that the origin of life on Earth was the result of organic chemistry that occurred in a significantly distinct ancient geochemical context. The crux of both of their arguments was that the origin of life depended on the supply of organic compounds from the environment undergoing chemical complexification that was analogous to biological natural selection.

This notion has subsequently been given names such as *chemical evolution* or *prebiotic chemistry* (Ponnamperuma and Chelaflores 1994). The first is a nonloaded term, which could perhaps be considered synonymous with the simple term *chemistry*, denoting transformation of matter over time. The second refines the notion to the chemical domain, though it implies the goal-directed transformation of chemicals ultimately leading to life (Cleaves 2012).

As we do not presently have a single compelling scenario for the origin of life, the prefix *pre-* in *prebiotic chemistry* is a necessarily speculative descriptor. A somewhat parallel term, *abiotic chemistry*, has also been used to describe chemistry, which occurs in the absence of life. For example, most petroleum is generally thought to be formed by abiotic transformation of biologically generated organic compounds, modified by nonbiological transformations such as heat, pressure, and mineral interactions (Killops and Killops 2013). It is not widely believed that such transformations lead to new independent origins of life, though there are some suggestions that deep Earth organic chemistry may still give rise to life (Gold 1992; Davies et al. 2009). Nevertheless, the two terms will be used almost interchangeably here, with the implied caveat that all of the described chemistry is *abiotic* and some of it *may* be *prebiotic*.

That the origin of life should depend on organic compounds is logical from two standpoints. First, contemporary organisms are overwhelmingly composed of organic compounds. Second, the complexity allowed by organic chemistry simply dwarfs that of inorganic chemistry. While some 4400 naturally occurring minerals are known (Hazen et al. 2008), the number of distinct relatively low-molecular-weight organic compounds is truly supra-astronomical; some estimates place the number on the order of 10^{60} compounds (Kirkpatrick and Ellis 2004).

Furthermore, covalent bonds between carbon atoms or between carbon and other atoms are intermediate in strength (300–850 kJ mol^{-1}) between what are considered weak bonds (such as hydrogen bonds [5–30 kJ mol^{-1}]) and ionic bonds (~750 to >2000 kJ mol^{-1}). Covalent bonds are thus stable at temperatures under which water, which is thought by many to be a requisite for life (Jones and Lineweaver 2010), is a liquid at normally encountered terrestrial surface pressures.

At the same time, the fact that these bonds are so stable means that biology has had to develop sophisticated mechanisms for making and breaking them, typically through enzyme-mediated catalysis. To break these bonds in the absence of catalysis typically requires very large amounts of energy, for example, high-energy radiation (e.g., ultraviolet (UV) light of ~3 of 9 eV, corresponding to wavelengths of approximately 140–400 nm) or high temperatures. As we will see, making these bonds in abiological systems typically depends on breaking bonds in high-energy regimes and then allowing the resulting excited ions or radicals to recombine during a quenching phase.

Presently, the idea of the spontaneous generation of life from abiotically or prebiotically supplied precursors remains an open question. Some authors have argued that this may be a question that is not amenable to experimental validation (Ourisson and Nakatani 1996), either because the initial conditions are too poorly constrained or because it is such an improbable event that it will likely never be replicated in the laboratory (see, e.g., the discussion in Orgel 1998a,b). Some are more optimistic, and still others consider such a demonstration beside the point, arguing that the goal of origin of life research is not to demonstrate the production of a living organism under laboratory conditions but to reconstruct a series of geochemically plausible steps that could have led to such a transformation (Eschenmoser 2007).

There are environmental limits to where life as we know it can exist, including limits based on temperature (from ~−12°C to +121°C), pressure (up to 1400 atm), and salinity (up to saturation of NaCl) (see, e.g., Cavicchioli 2002; Cleaves and Chalmers 2004; Harrison et al. 2013). However, these limits are consistently, though incrementally, moved outwards, and therefore, it may be presumptuous to define them too precisely.

5.2 WHAT IS LIFE?

Before we can tackle the question of what is needed to make life, it is prudent to touch up the question of what *life* is. Life is a notoriously difficult phenomenon to define (Lazcano 2008; Bedau and Cleland 2010). All extant terrestrial life is cellular, dependent on a deoxyribonucleic acid (DNA) information storage system and a ribonucleic acid (RNA)-based

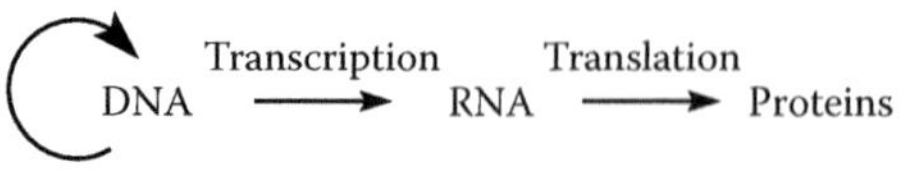

Transcription   Translation

DNA $\longrightarrow$ RNA $\longrightarrow$ Proteins

FIGURE 5.1   The central dogma of information flow in biology.

information translation system that converts the information stored in DNA to protein (the so-called *central dogma* of biology [Figure 5.1]) and largely dependent on protein-mediated catalysis. All life being cellular, some sort of boundary-forming mechanism appears necessary.

At a deeper level, it should be acknowledged that despite the various commonalities shared by all extant organisms, life is not necessarily materially defined (Gánti 2003;

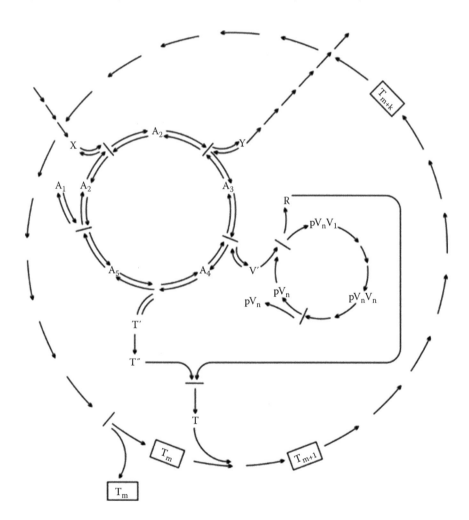

FIGURE 5.2   Tibor Gánti's chemoton model. Three cyclic subsystems, a boundary subsystem (T_m), a metabolic subsystem (A_n), and an informational control system (pV_n), are stoichiometrically linked to form an indivisible whole. (Reproduced from Gánti, T., *The Principles of Life*, Oxford University Press, Oxford, U.K., 2003, Figure 1.1 from p. 4. With permission.)

Louie 2009). Organisms exist that simply do not use many materials often considered universal to carry out their life processes. For example, archaebacteria do not use acylphospholipids to construct their cell membranes but instead use ether-based isoprenoid lipids (Matsumi et al. 2011), though these organisms also use DNA and the consensus genetic code (Freeland and Hurst 1998), suggesting the choice of lipid system was made after the natural selection of DNA and the genetic code (Peretó et al. 2004). But what then is fundamental to life, and which aspects observed in contemporary biochemistry should we consider essential, and which is necessary for an origin?

One theorist, Gánti (2003), has made the suggestion that life must be considered the union of three subsystems that cannot be separated without losing the integrity of the living system. These three subsystems are an informational subsystem, a metabolic subsystem, and a boundary subsystem (Figure 5.2).

It is especially significant in Gánti's model that the subsystems are linked. Each shares some component with at least one of the others. In this manner, the growth and replication of the subsystems are linked so that the entire organization can operate in a coordinated fashion, rather than being a simple collection of three independent systems. It is also significant in this model that while one can see natural cognates of the subsystems in biochemistry (e.g., lipids as the boundary subsystem, nucleic acids as the informational subsystem, and protein enzymes as components of the metabolic subsystem), the actual molecular identities of the subsystems are not defined, and in principle, many types of molecules could play any of these roles.

5.3 SITES FOR THE ORIGIN OF LIFE

It is generally believed that terrestrial life originated *on* Earth, though some arguments have been advanced based on various arguments that life began on Mars, or on comets or meteorite parent bodies (Kirschvink and Weiss 2002; Hoover 2008). We will not review here the early geochemical conditions on Mars or the arguments for interplanetary transport of life and the chemical conditions and types of organic compounds that could have been brought in from small solar system bodies such as comets and meteorites as this material is reviewed elsewhere in this volume. We will, however, briefly sketch the conditions that we think were necessary for the origin of life on Earth and review a few solar system bodies where there is at least circumstantial evidence for the minimal conditions thought to be necessary for the origin of life, namely, the presence of liquid water and organic compounds.

5.3.1 Earth

The Earth is believed to have formed some 4.55 Ga (giga-annum, or one billion years ago) based on isotopic dating of meteorites (Dalrymple 1994). There is evidence for life on Earth in the form of microfossils going as far back as ~3.5 Ga (Schopf 2008; Wacey et al. 2011) and evidence for light organic carbon (i.e., carbon depleted in ^{13}C, which is argued to be a marker of ancient biological carbon cycling) as far back as 3.8 Ga (Schidlowski 1988). This leaves open an approximately 0.7–1 Ga window for life to have originated on

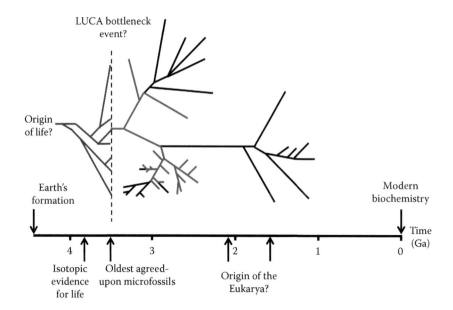

FIGURE 5.3 Timeline for major events in the history of life on Earth. Superposed earlier, the timeline is a schematic phylogenetic tree showing how the major classes of organisms can be grouped, though we cannot presently fix many of the branch points of this tree to the geological timeline with great accuracy. The origin of the Eukarya is represented by two arrows, showing the range of estimated dates.

Earth (Figure 5.3), which has been considered more than sufficient by some (Lazcano and Miller 1994), though others have argued it is impossible to estimate the pacing of this transition based on the available evidence (Orgel 1998). This is shown in Figure 5.3.

Precisely when conditions on the Earth's surface (here surface being used to denote upper crustal environments, including the ocean floor) became mild enough to support life, typically defined as when temperatures became cool enough to support standing bodies of liquid water, is a matter of considerable debate, and various studies continue to push the earliest possible date back further in time, now to ~4.3 Ga. For example, indications from ancient zircon grains found in some of Earth's earliest detrital deposits suggest that liquid water was present at the Earth's surface as early as 4.4–4.3 Ga (Mojzsis et al. 2001). It must be stressed that the temperature, pH, and solute composition of this water are unknown (Kempe and Degens 1985). Under a very thick atmosphere, the boiling point of water would be increased, meaning that the early oceans, as suggested by zircon evidence, may have been somewhat hotter than 100°C. Presently the upper limit of survival for terrestrial organisms is ~121°C (Kashefi and Lovley 2003), though these are of course highly adapted organisms with a host of repair mechanisms that allow them to cope with the stresses of such hot environments, and it is unclear what conditions the earliest life forms could have tolerated (Arrhenius 1999).

5.3.2 Mars

Exploration of Mars' surface by orbiters, landers, and rovers has now provided a wealth of data allowing for considerable reconstruction of Mars' earliest history. Apparent on

Mars' surface are depositional environments suggesting extensive standing bodies of liquid water, some perhaps hundreds of meters thick (Carr and Head 2003). The mineralogy of Mars' oldest surface rocks suggests weathering of basaltic materials at a neutral to slightly alkaline pH, not unlike that of the modern terrestrial ocean (pH ~ 8.1) (see, e.g., Ming et al. 2008). The composition of Mars' earliest atmosphere remains as contentious as the Earth's with arguments for both early oxidizing and reducing conditions having been put forward (see, e.g., Lammer et al. 2013). Of course, the same considerations for the early Earth's atmosphere likely also apply to the early Martian atmosphere: a highly or moderately reducing atmosphere could support the synthesis of abundant organic compounds via atmospheric synthesis mediated by various forms of impinging radiation, including solar UV, cosmic rays, and electric discharges, with more oxidized gas mixtures providing different types and quantities of compounds (Stribling and Miller 1987; Nna Mvondo et al. 2001). Of course, regardless of the atmospheric composition, extra-Martian infall of organic compounds to Mars' early surface is also likely.

The presence of hydrothermal environments driven by magmatism also seems possible on Mars, and various geological structures on Mars have been identified as hydrothermal vents (CIBA Foundation 2008).

5.3.3 Titan

Saturn's moon Titan remains one of the most intriguing bodies in the solar system from the standpoint of organic synthesis due to its reducing atmosphere composed of ~98% N_2 and 2% CH_4 with traces of other species such as H_2, CO, and light alkanes and nitriles (Niemann et al. 2005).

The presence of organic aerosols and low-molecular-weight organics in Titan's atmosphere has now been confirmed experimentally by the Huygens probe (Israël et al. 2005), and the composition of these in general compares well with those prepared in Earth-based laboratory simulations (Cleaves et al. 2014). Titan tholins are also remarkably similar to those formed under Miller–Urey-type conditions from a postulated early reducing terrestrial atmosphere (Cleaves et al. 2014).

Interestingly, it is now suspected that while Titan's surface is extraordinarily cold, with solid CH_4 and ethane possibly occurring on the surface (which at ~−180°C (Mitri et al. 2007) is well below the freezing point of water), Titan's subsurface may host liquid water and kept liquid by high concentrations of NH_3 (Thompson and Sagan 1992). Deeper still, temperatures may reach those of contemporary terrestrial environments that host living organisms. It has been postulated that the organic aerosols and other compounds that rain out from Titan's atmosphere may come into contact with this subsurface liquid water periodically. This could be due to either cryovolcanism or impact-induced heating, which could provide liquid water environments that could persist for thousands of years (ref). These are important points to keep in mind when considering that low-temperature hydrolysis of Titan tholins in concentrated NH_3 gives rise to a variety of organic compounds, including several important in terrestrial biochemistry, such as amino acids and nucleobases (Neish et al. 2010).

5.3.4 Enceladus

Jupiter's moon Enceladus is another intriguing solar system body from the perspective of organic chemistry. Enceladus remains one of only two solar system bodies besides the Earth with active volcanism (Manga and Wang 2007). A continual stream of icy material is ejected from beneath Enceladus' surface. This material has now been remotely sampled and found to include a number of low-molecular-weight compounds of interest to organic chemistry, including formaldehyde, hydrogen cyanide (HCN), ammonia, and methanol (Waite et al. 2009).

5.3.5 Europa

While there is presently no evidence for organic chemistry occurring on Europa's surface or in its subsurface, Europa remains intriguing due to the abundant evidence that it may host a large subsurface liquid water ocean, with a volume possibly exceeding that of the Earth's (Pappalardo 2010). This evidence includes surface features suggestive of ice sheets atop a liquid water matrix allowing their movement, as well as measurements of Europa's magnetic field. If Europa accreted from comet-like material, then it seems likely that it also contained some of the same organic compounds that comets are suspected to be composed of (Briggs et al. 1992; Sandford et al. 2006). How these have evolved or continue to evolve in Europa's subsurface ocean remains unknown, but it is possible that there is terrestrial-like submarine hydrothermal activity ongoing on Europa's ocean floor, which could be compatible with some of the types of chemistry purported to have been important for the chemical evolution, which led to the origin of life on Earth (Raulin et al. 2010).

5.3.6 Comets/Asteroids

In his seminal volume, Oparin suggested that extraterrestrial organic materials could have been important for the origin of life in his seminal volume (Oparin 1924). It has been known for well over 150 years that meteorites contain organic compounds, though their extraterrestrial origin was not conclusively confirmed until the 1970s (Kvenvold et al. 1970; Cronin and Moore 1971; Folsome et al. 1971; Oró et al. 1971).

The Earth is continuously bombarded with meteorites of various sizes, which are thought to be remnants of the material from which the solar system accreted (Lauretta and McSween 2006). These include meteorites that are almost entirely metallic, as well as various classes of stony meteorites. A subgroup of the latter, the carbonaceous chondrite meteorites, are known to include a significant fraction, up to ~2%–3% by weight, of indigenous organic material. There are several hundred specimens of this type of meteorite curated in various collections around the world. Their organic composition has been investigated extensively and remains a vibrant research activity in several laboratories.

In addition to a complex, kerogen-like high-molecular-weight organic material (known as insoluble organic material [IOM]), carbonaceous chondrite meteorites have been found to contain a complex host of organic materials including several classes of low-molecular-weight organic compounds representing compound classes that are important in contemporary terrestrial biochemistry, including amino acids, nucleobases, sugar-like derivatives, and various metabolic intermediates (Pizzarello et al. 2006; Schmitt-Kopplin et al. 2010; Cooper et al. 2011). These will be discussed in more detail in Section 5.4.

5.4 ANALOG STUDIES

5.4.1 Atmospheric Synthesis

Assuming life began on Earth, the composition of Earth's earliest atmosphere is of extreme importance, but as mentioned earlier, presently very poorly constrained due to the lack of geochemical evidence form Earth's earliest history. Harold Urey, the Nobel-prize-winning American chemist, first suggested that the early Earth's atmosphere would have been highly reducing (i.e., overwhelmingly dominated by molecular species bonded to hydrogen, for example, with carbon in the form of methane, CH_4) as opposed to oxidizing (e.g., with carbon largely in the form of species bonded to oxygen, such as carbon dioxide, CO_2) (Urey 1952). This conjecture was based on the observation of the abundance of hydrogen in the solar system compared with that of oxygen, for example, as in the atmospheres of Saturn and Jupiter. It was further conjectured, based on chemistry known at that time, that it should be easier to make organic compounds from reduced gases than it would be from oxidized gases.

Indeed, until the early 1950s, laboratory efforts to synthesize organic compounds from CO_2/H_2O gas mixtures mimicking an early CO_2-dominated steam atmosphere were found to be relatively inefficient at producing organic compounds (Garrison et al. 1951).

In 1952, Stanley Miller, then a graduate student, after seeing a lecture delivered by Urey on the topic of organic synthesis on the primitive Earth, proposed testing Urey's ideas as a thesis project. Shortly thereafter, he conducted the first deliberate test of Urey's atmospheric model in the laboratory. The experiment consisted of a gas-filled apparatus (Figure 5.4) simulating the early Earth atmosphere–ocean system acted upon by electrical discharges simulating lightning. It was a remarkable success, quickly producing a visible brownish organic polymer and upon further analysis also several amino acids important in biological proteins (Miller 1953).

Numerous experiments were conducted in the wake of this discovery testing the range of conditions under which such synthesis could take place, for example, examining the use of different types of energy sources and different gas compositions (Groth and Weyssenhoff 1957; Allen and Ponnamperuma 1967; Groth 1975; Miyakawa et al. 2002).

It was generally found that reducing gases produce not only a greater variety of organic compounds but also a greater yield (Stribling and Miller 1987). Among the types of compounds that have now been isolated from such systems include most of the major classes of biochemicals (amino acids, purines, pyrimidines, etc.), with modern analytical techniques continually finding new and more compounds, albeit in lower and lower yield (Levy 2000; Johnson et al. 2008).

Despite the success of such experiments, their relevance to the inventory of organic compounds on the early Earth has been questioned. Around the time Urey first proposed an early reducing atmosphere, other geochemists presented arguments for a relatively oxidizing CO_2/N_2-dominated atmosphere (Rubey 1951). This notion has been repeated

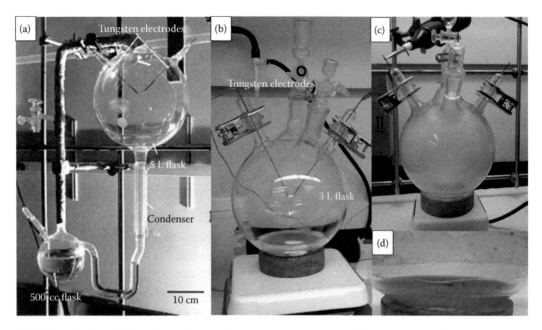

FIGURE 5.4 The Miller–Urey electric discharge apparatus. (a) The circulating flask of the type first used by Miller in 1953, (b) a room-temperature Miller–Urey electric discharge apparatus, (c) a room-temperature Miller–Urey electric discharge apparatus filled with methane and nitrogen gas after 48 h of sparking, and (d) close-up view of the organic materials produced in (c). (a: Reproduced from Lazcano, A. and Bada, J.L., *Orig. Life Evol. Biosph.*, 33, 235, 2003. With permission.)

over the years based on various lines of reasoning, including the oxidation state of various metals in ancient samples of what was presumably upper mantle material (Trail et al. 2011).

One important consideration that has received a great deal of attention is the so-called *dim young sun* paradox (Kasting and Siefert 2002), which can be summarized as follows: it is widely suspected, based on astronomical observations of stellar evolution, that the early Sun's energy output around the time of the origin of life was ~70% that of the present Sun's (Sagan and Mullen 1972). Given that there is evidence for liquid water during this period (e.g., from the measurements of ancient zircons described earlier), one solution to this paradox is to postulate a strong greenhouse atmosphere, which would raise Earth's surface temperature above the freezing point of water. Originally, gases such as CH_4 and NH_3 were proposed for this role, but their instability to photochemical destruction by UV radiation was soon pointed out (Ferris and Nicodem 1972; Kuhn and Atreya 1979), and the principal contender became CO_2. It has been calculated that at least several times the modern level of CO_2 is needed to generate the required warming, which would render the early atmosphere overwhelmingly neutral or oxidizing and pose limitations on the effectiveness of MU-type atmospheric organic synthesis. Various constraints have since been placed on the existence of a high CO_2 atmosphere, including the efficient removal of CO_2 by carbonatization of oceanic basalts (Sleep and Zahnle 2001), as well as various limitations imposed by the geological record (Kasting 1987).

The earliest atmosphere is largely thought to have been generated by the outgassing of the Earth's mantle; thus, the types of materials from which the Earth accreted and the rate at which these materials were outgassed or partitioned into the Earth's deep interior, removing them from interaction with the surface environment, become important considerations (Miller and Orgel 1974). Recently, measurements of ~4.3 Ga inclusions in zircon crystals have been used to argue that the upper mantle was essentially at its present oxidation state by that time and thus that the composition of gases released by early volcanism was similar to that that is presently observed (Trail et al. 2011). This leaves open the possibility of localized synthesis around volcanic settings (Johnson et al. 2008), as well as the possibility that the Earth passed through a more reducing phase still earlier in its evolution that might push the period of abundant atmospheric organic synthesis back to the very short period available before this (e.g., ~4.4 Ga) (Tian et al. 2005).

Alternatives to the high CO_2 atmosphere for greenhouse warming that rely on the presence of methane and organic aerosols have also been put forth recently (Wolf and Toon 2010). Such aerosols might have been produced from a mixture of CH_4 and CO_2 and could have shielded photolabile gases such as methane and ammonia from photolysis, allowing their concentrations to rise and then serve as greenhouse gases (Wolf and Toon 2010). The efficacy of abiotic organic synthesis from mixtures of CO_2, CO, and CH_4 remains relatively poorly explored, and it is possible that this offers a solution to the organic synthesis problem if the early atmosphere was not very reducing.

Other solutions to the problems this paradox poses to the supply of organics have been offered. First, organic compounds may have been supplied by extraterrestrial delivery (Pizzarello et al. 2006), and there are now a number of known mechanisms by which such materials could have been delivered to the early Earth and contributed to its organic inventory (Chyba and Sagan 1992). Such delivery seems almost certain, though there remains some debate about its relative importance. Second, as we explore in the next section, other endogenous mechanisms for organic synthesis may have contributed to the early Earth's organic inventory.

5.4.2 Hydrothermal Synthesis

In the late 1970s, after considerable doubt had been raised regarding the composition of the early atmosphere and its compatibility with MU-type synthesis, oceanographers working in small manned and unmanned submersibles discovered undersea hydrothermal systems near mid-ocean ridge spreading centers (Corliss et al. 1979). Not long after their discovery, these systems were suggested as potential sites for organic synthesis and the origin of life (Corliss et al. 1981). Among the benefits such sites potentially offer are protection from the abundant and potentially lethal UV radiation the surface likely received due to the lack of a significant ozone (O_3) layer (the modern ozone layer being largely the by-product of photochemical reactions of the O_2 produced by biological oxygenic photosynthesis) (Cleaves and Miller 1998), protection from the sterilizing effects of impactor bombardment (Maher and Stevenson 1988), and an abundant supply of energy in the form of chemical potential and heat.

These ideas were bolstered by the discovery of complex ecosystems supported in hydrothermal systems (for a discussion, see Van Dover 2000), a biology seemingly completely disconnected from sunlight as its primary energy source. Furthermore, it was later suggested based on the construction of phylogenetic trees of extant organisms that the last universal common ancestor (LUCA) was a thermophile or hyperthermophile (Forterre 1995; Stetter 1996; Di Giulio 2001), though this idea has also been challenged (see, e.g., Glansdorff et al. 2008). The obvious implication was that if LUCA was a thermophile, then the origin of life might also have occurred at high temperature, and by extension, hydrothermal sites were the cradle of life. There has been a considerable amount of discussion of the evidence for and against the earliest organisms being thermophiles and hydrothermal vents as the setting for the origin of life (Miller and Lazcano 1995; Arrhenius 1999; Di Giulio 2001).

First, it has been argued that even if hyperthermophiles are deeply branching groups of organisms, the organisms were already essentially modern organisms, complete with the modern genetic code, translation, and transcription apparatuses, as well as various biochemical innovations such as membrane-driven ATP synthesis, which depends on highly evolved molecular machinery (Islas et al. 2003). Therefore, a surmise is thus that while hyperthermophiles may be the most closely related to LUCA of all extant organisms, they are still highly evolved and thus may either be survivors of ocean-boiling impact events (Abramov and Mojzsis 2009) or, perhaps due to the increase in the rate of biochemical evolution in such environments, lower-temperature organisms may have colonized hydrothermal environments, then later reradiated into lower-temperature environments, displacing their ancestors (Arrhenius et al. 1999).

There are a number of geochemical ambiguities that do not strongly corroborate the hydrothermal origin model, for example, the fact that the earliest fossil organisms appear to be associated with tidal or shallow marine environments (Allwood et al. 2007), and the early light carbon isotope evidence is consistent with photosynthesis (Schidlowski 1988). Furthermore, it is notoriously difficult to calibrate molecular clocks so deeply back in time (Feng et al. 1997; Ayala 1999; Gaucher et al. 2003); thus, the age of the divergence from LUCA cannot be fixed with high confidence, and thus LUCA may well have appeared after the earliest known microfossils and putative period of ocean-sterilizing impact events (Cleaves 2013) (Figure 5.3).

5.4.3 Types of Modern Hydrothermal Environments

Since the discovery of the mid-ocean ridge-type hydrothermal vent environments, a variety of other lower-temperature vent environments have been discovered, including the so-called off-axis or carbonate-hosted systems (Früh-Green et al. 2003). A brief comparison of the two types is provided in Table 5.1 and a schematic showing their distribution and geological context is shown in Figure 5.5.

Bearing in mind that Table 5.1 contains data from many field measurements, it is readily evident that there is a wide range in the types of environments that may be found in any given vent, including the abundance of reduced gases such as methane and hydrogen and the relative oxidation state of sulfur, pH, and temperature. That the values differ so widely from those of seawater is evidence of the modification the fluids undergo upon contact with the host mineral assemblages.

Presently, there is some evidence for indigenous organic synthesis in such environments, with some corroboration coming from laboratory chemical simulations

TABLE 5.1   Some Representative Geochemical Data from Natural Submarine Hydrothermal Systems and of Ambient Modern Seawater, Showing the Variation in Conditions

Conditions	On-Axis Basalt-Hosted Vent System *Black Smoker*	Off-Axis Peridotite-Hosted Vent System *White Smoker*	Seawater
Temperature (°C)	185–370	3–75	7
pH	3.3–6.4	9–9.8	8.0
H_2 (mmol kg^{-1})	0.003–1.7	0.23–0.43	4×10^{-4}
CH_4 (mmol kg^{-1})	0.06–3.4	0.001–0.28	4×10^{-7}
H_2S (mmol kg^{-1})	1.3–9.3	0.064–2.1	0
SO_4^{2-} (mmol kg^{-1})	0–2	0.05–50	28.6

Source: Data from Kelley, D.S. et al., *Nature*, 412, 145, 2001.

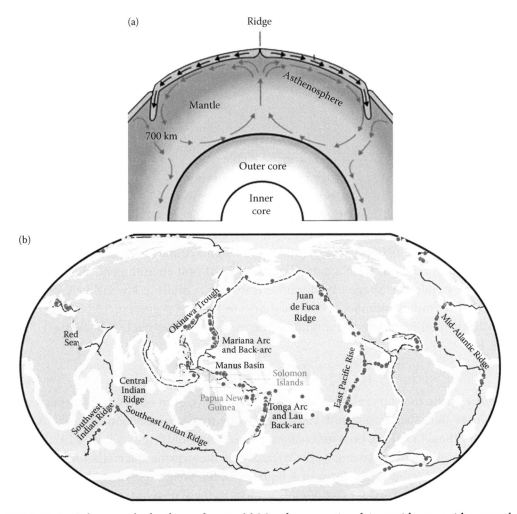

FIGURE 5.5   Submarine hydrothermal vents. (a) Mantle convection drives mid-ocean ridge spreading, (b) the global distribution of hydrothermal vent fields (yellow, suggested sites; red observed sites).

(*continued*)

(c)

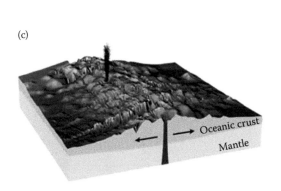

(d)

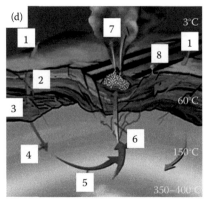

FIGURE 5.5 (continued) Submarine hydrothermal vents. (c) a black smoker atop a mid-ocean spreading ridge, and (d) schematic showing how hydrothermal convection drives vent dynamics. (1) Cold seawater is drawn into the high-temperature zone of the spreading ridge, where a variety of interactions with magmas and basalts (2–6) lead to significant changes in the composition of the fluids, including removal of dissolved oxygen and addition of various metals. Finally, these fluids are discharged into the ambient seawater, either directly (7) or via diffuse off-axis flow (8).

(Seewald et al. 2006; Proskurowski et al. 2008). For example, it has been shown that isotopically heavy low-molecular-weight hydrocarbons are found in the vent effluent of several systems, which are consistent with an abiological source and equilibration with the reduced mineralogy and chemistry of the vent environments (Proskurowski et al. 2008). The presence of more complex organics remains controversial, one reason for this being the ubiquity of biological materials in the seawater, which is the source of the vent effluent, and the abundance of microorganisms in the immediate vent environments (Bassez et al. 2009). To date, no unambiguous identification of compounds even as simple as amino acids has been returned from a natural hydrothermal vent environment, though various laboratory simulations suggest such synthesis may be possible in low yield (Aubrey et al. 2009). Furthermore, a number of computational techniques suggest such abiotic synthesis in hydrothermal environments should be thermodynamically viable (Shock 1990; Holm et al. 1992).

5.5 METEORITES AS TOUCHSTONES

We turn briefly to a survey of the compounds that have been observed in carbonaceous meteorites and other extraterrestrial materials that may have contributed to the organic inventory thought to have been necessary for a heterotrophic origin of life on Earth.

First, as mentioned previously, the majority of the organic materials found in such meteorites are bound in the form of a high-molecular-weight polymer (IOM). In addition to low-molecular-weight compounds present in these meteorites that are extractable by relatively low-temperature extraction, high-temperature (100°C–300°C) water or acid hydrolysis of this material gives rise to still more low-molecular-weight compounds, suggesting that many of these are bound in IOM macromolecular matrix (Pizzarello et al. 2006; Glavin et al. 2010).

TABLE 5.2   Organic Compounds Detected to Date in the Murchison Meteorite

Class	Concentration (ppm)	Number of Compounds Identified
Aliphatic hydrocarbons	>35	140
Aromatic hydrocarbons	22	87
Polar hydrocarbons	<120	10
Carboxylic acids	>300	48
Amino acids	60	74
Hydroxy acids	15	38
Dicarboxylic acids	>30	44
Dicarboximides	>50	2
Pyridine carboxylic acids	>7	7
Sulfonic acids	67	4
Phosphonic acids	2	4
N-heterocycles	7	31
Amines	13	20
Amides	n.d.	27
Polyols	30	19
Imino acids	n.d.	10

Source: Data from Pizzarello, S. et al., The nature and distribution of the organic material in carbonaceous chondrites and interplanetary dust particles, in: *Meteorites and the Early Solar System II*, University of Arizona Press, Tucson, AZ, in collaboration with Lunar and Planetary Institute, Houston, TX, 2006.

Note: n.d., not determined.

A variety of compounds have now been identified, including amino acids, nucleobases, sulfonic and phosphonic acids, fatty acids, polycyclic aromatic hydrocarbons, and straight- and branched-chain hydrocarbons, among others (Pizzarello et al. 2006). A summary of the results to date is presented in Table 5.2.

The mechanism of synthesis of these various compounds remains an active area of research, with the most parsimonious explanation being a combination of early solar nebula and postaccretionary processing. Briefly, the mode of accretion of these materials is summarized in Figure 5.6.

5.6 COMPOUND CLASSES

5.6.1 Lipids

Lipids are fundamental to biology as the compounds from which biological membranes are largely constructed. These can broadly be grouped into two classes, which are found in various types of organisms: the isoprenoid ether lipids and the fatty acid acyl lipids (Figure 5.7).

The fatty acid ester type (Figure 5.7b) is widely distributed across the eubacterial and eukaryotic domains of life and consists of the union of one or more straight saturated or unsaturated fatty acid chains. These chains typically contain 16–18 carbon atoms and are derived from the malonyl/acetyl-CoA biosynthetic pathways, joined to a polar head group, which is usually some derivative of glycerol, rendered more polar by the attachment

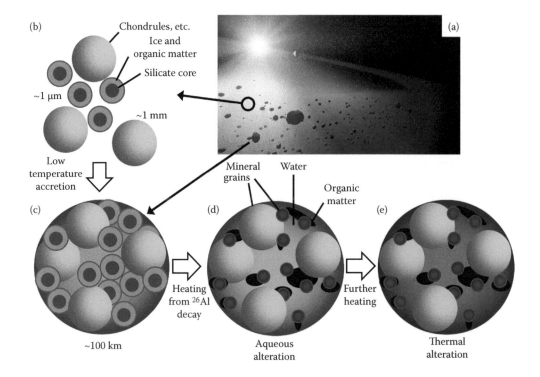

FIGURE 5.6 Suggested mechanisms of synthesis of organic compounds detected in carbonaceous meteorites. (a) shows an artists interpretation of the solar system during its early formation. (b) shows a close up of the tiny dust and ice grains which clumped together to form larger bodies such as asteroids. Both high temperature and energetic processing may contribute to organic synthesis. The dark brown regions in (d) and (e) represent organics derived from the release of simple precursors from the melting of the ice shown in (b) and (c). (a: Courtesy of NASA.)

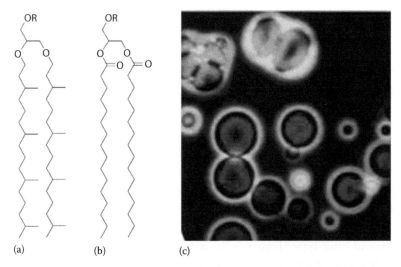

FIGURE 5.7 (a) Isoprenoid glycerol ether lipids common in Archaebacteria, (b) fatty acid glycerol ester lipids common in bacteria and eukaryotes, and (c) boundary-forming structures extracted from the Murchison meteorite. (c: Courtesy of NASA.)

of phosphate, choline, or some other low-molecular-weight moiety (represented as R in Figure 5.7a and b). The isoprenoid ether type (Figure 5.7a) is widely distributed among the archaebacteria (Matsumi et al. 2011), particularly in extremophilic species, perhaps rationalizable by their greater resistance to chemical degradation at extremes of salinity, pH, and temperature (Driessen and Albers 2007). These consist of one or more polyisoprenoid chains (derived from the isoprene biosynthetic pathways), attached by ether linkages to polar head groups such as substituted glycerol moieties.

Prebiotic syntheses of the straight-chain fatty acids have been suggested based on Fischer–Tropsch-type synthesis (FTT), for example, by the reactions of CO or CH_4 over suitable metal catalysts at high temperature (McCollom et al. 1999, 2010) (Figure 5.8).

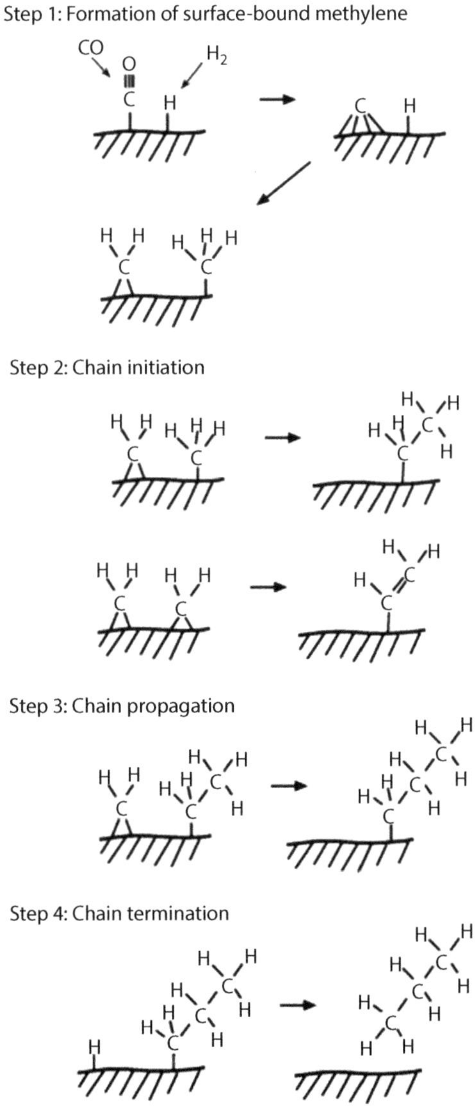

FIGURE 5.8   FTT synthesis of straight-chain amphiphiles from CO and H_2 on metal sulfide surfaces. (Reproduced from McCollom, T.M. et al., *Geochim. Cosmochim. Acta*, 74, 2717, 2010. With permission.)

The condensation of fatty acids with glycerol under drying conditions and phosphorylation of the resulting acyl derivatives have been demonstrated under conditions that might be expected in evaporating ponds or tidal environments (Rao et al. 1982).

The isoprenoid lipids have been suggested to be synthesizable from isobutene and formaldehyde by the Prins reaction (Ourisson and Nakatani 1994), and there is now some experimental evidence for this (Désaubry et al. 2003).

Carbonaceous meteorites have indeed been found to contain both straight-chain fatty acids and isoprenoids (Kvenvolden et al. 1970; Yuen and Kvenvolden 1973); however, analysis suggests the isoprenoids are terrestrial biological contaminants (Nooner and Oró 1967). Strong evidence for indigenous straight-chain fatty acids appears to be limited to those of C9 or shorter chains (Kvenvold et al. 1970).

Interestingly, it is known that straight-chain amphiphiles and various biological lipids will spontaneously self-assemble under the appropriate conditions of pH, temperature, and solute concentration to form aggregates known as micelles or vesicles, which in some cases are able to trap various solutes for extended periods of time (Deamer and Barchfeld 1982). These properties have led to suggestions that such materials may have been important for the self-assembly of the earliest protocells (Szostak 2001). Intriguingly, the straight-chain fatty acids found in carbonaceous chondrites to date are just at the threshold of the types of molecules that are known to display this behavior (Monnard et al. 2002), and even these are present in very low abundance (Yuen and Kvenvolden 1973). For example, the octanoic acid content of 1 g of the Murchison meteorite would need to be dissolved in a few tens of nanoliters of water in order to form vesicles (given a critical vesicle concentration [the concentration above which bilayered or higher aggregated lipid structures form spontaneously] of 130 mM [Apel et al. 2002] and a mean abundance of 0.01 μmol g^{-1} in a typical meteorite [Lawless and Yuen 1979]).

Nevertheless, organic solvent extracts of the Murchison meteorite show the presence of compounds that spontaneously assemble into boundary structures (Deamer 1985) (Figure 5.7c), though it is unlikely these have much compositional similarity to biological membranes.

5.6.2 Amino Acids

Amino acids are fundamental to biology as the structural units of the enzymes, which are responsible for the vast and varied catalytic repertoire of cells, and other cellular structural proteins. Terrestrial biology uses almost exclusively a set of 20 across all domains of life for the construction of coded proteins (Figure 5.9).

Amino acids were among the first biological compounds found in prebiotic organic synthesis experiments (Miller 1953), and since then, a variety of mechanisms have been found by which they can be produced abiotically.

Miller originally suggested the Strecker mechanism (Miller 1957) (Figure 5.10) for their synthesis under MU-type reaction conditions, and indeed, there is a good deal of evidence for the operation of this synthetic pathway.

The reaction is robust under certain conditions, namely, neutral to slightly basic conditions (Taillades and Commeyras 1974), and, depending on the ammonia concentration,

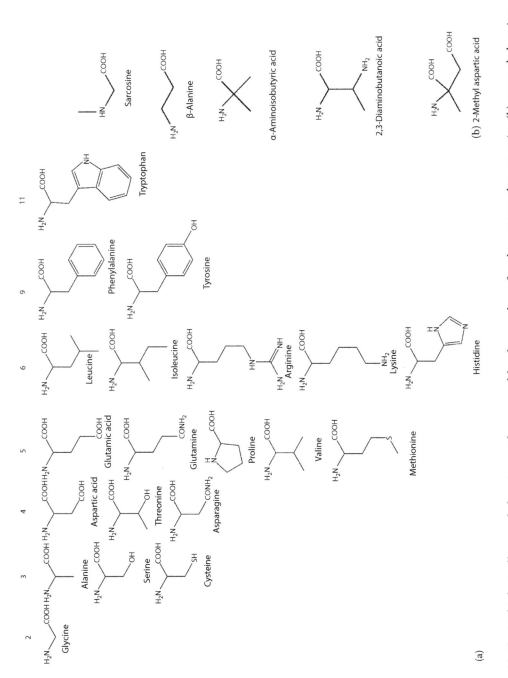

FIGURE 5.9 (a) The 20 biologically encoded amino acids arranged by the number of carbon atoms they contain. (b) noncoded amino acids also found in carbonaceous meteorites.

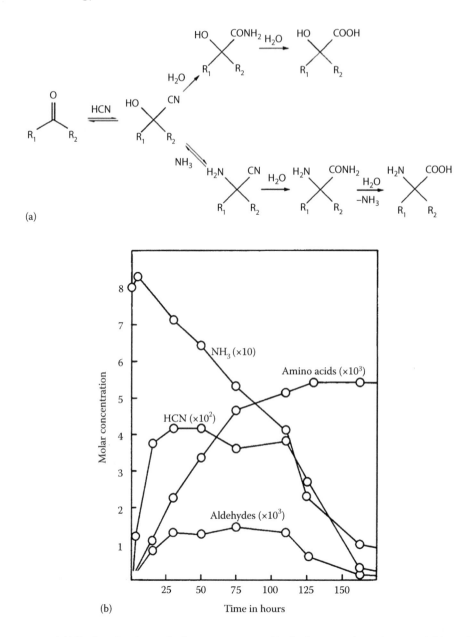

FIGURE 5.10 (a) The Strecker/cyanohydrin α-amino acid/α-hydroxy acid synthesis and (b) evidence for the operation of these synthetic pathways in the form of detection of the intermediates during the course of an electric discharge experiment. (b: Reproduced from Miller, S.L., *Biochim. Biophys. Acta*, 23, 480, 1957. With permission.)

also provides a mechanism to the cognate α-hydroxy acids (Peltzer et al. 1984). Depending on the starting conditions, ~12 of the 20 coded biological amino acids now have convincing prebiotic syntheses (Miller 1998). It is widely believed that the remainder were derived during the development of the biosynthetic pathways during biological evolution (Weber and Miller 1981; Cleaves 2010).

Other pathways also yield amino acids, for example, the hydrolysis of the polymer derived from the condensation of aqueous HCN gives rise to a variety of amino acids including serine, aspartic and glutamic acids, and α- and β-alanine (Ferris et al. 1978). The hydrolysis of various high-molecular-weight organic polymers (*tholins*) has also been found to liberate amino acids directly (Khare et al. 1986), suggesting that solution-phase conditions may be less important than previously thought if sufficiently reducing atmospheric conditions are available.

There is strong evidence for some of the aromatic amino acids (phenylalanine and tyrosine) in carbonaceous chondrites (Pizzarello and Holmes 2009); however, several biological amino acids such as histidine, tryptophan, arginine, and lysine remain difficult targets of prebiotic synthesis (Miller 1998).

It is unknown when proteins or simple peptides became integral parts of biochemistry; however, heating concentrated aqueous amino acid solutions or heating amino acids in the dry state can give rise to peptides of various molecular weights depending on the conditions of synthesis (Meggy 1954; Fox and Harada 1958; Cleaves et al. 2009).

In addition to the biological amino acids, abiotic synthesis may give rise to a variety of nonbiological amino acids, including N-substituted and β- and α,α-disubstituted amino acids, among other types, some of which are found in contemporary organisms (Figure 5.6) (Burton et al. 2012). It seems likely that abiotic synthesis provided some, but not all, of the coded amino acids in addition to many not found in coded proteins. Life's use of the canonical 20 coded amino acids is thus likely the result of a protracted period of biological evolution (Weber and Miller 1981; Cleaves 2010). That this occurred in the context of biological systems is also likely because of the difficulty of stringing amino acids together abiotically to form long polypeptide chains. Once sufficiently robust oligomerization mechanisms were available, life would have been free to explore the combinatorial catalytic peptide space this innovation allowed access to.

5.6.3 Nucleic Acids

The same year that Miller published his pioneering result regarding abiotic amino acid synthesis, the double-helical model for the structure of DNA was published (Watson and Crick 1953), which effectively clinched the role of nucleic acids in the inheritance of biological mutations. There are two types of nucleic acids important in biological systems, DNA and RNA, linked by the processes of transcription and biosynthesis (Figure 5.1).

Not long after Miller's and Watson and Crick's discoveries, the search for abiotic mechanisms for the synthesis of these important biochemicals began. Oró and Kimball showed that adenine, a biological purine, could be derived from the polymerization of aqueous HCN, of which it is formally a pentamer ($C_5H_5N_5$) (Oró and Kimball 1961). The mechanism of this synthesis was soon elucidated, and shortly thereafter, syntheses of other important purines including guanine, and the biosynthetic precursors xanthine and hypoxanthine, were elaborated (Ferris and Orgel 1965, 1966) (Figure 5.11).

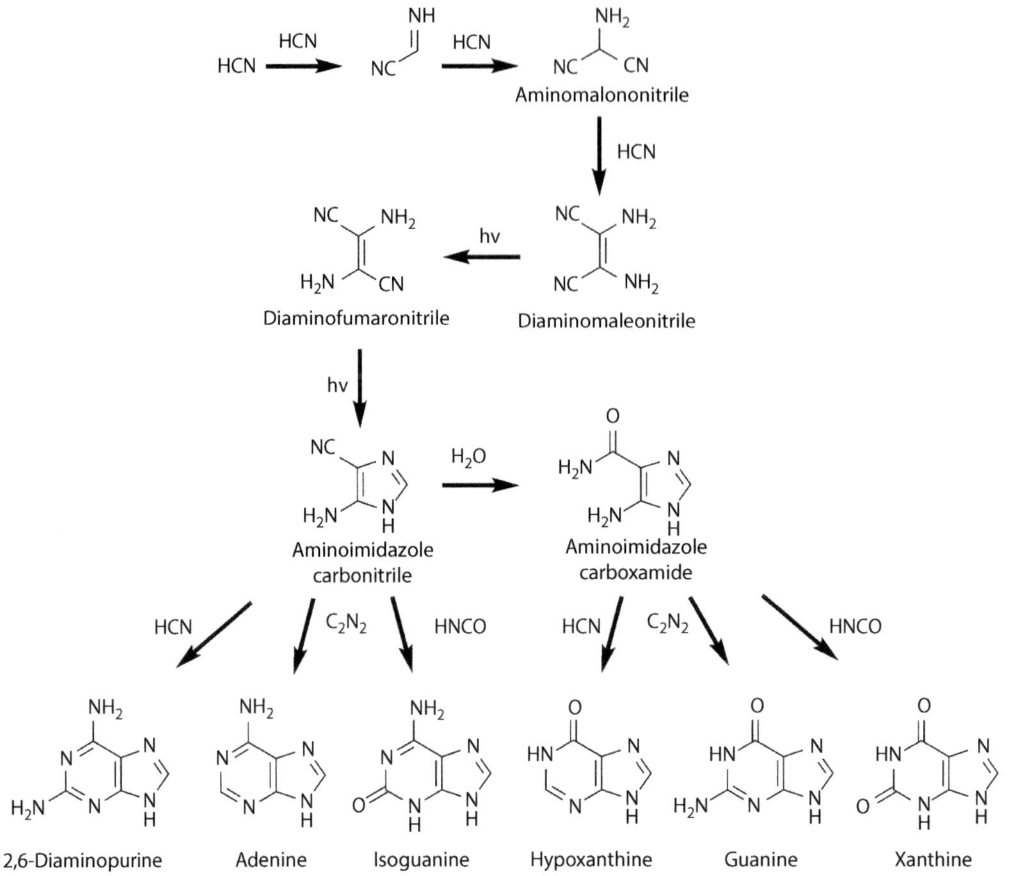

FIGURE 5.11 Demonstrated prebiotic mechanisms for the synthesis of the biological purines from HCN.

All of these purines have now been identified in carbonaceous chondrites (Stoks and Schwartz 1979; Martins et al. 2008; Callahan et al. 2011), possibly validating this mechanism making their widespread distribution in the early solar system extremely likely.

In the late 1960s, attention further turned to elucidating the abiotic synthesis of the other important class of biological nitrogen heterocycles, the pyrimidines (Ferris et al. 1968) (Figure 5.12).

Cyanoacetylene (HCCCN), which had already been found to be a major product of the action of electric discharges acting on reduced C- and N-containing gas mixtures and implicated as a precursor to the aspartic acid detected in such reactions (Stanley 1957; Sanchez et al. 1966), and cyanate could serve as precursors to the synthesis of cytosine and uracil in aqueous solution (Ferris et al. 1968). Further development of this synthesis starting from cyanoacetaldehyde ($OCHCH_2CN$) and urea or guanidine was found to give these and other pyrimidine derivatives (Robertson and Miller 1995; Robertson et al. 1996) (Figure 5.12).

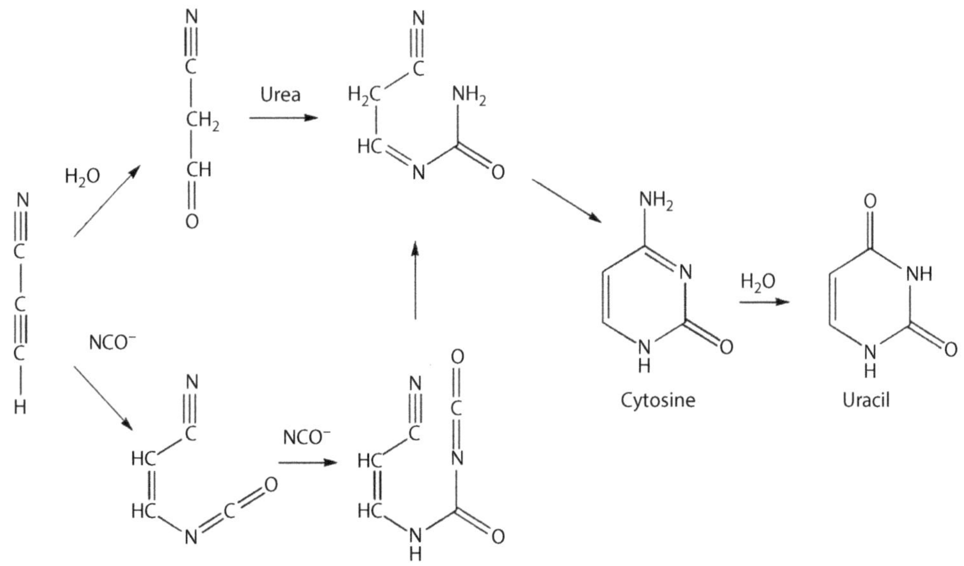

FIGURE 5.12 Demonstrated prebiotic mechanisms for the synthesis of the biological pyrimidines from plausible prebiotic starting materials.

5.6.4 Sugars

While sugars have a variety of important roles in biochemistry, we will explore them here as the important components of nucleic acid backbones in the form of ribose and deoxyribose. All canonical sugars share the empirical formula $(CH_2O)_n$, formally making them oligomers of formaldehyde (HCHO). Ribose, the sugar used in RNA, is but one of the isomeric pentamers where $n=5$, as each sugar may contain a number of stereo centers. Early in the development of organic chemistry as an empirical science, it was found that basic solutions of formaldehyde could give rise to a complex mixture of compounds, which included various sugars (Butlerow 1861). The mechanism of this synthesis has since been explored extensively (Breslow 1959) (Figure 5.13).

Early in its consideration as a prebiotic process for the production of ribose and other sugars, it was pointed out that the extreme diversity of products the reaction gives rise to, as well as the ultimate instability of sugars under the conditions of synthesis, may render this an implausible source of prebiotic carbohydrates (Reid and Orgel 1967; Larralde et al. 1995). Recently, there has been a resurgence of interest in this pathway as several new mechanisms have been discovered that produce a less diverse mixture, give rise to a higher yield of ribose, and importantly make the ultimate sugar derivatives considerably more stable than they are in the free form (Prieur 2001; Ricardo et al. 2004; Lambert et al. 2010). For example, conducting the reaction in the presence of borate selectively gives rise to a good yield of ribose–borate derivatives (Ricardo et al. 2004).

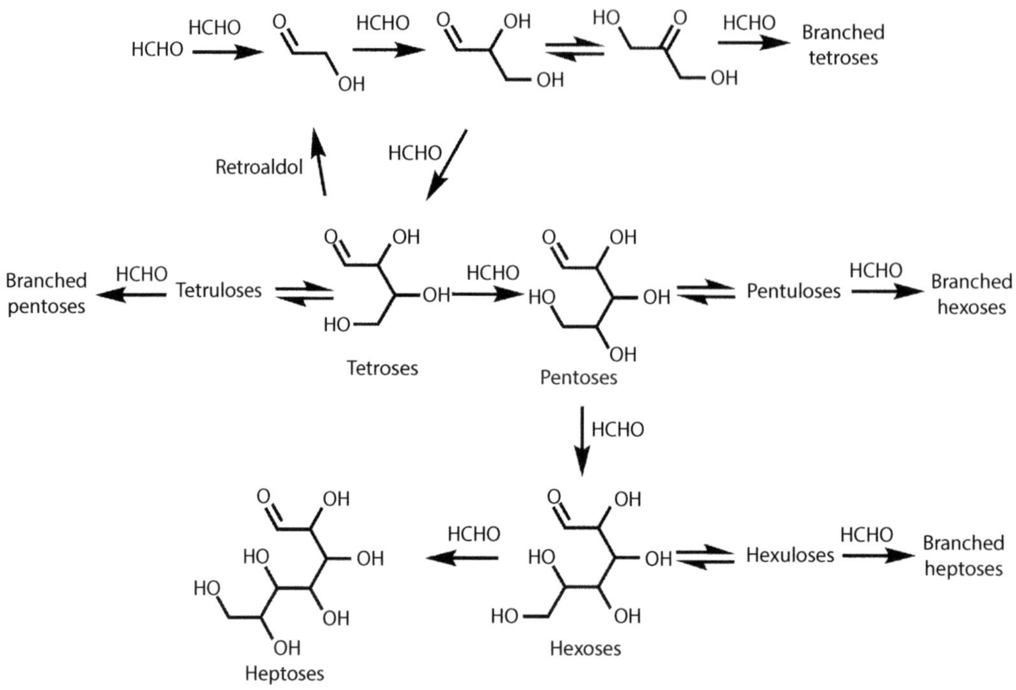

FIGURE 5.13 The general course of the formose or Butlerow reaction.

5.6.5 Nucleosides and Nucleotides

The construction of nucleosides depends on the union of a nitrogenous base, via the correct linkage, with a sugar derivative. Some success in prebiotic synthesis has been achieved in this area. For example, it has been found that heating pure ribose with purines gives rise to small yields of purine ribosides, though a variety of isomers are produced (Fuller et al. 1972) (Figure 5.14a). The equivalent reaction using pyrimidines does not work well, however. Recently, though, it has been found that a nonnatural pyrimidine can be linked under the same condition to give a pyrimidine nucleoside (Bean et al. 2007).

The phosphorylation of nucleosides to give nucleotides has been accomplished in a variety of manners that are conceivably prebiotic. Dry heating various mixtures of nucleosides in the presence of ammonium salts and orthophosphate or apatite and cyanate gives decent yields of pyrimidine nucleotides (Bishop et al. 1972; Schwartz 1972; Schwartz et al. 1973).

The limitations of the synthesis of pyrimidine nucleosides led Orgel and coworkers to examine other disconnects to find less obvious, unorthodox syntheses, which have been elucidated and explored further more recently (Sanchez and Orgel 1970; Powner et al. 2007) (Figure 5.14b). This approach is novel in that the sugar and nitrogenous base are constructed simultaneously and phosphate is incorporated prior to the completion of the nucleoside's synthesis. This has proven effective for the pyrimidines, but to date, a complete analogous synthesis for the purine nucleotides has proven elusive (Powner et al. 2010).

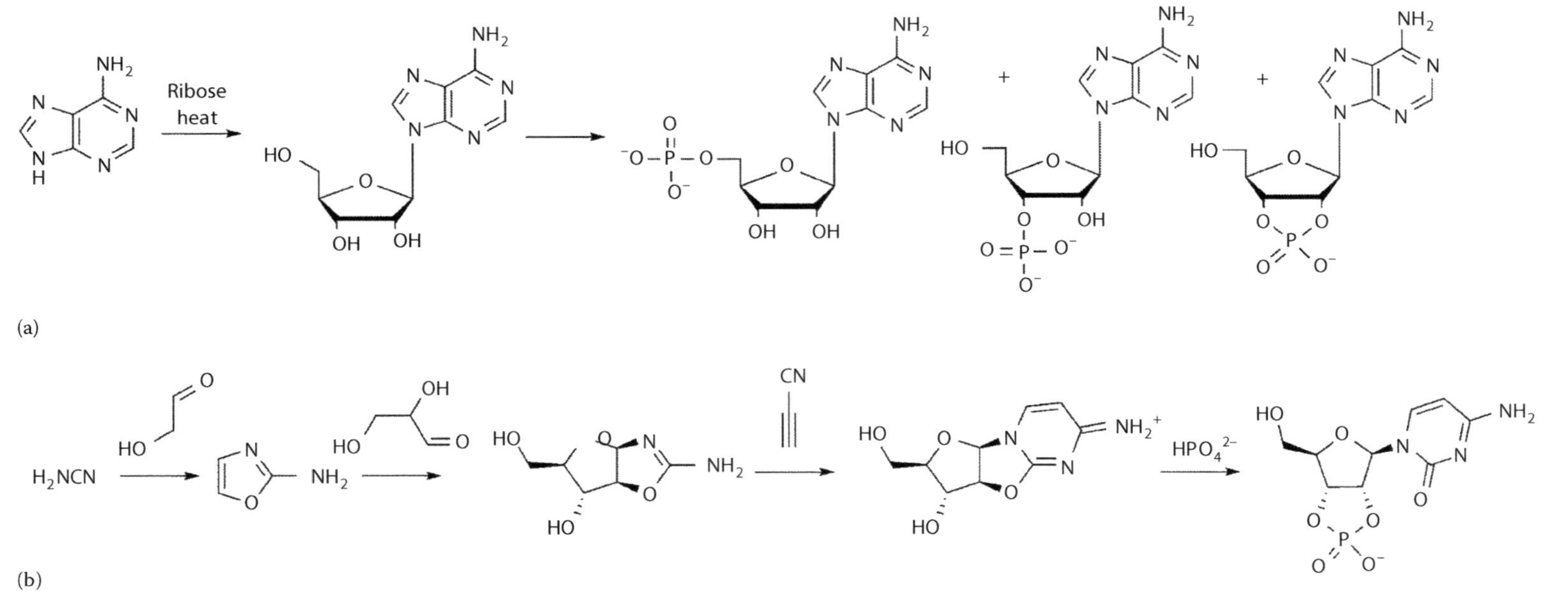

FIGURE 5.14 Some reactions that have been investigated in the context of the prebiotic synthesis of nucleosides and nucleotides. (a) Dry heating of purines with ribose gives low yields of purine ribosides; the equivalent reaction does not work with the pyrimidines. Dry heating ribosides with inorganic phosphate can give rise to mixed nucleotide isomers, including 3′,5′- and 2′,3′-cyclic species. (b) More recent investigations have extended previous work on the synthesis of activated pyrimidine nucleotides, showing that sequential addition of reagents can selectively give rise to activated pyrimidine nucleotides. In this case, an equivalent set of reactions has not been demonstrated for the purine nucleotides.

5.7 SUMMARY AND FUTURE DIRECTIONS

There are a variety of ways in which many of the organic compounds used in modern biochemistry could have been formed abiologically in plausible early solar system environments, including some beyond Earth. Not all of these compounds have proven equally easy to make, and some have resisted synthesis altogether. We have not reviewed here all of the compounds whose synthesis has been investigated, and it is likely that much remains to be discovered in this area.

There has been extensive work attempting to make higher-order structures, such as polypeptides and polynucleotides, from these relatively simple biological monomers, with mixed degrees of success (see, e.g., Cleaves 2008 and other contributions in that volume). Thus, Ganti's dream of a minimal integrated set of systems remains well beyond the rather reductionist approaches explored by prebiotic chemists thus far.

It should be kept in mind that we do not know if the first living systems were composed of entirely different compounds and gradually adopted the modern components during the course of biological evolution. Indeed, analysis of carbonaceous meteorites and laboratory simulations has shown that often many thousands of compounds that are not important in modern biochemistry occur and are produced alongside those important in biochemistry (Schmitt-Kopplin et al. 2010; Vuitton et al. 2010), and this has been pointed out as a serious potential flaw in many models for the origin of life (Shapiro 1984, 1987, 1988, 1995). It is also becoming increasingly clear that a variety of other molecules can carry out the functions of modern biopolymers (Egholm et al. 1992; Pinheiro et al. 2012), and it thus remains unclear whether biology has chosen the best solutions or perhaps only those that it managed to access during its evolutionary development. Clearly a significant amount of work remains to be done in this most enigmatic aspect of biology.

GLOSSARY

Acyl phospholipid: A lipid in which typically glycerol is linked to two straight-chain fatty acids via acyl (ester) linkages. These are common components of bacterial and eukaryotic cell membranes.

Amino acid: An organic compound containing both an amino (–NH$_2$) group and a carboxylic acid (–COOH) group. Biology uses a consistent set of these to make its coded proteins.

Archaebacteria: Unicellular microorganisms similar to bacteria in size and simplicity but evolutionarily more similar to eukaryotes. Many of them are extremophiles, and they exhibit a wide diversity of metabolisms.

Central dogma: An explanation of the flow of genetic information in biological systems. It holds that information flow in biological systems deals with the detailed residue-by-residue transfer of information in biopolymers. It holds that information cannot be transferred back from protein to either protein or nucleic acid.

Detrital: Referring to grains or fragments of rocks that have been eroded from earlier rocks.

Isoprenoid: Any of a member of a large and diverse class of naturally occurring organic chemicals, also called terpenoids, biosynthetically derived from five-carbon isoprene units. They are the largest group of natural products.

Nucleobase: A nitrogen-containing heterocycle used as a base-paring element in a nucleic acid. These can be pyrimidines or purines.

Nucleoside: A monomeric structural element of a nucleic acid consisting of a sugar (in biological nucleic acids, either ribose or deoxyribose) and a nucleobase.

Nucleotide: A monomeric structural element of a nucleic acid consisting of a sugar, a nucleobase, and a phosphate moiety.

Oxidizing atmosphere: A planetary atmosphere containing an excess of oxygen over hydrogen. Examples of planets in our solar system that have oxidizing atmospheres include Venus, Earth, and Mars.

Ozone: An inorganic compound with the chemical formula O_3. It is an allotrope of oxygen that is much less stable than O_2. Ozone is formed from O_2 by the action of UV light and atmospheric electrical discharges. Ozone plays a significant role in blocking ionizing UV radiation from the Sun.

Peridotite: A dense igneous rock, consisting mostly of olivine and pyroxene, containing less than 45% silica. It is high in Mg, with appreciable Fe. Peridotite is the dominant rock of the Earth's upper mantle.

Purine: A nitrogen-containing aromatic organic compound consisting of a fused six-membered pyrimidine and five-membered imidazole ring system with variable exocyclic groups.

Pyrimidine: A nitrogen-containing aromatic organic compound consisting of a six-membered aromatic ring containing two nitrogen atoms with variable exocyclic groups.

Reducing atmosphere: A planetary atmosphere containing an excess of hydrogen over oxygen. Examples of planets in our solar system that have reducing atmospheres include Jupiter and Saturn.

Zircon: A silicate mineral with the chemical formula $ZrSiO_4$. Zircon crystals are highly resistant to weathering and often include various other rare earth elements that give clues to their conditions of formation. As such, they can be used as markers of ancient geological processes.

REVIEW QUESTIONS

1. What factors would have affected the supply of organic compounds on the primitive Earth?

2. What are the major classes of biochemical compounds? Do you think all of these were required for the origin of life? If not, which ones do you think were crucial?

3. Based on the information presented here, what do you think would be the most amenable environment for the origin of life?

4. Based on the information presented here, what do you think would be the most amenable location in our solar system (besides Earth) to look for evidence of life?

5. What traits or characteristics do you consider to be required for life?

6. Describe the Miller–Urey experiment and explain why it was critical for astrobiology.

7. What are hydrothermal vents? What type of chemistry is going on there?

8. Describe some of the classes of compounds that were found on Murchison meteorite.

9. What are the difficulties in making sugars prebiotically?

10. Which of the syntheses presented is the most robust in your opinion?

REFERENCES

Abramov, O. and S. J. Mojzsis (2009). Microbial habitability of the Hadean Earth during the late heavy bombardment. *Nature* **459**(7245): 419–422.

Allen, W. V. and C. Ponnamperuma (1967). A possible prebiotic synthesis of monocarboxylic acids. *Currents in Modern Biology* **1**(1): 24–28.

Allwood, A. C., M. R. Walter et al. (2007). 3.43 billion-year-old stromatolite reef from the Pilbara Craton of Western Australia: Ecosystem-scale insights to early life on Earth. *Precambrian Research* **158**(3–4): 198–227.

Apel, C. L., D. W. Deamer et al. (2002). Self-assembled vesicles of monocarboxylic acids and alcohols: Conditions for stability and for the encapsulation of biopolymers. *Biochimica et Biophysica Acta (BBA)—Biomembranes* **1559**(1): 1–9.

Arrhenius, G., J. L. Bada et al. (1999). Origin and ancestor: Separate environments. *Science* **283**(5403): 792–792.

Aubrey, A. D., H. J. Cleaves et al. (2009). The role of submarine hydrothermal systems in the synthesis of amino acids. *Origins of Life and Evolution of the Biosphere* **39**(2): 91–108.

Ayala, F. J. (1999). Molecular clock mirages. *Bioessays* **21**(1): 71–75.

Bassez, M.-P., Y. Takano et al. (2009). Organic analysis of peridotite rocks from the Ashadze and Logatchev hydrothermal sites. *International Journal of Molecular Sciences* **10**(7): 2986–2998.

Bean, H. D., Y. Sheng et al. (2007). Formation of a beta-pyrimidine nucleoside by a free pyrimidine base and ribose in a plausible prebiotic reaction. *Journal of the American Chemical Society* **129**(31): 9556–9557.

Bedau, M. A. and C. E. Cleland (2010). *The Nature of Life*. Cambridge University Press, Cambridge, U.K.

Bishop, M. J., R. Lohrmann et al. (1972). Prebiotic phosphorylation of thymidine at 65 degrees C in simulated desert conditions. *Nature* **237**(5351): 162–164.

Breslow, R. (1959). On the mechanism of the Formose reaction. *Tetrahedron Letters* **21**: 22–26.

Briggs, R., G. Ertem et al. (1992). Comet Halley as an aggregate of interstellar dust and further evidence for the photochemical formation of organics in the interstellar medium. *Origins of Life and Evolution of the Biosphere* **22**(5): 287–307.

Burton, A. S., J. C. Stern et al. (2012). Understanding prebiotic chemistry through the analysis of extraterrestrial amino acids and nucleobases in meteorites. *Chemical Society Reviews* **41**(16): 5459–5472.

Butlerow, A. (1861). Formation synthetique d'une substance sucree. *Comptes Rendus des Seances de l'a Academie des Sciences* **53**: 145–147.

Callahan, M. P., K. E. Smith et al. (2011). Carbonaceous meteorites contain a wide range of extraterrestrial nucleobases. *Proceedings of the National Academy of Sciences of the United States of America* **108**(34): 13995–13998.

Carr, M. H. and J. W. Head (2003). Oceans on Mars: An assessment of the observational evidence and possible fate. *Journal of Geophysical Research: Planets* **108**(E5): 5042.

Cavicchioli, R. (2002). Extremophiles and the search for extraterrestrial life. *Astrobiology* **2**(3): 281–292.

Chyba, C. and C. Sagan (1992). Endogenous production, exogenous delivery and impact-shock synthesis of organic molecules: An inventory for the origins of life. *Nature* **355**: 125–132.

CIBA Foundation (2008). *Evolution of Hydrothermal Ecosystems on Earth (and Mars?)*. Wiley, New York.

Cleaves, H. J. (2008). Prebiotic chemistry, the primordial replicator, and modern protocells. In: *Protocells: Bridging Nonliving and Living Matter*, edited by M. A. Bedau, S. Rasmussen, L. Chen et al. The MIT Press, Cambridge, MA, pp. 583–614.

Cleaves, H. J., 2nd (2010). The origin of the biologically coded amino acids. *Journal of Theoretical Biology* **263**(4): 490–498.

Cleaves, H. J. (2012). Prebiotic chemistry: What we know, what we don't. *Evolution: Education and Outreach* **5**(3): 342–360.

Cleaves, H. J. (2013). Prebiotic chemistry: Geochemical context and reaction screening. *Life* **3**(2): 331–345.

Cleaves, H. J., A. D. Aubrey et al. (2009). An evaluation of the critical parameters for abiotic peptide synthesis in submarine hydrothermal systems. *Origins of Life and Evolution of the Biosphere* **39**(2): 109–126.

Cleaves, H. J., 2nd and J. H. Chalmers (2004). Extremophiles may be irrelevant to the origin of life. *Astrobiology* **4**(1): 1–9.

Cleaves, H. J. and S. L. Miller (1998). Oceanic protection of prebiotic organic compounds from UV radiation. *Proceedings of the National Academy of Sciences of the Unites States of America* **95**(13): 7260–7263.

Cleaves, H. J., C. Neish, M. P. Callahan, E. Parker, F. M. Fernández, and J. P. Dworkin. (2014, in press) Amino acids generated from hydrated Titan tholins: Comparison with Miller–Urey electric discharge products. *Icarus*.

Cooper, G., C. Reed et al. (2011). Detection and formation scenario of citric acid, pyruvic acid, and other possible metabolism precursors in carbonaceous meteorites. *Proceedings of the National Academy of Sciences of the United States of America* **108**(34): 14015–14020.

Corliss, J. B., J. Baross et al. (1981). An hypothesis concerning the relationship between submarine hot springs and the origin of life on Earth. *Oceanologica Acta* **4**(Supplement): 59–69.

Corliss, J. B., J. Dymond et al. (1979). Submarine thermal springs on the Galapagos Rift. *Science* **203**(4385): 1073–1083.

Cronin, J. R. and C. B. Moore (1971). Amino acid analyses of the Murchison, Murray, and Allende carbonaceous chondrites. *Science* **172**(3990): 1327–1329.

Dalrymple, G. B. (1994). *The Age of the Earth*. Stanford University Press, Stanford, CA.

Davies, P. C., S. A. Benner et al. (2009). Signatures of a shadow biosphere. *Astrobiology* **9**(2): 241–249.

Deamer, D. W. (1985). Boundary structures are formed by organic components of the Murchison carbonaceous chondrite. *Nature* **317**(6040): 792–794.

Deamer, D. W. and G. L. Barchfeld (1982). Encapsulation of macromolecules by lipid vesicles under simulated prebiotic conditions. *Journal of Molecular Evolution* **18**(3): 203–206.

Désaubry, L., Y. Nakatani et al. (2003). Toward higher polyprenols under 'prebiotic' conditions. *Tetrahedron Letters* **44**(36): 6959–6961.

Di Giulio, M. (2001). The universal ancestor was a thermophile or a hyperthermophile. *Gene* **281**(1): 11–17.

Driessen, A. J. and S.-V. Albers (2007). Membrane adaptations of (hyper) thermophiles to high temperatures. In: *Physiology and Biochemistry of Extremophiles*, edited by C. Gerday and N. Glansdorff. ASM Press, Washington, DC, pp. 104–116.

Egholm, M., O. Buchardt et al. (1992). Peptide nucleic acids (PNA). Oligonucleotide analogs with an achiral peptide backbone. *Journal of the American Chemical Society* **114**(5): 1895–1897.

Eschenmoser, A. (2007). The search for the chemistry of life's origin. *Tetrahedron* **63**(52): 12821–12844.

Farley, J. (1977). *The Spontaneous Generation Controversy from Descartes to Oparin*. The Johns Hopkins University Press, Baltimore, MA.

Feng, D.-F., G. Cho et al. (1997). Determining divergence times with a protein clock: Update and reevaluation. *Proceedings of the National Academy of Sciences of the United States of America* **94**(24): 13028–13033.

Ferris, J. P., P. C. Joshi et al. (1978). Chemical evolution 30. HCN—Plausible source of purines, pyrimidines and amino acids on primitive Earth. *Journal of Molecular Evolution* **11**(4): 293–311.

Ferris, J. P. and D. E. Nicodem (1972). Ammonia photolysis and the role of ammonia in chemical revolution. *Nature* **238**(5362): 268–269.

Ferris, J. P. and L. E. Orgel (1965). Aminomalononitrile and 4-amino-5-cyanoimidazole in hydrogen cyanide polymerization and adenine synthesis. *Journal of the American Chemical Society* **87**(21): 4976–4977.

Ferris, J. P. and L. E. Orgel (1966). Studies on prebiotic synthesis. I. Aminomalononitrile and 4-amino-5-cyanoimidazole. *Journal of the American Chemical Society* **88**(16): 3829–3831.

Ferris, J. P., R. A. Sanchez et al. (1968). Studies in prebiotic synthesis. 3. Synthesis of pyrimidines from cyanoacetylene and cyanate. *Journal of Molecular Biology* **33**(3): 693–704.

Folsome, C. E., Lawless, J., Romiez, M., and Ponnamperuma, C. (1971). Heterocyclic compounds indigenous to the Murchison meteorite. *Nature* **232**: 108–109.

Forterre, P. (1995). Looking for the most "primitive" organism(s) on Earth today: The state of the art. *Planetary and Space Science* **43**(1–2): 167–177.

Fox, S. W. and K. Harada (1958). Thermal copolymerization of amino acids to a product resembling protein. *Science* **128**(3333): 1214.

Freeland, S. J. and L. D. Hurst (1998). The genetic code is one in a million. *Journal of Molecular Evolution* **47**(3): 238–248.

Früh-Green, G. L., D. S. Kelley et al. (2003). 30,000 years of hydrothermal activity at the lost city vent field. *Science* **301**(5632): 495–498.

Fry, I. (2000). *Emergence of Life on Earth: A Historical and Scientific Overview.* Rutgers University Press, New Brunswick, NJ.

Fuller, W. D., R. A. Sanchez et al. (1972). Studies in prebiotic synthesis: VII. Solid-state synthesis of purine nucleosides. *Journal of Molecular Evolution* **1**(3): 249–257.

Gánti, T., Ed. (2003). *The Principles of Life.* Oxford University Press, Oxford, U.K. www.oup.com.

Garrison, W. M., D. Morrison et al. (1951). Reduction of carbon dioxide in aqueous solutions by ionizing radiation. *Science* **114**(2964): 416–418.

Gaucher, E. A., J. M. Thomson et al. (2003). Inferring the paleoenvironment of ancient bacteria on the basis of resurrected proteins. *Nature* **425**(6955): 285–288.

Geison, G. L. (1997). *The Private Science of Louis Pasteur.* Princeton University Press, Princeton, NJ.

Glansdorff, N., Y. Xu et al. (2008). The last universal common ancestor: Emergence, constitution and genetic legacy of an elusive forerunner. *Biology Direct* **3**(29): 56–125.

Glavin, D. P., M. P. Callahan et al. (2010). The effects of parent body processes on amino acids in carbonaceous chondrites. *Meteoritics & Planetary Science* **45**(12): 1948–1972.

Gold, T. (1992). The deep, hot biosphere. *Proceedings of the National Academy of Sciences of the United States of America* **89**: 6045–6049.

Groth, W. (1975). Photochemical formation of organic compounds from mixtures of simple gases simulating the primitive atmosphere of the earth. *Biosystems* **6**(4): 229–233.

Groth, W. and H. Weyssenhoff (1957). Photochemische Bildung von Aminosäuren aus Mischungen einfacher Gase. *Naturwissenschaften* **44**(19): 510–511.

Haldane, J. B. S. (1929). The origin of life. *The Rationalist Annual* **148**: 3–10.

Harrison, J. P., N. Gheeraert et al. (2013). The limits for life under multiple extremes. *Trends in Microbiology* **21**(4): 204–212.

Hazen, R. M., D. Papineau et al. (2008). Mineral evolution. *American Mineralogist* **93**(11–12): 1693–1720.

Holm, N. G., A. G. Cairns-Smith et al. (1992). Marine hydrothermal systems and the origin of life: Future research. *Origins of Life and Evolution of the Biosphere* **22**(1–4): 181–242.

Hoover, R. B. (2008). Comets, carbonaceous meteorites, and the origin of the biosphere. In: *Biosphere Origin and Evolution*, edited by N. Dobretsov, N. Kolchanov et al. Springer, Berlin, Germany, pp. 55–68.

Islas, S., A. M. Velasco et al. (2003). Hyperthermophily and the origin and earliest evolution of life. *International Microbiology* **6**(2): 87–94.

Israël, G., C. Szopa et al. (2005). Complex organic matter in Titan's atmospheric aerosols from in situ pyrolysis and analysis. *Nature* **438**(7069): 796–799.

Johnson, A. P., H. J. Cleaves et al. (2008). The Miller volcanic spark discharge experiment. *Science* **322**(5900): 1.

Jones, E. G. and C. H. Lineweaver (2010). To what extent does terrestrial life "follow the water"? *Astrobiology* **10**(3): 349–361.

Kashefi, K. and D. R. Lovley (2003). Extending the upper temperature limit for life. *Science* **301**(5635): 934–934.

Kasting, J. F. (1987). Theoretical constraints on oxygen and carbon dioxide concentrations in the Precambrian atmosphere. *Precambrian Research* **34**: 205–229.

Kasting, J. F. and J. L. Siefert (2002). Life and the evolution of Earth's atmosphere. *Science* **296**(5570): 1066–1068.

Kelley, D. S., J. A. Karson et al. (2001). An off-axis hydrothermal vent field near the Mid-Atlantic Ridge at 30[deg] N. *Nature* **412**(6843): 145–149.

Kempe, S. and E. T. Degens (1985). An early soda ocean? *Chemical Geology* **53**(1–2): 95–108.

Khare, B. N., C. Sagan et al. (1986). Amino acids derived from Titan tholins. *Icarus* **68**(1): 176–184.

Killops, S. D. and V. J. Killops (2013). *Introduction to Organic Geochemistry*, Wiley, New York.

Kirkpatrick, P. and C. Ellis (2004). Chemical space. *Nature* **432**(7019): 823.

Kirschvink, J. L. and B. P. Weiss (2002). Mars, panspermia, and the origin of life: Where did it all begin. *Palaeontologia Electronica* **4**(2): 8–15.

Kuhn, W. R. and S. K. Atreya (1979). Ammonia photolysis and the greenhouse effect in the primordial atmosphere of the earth. *Icarus* **37**(1): 207–213.

Kvenvold, K., J. Lawless et al. (1970). Evidence for extraterrestrial amino-acids and hydrocarbons in Murchison meteorite. *Nature* **228**(5275): 923–926.

Lambert, J. B., S. A. Gurusamy-Thangavelu et al. (2010). The silicate-mediated formose reaction: Bottom-up synthesis of sugar silicates. *Science* **327**(5968): 984–986.

Lammer, H., E. Chassefière et al. (2013). Outgassing history and escape of the martian atmosphere and water inventory. *Space Science Reviews* **174**(1–4): 113–154.

Larralde, R., M. P. Robertson et al. (1995). Rates of decomposition of ribose and other sugars— Implications for chemical evolution. *Proceedings of the National Academy of Sciences of the United States of America* **92**(18): 8158–8160.

Lauretta, D. S. and H. Y. McSween (2006). *Meteorites and the Early Solar System II*, University of Arizona Press, Tucson, AZ. In collaboration with Lunar and Planetary Institute, Houston, TX.

Lawless, J. G. and G. U. Yuen (1979). Quantification of monocarboxylic acids in the Murchison carbonaceous meteorite. *Nature* **282**(5737): 396–398.

Lazcano, A. (2008). What is life? A brief historical overview. *Chemistry & Biodiversity* **5**(1): 1–15.

Lazcano, A. and J. L. Bada (2003). The 1953 Stanley L. Miller experiment: Fifty years of prebiotic organic chemistry. *Origins of Life and Evolution of the Biosphere* **33**(3): 235–242.

Lazcano, A. and S. L. Miller (1994). How long did it take for life to begin and evolve to cyanobacteria? *Journal of Molecular Evolution* **39**(6): 546–554.

Levy, M., S. L. Miller et al. (2000). Prebiotic synthesis of adenine and amino acids under Europa-like conditions. *Icarus* **145**(2): 609–613.

Louie, A. H. (2009). *More Than Life Itself: A Synthetic Continuation in Relational Biology.* De Gruyter, Berlin, Germany.

Lunine, J. I. and D. J. Stevenson. (1987). Clathrate and ammonia hydrates at high pressure: Application to the origin of methane on Titan. *Icarus* **70**(1): 61–77.

Maher, K. A. and D. J. Stevenson (1988). Impact frustration of the origin of life. *Nature* **331**(6157): 612–614.

Manga, M. and C. Y. Wang (2007). Pressurized oceans and the eruption of liquid water on Europa and Enceladus. *Geophysical Research Letters* **34**(7): L07202.

Martins, Z., O. Botta et al. (2008). Extraterrestrial nucleobases in the Murchison meteorite. *Earth and Planetary Science Letters* **270**(1–2): 130–136.

Matsumi, R., H. Atomi et al. (2011). Isoprenoid biosynthesis in Archaea—Biochemical and evolutionary implications. *Research in Microbiology* **162**(1): 39–52.

McCollom, T. M., G. Ritter et al. (1999). Lipid synthesis under hydrothermal conditions by Fischer-Tropsch-type reactions. *Origins of Life and Evolution of the Biosphere* **29**(2): 153–166.

McCollom, T. M., B. S. Lollar et al. (2010). The influence of carbon source on abiotic organic synthesis and carbon isotope fractionation under hydrothermal conditions. *Geochimica et Cosmochimica Acta* **74**(9): 2717–2740.

Meggy, A. B. (1954). The free energy of formation of the amide bond in polyamides. *Journal of Applied Chemistry* **4**(4): 154–159.

Miller, S. L. (1953). A production of amino acids under possible primitive Earth conditions. *Science* **117**(3046): 528–529.

Miller, S. L. (1957). The mechanism of synthesis of amino acids by electric discharges. *Biochimica et Biophysica Acta* **23**(3): 480–489.

Miller, S. L. (1998). The endogenous synthesis of organic compounds. In: *The Molecular Origins of Life: Assembling Pieces of the Puzzle,* edited by A. Brack. Cambridge University Press, Cambridge, U.K.

Miller, S. L. and A. Lazcano (1995). The origin of life—Did it occur at high temperatures? *Journal of Molecular Evolution* **41**: 689–692.

Miller, S. L. and L. E. Orgel (1974). *The Origins of Life on the Earth.* Prentice-Hall, Englewood Cliffs, NJ.

Ming, D. W., R. V. Morris et al. (2008). Aqueous alteration on Mars. In: *The Martian Surface: Composition, Mineralogy, and Physical Properties,* edited by J. F. Bell. Cambridge University Press, Cambridge, U.K., pp. 519–540.

Mitri, G., A. P. Showman et al. (2007). Hydrocarbon lakes on Titan. *Icarus* **186**(2): 385–394.

Mitri, G., A. P. Showman, J. I. Lunine, and R. Lopes (2008). Resurfacing of Titan by ammonia–water cryomagma. *Icarus* **196**(1): 216–224.

Miyakawa, S., H. Yamanashi et al. (2002). Prebiotic synthesis from CO atmospheres: Implications for the origins of life. *Proceedings of the National Academy of Sciences of the United States of America* **99**(23): 14628–14631.

Mojzsis, S. J., T. M. Harrison et al. (2001). Oxygen-isotope evidence from ancient zircons for liquid water at the Earth's surface 4,300 Myr ago. *Nature* **409**(6817): 178–181.

Monnard, P. A., C. L. Apel et al. (2002). Influence of ionic inorganic solutes on self-assembly and polymerization processes related to early forms of life: Implications for a prebiotic aqueous medium. *Astrobiology* **2**(2): 139–152.

Neish, C. D., Á. Somogyi et al. (2010). Titan's primordial soup: Formation of amino acids via low-temperature hydrolysis of tholins. *Astrobiology* **10**(3): 337–347.

Niemann, H., S. Atreya et al. (2005). The abundances of constituents of Titan's atmosphere from the GCMS instrument on the Huygens probe. *Nature* **438**(7069): 779–784.

Nna Mvondo, D., R. Navarro-González et al. (2001). Production of nitrogen oxides by lightning and coronae discharges in simulated early Earth, Venus and Mars environments. *Advances in Space Research* **27**(2): 217–223.

Nooner, D. W. and J. Oró (1967). Organic compounds in meteorites—I. Aliphatic hydrocarbons. *Geochimica et Cosmochimica Acta* **31**(9): 1359–1394.

Oparin, A. I. (1924). *The Origin of Life*. Izd. Moskovshii Rabochii, Moscow, Russia.

Orgel, L. E. (1998a). The origin of life—A review of facts and speculations. *Trends in Biochemical Science* **23**(12): 491–495.

Orgel, L. E. (1998b). The origin of life—How long did it take? *Origins of Life and Evolution of the Biosphere* **28**(1): 91–96.

Oró, J., J. Gibert et al. (1971). Amino acids, aliphatic and aromatic hydrocarbons in the Murchison meteorite. *Nature* **230**(5289): 105–106.

Oró, J. and A. P. Kimball (1961). Synthesis of purines under possible primitive Earth conditions. I. Adenine from hydrogen cyanide. *Archives of Biochemistry and Biophysics* **94**: 217–227.

Ourisson, G. and Y. Nakatani (1994). The terpenoid theory of the origin of cellular life: The evolution of terpenoids to cholesterol. *Chemistry & Biology* **1**(1): 11–23.

Ourisson, G. and Y. Nakatani (1996). Can the molecular origin of life be studied seriously? *Comptes Rendus de l'Académie des Sciences. Série II, Mécanique, Physique, Chimie, Astronomie* **322**(4): 323–334.

Pappalardo, R. T. (2010). Seeking Europa's ocean. *Proceedings of the International Astronomical Union* **6**(S269): 101–114.

Peltzer, E. T., J. L. Bada et al. (1984). The chemical conditions on the parent body of the Murchison meteorite: Some conclusions based on amino, hydroxy and dicarboxylic acids. *Advances in Space Research* **4**(12): 69–74.

Peretó, J., P. López-García et al. (2004). Ancestral lipid biosynthesis and early membrane evolution. *Trends in Biochemical Sciences* **29**(9): 469–477.

Pinheiro, V. B., A. I. Taylor et al. (2012). Synthetic genetic polymers capable of heredity and evolution. *Science* **336**(6079): 341–344.

Pizzarello, S., G. W. Cooper et al. (2006). The nature and distribution of the organic material in carbonaceous chondrites and interplanetary dust particles. In: *Meteorites and the Early Solar System II*, edited by D. S. Lauretta and H. Y. McSween, University of Arizona Press, Tucson, AZ. In collaboration with Lunar and Planetary Institute, Houston, TX, pp. 625–651.

Pizzarello, S. and W. Holmes (2009). Nitrogen-containing compounds in two CR2 meteorites: N-15 composition, molecular distribution and precursor molecules. *Geochimica et Cosmochimica Acta* **73**(7): 2150–2162.

Ponnamperuma, C. and J. Chelaflores (1994). Chemical evolution—Structure and model of the first cell—Preface. *Journal of Biological Physics* **20**(1–4): R11–R11.

Powner, M. W., C. Anastasi et al. (2007). On the prebiotic synthesis of ribonucleotides: Photoanomerisation of cytosine nucleosides and nucleotides revisited. *Chembiochem* **8**(10): 1170–1179.

Powner, M. W., J. D. Sutherland et al. (2010). Chemoselective multicomponent one-pot assembly of purine precursors in water. *Journal of the American Chemical Society* **132**(46): 16677–16688.

Prieur, B. E. (2001). Étude de l'activité prébiotique potentielle de l'acide borique. *Comptes Rendus de l'Académie des Sciences—Series IIC—Chemistry* **4**(8–9): 667–670.

Proskurowski, G., M. D. Lilley et al. (2008). Abiogenic hydrocarbon production at lost city hydrothermal field. *Science* **319**(5863): 604–607.

Rao, M., M. R. Eichberg et al. (1982). Synthesis of phosphatidylcholine under possible primitive earth conditions. *Journal of Molecular Evolution* **18**(3): 196–202.

Raulin, F., K. P. Hand et al. (2010). Exobiology and planetary protection of icy moons. *Space Science Reviews* **153**(1–4): 511–535.

Reid, C. and L. E. Orgel (1967). Synthesis in sugars in potentially prebiotic conditions. *Nature* **216**(5114): 455.

Ricardo, A., M. A. Carrigan et al. (2004). Borate minerals stabilize ribose. *Science* **303**(5655): 196.

Robertson, M. P., M. Levy et al. (1996). Prebiotic synthesis of diaminopyrimidine and thiocytosine. *Journal of Molecular Evolution* **43**(6): 543–550.

Robertson, M. P. and S. L. Miller (1995). An efficient prebiotic synthesis of cytosine and uracil. *Nature* **375**(6534): 772–774.

Rubey, W. W. (1951). Geologic history of sea water. *Geological Society of America Bulletin* **62**(9): 1111–1148.

Sagan, C. and G. Mullen (1972). Earth and Mars: Evolution of atmospheres and surface temperatures. *Science* **177**(4043): 52–56.

Sanchez, R. A., J. P. Ferris et al. (1966). Cyanoacetylene in prebiotic synthesis. *Science* **154**(750): 784–785.

Sanchez, R. A. and L. E. Orgel (1970). Studies in prebiotic synthesis. V. Synthesis and photoanomerization of pyrimidine nucleosides. *Journal of Molecular Biology* **47**(3): 531–543.

Sandford, S. A., J. Aleon et al. (2006). Organics captured from comet 81P/Wild 2 by the Stardust spacecraft. *Science* **314**(5806): 1720–1724.

Schidlowski, M. (1988). A 3,800-million-year isotopic record of life from carbon in sedimentary rocks. *Nature* **333**(6171): 313–318.

Schmitt-Kopplin, P., Z. Gabelica et al. (2010). High molecular diversity of extraterrestrial organic matter in Murchison meteorite revealed 40 years after its fall. *Proceedings of the National Academy of Sciences of the Unites States of America* **107**(7): 2763–2768.

Schopf, J. W. (2008). *Cradle of Life: The Discovery of Earth's Earliest Fossils.* Princeton University Press, Princeton, NJ.

Schwartz, A. W. (1972). Prebiotic phosphorylation-nucleotide synthesis with apatite. *Biochimica et Biophysica Acta* **281**(4): 477–480.

Schwartz, A. W., M. van der Veen et al. (1973). Prebiotic phosphorylation. II. Nucleotide synthesis in the reaction system apatite-cyanogen-water. *Currents in Modern Biology* **5**(3): 119–122.

Seewald, J. S., M. Y. Zolotov et al. (2006). Experimental investigation of single carbon compounds under hydrothermal conditions. *Geochimica et Cosmochimica Acta* **70**(2): 446–460.

Shapiro, R. (1984). The improbability of prebiotic nucleic acid synthesis. *Origins of Life* **14**(1–4): 565–570.

Shapiro, R. (1987). *Origins: A Skeptic's Guide to the Creation of Life on Earth.* Bantam Books, New York.

Shapiro, R. (1988). Prebiotic ribose synthesis: A critical analysis. *Origins of Life and Evolution of the Biosphere* **18**(1–2): 71–85.

Shapiro, R. (1995). The prebiotic role of adenine: A critical analysis. *Origins of Life and Evolution of the Biosphere* **25**(1–3): 83–98.

Shock, E. L. (1990). Do amino acids equilibrate in hydrothermal fluids? *Geochimica et Cosmochimica Acta* **54**: 1185–1189.

Sleep, N. H. and K. Zahnle (2001). Carbon dioxide cycling and implications for climate on ancient Earth. *Journal of Geophysical Research: Planets (1991–2012)* **106**(E1): 1373–1399.

Stanley L. M. (1957). The mechanism of synthesis of amino acids by electric discharges. *Biochimica et Biophysica Acta* **23**(0): 480–489.

Stetter, K. O. (1996). Hyperthermophilic prokaryotes. *FEMS Microbiology Reviews* **18**(2–3): 149–158.

Stoks, P. G. and A. W. Schwartz (1979). Uracil in carbonaceous meteorites. *Nature* **282**(5740): 709–710.

Stribling, R. and S. L. Miller (1987). Energy yields for hydrogen cyanide and formaldehyde syntheses—The HCN and amino acid concentrations in the primitive ocean. *Origins of Life and Evolution of the Biosphere* **17**(3–4): 261–273.

Szostak, J. W., D. P. Bartel et al. (2001). Synthesizing life. *Nature* **409**(6818): 387–390.

Taillades, J. and A. Commeyras (1974). Systemes de strecker et apparentes—II: Mécanisme de formation en solution aqueuse des α-alcoylaminoisobutyronitrile à partir d'acétone, d'acide cyanhydrique et d'ammoniaque, methyl ou diméthylamine. *Tetrahedron* **30**(15): 2493–2501.

Thompson, W. R. and C. Sagan (1992). *Organic Chemistry on Titan: Surface Interactions. Symposium on Titan.* European Space Agency, Noordwijk, the Netherlands.

Tian, F., O. B. Toon et al. (2005). A hydrogen-rich early Earth atmosphere. *Science* **308**(5724): 1014–1017.

Trail, D., E. B. Watson et al. (2011). The oxidation state of Hadean magmas and implications for early Earth's atmosphere. *Nature* **480**(7375): 79–82.

Urey, H. C. (1952). *The Planets: Their Origin and Development.* Yale University Press, New Haven, CT.

Van Dover, C. (2000). *The Ecology of Deep-Sea Hydrothermal Vents.* Princeton University Press, Princeton, NJ.

Vuitton, V., J.-Y. Bonnet et al. (2010). Very high resolution mass spectrometry of HCN polymers and tholins. *Faraday Discussions* **147**: 495–508.

Wacey, D., M. R. Kilburn et al. (2011). Microfossils of sulphur-metabolizing cells in 3.4-billion-year-old rocks of Western Australia. *Nature Geoscience* **4**(10): 698–702.

Waite Jr, J. H., W. Lewis et al. (2009). Liquid water on Enceladus from observations of ammonia and 40Ar in the plume. *Nature* **460**(7254): 487–490.

Watson, J. D. and F. H. C. Crick (1953). Molecular structure of nucleic acids: A structure for deoxyribose nucleic acid. *Nature* **171**(4356): 737–738.

Weber, A. and S. Miller (1981). Reasons for the occurrence of the 20 coded protein amino-acids. *Journal of Molecular Evolution* **17**(5): 273–284.

Wolf, E. and O. Toon (2010). Fractal organic hazes provided an ultraviolet shield for early Earth. *Science* **328**(5983): 1266–1268.

Yuen, G. U. and K. A. Kvenvolden (1973). Monocarboxylic acids in Murray and Murchison carbonaceous meteorites. *Nature* **246**(5431): 301–303.

CHAPTER 6

Biochemical Pathways as Evidence for Prebiotic Syntheses

Gene D. McDonald

CONTENTS

6.1 INTRODUCTION

At some point in the history of life on Earth, prebiological chemistry became biological chemistry. We know a great deal about the metabolic basis of life on Earth today, although details continue to emerge from active research. In contrast, our understanding of the transition that must have occurred between purely abiotic chemistry and a chemical system (or systems) of sufficient complexity and novelty to be considered metabolically biological is sorely lacking.

One of the first scientists to ponder this question, at least in print, was the biochemist Norman Horowitz. In 1945, he proposed a general scenario in which a primitive metabolic system on the early Earth initially used only those abiotically produced organic compounds that were available in its environment. As abiotic production diminished and each of those compounds became depleted, the system began to accumulate their immediate chemical precursors and then to evolve catalytic capabilities for converting those precursors into the needed compounds. In this way, Horowitz imagined that metabolic pathways evolved in reverse, from end products back to the initial precursors used in modern metabolism.

Our understanding of metabolism today is orders of magnitude greater than that in 1945. We realize now that the origin of biological metabolism, on the early Earth or in an extraterrestrial environment, must have been more complicated and convoluted than Horowitz's scheme. To begin to piece together plausible scenarios for this transition, we must first look at the flow of mass and energy through contemporary biochemical systems and the catalysts and accessory compounds that enable this flow. We can then try to use the components of metabolism that are common to all life on Earth today to infer the metabolic pathways and components that existed in the last common ancestor of life on Earth. From there, we can begin to at least speculate about the composition of the first metabolic systems on Earth and how those putative systems fit into various models of prebiotic chemistry that have been proposed. Finally, we can try to use our understanding of past and present metabolism on Earth as a guide in the search for life elsewhere in the universe.

6.1.1 Overview of Central Metabolism Today

Before we explore metabolism in ancient ancestors of life on Earth, and what that metabolism might tell us about the prebiotic Earth, it is necessary to point out a few features of metabolism today that will be relevant to the ensuing discussion. There is not space here for a complete review of modern biochemical metabolism; the reader unfamiliar with the topic is referred to the numerous biochemistry textbooks available (see the reference list at the end of the chapter for some examples). The purpose of this section is to highlight certain features of metabolism that the reader needs to understand in general terms in order to be able to refer to them later in the chapter.

6.1.2 Logic of Central Metabolism

In general, each metabolic pathway functions primarily in either catabolism or anabolism (Figure 6.1). Catabolism is the breakdown of energy-containing molecules in order to convert that energy into a form that can be used for biological processes. Catabolic metabolism generally involves the use of many different related nutrient molecules, which are broken down into a relatively small number of intermediate products. Anabolism is the use of catabolic products for the synthesis of biomolecules such as amino acids, nucleic acid bases, and lipids. Anabolic pathways generally take a small number of building blocks, many of them products of catabolism, and assemble them into a wide variety of biomolecules. Some metabolic pathways, such as the tricarboxylic acid (TCA) cycle, are amphibolic, meaning they have both catabolic and anabolic functions.

In general, catabolic pathways are exergonic (net release of energy) while anabolic pathways are endergonic (net consumption of energy). The chemical energy of nutrient

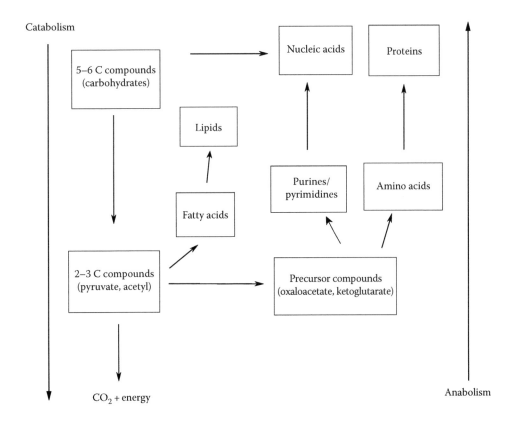

FIGURE 6.1 Schematic of catabolic/anabolic relationships in modern organisms. In general, catabolic (energy-harnessing) processes proceed from top to bottom in the figure, while anabolic (biomolecule-synthesizing) processes proceed from bottom to top. C, carbon atoms.

molecules is transferred by catabolic pathways to a set of energy carriers. Some of these are metabolic intermediates themselves, while others are strictly carriers that are recycled constantly and reused. These carriers then make possible the energy-consuming anabolic processes of biosynthesis. The most common intermediate energy carrier is adenosine triphosphate, or ATP. The phosphodiester bond of the terminal phosphate group in ATP can be hydrolyzed to yield over 30 kJ/mol of free energy. Other phosphorylated intermediates such as phosphoenolpyruvate and 1,3-bisphosphoglycerate can also carry energy via phosphoester bonds. Oxidation–reduction (redox) cofactors such as nicotinamide adenine dinucleotide (NADH) and flavin adenine dinucleotide (FADH$_2$) carry reducing power (i.e., extra electrons), which is another intermediate form of chemical energy.

Both catabolism and anabolism use primarily two-carbon units. The major reason for this lies in the chemistry of carbon-containing compounds. It is mechanistically rather difficult, although not impossible, to form a carbon–carbon bond between 2 one-carbon molecules such as carbon dioxide (CO$_2$), carbon monoxide (CO), or methane (CH$_4$). With a two-carbon unit, on the other hand, one of the carbons can be oxidized to a keto (C=O) group, which is susceptible to nucleophilic attack by another sufficiently activated carbon, forming a C—C bond. It is also easier to break a long hydrocarbon chain into two-carbon units than one-carbon units. By oxidizing alternating carbons to keto groups, β-keto structures can be formed, which are relatively labile to C—C bond breakage. This breakage results in an acetyl (CH$_3$C=O) group, which can be transferred to carrier molecules with compatible functional groups, such as thiol (—SH) and amino (—NH$_2$) groups, to form esters or amides.

6.1.3 Catalysts and Cofactors: The Tools of Metabolism

A catalyst is a substance that increases the rate of a chemical reaction without being consumed in the process. Metabolism requires catalysts, because many catabolic and anabolic reactions have very slow reaction rates under biological conditions. These catalysts are protein (for the most part) enzymes, which are composed of linear chains of amino acids, molecules containing both carboxylic acid (—COOH) and amino (—NH$_3$) groups. The sequence of these amino acids is encoded in the genome of the organism. Some of these amino acids have reactive functional groups that participate in the catalytic mechanism of the enzyme. Others are relatively inert and serve mostly structural purposes. All known enzymes can be grouped into one of six major classes, depending on the type of chemical reaction they catalyze (Table 6.1).

TABLE 6.1 Enzyme Classes as Defined by Enzyme Commission, International Union of Biochemistry

Enzyme Class	Reaction Type Catalyzed
Oxidoreductases	Redox
Transferases	Functional group transfer
Hydrolases	Bond hydrolysis
Lyases	Addition across double bonds
Isomerases	Isomerization
Ligases	Bond formation with ATP cleavage

TABLE 6.2   Prebiotically Relevant Enzyme Cofactors

Function	Cofactor	Active Structural Element
Electron transfer	Nicotinamide adenine dinucleotide (NADH)	Nicotinamide
	Flavin adenine dinucleotide (FADH$_2$)	Riboflavin
	Metal ions	Cu, Ni, Mo, W
	Metal–nonmetal clusters	Fe$_2$S$_2$, Fe$_2$S$_4$, Fe$_4$S$_4$
Chemical group transfer	CoA	Thiol
	Folic acid	Pterin
	Lipoic acid	Thiol
	Pantothenic acid	Thiol
	Cobalamin	Tetrapyrrole

Many metabolic enzymes require nonprotein molecules known as coenzymes or cofactors. These molecules are either synthesized by other enzymes also encoded in the genome or taken up from the organism's environment or food sources (the latter are known as vitamins). The major role of coenzymes is to provide chemical functional groups that the 20 amino acids encoded by the genetic code do not possess. These functional groups are mainly involved in the transfer of other functional groups, or of electrons, between substrate and product molecules (Table 6.2).

6.1.4 Catabolism

Glycolysis, also known as the Embden–Meyerhof pathway, is the central metabolic pathway in all organisms for the initial harvesting of chemical energy from carbohydrates. The major starting substrate for glycolysis is glucose, which comes directly from the cellular environment or from hydrolysis of storage materials such as starch. Other monosaccharide sugars, such as fructose, galactose, and mannose, can also be used via conversion to glucose or other intermediates in the pathway. Glycolysis breaks down sugars in a multistep process into pyruvate, with a small net yield of chemical energy. This molecule can then be used for synthesis of biomolecules such as nucleic acids and proteins (anabolic metabolism, discussed in the following). Pyruvate can also be activated by one of the coenzymes, coenzyme A (CoA), to be used for further generation of chemical energy. When carbohydrates or other energy-rich substrates are unavailable, biochemical systems can break down fatty acids into acetyl groups and amino acids into keto acids, which can then be fed into catabolic pathways to recover some energy.

There are two phases in glycolysis. In the first phase (Figure 6.2a), glucose is activated by phosphorylation and then broken into 2 three-carbon intermediate products. This phase actually requires the expenditure of chemical energy in the form of two ATP molecules. The terminal phosphate groups from ATP, with their high-energy phosphoester bonds, are transferred to these intermediates.

In the second phase (Figure 6.2b), the three-carbon intermediates are oxidized to pyruvate. The high-energy phosphate ester bonds formed in the first phase are transferred to adenosine diphosphate (ADP) to form four ATP molecules, while the electrons from the oxidation are captured by NADH for use in further ATP generation by other pathways. The direct generation of biochemical carrier energy, then, occurs only in the second half of the glycolytic pathway.

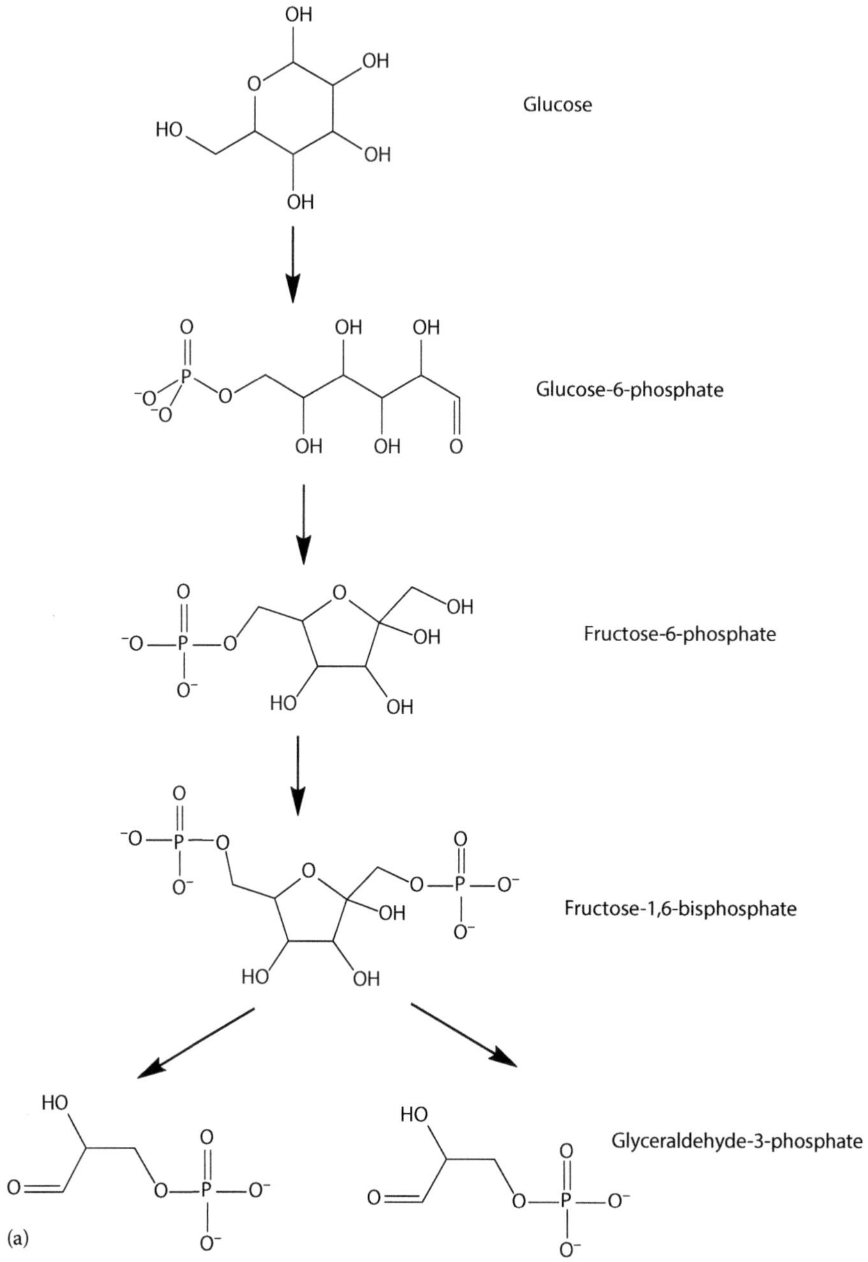

FIGURE 6.2   Glycolysis: (a) the first half, converting glucose to glyceraldehyde-3-phosphate and consuming energy in the form of two ATP molecules.

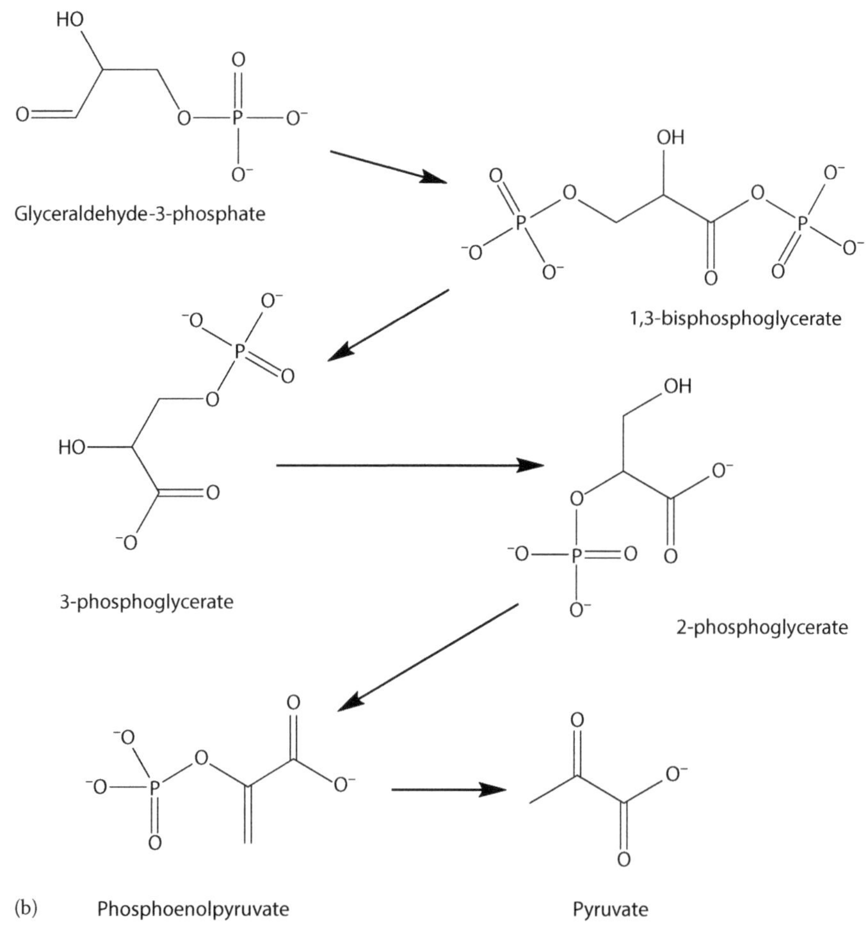

Glyceraldehyde-3-phosphate

1,3-bisphosphoglycerate

3-phosphoglycerate

2-phosphoglycerate

(b)   Phosphoenolpyruvate

Pyruvate

FIGURE 6.2 (continued)   Glycolysis: (b) the second half, converting glyceraldehyde-3-phosphate to pyruvate and generating energy in the form of four ATP molecules.

In addition to glycolysis, glucose can be utilized via the pentose phosphate pathway. This pathway converts glucose-6-phosphate to ribose-6-phosphate, with the accompanying production of NADPH (a cofactor similar to NADH). The reducing power of NADPH is used as energy to drive several anabolic reactions. Ribose-6-phosphate is used to synthesize ribonucleotides for incorporation into RNA and deoxyribonucleotides for incorporation into DNA.

The TCA cycle, also known as the citric acid cycle or the Krebs cycle, is the second major catabolic pathway common to all organisms (Figure 6.3). The major input to the TCA cycle is acetyl-CoA, which is produced from the pyruvate product of glycolysis. Pyruvate is decarboxylated to form an acetyl group plus CO_2. CoA then forms a thioester with the acetyl group, activating it for insertion into the TCA cycle. Acetyl-CoA can also be generated by fatty acid oxidation and some amino acid degradation pathways. The concentration of acetyl-CoA is one of the major indicators of the metabolic state of a cell.

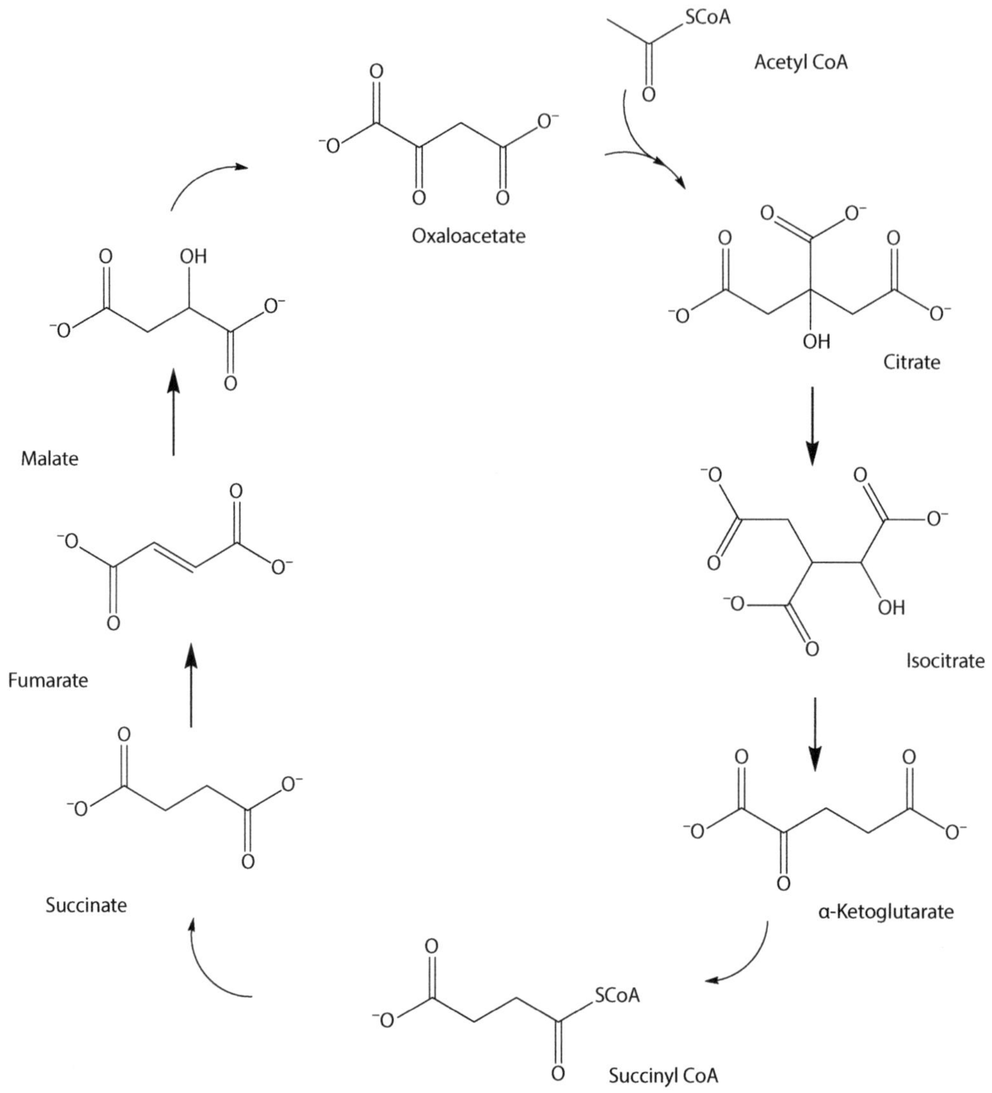

FIGURE 6.3 TCA cycle. The acetyl group from acetyl-CoA is added to oxaloacetate to form citrate; subsequent oxidations and group transfers generate the reduced cofactor NADH, which can be used by other pathways to form ATP. Two carbons per cycle are lost as carbon dioxide.

The main principle of the TCA cycle is the addition of the two-carbon acetyl group from acetyl-CoA to the four-carbon intermediate oxaloacetate to generate the six-carbon compound citrate. Subsequent reactions in the cycle oxidize citrate, removing two carbons as CO_2 and generating electrons that are carried away by NADH for use in ATP generation. The end result is oxaloacetate, which is joined with another acetyl group to repeat the cycle. The TCA cycle can also receive other products of amino acid degradation pathways. These include TCA intermediates oxaloacetate, fumarate, succinyl-CoA, and α-ketoglutarate, as well as TCA precursors pyruvate and acetyl-CoA.

In contrast to glycolysis, in the TCA cycle, there is no activation of intermediates by phosphorylation and thus no direct ATP synthesis. The TCA cycle only generates energy indirectly, if coupled to other pathways such as oxidative phosphorylation. Oxidative phosphorylation, as the name implies, is only completely active in the presence of molecular oxygen. Basically, it involves the transfer of electrons supplied primarily by NADH along a chain of carrier molecules. The energy of this electron flow is used by many protein complexes to generate proton gradients across lipid membranes, which are then used to generate ATP. The details of oxidative phosphorylation need not concern us here, since it probably arose on Earth only after the advent of oxygenic photosynthesis and thus after the protobiotic period had past. It is important for the coming discussion, however, to note that several of the proteins involved in the electron transport chain have flavin or iron–sulfur (Fe–S) cofactors.

6.1.5 Anabolism

Microorganisms can generally synthesize all amino acids, purine and pyrimidine bases, fatty acids, and enzyme cofactors from one- or two-carbon starting materials. Autotrophs, such as plants and some microorganisms, can synthesize all needed compounds from one-carbon inorganic sources such as CO_2, CH_4, or bicarbonate (HCO_3^-). Heterotrophs must obtain energy and anabolic starting materials from compounds already synthesized by other organisms. Higher organisms such as vertebrates often must depend on either diet or symbiotic microorganisms for so-called essential amino acids and cofactors (vitamins).

Purine and pyrimidine bases for use in nucleic acids are synthesized primarily from amino acids and one-carbon molecules (mediated mostly by folate-dependent enzymes). One-carbon chemistry, in contrast, is rare in amino acid biosynthetic pathways. Amino acids are synthesized primarily from TCA cycle intermediates through a number of branched anabolic pathways. Each pathway starts with one compound and leads to several amino acids, which are often referred to collectively as an amino acid family. Many of the TCA intermediates are either hydroxy acids or keto acids, both of which are relatively easily converted into amino acids by addition of an amino group and in some cases a round of oxidation. Fatty acid synthesis, for incorporation into lipids in membranes and other uses, involves the sequential addition of acetyl groups, from acetyl-CoA, onto a growing hydrocarbon chain (Figure 6.1).

6.2 METABOLISM IN THE LAST COMMON ANCESTOR

It is generally held that all life on Earth today is descended from a common ancestor. The similarity of the biochemical machinery and the genetic material among all known organisms lends powerful support to this theory, even though 3 billion years worth of evolution has resulted in a wide variety in the details of genetic organization and biomolecular structure and function. Before we can look at the possible metabolic functions of the earliest biochemical systems on Earth, we need to consider what we know about the last common ancestor of modern biochemistry.

The composition of the last universal common ancestor of all life on Earth, which we will refer to as LUCA, can only be inferred from available current data. Those data include primarily phylogenetic, but also some geochemical, evidence. The phylogenetic evidence lies in the evolutionary trees that can be constructed using the similarity (homology) between nucleotide sequences of genes, or amino acid sequences of the proteins coded for by genes, in modern species both closely and distantly related. The geochemical evidence is contained, for the most part, in stable isotope ratios of biologically important elements in ancient rocks. In both cases, the evidence is somewhat fragmentary and difficult to interpret. Our picture of the metabolic functions of LUCA is thus really only a model, or a set of models, based on inference.

6.2.1 Size of the Genome

The overall size of model LUCA genomes varies quite a bit. Since the divergence of the three domains of life (Bacteria, Archaea, and Eukaryota) is generally considered to be an ancient event, comparisons of archaeal genomes with those of bacteria and eukaryotes have been used by numerous researchers to try to estimate the number of genes that LUCA might have contained. Some studies have suggested that the genome could have been as small as 92 genes, while others envision genomes as large as 1600–1800 genes.

The genome of the microorganism *Methanococcus* was the first archaeal genome to be fully sequenced, and comparison with bacterial and eukaryotic genomes led to an estimate of approximately 300 universal genes that might have made up the LUCA genome. Another comparison of orthologous gene groups in a range of widely diverged genomes suggested approximately 600 genes in a LUCA model genome. Using an approach based on the coevolution of interacting proteins, the LUCA genome size was estimated by yet another study at 743 proteins.

A compilation of data from a number of these studies resulted in a *mean* LUCA genome size of approximately 400 genes. For comparison, the smallest known modern genome, that of the parasitic microorganism *Mycoplasma*, has around 500 genes. The overall picture from these studies seems to be that the size of the LUCA genome was probably comparable to the simplest extant organisms on Earth today.

6.2.2 Metabolic Functions

The percentage of inferred LUCA genes that code for metabolic enzymes varies from study to study. One analysis comparing genomes of bacteria, archaea, and yeast resulted in a putative LUCA genome in which 45 out of 115 genes are metabolic. Thirty-three of these 45 genes involve either sugar or nucleotide metabolism. Another phylogenetic study identified approximately 300 highly conserved protein-coding genes, with about 100 of those coding for metabolic enzymes. Of these, the metabolically related genes were again primarily involved in sugar and nucleotide metabolism.

Consensus LUCA genomes based on compilation of data from several studies that took sequence-, structure-, and function-based approaches included several classes of enzymes. One such class is oxidoreductases, which catalyze redox reactions. Also included are glycosyl transferases, which act on sugars; phosphotransferases, which catalyze addition of phosphate groups; and amino acid ligases. Analysis of the occurrence of various protein folding domains across modern species resulted in the identification of five particularly ancient domains. These include domains that hydrolyze nucleoside triphosphates, bind nucleic acids, bind nucleotide cofactors, and interact with Fe–S clusters.

It has also been pointed out that the glycerol derivatives used in both bacteria and archaea to synthesize membrane phospholipids are made from dihydroxyacetone phosphate (DHAP), a key intermediate in glycolysis. This suggests, assuming LUCA had phospholipid membranes similar to those of modern bacteria and eukaryotes, that DHAP was available within the cell. This, in turn, indirectly suggests that most of the glycolytic pathway was operational in LUCA. Overall, the bulk of central metabolism has been inferred to have been strongly conserved in most model LUCA genomes. This includes amino acid and nucleotide synthesis, lipid and coenzyme synthesis, as well as most glycolysis and TCA cycle enzymes.

The results of phylogenetic comparisons like these are not always unambiguous, however. In one putative genome, for example, the glycolytic enzymes enolase, pyruvate kinase, and phosphoglycerate kinase were all present, while phosphoglyceromutase and glyceraldehyde-3-phosphate dehydrogenase were absent. Phosphoglyceromutase is the enzyme that sits in the middle of the second phase of glycolysis (the energy-producing phase). How LUCA would have managed to carry out glycolysis without this step in the pathway is unclear.

6.2.3 Nitrogen Fixation in LUCA

A particularly interesting question about the functionalities present in the LUCA genome is whether nitrogen fixation genes were present. Nitrogen fixation, the reduction of molecular nitrogen (N_2) to the oxidation level of ammonia (NH_3), is a very energy-intensive process and is active today in only certain microorganisms. However, since the current Earth environment is relatively deficient in reduced nitrogen, all life on Earth depends on these few species.

Did LUCA contain the machinery for nitrogen fixation? Perhaps the early Earth environment contained sufficient supplies of at least partially reduced nitrogen species (NO, N_2O, etc.) to allow the biosynthesis of amino acids, purine and pyrimidine bases, and other essential nitrogen-containing biomolecules without the reduction of N_2. An examination of the molecular phylogeny of nitrogen fixation (*nif*) genes led to the conclusion that sequence similarity in all *nif* genes suggests an ancient single ancestor gene. Two scenarios were noted as possible: (1) LUCA was a nitrogen-fixing organism, and the *nif* genes were lost in eukaryotes after the three domains diverged; (2) LUCA was not a nitrogen-fixing organism, and the *nif* ancestor gene arose in methanogenic archaea and was then duplicated and spread by horizontal gene transfer into bacteria and archaea.

It has been proposed that the ancestor of several *nif* genes could have been present in the LUCA genome, but not necessarily as part of a nitrogen fixation pathway. If the only nitrogen source present in LUCA's environment was N_2, then at least the ancestor *nif* gene would have been necessary for nitrogen fixation. However, an alternate scenario would have the ancestral *nif* gene product functioning as a cyanide detoxifying enzyme, in a more reducing environment where nitrogen was available as NO_3^-, NO_2^-, or even NH_4.

Phylogenetic analysis of *nif* genes has, in at least one case, led to the conclusion that Mo-dependent nitrogenase is unlikely to have been part of the LUCA genome. This study concluded that the *nif* ancestor gene arose around 1.5–2.2 billion years ago (Ga), after the rise of atmospheric oxygen levels (around 2.5 Ga).

Molybdenum, tungsten, and copper are key metal cofactors in a number of enzymes, including nitrogenase. It has been observed that copper and molybdenum have low solubilities in a low-oxygen environment, such as that in which LUCA probably lived, while the solubility of tungsten is somewhat better. At least some subfamilies of complex Fe–S molybdoenzymes, including reductases for carbon dioxide, polysulfide, and nitrate and oxidases for arsenite, appear to have been present in LUCA based on molecular phylogenetic analysis. Perhaps ancestral enzymes used W instead of Mo (the two elements have similar chemistry) or perhaps obtained Mo from alkaline environments like hydrothermal vents.

Nitrogen stable isotope ratios in sedimentary organic matter reach modern values in 2.1 Ga rocks but are lighter in rocks from 3.5 to 2.1 Ga time range. This suggests that the presence of biological N_2 fixation pathways in the Archaean era is possible, but not clearly indicated. Nitrogen isotope values in 2.5 Ga shale suggest that biological nitrifying and denitrifying pathways were in existence prior to O_2 accumulation in the atmosphere. Kerogen from 2.67 Ga shale has nitrogen isotope ratios that indicate presence of coupled nitrification and denitrification pathways. Nitrogen limitation may have been a problem for organisms after oxygenic photosynthesis arose, perhaps around 2.6 Ga.

One hypothesis is that reduced nitrogen was supplied completely by abiotic nitrogen fixation, that is, the conversion of N_2 to NO by lightning-induced atmospheric chemistry in the period from approximately 4 Ga to about 2.2 Ga. The decrease in output from this chemistry as the atmospheric oxygen level rose could have resulted in a *nitrogen crisis* that could have spurred the evolution of biological nitrogen fixation. The overall picture, then, seems to be that nitrogen fixation was probably not part of the metabolism of LUCA.

6.2.4 Was LUCA a *Real* Species?

Carl Woese, one of the pioneers of molecular phylogenetic analysis, proposed that LUCA was not in fact one discreet species as we understand that concept today. His idea was that what we call LUCA was in fact a consortium of protocells, with different genes and functions, that acted as a sort of superorganism in order to carry out all the functions needed for survival and reproduction. In this scenario, there would have been much swapping of genetic material between protocells, a process we now call lateral (or horizontal) gene transfer. This lateral gene transfer would have, over time, allowed each subpopulation of protocells to accumulate a different set of genes for useful functions and thus slowly become a recognizable, genetically distinct species that could live independently. In Woese's scenario, true species did not appear until after the divergence of the three domains of life—Archaea, Bacteria, and Eukaryota.

6.3 METABOLIC PREDECESSOR TO LUCA

In constructing a model for the metabolic functions of the LUCA of all current life on Earth, molecular phylogeny of existing species at least provides some guidance. In contrast, for the biological or protobiological systems that existed before LUCA, the phylogenetic methods discussed earlier cannot be used, since by definition any system predating LUCA is not related to all extant species. The only available data from which to work are putative chemical traces, as deciphered from Archaean geochemical data, and educated guesses about the minimum number and type of metabolic pathways needed for a chemical system to function as a metabolic one. We can, however, assume that any complex biochemical function or pathway that was (probably) not present in LUCA was also (probably) not present in its (presumably simpler) predecessors. Nitrogen fixation, which we have already more or less ruled out in LUCA, would be in this category. To look back before LUCA, then, involves mostly inferences from basic chemical and biochemical principles about what metabolic pathways might have been possible.

6.3.1 Use of Terms

When discussing the emergence of life on Earth (or elsewhere for that matter), the terminology used can be problematic. The phrase *origin of life*, which has been used for many years to describe this general field of study, seems to suggest a clear dividing line between nonliving and living systems, a concept that is not totally consistent with all major theories in the field. The definition of *life* itself has been vigorously debated, without the emergence of a clear consensus. The terms *species*, *cell*, and *organism* also carry specific connotations about the genetic and/or structural makeup of the system being described.

In this discussion, we will try to avoid the preconceptions that these terms carry by using the (admittedly somewhat awkward) term first metabolic system (FIMS). The definition of FIMS we will use is that of one or more chemical systems that carried out at least a significant subset of the general metabolic functions that are common to all biological cells today. These would have included at least some anabolic reactions and perhaps some catabolic ones as well. In keeping with the overall theme of the chapter, we will not dwell on the nature of the genetic material of FIMS, nor will we ponder the thorny question of whether FIMS was capable of *true* Darwinian evolution.

6.3.2 Relationship between LUCA and FIMS

The relationship between LUCA and FIMS was both temporal and compositional. There are three possibilities. The first is identity, that is, the FIMS on Earth was also the last common ancestor of all extant life. The hypothesis that FIMS was also LUCA seems rather unlikely. As we have seen, even the most conservative models of the composition of LUCA paint it as a quite complex system, a true organism. A system like this would be very difficult to imagine arising directly from purely prebiotic chemical reactions.

The second possibility is that of close relationship, that is, the emergence of FIMS led rapidly to LUCA. A rapid transition from FIMS to LUCA, with only a few intermediate steps, is perhaps a bit more conceivable. It has been pointed out that on a geological time scale, the *origin of life* was probably a relatively swift event (or series of events). Of course, *swift* in geological terms could still refer to a period of up to a few hundred million years.

The third possibility is that of a distant relationship, that is, there were numerous iterations of FIMS before a system emerged that was sufficiently *biological* to have been the ancestor of all life on Earth. This is probably the most likely of the three possibilities and the one that informs most current thinking. In this scenario, the existence of a number of increasingly complex and capable metabolic (and possibly genetic) systems would have, over some time period, ultimately led to LUCA. If FIMS and LUCA were only distantly related, then the metabolic pathways present in LUCA may not have all been present in FIMS. It is possible that some pathways present in some species today were also present in FIMS but were lost on the evolutionary road to LUCA, only to be *reinvented* later by some of LUCA's descendants. It is also possible that there existed simultaneously multiple distinct versions of FIMS, some of which eventually contributed components to LUCA and others of which did not.

6.3.3 Minimum Requirements for a Metabolic System

The minimal set of functions needed by any primitive metabolic system would be heavily influenced by the environmental conditions in which the system operates. If significant amounts of abiotically produced small metabolic precursors are available, then the minimum required anabolism would mostly involve polymerization or condensation reactions to form larger molecules. Abiotic amino acids might be condensed into peptides, for example, or purine and pyrimidine bases incorporated into nucleic acid analogs. If soluble organic compounds with significant energy content (e.g., simple carbohydrates or organophosphates) are available abiotically, then catabolism would only need to be heterotrophic, breaking up molecules to harness their chemical energy. Autotrophic carbon fixation would not be needed.

Conversely, if abiotic organics are not readily available from the environment, then anabolic processes would have to start with inorganic carbon, probably CO_2 but perhaps CO or CH_4, and would have to synthesize any soluble organic compounds needed by the system. These compounds could include molecular catalysts, information-carrying molecules, and components of compartment structures. In this case, heterotrophic catabolic processes would be limited to scavenging of organic compounds, such as unneeded side products or damaged molecules, initially produced by autotrophic carbon fixation.

Much of our understanding of the minimum requirements for a metabolic system comes from endosymbiotic microorganisms that inhabit the blood or digestive tract of higher organisms. These species, including members of the genera *Mycoplasma*, *Rickettsia*, and *Chlamydia*, absorb many necessary compounds from their hosts. They have around 150–180 essential genes that are common to them. These include genes for enzymes in the glycolysis and pentose phosphate pathways, as well as enzymes that synthesize nucleotides, phospholipids, and a number of coenzymes. However, these organisms do not synthesize fatty acids, amino acids, or most purine and pyrimidine bases. These compounds are obtained from the environment inside the host.

So are these endosymbionts a viable model for FIMS? Perhaps, if the environment in which FIMS existed was relatively rich in abiotically produced organic compounds such as amino acids and fatty acids. An environment like that would be similar to the one in which these organisms now live, and so we could infer that the metabolic capabilities that are sufficient for them would also have been sufficient for FIMS. Conversely, if the environment was significantly poorer in abiotic organics than the bloodstream of a vertebrate animal, for example, then FIMS would probably have needed more anabolic capability than the endosymbionts currently possess. Of course, if FIMS were, for example, a system that used ribozymes as catalysts instead of protein enzymes, amino acids would not have been vital. The same would have been true for phospholipids, if FIMS did not contain a lipid membrane.

It has also been pointed out that not all extant metabolic pathways are chemically minimal. In other words, in some cases, it is possible to envision pathways from one intermediate to another that would be chemically possible but contain fewer steps than those that have evolved over time. A few such *shortcut* pathways exist in some modern organisms, in fact. One example is the Entner–Doudoroff pathway. This alternate pathway is activated in some microorganisms under certain conditions and converts glucose-6-phosphate to glyceraldehyde phosphate and pyruvate in only four steps. This is in contrast to glycolysis, which uses nine steps for the same transformation. It seems possible, then, that FIMS could have carried out the same general catabolic and anabolic functions we see in biology today but with fewer pathway steps and thus fewer required enzymes. However, any metabolic system that we would recognize as such would appear to require roughly 100 genes/enzymes/pathway steps in order to function.

On the issue of nitrogen fixation, it seems relatively safe to assume, at least, that the complex proteins coded for by the *nif* genes were not part of FIMS. It also seems relatively safe to assume that the early Earth environment at the time of FIMS was at least as reducing, if not more so, than at the time of LUCA. Since we have tentatively concluded that LUCA probably did not fix molecular nitrogen because it did not need to, it is logical to extend that conclusion to FIMS as well.

6.3.4 Alternate Metabolic Pathways

There are a small number of alternate metabolic pathways in existence today in some organisms that have been suggested as possibly primitive (Table 6.3). One possible FIMS pathway is the reductive TCA (rTCA) cycle. This pathway is used for autotrophic carbon fixation in a few bacterial species today. This cycle reduces CO_2 to various anabolic building blocks, in a general reversal of the overall scheme of the regular (or oxidative) TCA cycle.

Although most of the intermediates are shared between the oxidative and rTCA cycles, different enzymes are required at several key steps in the reductive cycle that are irreversible in the oxidative pathway. A FIMS using the rTCA cycle would have been reductively autotrophic, carrying out functions such as redox, carboxyl transfer, hydrolysis/dehydration, phosphorylation/dephosphorylation, amination (addition of an amino

TABLE 6.3 Proposed Alternate Pathways for Protometabolism

Pathway	Starting Materials	Products
Reductive TCA	Oxaloacetate, $2CO_2$	Citric acid and other TCA intermediates
Acetyl-CoA	CoA, $2CO_2$	Acetyl-CoA
3-Hydroxypropionate/4-hydroxybutyrate	Acetyl-CoA, CO_2	Two acetyl-CoA
Dicarboxylate/4-hydroxybutyrate	Acetyl-CoA, CO_2	Two acetyl-CoA

group to a molecule), and acylation (addition of an acyl group to a molecule). The intermediates in the rTCA cycle are used mostly to synthesize anabolic cofactors, amino acids, nucleotides, and other intermediates in anabolic pathways.

The acetyl-CoA, or Wood–Ljungdahl, pathway occurs in a sizeable number of prokaryotic species today and involves the conversion of two carbon dioxide molecules to an acetyl group attached to CoA. The first CO_2 is reduced initially to formate, then in subsequent steps to the oxidation level of a methyl group. The enzymes that catalyze these steps all have folate cofactors. The second CO_2 fixed is reduced to CO (via a Ni-cofactor enzyme) and attached to the methyl group to form acetyl-CoA. The acetyl-CoA can then be used to generate ATP. This pathway is exergonic overall. In addition to the folate and Ni cofactors, several of its steps are catalyzed by enzymes containing Fe–S clusters. The early Earth settings proposed for the emergence of this pathway are mostly hydrothermal systems, which contain abundant molecular hydrogen and sulfides. The H_2 could have served as an electron donor and CO_2 as an electron acceptor for an early metal-catalyzed version of the acetyl-CoA pathway.

The 3-hydroxypropionate/4-hydroxybutyrate pathway and dicarboxylate/4-hydroxybutyrate pathway are two related carbon-fixation pathways that are used by a few autotrophic archaea. Each pathway starts with one acetyl-CoA and fixes two carbons, from CO_2 and/or bicarbonate (HCO_3^-). The dicarboxylate/4-hydroxybutyrate pathway uses a series of dicarboxylic acid intermediates to form succinyl-CoA, while the 3-hydroxypropionate/4-hydroxybutyrate pathway uses 3-hydroxypropionate and related intermediates to form the same compound. Both pathways then convert the succinyl-CoA into acetoacetyl-CoA via 4-hydroxybutyrate and related compounds. The acetoacetyl-CoA then is split (with a second CoA molecule) to yield two acetyl-CoA molecules.

The carbon stable isotope signatures generated by these alternate carbon-fixation pathways have been determined experimentally. The rTCA and 3-hydroxypropionate (3-HP) pathways result in $\delta^{13}C$ values of 0 to −20 per mil, while the acetyl-CoA pathway generates values of −10 to −45 per mil. Carbon in rocks from the 3.8 Ga Isua formation shows fractionations consistent with rTCA/3HP, although some ambiguities in these $\delta^{13}C$ measurements exist. South African rocks from 3.5 Ga are more consistent in $\delta^{13}C$ with the reductive pentose phosphate pathway (Calvin cycle) as in modern photosynthesis.

6.3.5 Overall Picture of FIMS

The alternate metabolic pathways discussed earlier are all essentially autotrophic. In the event that the early Earth environment was relatively poor in abiotically-generated reduced organic compounds, autotrophy of some type would have probably been necessary for any system that could be considered metabolic to sustain itself. The pathways involved would have been mostly anabolic. On the other hand, if there were significant amounts of reduced organics supplied to the early Earth by either endogenous or exogenous sources, then FIMS could have existed on heterotrophic reactions. These pathways would certainly have included catabolism, to harvest the energy contained in carbon–carbon bonds. Some anabolic pathways would probably have also been necessary, to generate specific compounds not produced in adequate amounts by abiotic processes.

The molecular composition of FIMS, in terms of small molecules that we would identify as metabolic intermediates and anabolic building blocks, is of key importance in relating early metabolism to the prebiotic environment. Researchers have examined the connectivity of pathways in *Escherichia coli* and found 12 metabolites with the highest values. These were presumed to be ancient. They include the amino acids glutamate, glutamine, aspartate, and serine. Also in this set of compounds are the central metabolic intermediates pyruvate, α-ketoglutarate, succinate, and 3-phosphoglycerate and the enzyme cofactors CoA and tetrahydrofolate. It has been suggested that metal ions could have been used as cofactors by primitive enzymes, regardless of whether those enzymes were proteins or nucleic acids. The coevolution of genetic material (RNA) and metabolic pathways would seem to require abiotic synthesis of amino acids, purine and pyrimidine bases, and two- and three-carbon metabolic intermediates.

The FIMS could also have been chemolithoautotrophic, that is, deriving energy from mineral sources. It has been suggested that such a system could have used the reaction of FeS and H_2S to produce FeS_2 (pyrite) and H_2. A primitive hydrogenase enzyme could then have used the generated H_2 to produce a transmembrane proton gradient that an equally primitive ATPase would have used to make ATP. Anabolic pathways would have been fed by fermentation of environmentally available carbon compounds. A number of redox proteins—molybdopterin-containing reductases and oxidases, some FeS-containing hydrogenases and reductases, and NiFe redox proteins—appear from molecular phylogeny and protein structural similarity to have diverged before the clear appearance of LUCA. Molecular hydrogen and perhaps arsenite could have been used as electron donors by these early redox enzymes.

6.4 CLUES TO THE PREBIOTIC CHEMICAL ENVIRONMENT FROM METABOLISM THEN AND NOW

The number and variety of hypotheses that have been put forward for the transition from prebiotic chemistry to protometabolism, that is, FIMS, are almost staggering. Many of these are discussed in detail elsewhere in this book. We will now examine a few of the major hypotheses in light of our previous discussion of the possible metabolic constituents of FIMS, LUCA, and metabolism today. Our primary goal is to see if similarities can be found between the chemistry postulated for each model and that seen in biochemical metabolism.

6.4.1 Hypotheses Involving Iron and Sulfur

The *Fe–S world* model is a set of hypotheses about the origin of life that postulate a central role for Fe–S and/or nickel–sulfur (Ni–S) minerals. They are autotrophic, generally speaking, because they all involve fixation of carbon from inorganic sources to form simple organic molecules. One such scenario, proposed by Gunter Wachtershauser and also referred to as the surface metabolism model, requires no initial input of abiotically produced organic compounds. In this model, the surface of Fe–S or Ni–S minerals such as pyrrhotite serves as a catalyst for the reduction of carbon monoxide and/or hydrogen cyanide to simple carboxylic acids and amino acids. The reducing power is provided by molecular hydrogen from submarine hydrothermal systems, and the conversion of pyrrhotite to pyrite (FeS_2) furnishes the chemical energy required.

A number of enzymes contain Fe–S clusters that are part of the catalytic mechanism of the enzyme. These clusters can be as simple as one iron bound to the sulfurs of four cysteines that are part of the protein structure in rubredoxin or as complex as Fe_3S_4, Fe_4S_4, and Fe_6S_6 clusters in proteins such as ferredoxin, aconitase, nitrogenase iron protein, and high-potential Fe–S protein. Most of the Fe–S proteins serve as electron-transfer catalysts, in electron-transfer chains such as photosynthesis and oxidative phosphorylation. Also of note is the presence of the Fe–S enzyme aconitase in both the oxidative and rTCA cycles; it catalyzes one of the low-energy, reversible steps that is the same in both directions of the pathway.

It is interesting to note that, in experimental tests of the surface metabolism model, nickel sulfides seem to be more efficient than iron sulfides at catalyzing carbon fixation from CO or HCN. This is notable because the enzyme acetyl-CoA synthase, which contains both Fe and Ni cofactors, is involved in the acetyl-CoA pathway discussed previously. Also of interest is the observation that, while only one enzyme in modern higher organisms uses Ni as a cofactor, several enzymes in prokaryotes are Ni dependent, and many of these are involved in one-carbon metabolism (i.e., carbon fixation from CO_2) or production of energy from H_2.

The *thioester world* model of Christian de Duve envisions abiotically produced thioesters of protobiotic monomers, such as amino acids, as the activated building blocks for polymer (e.g., peptide) synthesis. According to the model, thioesters of inorganic orthophosphate could have transferred the chemical energy of the thioester bond to the synthesis of pyrophosphate, which could have served a function analogous to ATP in modern metabolism. The model also suggests a transition from a protometabolism based only on thioesters and random peptides to an RNA world-type system. This proposed transition would be consistent with the nucleotide-containing structure of CoA.

The enzyme cofactors CoA (Figure 6.4) and lipoic acid (Figure 6.5) have thiol groups that form thioesters in the process of acyl group transfer and activation. The central place of CoA in both catabolic and anabolic metabolism in all known species on Earth is a persuasive data point supporting the contention that thiol chemistry has been involved in metabolism since at least FIMS. All the alternate pathways for anabolism in FIMS previously discussed make use of CoA. But CoA is a complex molecule, comprised of ADP and phosphopantetheine structures. Generation of such a complex compound in FIMS, much less in a protometabolic system, is difficult to imagine.

Lipoic acid, on the other hand, consists of a cyclic disulfide group, a short hydrocarbon chain, and a carboxylic acid group, which in modern proteins is coupled to a lysine side chain. The synthesis of lipoic acid or a close analog in a protometabolic thioester or surface metabolism scenario is much easier to picture. Modern enzymes containing lipoic acid include pyruvate dehydrogenase and α-ketoglutarate dehydrogenase, which transfer acyl groups to CoA in glycolysis and the TCA cycle. This relationship suggests the interesting idea that perhaps lipoic acid or a similar simple thiol might have been the original

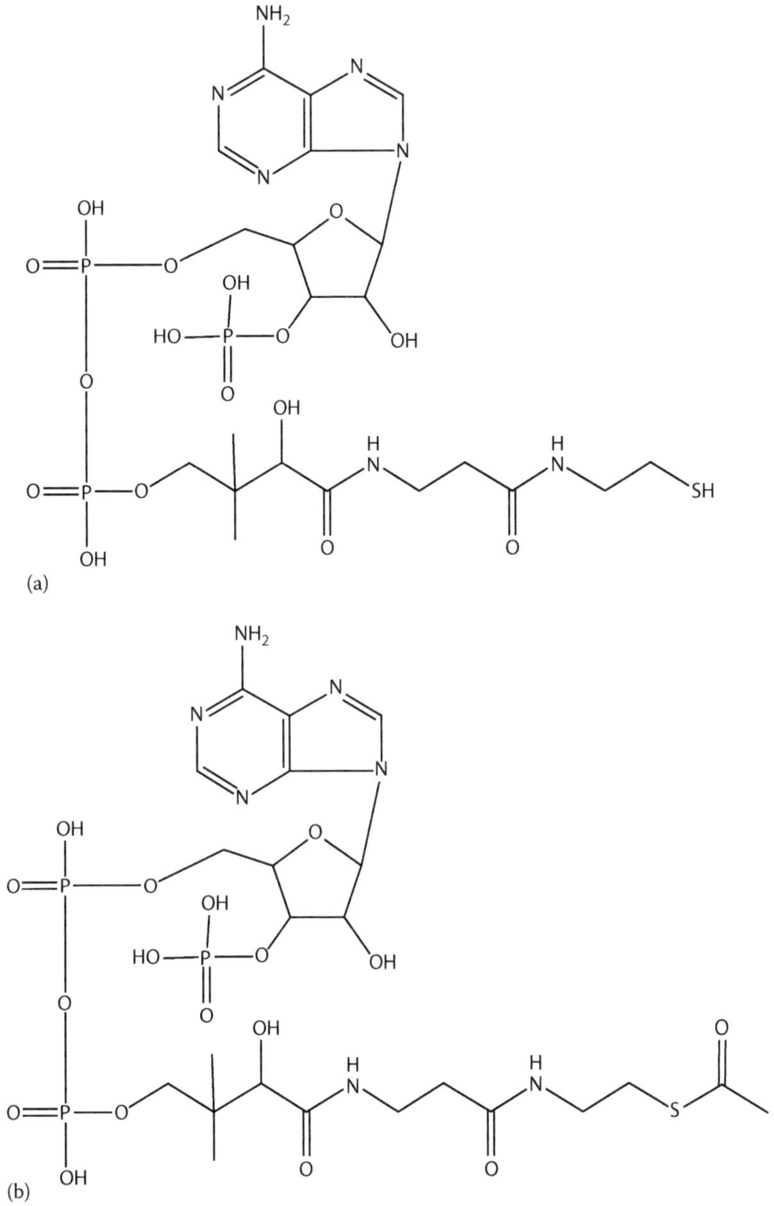

FIGURE 6.4   CoA. (a) The free form without any acetyl group attached to the terminal thiol (–SH) group. (b) The free form with an acetyl group bound to the terminal thiol.

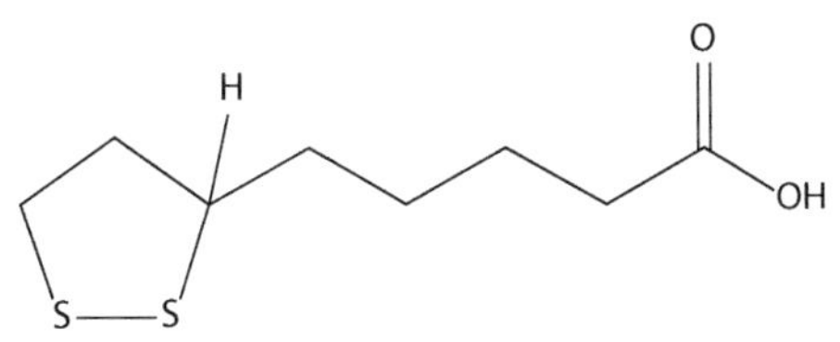

FIGURE 6.5   Lipoic acid, oxidized form. The disulfide bond between the two sulfur atoms in the ring can be broken to form two thiol groups, which can then form thioesters and transfer acyl groups.

activating coenzyme for acyl groups in a protometabolic system or perhaps even in FIMS. CoA, with its nucleotide components, could have been added into growing pathways as nucleic acid biosynthesis became common, eventually becoming the single most important activating compound in central metabolism.

6.4.2 RNA World Hypotheses

The *RNA world* model put forth by Gilbert and others proposes that the FIMS was composed almost entirely of nucleic acids. These molecules would have served as both catalysts and information carriers. One particular version put forth by Steven Benner envisions three stages. The first stage would have involved an RNA-only system without significant protein component. This would have then been followed by a more complex system containing some simple proteins synthesized by RNA. Finally, LUCA would have appeared, with a protein–DNA biochemistry similar to that of modern organisms.

The modern enzyme cofactors NAD^+, $NADP^+$, FAD, and CoA all have ribonucleotide subunits in their structures. The Fe–S cluster protein ferredoxin also interacts with both $NADP^+$ and FAD cofactors in the electron transport chain in contemporary photosynthesis. This has led to proposals of a linkage between metal–sulfur/thioester protometabolism and RNA- or RNA-analog-based systems. It has also been inferred from phylogeny that at some stage of the RNA world tetrapyrrole synthesis evolved, leading to the early use of cobalamin as an enzyme cofactor for one-carbon transfer reactions.

It is not clear, however, that a protobiotic system based on catalytic nucleic acids would have necessarily used ribose as the sugar in the backbone, since the prebiotic availability of ribose is somewhat problematic. For linkage between the *RNA world* model and contemporary metabolism, we are probably best served looking at the modern biosynthetic pathways for purine and pyrimidine bases. The precursors for de novo synthesis of purines are aspartate, glycine, formate, glutamine, and CO_2. With the exception of glutamine, all these compounds are quite plausible components of the prebiotic environment, either from exogenous delivery or endogenous production. The only part of glutamine that is used in purine synthesis is the amide nitrogen, so simpler and more prebiotically plausible analogs such as acetamide can be envisioned as substitutes. The enzymes in this pathway use folate coenzymes at two steps, with Fe involved in one step, but not thiol coenzymes such as CoA. De novo pyrimidine synthesis involves three precursors, aspartate, CO_2, and glutamine (the amide nitrogen). Enzymatically, one step requires both NAD^+ and FAD cofactors, as well as an Fe_4S_4 cluster in the relevant enzyme. A couple of other steps in this pathway have at least some dependence on metal ions such as Zn^{2+}, Co^{2+}, or Cd^{2+}.

6.4.3 Low-Temperature/Atmospheric/Extraterrestrial Prebiotic Synthesis Hypotheses

The classical view of prebiotic chemistry, in broad overview, involves the abiotic synthesis of small organic molecules by atmospheric and low-temperature aqueous processes, combined with the delivery to the Earth's surface from meteorites and interplanetary dust particles. Although a number of biologically relevant classes of compounds have been synthesized or detected from these sources, the amino acids have always been a major focus.

They are relatively abundant in carbonaceous meteorites and are easily synthesized in a variety of prebiotic simulation experiments with a wide range of initial reactants.

Of the 20 amino acids encoded by the genetic code in modern organisms, however, not all have been synthesized prebiotically or identified in extraterrestrial material. The basic amino acids, lysine, arginine, and histidine, are notably absent, along with the aromatic amino acids phenylalanine, tyrosine, and tryptophan. It is at least conceivable that the earliest peptides and proteins did not contain these amino acid groups and could have still carried out at least some catalytic and structural functions.

Looking to modern amino acid anabolic pathways, we see that several of the most abundant prebiotic amino acids are not only the most abundant in modern proteins but are also near the beginning of branched pathways that lead to multiple other amino acids. Aspartic acid is the precursor for the amino acids asparagine, methionine, threonine, isoleucine, and in some organisms lysine. Glutamic acid is the first amino acid made in the anabolic pathway beginning with α-ketoglutarate (a TCA cycle intermediate); other amino acids downstream in this pathway are glutamine, proline, arginine, and in some cases lysine. Alanine is the first product of the pyruvate anabolic pathway, which also leads to valine and leucine.

These metabolic arrangements are certainly in part due simply to chemical structure; the most common prebiotic amino acids are also some of the smallest and most structurally simple, so they are natural precursors of the more complex amino acids. They are also easily synthesized from central metabolic intermediates such as pyruvate and α-ketoglutarate, which would be abundant in both modern biochemical systems and any protobiotic systems that involved the rTCA cycle or similar alternate anabolic pathways. But the same principle also should govern abiotic synthesis; it is usually the case that the simplest members of any compound class are produced in highest yield by nonspecific organic chemical reactions. So while it is certainly possible that the position of certain amino acids in modern metabolism indicates that they were readily available from abiotic sources on the early Earth, it is difficult to conclude this definitively from metabolic arrangements alone.

The established prebiotic sources of purine bases, such as adenine and guanine, are in general more robust than those of pyrimidines, such as cytosine, thymine, and uracil. Purines have been easier to make in prebiotic simulations and have also generally been found to be more abundant in meteorites. It is interesting to note that, in the modern anabolic pathways leading to nucleotides, purine bases are assembled one or two carbons at a time on the ribose structure; free purine bases only exist in modern biochemistry as a result of nucleic acid degradation. In contrast, pyrimidine bases are synthesized as free compounds, with aspartate and bicarbonate as starting materials, and then attached to ribose.

This suggests an interesting possibility for a FIMS version containing nucleic acids. If purines were available abiotically to FIMS, but pyrimidines were not, then it would have been necessary for FIMS to evolve a pyrimidine synthesis pathway before one for purine synthesis. If ribose had not yet been incorporated as the backbone sugar of nucleic acids, it would stand to reason that the pyrimidine synthesis pathway would have evolved independently of the specific sugar involved, as we see today. If purine biosynthesis, on the other hand, evolved later, after abiotic purines had been depleted in the environment and also after ribose had been *locked in* to nucleic acid structure, it would also make sense that the purine pathway would have evolved to build the base onto the ribose molecule rather than initially synthesizing it in free base form.

The implications of modern carbohydrate catabolism for the possible prebiotic chemistry of carbohydrates are also interesting. Most prebiotic carbohydrate synthesis studies have focused on ribose, in the context of the RNA world model. However, the chemistry of formaldehyde in aqueous solution (often referred to as the formose reaction) has been shown to yield a range of carbohydrates of varying sizes. Glyceraldehyde is one of these compounds. This is potentially interesting from the metabolic point of view because phosphorylated glyceraldehyde is the intermediate at what might be thought of as the midpoint of glycolysis. The glycolytic reactions upstream from glyceraldehyde-3-phosphate (Figure 6.2a) consume energy, while those downstream from it (Figure 6.2b) generate two ATP molecules plus one NADH molecule per molecule of glyceraldehyde-3-phosphate. It is possible to envision a protometabolism in which abiotically produced glyceraldehyde was used as an energy source via reactions similar to the *second half* of glycolysis. The pathway from glucose to glyceraldehyde-3-phosphate could have evolved much later, as monosaccharides became available through the advent of photosynthesis.

6.4.4 Vesicle-World Hypotheses

A distinct set of alternatives to the prebiotic/protobiotic models already discussed is what might be generally termed the vesicle-world models. The simplest version involves the accumulation of soluble organic compounds in vesicles formed by abiotically produced amphiphilic compounds such as fatty acids. An autotrophic version invokes simple chromophores such as polycyclic aromatic hydrocarbons to harvest photons and generate ion gradients. These would be coupled to the synthesis of amphiphilic molecules, which would form lipid bilayer vesicles in which the chromophores are embedded. Reproduction would occur by physical division of the vesicles as they reached a certain size, with the molecular components of the metabolic system partitioned between the resultant vesicles by physical diffusion.

Fatty acid synthesis in modern biochemistry is a very complex process. It involves large multienzyme complexes that build the fatty acid chains by the addition of two-carbon acetyl groups. As one might guess, CoA is heavily involved in this process, as is the phosphopantetheine cofactor of acyl carrier protein. The role of these two thiol cofactors could be taken as suggestive of some type of hybrid protometabolism incorporating elements of the thioester and vesicle models.

6.4.5 Overall Picture of Metabolism and Prebiotic Chemistry

A number of modern metabolic components share chemistry with some, but not all, of the prebiotic scenarios. The cofactors NAD/NADH and FAD/FADH$_2$ have ribonucleotide structural components. The Fe–S clusters in numerous enzymes have similar structures to minerals. Many of the simpler metabolic intermediates, as well as some products such as amino acids and purine/pyrimidine bases, also appear to various degrees in multiple prebiotic schemes.

The one component of biochemical metabolism today that seems to be at least some-what related to all the prebiotic hypotheses discussed earlier is CoA. It is a thiol that forms thioesters, its structure contains a ribonucleotide, its pantothenic acid component has had a low-temperature amino acid–based prebiotic synthesis suggested for it, and it is involved in the synthesis of fatty acids. In short, it is the one key metabolic compound that could fairly easily be pictured as a *descendant* of any of the chemistries involved in any of the major prebiotic theories that are operational today. That does not mean, of course, that CoA as we know it now was present in the earliest metabolic systems; it only suggests that at least some of its structural and functional groups would have been a plausible part of them. In fact, the varied components of CoA might be a hint that multiple FIMS with dif-ferent chemistries coexisted for a time and then merged, folding several discrete compo-nents into a single central metabolic molecule for the sake of efficiency.

Before we finish our discussion of the implications of metabolism for prebiotic chemis-try, we need to consider nitrogen fixation one more time. Most of the biomolecules we have examined as either prebiotic or protometabolic components contain nitrogen atoms. These nitrogen atoms are all at lower oxidation states than those in molecular nitrogen (N_2). The reason it is important for prebiotic chemistry that (apparently) neither LUCA nor FIMS contained the capability to reduce molecular nitrogen is that this tells us that the prebi-otic environment contained at least some reduced nitrogen compounds. Otherwise, these metabolic systems would not have been able to synthesize nitrogen-containing biomol-ecules. Whether those reduced nitrogen compounds were generated primarily by atmo-spheric processes, such as lightning-driven chemistry, or by mineral-catalyzed reactions in hydrothermal systems is unclear. What does seem relatively clear is that N_2 was not the only form of inorganic nitrogen on the early Earth.

6.5 SEARCHING FOR EVIDENCE OF METABOLISM IN EXTRATERRESTRIAL ENVIRONMENTS

The environment of the early Earth has long since been replaced by the planetary envi-ronment in which we live today. In addition to a small amount of geochemical evidence from that period, and laboratory simulations of possible early Earth chemistry, our best chance of truly understanding the advent of metabolism on Earth may be to search for it on other planets. There is a certain paradoxical nature to such an effort. We search for extraterrestrial forms of metabolism in order to define what is truly necessary for the sim-plest such systems to exist. But in order to intelligently design and carry out such a search, we must make assumptions about the nature of the basic metabolic systems that we seek to find. This paradox has led to the proposal and/or application of a number of different approaches to the detection of extraterrestrial metabolism.

6.5.1 Anabolism or Catabolism of Carbon

Probably the most specific approach to the search for metabolism in an extraterres-trial environment is the attempt to directly detect catabolic or anabolic processes. This approach is exemplified by two of the Viking life detection experiments, the labeled release and pyrolytic release experiments. The labeled release experiment looked for catabolism by heterotrophic organisms by supplying ^{14}C-labeled nutrient compounds to Martian soil and monitoring for release of radioactive CO_2 as a product of pathways similar to glycolysis and the TCA cycle. The pyrolytic release experiment looked for anabolism by autotrophic organisms by supplying radiolabeled CO_2, allowing any organisms present to fix that carbon into high-molecular-weight compounds, and then heating the sample to convert those compounds back into detectable ^{14}C-containing gaseous molecules.

This approach has the major limitation that it can only detect active metabolic processes in extant organisms. It assumes that an extraterrestrial metabolism is carbon-based, although this assumption is probably a safe one. It also assumes that the metabolic rates of extraterrestrial organisms are similar to terrestrial ones, an assumption that may not be valid given the generally lower-temperature regimes relative to Earth on most solar system bodies that might harbor life. Even more problematic is the potential for biological metabolic processes that involve redox reactions to be mimicked by abiotic redox chemistry. This appears to have been the case for the Viking experiments, although that conclusion is still the subject of some controversy.

6.5.2 Chemical Disequilibrium

A less specific metabolism detection approach is the search for chemical disequilibrium. Metabolism can be thought of as the harnessing of energy by biological systems to maintain themselves in disequilibrium with their environment. A number of different approaches to disequilibrium detection have been either proposed or deployed. These include the Viking gas exchange experiment and the proposed spectroscopic searches of extrasolar planetary atmospheres. Both of these approaches involve attempts to detect nonequilibrium gaseous products of metabolism. Biogenic minerals, those formed as byproducts of microbial metabolic processes but not by abiotic geochemical reactions, have also been investigated as potential indicators of life.

These chemical disequilibrium approaches also primarily detect extant metabolism, although some biogenic minerals may be stable enough to serve as fossil biomarkers. While the search for disequilibrium has fewer inherent assumptions than the search for specific carbon metabolism, it also has more possible abiotic mimics. These include radiation-driven chemical processes in atmospheres and on icy surfaces, which can produce O_2, nitrogen oxides, and other molecules that can also be products of metabolism. Other concerns include geochemical reactions that have kinetic barriers that greatly slow, or prevent entirely, the attainment of equilibrium in mineral systems.

6.5.3 Metabolic Product Distribution

An alternative approach to the direct attempt to detect metabolism is the search for distinctive distributions of metabolic products. The gas chromatography/mass spectrometry instruments on the Viking and Curiosity Mars spacecraft, as well as the analysis of Mars meteorites and other extraterrestrial samples for organic compounds, are examples of this approach. The metabolic synthesis of discrete classes of compounds generally results in only some of the chemically possible homologs and isomers in that class. Abiotic synthesis, on the other hand, typically results in all possible compounds within a class, in abundances that generally decrease logarithmically with the number of carbons (and perhaps nitrogens) in the molecule. The comparison of biological amino acid compositions with the abiotically produced contents of carbonaceous meteorites is a notable example.

In the case of chiral biomolecules such as amino acids, an enantiomeric excess can also be indicative of biological metabolism. It is generally held that any complex biochemical system would require the use of enantiomerically pure monomers, such as amino acids, for the synthesis of polymers such as proteins that have catalytic activity based on 3D binding specificity. Abiotic synthesis, on the other hand, almost always produces equal amounts of enantiomers (a racemic mixture). Stable isotope ratios can also be used to identify products of metabolism. Enzymatically catalyzed synthesis reactions often fractionate carbon and nitrogen atoms by mass due to the differing bond strengths associated with different isotopes. This is particularly true for photosynthesis but also is seen in amino acid and other biosynthetic pathways. Abiotic synthetic reactions, particularly those that take place at low temperatures, have very different (and generally much heavier) stable isotope signatures than metabolic products.

These approaches also assume carbon-based metabolism, but are not automatically restricted to extant and active systems. Their ability to detect extinct metabolic systems, however, can be limited by the degradation of organic compounds in some environments over time. This degradation can be complete, as in oxidation of organics to volatiles such as CO_2 by environmental oxidants. It can also be partial, resulting not in destruction of metabolic products but in their alteration to other molecules that may in some cases resemble products of abiotic chemistry. It has been suggested that the suite of polycyclic aromatic hydrocarbons detected in several meteorites from Mars may be an example of this problem.

GLOSSARY

Note: All standard biochemical terms are defined/explained in the text as they were encountered first.

Abiotic: Produced by or from nonbiological chemical or other processes.

Amphibolic: Metabolic pathway that has both synthetic (anabolic) and energy-producing (catabolic) functions.

Amphiphilic: Molecule with both hydrophobic or nonpolar and hydrophilic or polar functional groups.

Anabolic: Pathway that functions to synthesize biomolecules.

Archaea: One of the three domains of life; single-celled microorganisms with no cell nucleus; most live in extreme environments.

Archaean era: Period in Earth's history from 4 to 2.5 billion years ago; period during which the origin of life on Earth is generally thought to have occurred.

Autotrophic: An organism deriving all biomolecules from simple inorganic compounds present in the environment and all energy from light or chemical reactions in the environment.

Bacteria: One of the three domains of life; single-celled microorganisms with no cell nucleus; most live in common environments.

Biogenic mineral: Mineral that is produced by biologically driven chemical reactions and does not normally occur in the absence of biology.

Biomarker: Chemical compound or mineral that forms as the result of biological processes and is used as evidence for the presence of biology in a given environment.

Carbon fixation: The process of incorporating one-carbon molecules such as carbon dioxide into multicarbon biomolecules such as carbohydrates.

Carbonaceous meteorite: Meteorite containing unusually high levels of abiotically synthesized organic material.

Catabolic: Pathway that functions to break down organic molecules in order to release chemical energy.

Chemolithoautotrophic: Organism that derives energy from mineral compounds or materials.

Chromophore: Molecule that absorbs light in the visible or ultraviolet spectral regions.

Coenzyme (or cofactor): Molecule that has a role in the catalytic mechanism of an enzyme, but that is not part of the protein structure of the enzyme itself.

Delta C-13 (δ^{13}C) value: Measurement of the $^{12}C/^{13}C$ ratio in a molecule or sample, relative to a standard carbonate mineral.

Enantiomer: One of two non-superimposable mirror-image isomers of a compound with a chiral center.

Endergonic: Chemical reaction in which products have higher total energy than reactants.

Endosymbiont: Organism that lives within the body or cells of another organism and derives nutrients or other advantages from the association.

Eukaryota: One of the three domains of life; single- or multicellular organisms with a cell nucleus.

Exergonic: Chemical reaction in which reactants have higher total energy than products.

FIMS: First metabolic system, the earliest chemical system on Earth that contained either catabolic or anabolic pathways.

Ga: Giga-annum, billion years before present.

Glycolysis: The metabolic pathway in all extant organisms that converts glucose and other sugars into pyruvate, with the production of chemical energy.

Heterotrophic: An organism deriving all biomolecules and all energy from organic compounds synthesized by other organisms.

Homologous: Genes or proteins in different biological species having shared ancestry.

Horizontal (or lateral) gene transfer: Transfer of genes between organisms via nonreproductive mechanisms.

Hydrothermal vents: Submarine geological systems in which seawater is heated by volcanic activity, often accompanied by high concentrations of dissolved minerals.

Iron–sulfur world: A set of hypotheses for the origin of life on Earth in which carbon–carbon bond formation and other prebiotic chemistry are catalyzed by Fe–S minerals such as pyrite.

Kerogen: High-molecular-weight organic material formed by polymerization and alteration of smaller organic compounds under elevated temperatures and pressures.

LUCA: Last universal common ancestor, the species that is the common genetic ancestor of all extant and extinct species of life on Earth.

Metabolism: A set of chemical reactions, usually occurring in living cells, that allows organisms to harness chemical energy and synthesize necessary biomolecules.

Molecular phylogeny: The evolutionary relationships between species as indicated by their hereditary similarities and differences in genetic material.

Nitrification: Oxidation of ammonia into nitrates and nitrites; usually biologically mediated.

Nitrogen fixation: Reduction of molecular nitrogen into ammonia or organic amine compounds.

Nucleophilic: Chemical group that is attracted to electron-deficient regions of a molecule.

Orthologous: Homologous genes descended from the same ancestral sequence separated by a speciation event.

Phosphoester bond: Bond formed between a phosphate group and a hydroxyl group; a phosphodiester bond (as in DNA and RNA) contains one phosphate participating in two phosphoester bonds.

Phospholipid: Molecule composed of two fatty acids or other nonpolar molecules, joined by a polar molecule containing a phosphate group; major component of biological membranes.

Phosphorylation: Formation of chemical bond between a phosphate group and another chemical group such as a hydroxyl group.

Phylogenetic: Referring to the evolutionary relationships among groups of organisms, genes, or gene products.

Prebiological: Related to the chemistry that preceded and led to the origin of life on Earth.

Protobiological: Related to the chemistry of systems on the early Earth that shared at least some characteristics with modern biochemistry, but were not unambiguously biological.

Protometabolism: Groups of chemical reactions that occurred in protobiological systems and served metabolic functions.

Purine and pyrimidine bases: Cyclic organic molecules containing nitrogen that are part of nucleic acids; their sequence in DNA carries the genetic code.

Racemic mixture: A sample containing equal amounts of two enantiomers of a chiral compound.

Redox: Abbreviation for oxidation–reduction; commonly used to identify enzymes that catalyze electron-transfer reactions.

Ribonucleotide: Nucleotide containing the sugar ribose.

Ribozyme: A molecule of RNA that is capable of catalyzing a chemical reaction in a similar fashion to a protein enzyme.

RNA world: A set of hypotheses for the origin of life on Earth in which RNA was the first true biomolecule and had both catalytic and genetic functions.

Stable isotope ratio: The ratio in a sample or compound of the stable (nonradioactive) isotopes of a given element.

Tetrapyrrole: Nitrogen-containing cyclic organic molecule that can bind to metal ions such as iron or cobalt; serves as cofactor in numerous enzymes.

Thioester world: A subset of the Fe–S world hypotheses for the origin of life on Earth in which chemical energy is stored in thioester bonds between organic acids and thiol compounds.

Tricarboxylic acid cycle: The metabolic pathway in all extant organisms that uses acetyl groups generated by glycolysis or other pathways along with organic acids to generate chemical energy.

Vesicle: Spherical bilayer structure formed in aqueous solutions from abiotically produced amphiphilic compounds such as fatty acids.

Vesicle-world hypothesis: A set of hypotheses for the origin of life on Earth in which the earliest protometabolic systems were vesicles that carried only catalytic and energy-harvesting components without any genetic material.

REVIEW QUESTIONS

1. Describe the basic differences between catabolic and anabolic pathways.

2. Which type of pathway generally involves the oxidation of carbon? The reduction of carbon?

3. Compare the maximum and minimum sizes of model LUCA genomes with the smallest known modern genomes.

4. What is the smallest known modern genome in a free-living (nonparasitic) organism?

5. Rank the following nitrogen-containing molecules in the order of decreasing nitrogen oxidation state: NO_2, NH_3, N_2, NO_2. What is the source of each in the Earth's atmosphere today?

6. All the major alternate metabolic pathways suggested for FIMS are autotrophic. Why?

7. In the alternate pathways, to what form is CO_2 reduced? What is typically the carrier of this molecule? Are there other plausible carriers for this molecule?

8. If several protometabolic systems, using different alternate pathways, existed at the same time on the early Earth, would they have competed with each other for resources? If so, for which resources?

9. What type of change in environmental conditions might have caused one of these protometabolic systems to outcompete the others and thus form the basis for the metabolism of LUCA?

10. What chemical structures or functional groups would a precursor of CoA require in order to function? Are there other plausible prebiotic molecules that have the required structures or groups?

11. Are there functional groups in the structure of RNA that, without modification, could serve any of the functions now carried out by enzyme cofactors? What modifications to RNA would be needed for it to fill all these functions?

SUGGESTED READING

Benner, S. A., A. D. Ellington, and A. Tauer. Modern metabolism as a palimpsest of the RNA world. *Proc. Natl. Acad. Sci. USA* 86 (1989): 7054–7058.

Cody, G. D. and J. H. Scott. The roots of metabolism. In *Planets and Life: The Emerging Science of Astrobiology*, W. T. Sullivan and J. A. Baross (eds.), pp. 174–186. Cambridge, U.K.: Cambridge University Press, 2007.

de Duve, C. Clues from present-day biology: The thioester world. In *The Molecular Origins of Life: Assembling Pieces of the Puzzle*, A. Brack (ed.), pp. 219–236. Cambridge, U.K.: Cambridge University Press, 1998.

Fuchs, G. Alternative pathways of carbon dioxide fixation: Insights into the early evolution of life? *Annu. Rev. Microbiol.* 65 (2011): 631–658.

Garrett, R. H. and C. M. Grisham. *Biochemistry* (3rd edn.). Belmont, CA: Thomson Brooks/Cole, 2007.

Horowitz, N. H. On the evolution of biochemical syntheses. *Proc. Natl. Acad. Sci. USA* 31 (1945): 153–157.

Klein, H. P. The Viking biology experiments: Epilogue and prologue. *Origins Life Evol. Biosph.* 21 (1992): 255–261.

Lazcano, A. and S. L. Miller. On the origin of metabolic pathways. *J. Mol. Evol.* 49 (1999): 424–431.

McDonald, G. D. and M. C. Storrie-Lombardi. Amino acid distribution in meteorites: Diagenesis, contamination, and standard metrics in the search for extraterrestrial biosignatures. *Astrobiology* 6 (2006): 17–33.

McDonald, G. D. and M. C. Storrie-Lombardi. Biochemical constraints in a protobiotic Earth devoid of basic amino acids: The "BAA-World". *Astrobiology* 10 (2010): 989–1000.

Mirkin, B. G., T. I. Fenner, M. Y. Galperin, and E. V. Koonin. Algorithms for computing parsimonious evolutionary scenarios for genome evolution, the last universal common ancestor and dominance of horizontal gene transfer in the evolution of prokaryotes. *BMC Evol. Biol.* 3 (2) (2003). http://www.biomedcentral.com/1471-2148/3/2 (accessed February 12, 2013).

Morowitz, H. J., B. Heinz, and D. W. Deamer. The chemical logic of a minimum protocell. *Origins Life Evol. Biosph.* 18 (1988): 281–287.

Navarro-Gonzalez, R., C. P. McKay, and D. Nna Mvondo. A possible nitrogen crisis for Archaean life due to reduced nitrogen fixation by lightning. *Nature* 412 (2001): 61–64.

Nelson, D. L. and M. M. Cox. *Lehninger Principles of Biochemistry* (3rd edn.). New York: Worth, 2000.

Noor, E., E. Eden, R. Milo, and U. Alon. Central carbon metabolism as a minimal biochemical walk between precursors for biomass and energy. *Mol. Cell* 39 (2010): 809–820.

Scott, J. H., D. M. O'Brien, D. Emerson et al. An examination of the carbon isotope effects associated with amino acid biosynthesis. *Astrobiology* 6 (2006): 867–880.

Wachtershauser, G. Origin of life in an iron–sulfur world. In *The Molecular Origins of Life: Assembling Pieces of the Puzzle*, A. Brack (ed.), pp. 206–218. Cambridge, U.K.: Cambridge University Press, 1998.

Woese, C. The universal ancestor. *Proc. Natl. Acad. Sci. USA* 95 (1998): 6854–6859.

Zerkle, A. L., C. H. House, and S. L. Brantley. Biogeochemical signatures through time as inferred from whole microbial genomes. *Am. J. Sci.* 305 (2005): 467–502.

CHAPTER 7

Roles of Silicon in Life on Earth and Elsewhere

Joseph B. Lambert and Senthil Andavan Gurusamy-Thangavelu

CONTENTS

7.1 CARBON AND LIFE

The science fiction media have a penchant for referring to life on Earth as being carbon based. Indeed, the chemical materials that constitute living substances are universally based on carbon-containing molecules that we refer to as organic compounds. Molecules that make up living tissue, that provide genetic information, that generate energy for life, and that catalyze chemical reactions within organisms share the universality of carbon as the fundamental atomic building block. In infinite variety, carbon atoms bond covalently to other carbon atoms to form chains and rings, in which the carbon atoms usually are capped with hydrogen atoms and occasionally are interspersed with other atoms, referred to as heteroatoms, such as oxygen, nitrogen, sulfur, and phosphorus.

Carbon is the ideal atom for the formation of varieties of stable, complex substances for several reasons. Because of its electronic structure and placement in the periodic table, carbon has a valence of four and hence can form four different bonds. Similarly, nitrogen can form only three bonds in its neutral forms, and oxygen only two. In a numbers game, two carbons singly bonded to each other still have six opportunities for further bonds, whereas two nitrogens have four and two oxygens only two. Carbon thus has an advantage in terms of variety and complexity.

Those four bonds of carbon have to be as far apart as possible to avoid repulsions between the electrons in the C—C bonds. Therefore, the four bonds of carbon protrude into three dimensions, as in **1**,

1   **2**

rather than remaining in a plane, as in **2**. The three-dimensionality of carbon bonding (its *stereochemistry*) generates important ramifications that are beyond the scope of this presentation. The four bonds of the central carbon are fully illustrated in **1**. For visual simplicity, the bonds of the outer carbons are not. They could be bonded to hydrogen atoms, to other carbon atoms, or to heteroatoms. In the tetrahedral arrangement of **1**, each C—C bond is as far as possible from any other C—C bond, resulting in C—C—C angles of about 109°. In **2**, the angles are 90°, so the bonds are closer and electronic repulsion between the bonds is larger. Intermediate or smaller angles also offer greater repulsions than the optimal stereochemistry of **1**.

The C—C bonds illustrated in **1** are called single or sigma (σ) bonds, and the system is said to be *saturated*. Carbon also has the ability to form multiple bonds, which take the forms of, for example, the molecular fragments in **3** and **4**.

3   **4**

Again, the bonds of the outer carbons are left to the imagination. Such systems are said to be *unsaturated*. In **3**, two central carbon atoms are doubly bonded to each other. Although these carbons still have four bonds, they are attached to only three other atoms. For doubly bonded carbon, C—C—C angles of 120° provide the least repulsion between electrons, that is, the C—C bonds are as far apart as possible. Bond angles of 120° require the resulting six-atom piece of **3** to be planar. The two central carbons in **3** are shown to be bonded to other carbon atoms, but alternatively, they could be bonded to hydrogen atoms (H, as in >C=CH$_2$), which are univalent and terminate the chain. In **4**, the central carbon atom is doubly bonded to an oxygen atom to form the carbonyl group. Since neutral oxygen is divalent, it has no further bonds. The additional electrons in the atomic structure of oxygen, needed to achieve an octet, reside in nonbonding orbitals, or lone pairs, as depicted by the two pairs of dots in **4**.

Carbon also can form triple bonds (—C≡C— or —C≡N:), arrangements that are extremely stable and are found in interstellar space. The angle between the bonds attached to the triple bond is 180° in order to achieve the greatest distance and the least electronic repulsion. As a result, a triply bonded system is linear (compared with planar for doubly bonded and tetrahedral for singly bonded). The carbon–nitrogen triple bond, called the nitrile or cyano group, is a biochemical precursor for the nucleobase adenine, a common constituent of nucleic acids.

Although nitrogen and oxygen also can bond to each other and catenate (form chains), their lone pairs repel each other strongly. The resulting molecules, such as **5**

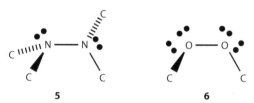

(a hydrazine) and **6** (a peroxide), are relatively unstable because of repulsions between adjacent lone pairs. Further bondings are shown to carbon, but they also could be to hydrogen (as in HOOH, hydrogen peroxide, and NH_2NH_2, parent hydrazine). The additional bondings of the illustrated carbon atoms are omitted in **5** and **6**. Note that nitrogen, bonded to three other atoms, completes its octet with a single lone pair, and the oxygens, bonded to two other atoms, have two lone pairs. Thus, hydrazines and peroxides, the respective building blocks of nitrogen-based and oxygen-based systems, are highly energetic and subject sometimes to explosive decomposition. Both have been used as rocket fuels.

Bonds formed by carbon are quite strong. *Bond energy* (the heat required to decompose 1 mol of a molecule into its constituent atoms) is one measure of bond strength (Rappoport and Apeloig, 1998). Carbon–carbon single-bond energies average about 83 kcal mol^{-1}, and carbon–hydrogen bonds are even stronger (ca. 100 kcal mol^{-1}). In contrast, the oxygen–oxygen bond energy is only about 40 kcal mol^{-1} because of the repulsions between lone pairs. Since less energy is required to break O—O and N—N bonds, hydrocarbons are more stable than hydrazines and peroxides. Moreover, C—C bonds are remarkably inert to attack by acids and bases, two of the most common reaction-initiated species. Thus, single bonds between carbons have good thermodynamic strength (inherent stability) and low kinetic reactivity (propensity to undergo chemical reactions).

Bending back upon themselves, saturated carbon chains can form 3D rings

(**7**, a molecule known as cyclohexane). Unsaturated carbon chains can form 2D rings (**8**, benzene, which often is represented as **9** to emphasize the equivalence of all the ring C—C bonds). In the representations of **7** and **8**, carbon atoms are present wherever two lines (representing bonds) come together. The atomic letters are omitted to make the structures less crowded. In common representations, even the hydrogen atoms and the bonds to them are omitted.

In summary, saturated carbon–based molecules are constructed in three dimensions from C—C single bonds (**1**), interspersed occasionally with oxygen atoms to form ethers (C—O—C), or with nitrogen atoms to form amines (C—N—C, plus one more bond to N, not shown). Planar structures are formed from double bonds (**3**), in which one of the atoms occasionally is replaced by oxygen to form the carbonyl group (**4**) or by nitrogen to form the imine group (>C=N:). Directly bonded oxygen and nitrogen atoms generally are avoided (**5** and **6**). Saturated and unsaturated rings (**7**, **8**, and **9**) add complex variations.

From these simple rules, the molecules of earthly life are generated. Carbohydrates have saturated rings of five or six atoms, of which one always is oxygen. Proteins are chains containing repeated arrangements of carbonyl groups directly bonded to carbon on one side and to nitrogen on the other (**10**, in which the wiggly lines indicate multiple repetition of the same structure on both sides).

10

The letter R represents an unspecified organic entity. Nucleic acids are composed of five-membered carbohydrate rings on which an unsaturated ring (the nucleobase) derived from benzene (**8** and **9**) of varying ring size is substituted and with carbons often replaced by nitrogen or carbonyl groups. The carbohydrate rings in nucleic acids are connected through phosphate linkages at two points to generate complex chains. Carbohydrates, proteins, and nucleic acids are probably the most important classes of molecules required for life, but there are many others, including lipids such as steroids and prostaglandins, all made from carbon-based building blocks.

7.2 BONDING PROPERTIES OF SILICON

The basic tenets of the periodic table of elements dictate that atoms in a column have similar properties. Group 16 starts with oxygen and continues down with sulfur, selenium, and tellurium. The most common bonding arrangement for all these elements, when neutral, is bivalency, as in ethers (R—O—R, in which R represents an organic grouping). There is a trend toward higher valencies (numbers of bonds) in the lower regions of the periodic table, as numbers of electrons accumulate. The atoms in group 15 are nitrogen, phosphorus, arsenic, antimony, and bismuth, with trivalency the dominant pattern in the lighter elements, as in amines (R_3N:). Carbon tops group 14, followed by silicon, germanium, tin, and lead. Here, the preferred arrangement is tetravalency for neutral species, and again, higher valencies are found increasingly with the heavier elements. Thus, the simplest organic compound (**11**, methane) has a strong resemblance to the simplest organosilicon compound (**12**, silane).

At first glance, it would seem that silicon could slip into the same roles as carbon and provide silicon analogues of all the molecules of life. In this fashion, silicon would replace carbon in silicon-based carbohydrates, proteins, and nucleic acids. Nature has given silicon an enormous advantage in terms of availability. On Earth's lithosphere or crust, silicon is the second most abundant element (27%–28%), after oxygen. In contrast, despite the richness of life, carbon is present in less than 0.2% of the Earth's crust. earthlike objects (our Moon, Mars, Venus, Mercury) have similar crustal composition to Earth's. Thus, silicon would seem to have two factors that should be favorable to its use as the building block of life—similar valencies to carbon and wide availability.

Despite these favorable factors, reality is quite different. On Earth, silicon has never been identified in a single compound in a living system. The bonding properties of silicon in fact militate against its use in biochemical molecules. A common aspect of all such molecules is the presence of unsaturation. Carbohydrates are formed from aldehydes, which are defined by the presence of a carbonyl group attached to hydrogen (HC=O). Proteins require the peptide linkage illustrated in **10**, which contains a carbonyl group, in this case part of the amide functionality. Nucleic acids contain carbohydrate rings, which again require the carbonyl group. Attached to the carbohydrate rings in nucleic acids are side rings related to benzene (the nucleobase). They contain C=C, C=O, and C=N double bonds. Photosynthesis is the basis of life for many organisms, whereby carbon dioxide in the atmosphere is converted to carbohydrate rings. Carbon dioxide (O=C=O) contains two double bonds. Thus, unsaturation is a fundamental structural motif in carbon-based biological molecules.

Whereas unsaturation is common in the world of carbon, it is nearly absent in the world of silicon. When organic chemists started creating compounds in which carbon is replaced by silicon, they quickly synthesized a host of saturated organosilicon compounds with Si—Si, Si—C, and Si—O single bonds, but none with double bonds to a silicon atom.

Eventually, it was realized that multiple bonding with silicon is much weaker than with carbon, for two reasons. First, bond lengths increase with increasing atomic number, so that elements lower in the periodic table have longer bonds. Thus, Si—Si bonds are about 2.33 Å, C—Si bonds about 1.82 Å, and C—C bonds about 1.54 Å. When double, or pi (π), bonds form, the p orbitals must be able to interact through side-by-side overlap, as illustrated in **13**

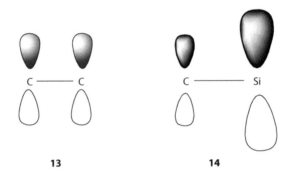

for the C=C bond. This is the main distinction between σ bonds, in which orbitals interact face on, and π bonds, in which the orbitals interact side by side. The corresponding orbitals for a C=Si bond are considerably farther apart (**14**). As a result, they interact more weakly, and the resulting double bond is weaker. For the Si=Si bond, the situation is even worse. There is a second factor working against the C=Si bond, because the interacting orbitals are of considerably different sizes (resulting from the atomic sizes). The carbon p orbital fits poorly with the silicon p orbital in **14**, so that overlap is further diminished. Organosilicon chemists were frustrated for many decades in their efforts to synthesize unsaturated silicon systems. When they finally were successful, the compounds proved to be relatively unstable. To this day, the silicon–oxygen double bond (Si=O) has not been prepared in a stable compound. Thus, the silicon analogue of the carbonyl group is singularly absent even in the synthetic world, so that the analogues of carbohydrates, proteins, and nucleic acids cannot exist. The basic building blocks of carbon-based life are absent in the silicon world. To make matters worse, the silicon analogue of carbon dioxide (O=Si=O) contains two unstable double bonds and has never been prepared, so there could not be silicon-based photosynthesis.

When unsaturated compounds are removed from biochemistry, only molecules with single bonds are left. Still, nature seems to ignore silicon even in this context. Concatenations of C—C single bonds are called alkanes. In the living world, lipids including fats, waxes, and steroids contain significant segments of saturated C—C bonds, but equally important are their unsaturated components, for which no silicon analogue would be stable. The silicon analogues of alkanes have long chains of silicon atoms (—Si—Si—Si—Si—) and are called *polysilanes*. Like alkanes, they also can be branched or cyclic. Just as a biochemistry based on alkanes alone is hopelessly limited, a polysilane biology would be even more limited. Life requires metabolic reactions and the ability to reproduce. These functions involve the reactions of unsaturated molecules and are beyond the capabilities of saturated molecules.

Moreover, organosilanes generally are less stable than alkanes, both thermodynamically (inherent stability) and kinetically (propensity to react with other molecules). The Si—Si bond energy is about 75 kcal mol^{-1}, some 10 kcal mol^{-1} less than that of C—C bonds. In addition, Si—Si bonds are reactive with acids, bases, and light, whereas C—C bonds are much less so. Organosilanes have neither the ability nor the stability to serve as the start for a silicon-based biochemistry. Bonds between silicon and carbon (Si—C) are slightly less stable (ca. 80 kcal mol^{-1}) than those between carbons, and such bonds are particularly sensitive to attack by bases. Alkanes are composed of alkyl groups, of which the simplest is methyl (CH_3), followed by ethyl (CH_3CH_2), and so on. The Si—H bond is extremely reactive under numerous conditions, in stark contrast with the C—H bond. Thus, the silicon analogue of the methyl group (SiH_3) is not a practical constituent of biological molecules and similarly for other alkyl analogues.

Thus, silicon bonds in nature seem to be frustrated at every approach, with serious problems arising with Si=Si, Si=O, Si=C, Si—Si, Si—C, and Si—H bonds. There is an exception, however, in that bonds between silicon and oxygen (Si—O) are extremely strong (ca. 110 kcal mol^{-1}). Pauling (1980) attributed this unusual strength to higher ionicity and partial double bonding. Because silicon is electropositive and oxygen is electronegative, there is considerable positive charge buildup on silicon and negative charge buildup on oxygen, resulting in the resonance hybrid of Si—O $\leftrightarrow$ Si$^+$ O$^-$. The atoms attract each other and create a stronger bond. Pauling's double bonding may be illustrated by the resonance structures of a four-atom fragment, Si—O—Si—O $\leftrightarrow$ Si$^+$ O=Si$^-$ O, but the contribution may be less important than Pauling thought. The Si—O bond still can be characterized as covalent rather than ionic. Molten sand, made up of many Si—O bonds, for example, is not a conductor. The high availability of silicon and oxygen in the Earth's crust and the strength of the bond between them result in the widespread occurrence of silicate minerals, which constitute 90% of the Earth's crust. Silica or sand is given the formula of SiO_2, but this formula does not indicate structure but rather atomic ratios. On average, there are two oxygens for every silicon because of their difference in valence. Much of the field of mineralogy in fact is a study of silicate species. The very strength of the Si—O bond, however, works against its use as a biochemical building block, because the bond is so stable. The silicate minerals of terrestrial planets and moons result from the thermodynamic sink of the Si—O bond.

Another reflection of the poor bonding characteristics of silicon is its low occurrence in interstellar space (Pascoli and Comeau, 1995; Ziurys, 2006). Whereas dozens of molecules containing carbon have been identified spectroscopically, there is only a handful of silicon molecules, including SiC, SiN, SiO, SiS, SiCN, SiNC, SiC_2, SiC_3, and SiH_4, five of which also contain carbon.

7.3 SILICON AND LIFE ON EARTH

Although there has not been a single organosilicon compound found indisputably in living organisms, inorganic silicon in the form of silica (SiO_2) is not uncommon. Silica provides support structure to certain marine organisms such as diatoms, radiolarians, and sponges, playing a role similar to calcium carbonate in shells and hydroxyapatite in mammalian bones. Silica occurs in many other organisms, also providing a structural role (Epstein, 1999; Pryer et al., 2004; Liu et al., 2013). Rice husks, for example, contain 20% or more of silica, which is being exploited as a starting material for elemental silicon used in the electronics industry. The ancient vascular plant called the horsetail (genus *Equisetum*) also contains a relatively large amount of silica and has found uses as an abrasive. Silica-containing plants leave a residue of silica in the ash both after combustion and following natural decomposition. In the nineteenth century, some of these residues were named *phytoliths* (Greek for *plant stone*). Today, archaeologists and paleoecologists use the morphology of phytoliths left after natural plant decay to identify species in the ancient biological environment (Fritz, 2005; Piperno, 2005).

The fact that many organisms can extract natural silica from the sea and the ground and rework it as a biological material implies that there is a biochemical process for uptake and conversion. Surely, these processes involve the formation of some types of silicon-containing intermediates, but to date, they have not been found. The enzymes involved in these processes have been termed *silicateins*, by analogy with proteins, which process carbon-based materials. The details of these biochemical processes remain murky, but it is reasonable that Si—O bonds remain intact and it is possible that Si—C bonds never form (Shimizu et al., 1998; Müller et al., 2013).

There are many different structural possibilities for materials containing Si—O or Si—C bonds. First, in the inorganic world without carbon (when all the heavy atoms are silicon and all the silicon atoms are attached to four oxygen atoms), the substance is a silicate (silica, sand, or quartz minerals). When some silicon atoms in silicate are replaced by aluminum as aluminosilicates, still in the absence of carbon components, the material is clay. The complex chemistry of silicates may provide opportunities for roles of silicon in life, and Cairns-Smith and Graham (1990) even have proposed that Earth's first organisms were based on clay minerals. Such rudimentary organisms, even if they could exist, would lack the enzymatic and genetic capabilities required to generate more advanced functions.

When one Si—O linkage of each silicon in silicate is replaced with a Si—C linkage, the resulting materials are called *silsesquioxanes*. When two Si-O linkages of each silicon are replaced by Si—C linkages, complex polymers are formed that are named *silicones*. These materials have a long chain with the structure —Si—O—Si—O—, in which each silicon is bonded to two carbons (not shown) and to two oxygens. The silicone industry is vast, but the materials are not so stable as carbon polymers. When three Si—O linkages of each silicon are replaced with Si—C linkages, the result is a simple organosilicon compound, without a special, collective name, for example, $(CH_3)_3SiOR$. All of these structures have Si—C bonds, which are unstable under many conditions. In these simplest compounds, the oxygen on silicon could in turn be part of a hydroxy group (HO—Si, hydroxysilanes) or an alkoxy group (RO—Si, alkoxysilanes). The former are unstable with respect to dimerization to a Si—O—Si (siloxanes) linkage ($R_3SiOH \rightarrow R_3SiOSiR_3$), and the latter also are easily decomposed.

If Si—Si, Si—C, and Si—H linkages are excluded from roles in a silicon-based biochemistry, what about other functionalities that take advantage of the strength of the Si—O bond? Such would involve a mixture of silicon and carbon. These functionalities would include Si—O—C groupings, by which silicon is involved without having to form unstable bonds directly with silicon, carbon, or hydrogen. If we are to find an island of stable activity for silicon, it seems that the Si—O—C bond would be a very likely candidate. As each silicon still is bonded to four oxygens, such molecules are called *organosilicates*, in which one or more of the oxygen atoms are capped with a carbon atom. These sorts of structures may be involved, for example, in the biological uptake of silica. Only a few studies have examined the role of the Si—O—C bond, so more background is needed first.

There have been numerous studies of so-called prebiotic formulas for creating the molecules that are essential for carbon-based biochemistry. Miller (1953) reported experiments with simple reagents and conditions that generated essential amino acids, and Miller and Urey (1959) expanded the yield to include other biomolecules. Even earlier, the Russian Butlerov discovered the formose reaction, whereby small aldehydes under basic conditions oligomerize to C5 and C6 carbohydrates. This reaction provides a feasible pathway for generation of the carbohydrates necessary for metabolism and genetics (Gabel and Ponnamperuma, 1967; Reid and Orgel, 1967). This formose reaction, however, suffers from the fatal drawback that the carbohydrate products decompose under the required basic conditions. Lambert and coworkers (2010) demonstrated that the formose reaction in the presence of sodium silicate, however, provides stable sugar products. The sugars, upon formation, react spontaneously with sodium silicate to form sugar silicates, which are stable to base. These authors provided spectroscopic evidence that these sugar complexes possess Si—O—C bonds. Moreover, the species have low volatility. In this silicate-mediated formose reaction, sugars are sequestered as silicates according to structure and stereochemistry. These experiments provided a model whereby silicon, exploiting the strength of the Si—O bond, can guide the synthesis of complex carbohydrates as stable, nonvolatile products. Volatility could be important in the context of an astronomical body lacking an atmosphere.

Another drawback to the formose reaction as the prebiotic process to form carbohydrates is the need for conditions of strong base. The reaction of sugars with silicates requires a pH of about 10. Although extreme environments might exist that could allow the reaction, they are rare and an alternative form of catalysis would be better. Lambert and coworkers also have shown that natural clays can provide the same role as strong base. Figure 7.1 demonstrates that the yield of carbohydrates under base-free conditions increases with temperature in the presence of many clays but is negligible in the absence of clay (points with dark green circles at all temperatures). Thus, the aluminosilicates of clay might have played a role in the prebiotic synthesis of carbohydrates.

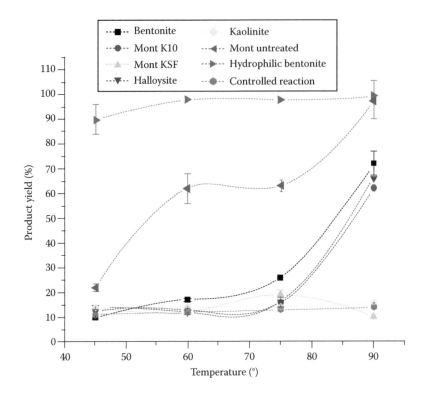

FIGURE 7.1 The total product yield of sugars as a function of temperature (°C) for a variety of clay catalysts listed in the insert at top right. The control reaction, lacking any clay catalyst and represented by circular points in olive drab, is found to provide a level yield of about 10%. Increases above this level indicate the presence of clay catalysis.

7.4 SUMMARY

Because of its structural limitations, silicon cannot be the primary atomic basis for an alternative biochemistry needed as the basis of life. First, almost all bonds to silicon are unstable to acid and base. Second, unsaturated bonds involving silicon are of high energy. Third, the silicon analogue of carbon dioxide does not exist. These insurmountable drawbacks preclude silicon-based life. Nonetheless, silicon can assist in carbon-based life processes (CIBA Foundation Symposium, 2007; Epstein, 2007). The highly stable Si—O bond dominates most of the rocks of terrestrial planets and moons, provides structural materials for organism such as sponges and horsetails, and may be involved in the abiotic creation of carbon-based biological molecules such as carbohydrates.

GLOSSARY

Bond energy: The heat required to break 1 mol of a molecule into its individual atoms. It is a measure of the strength of a chemical bond and is related to the distance between the two atoms composing the bond (the bond length). The units may be kcal mol^{-1} or kJ mol^{-1}.

Carbonyl group: A molecular unit consisting of a carbon atom doubly bonded to an oxygen atom. The carbon atom carries two other substituents, but the oxygen atom has no further bonds (>C=O). The carbonyl group is the integral component of aldehydes, ketones, carboxylic acids, esters, and other organic functionalities.

Covalent bond: The chemical bond between two atoms that involves sharing of electron pairs, thus enabling the constituent atoms to achieve a stable octet of electrons. The result is achieved by overlap of electronic orbitals on the atoms through a balance of attractive and repulsive forces.

Heteroatoms: In organic compounds, heavy atoms other than carbon, usually oxygen, nitrogen, sulfur, phosphorus, and halogens.

Ionic bond: The chemical bond between two atoms that results primarily from the electrostatic attraction between ions of opposite charge. The positive charge of the cation attracts the negative charge of the anion.

Lone pair: A pair of electrons on an atom that is necessary to achieve the stable octet but is not involved in a covalent bond with another atom. Typically in neutral systems, nitrogen has one lone pair, oxygen two lone pairs, and halogens three lone pairs. Also called nonbonding electrons.

Phytolith: Microscopic plant remains composed of silica, derived from siliceous deposits within intracellular and extracellular structures. Phytoliths are revealed and deposited during the natural decay of the organic portions of the plant.

Pi bond: A covalent bond between adjacent atoms involving only p (pi) orbitals. Because of the shape of π orbitals (**13**), there is a nodal plane of zero electron density that passes through both constituent atoms. The geometry is a side-by-side overlap that is weaker than sigma overlap.

Saturation: Characteristic of a molecule or molecular component that contains only sigma bonds.

Sigma bond: A covalent bond between adjacent atoms involving only orbitals with s (sigma) character. The overlap is head on between orbitals and is stronger than the π overlap. The overlap can be between pure s orbitals (**15**) or between mixtures (hybrids) of s and p orbitals (**16**) (the lines between atoms in these two figures represent orbital overlap rather than chemical bonds).

15	**16**

Silacatein: An enzyme used by marine organisms such as sponges to take up silica and build silica support systems.

Silicone: Polymeric materials consisting of long chains of alternating oxygen and silicon atoms, $[-(R_2Si)-O-]_n$, in which the R side groups may be alkyl such as methyl, ethyl, phenyl, or others.

Silsesquioxane: A compound with the empirical formal $RSiO_{1.5}$, in which R is an organic group. In practice, the compounds form complex cages composed of silicon atoms, each of which is bonded to three oxygen atoms and one organic group.

Thermodynamic sink: The informal term for a particularly stable reaction product. The analogy is the ability of water to flow to its lowest available point. Interconverting molecules eventually favor the forms with the lowest energy, the thermodynamic sink.

Unsaturation: Characteristic of a molecule or molecular component that contains π bonds. In a more general sense, unsaturation corresponds to the presence of multiple bonds and rings in a molecule.

Valence: The number of bonds an atom can accept.

REVIEW QUESTIONS

1. What is the normal valency (number of bonds) for neutral carbon? For neutral sulfur? For neutral nitrogen?

2. Why does saturated carbon have a tetrahedral arrangement of bonded atoms around it, as in Figure 7.1?

3. What is the difference between a σ and a π bond?

4. Give examples of double bonds in the principal molecules of life (proteins, carbohydrates, nucleic acids)?

5. Are Si–Si or C–C bonds generally stronger? Which is more reactive?

6. What organic compounds containing silicon have been found in living systems?

7. Why are silicate minerals so common on Earth?

8. What generally are the arguments against the existence of silicon-based life forms?

REFERENCES

General introduction to organosilicon chemistry

Rappoport, Z. and Y. Apeloig, eds. *The Chemistry of Organic Silicon Compounds*. Chichester, U.K.: John Wiley, Vols. 1 and 2, 1998; Vol. 3, 2003. Chapters 1 and 2 provide general introduction to structure and tables of bond energies.

Stability of the silicon–oxygen bond

Pauling, L. 1980. The nature of silicon–oxygen bonds. *American Mineralogist* 65: 321–323.

Molecules in space

Pascoli, G. and M. Comeau. 1995. Silicon carbide in circumstellar environment. *Astrophysics and Space Science* 226: 149–163.

Ziurys, L. M. 2006. The chemistry in circumstellar envelopes of evolved stars: Following the origin of the elements to the origin of life. *Proceedings of the National Academy of Sciences* 103: 12274–12279.

Silica in plants

Epstein, E. 1999. Silicon. *Annual Review of Plant Physiology and Plant Molecular Biology* 50: 641–664.

Liu, N., K. Huo, M. T. McDowell, J. Zhao, and Y. Cui. 2013. Rice husks as a sustainable source of nanostructured silicon for high performance Li-ion battery anodes. *Nature Scientific Reports* 3: Article number 1919.

Pryer, K. M., E. Schuettpelz, P. G. Wolf, H. Schneider, A. R. Smith, and Cranfill, R. 2004. Phylogeny and evolution of ferns (monilophytes) with a focus on the early leptosporangiate divergences. *American Journal of Botany* 91: 1582–1598.

Phytoliths

Fritz, G. J. 2005. Paleoethnobotanical methods and applications. In: Maschner, H. D. G. and C. Chippindale, eds. *Handbook of Archaeological Methods*. Lanham, MD: Altamira Press, Vol. II, pp. 816–819.

Piperno, D. R. 2005. *Phytolith Analysis: An Archaeological and Geographical Perspective*. Lanham, MD: Altamira Press.

Silicateins

Müller, W. E., H. C. Schröder, Z. Burghard, D. Pisignano, and X. Wang. 2013. Silicateins—A novel paradigm in bioinorganic chemistry: Enzymatic synthesis of inorganic polymeric silica. *Chemistry* 19: 5790–5804.

Shimizu, K., J. Cha, G. D. Stucky, and D. E. Morse. 1998. Silicatein alpha: Cathepsin L-like protein in sponge biosilica. *Proceedings of the National Academy of Sciences of the United States of America* 95: 6234–6238.

Origin of life

Cairns-Smith, A. and A. Graham. 1990. *Seven Clues to the Origin of Life*. Cambridge, U.K.: Cambridge University Press.

Gabel, N. W. and C. Ponnamperuma. 1967. Model for origin of monosaccharides. *Nature* 216: 453–455.

Lambert, J. B., S. A. Gurusamy-Thangavelu, and K. Ma. 2010. The silicate-mediated formose reaction: Bottom-up synthesis of sugar silicates. *Science* 327: 984–986.

Miller, S. L. 1953. Production of amino acids under possible primitive Earth conditions. *Science* 117: 528–529.

Miller, S. L. and H. C. Urey. 1959. Organic compound synthesis on the primitive Earth. *Science* 130: 245–251.

Reid C. and L. E. Orgel. 1967. Model for origin of monosaccharides: Synthesis of sugars in potentially prebiotic conditions. *Nature* 216: 455–457.

Silicon biochemistry

CIBA Foundation Symposium. 2007. *Silicon Biochemistry*. Novartis Foundation Symposia No. 256.

Epstein, E. 2007. The anomaly of silicon in plant biology. *Annals of Botany* 100: 1383–1389.

CHAPTER 8

Fossil Records for Early Life on Earth

David Wacey

CONTENTS

8.1 INTRODUCTION

The search for life elsewhere in our universe and the task of ratifying the evidence collected in this search must be informed by data gathered from the rock record here on Earth. The return of rock samples from Mars, or even Saturn's moon Titan or Jupiter's moon Europa, is getting ever closer, so clear criteria for unambiguously identifying life in such samples are now required. Earth's early fossil record provides our best chance of doing this. However, there are numerous challenges associated with the study and interpretation of this early fossil record. First, we must identify suitable targets where life may have been preserved, that is, rocks that were not only deposited in a setting habitable for life but have also largely escaped postdepositional contamination, plus metamorphic heating and deformation that would alter or destroy former signs of life. Second, Earth's earliest life would have been very small and only subtly different from prebiotic or co-occurring abiotic artifacts. Hence, specialized techniques must be developed and employed in the study of this earliest potential fossil record. Third, the scientific community needs to reach a consensus, using well-defined and robust criteria, on what is and what is not a sign of life in the early rock record.

This chapter provides some guidelines for those wishing to study Earth's earliest life, together with a summary of what we currently know about life on Earth in excess of 3 billion years ago. It will become clear to the reader that not all of the challenges and questions have been resolved; however, great advances have been made over the last decade or so, and a consensus is emerging that life was abundant and diverse in multiple environments back to about 3.5 billion years ago.

8.2 WHAT ARE THE TYPES OF EVIDENCE FOR EARLY LIFE ON EARTH?

There are three main ways that Earth's earliest life may reveal itself in the rock record: body fossils, trace fossils, and chemical fossils.

Body fossils are morphological remains of an organism preserved within a rock. Body fossils can be of varied form, for example, bones or shells. But in the early Archean, organisms with hard body parts did not exist, nor did complex cells with nuclei, so we may only expect to find very simple and very small (a few micrometers [μm] to a few tens of μm) microbial cells. Such small and delicate structures are prone to decay so one might expect the potential for them to be preserved for over billions of years in the rock record to be very low. This is the case in general, but in some circumstances, even delicate cells can be remarkably preserved (Figure 8.1). Remarkable preservation usually occurs where organisms are rapidly encased in a mineral or rapidly buried to prevent oxidative degradation. The lack of oxygen in the early Archean may fortuitously have facilitated a relatively greater fidelity of microbial preservation compared to the younger rock record. The quality of preservation also depends on which mineral first comes in contact with the organism, with silica (SiO_2), pyrite (FeS_2), calcium phosphate ($CaPO_4$), and some clay minerals being particularly good preservation media for microfossils. The subsequent geological history of the rock is also important in the search for early life, with good preservation favored by rapid occlusion of sediment pore space, followed by minimal postdepositional deformation and metamorphism that would likely destroy delicate cellular body fossils. Certain parts of cells are also more likely to survive in the rock record than others. For example, RNA and DNA molecules, plus other cell contents, are unstable and will be rapidly degraded. The cell membrane is more stable, being made of phospholipids and fatty acids, but is still rather weak so has a rather low chance of preservation. The cell wall, made of stronger peptidoglycan polymer, probably has the best chance of preservation. Extracellular polymeric mucus–like substances that often form envelopes around cells also have a reasonable chance of being preserved, since they contain functional groups (e.g., carboxyl) on which mineral ions can rapidly nucleate.

Trace fossils are non-body remains that in the broadest sense indicate the activity of an organism. An obvious example would be dinosaur footprints. Early Archean trace fossils are much smaller and less complex and are difficult to unambiguously attribute to a

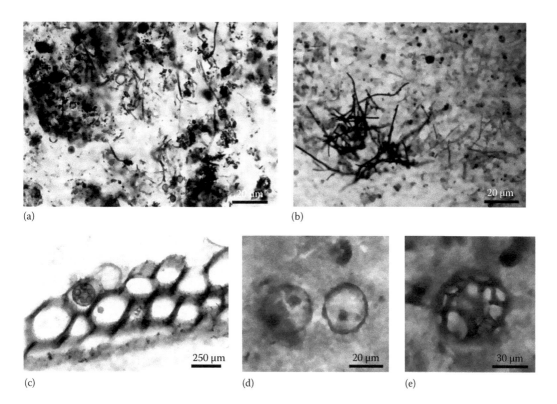

FIGURE 8.1   Examples of remarkable body fossil preservation in the rock record. (a) Carbonaceous coccoid and filamentous bacterial body fossils preserved in chert from the 1900 Ma Gunflint Formation, Canada. (b) Carbonaceous filamentous bacteria replicated by pyrite (black) and preserved in chert from the 1900 Ma Gunflint Formation, Canada. (Modified from Wacey, D. et al., *Proc. Natl. Acad. Sci. USA*, 110, 8020, 2013b.) (c) Plant cells preserved in chert from the 410 Ma Rhynie Chert, Scotland. (d, e) Probable eukaryote cells preserved in phosphate from the 1000 Ma Torridon Group, Scotland. (e: Modified from Wacey, D., In situ morphologic, elemental and isotopic analysis of Archean life, in: Dilek, Y. and Furnes, H., eds., *Evolution of Archean Crust and Early Life, Modern Approaches in Solid Earth Sciences*, Vol. 7, Springer, Dordrecht, the Netherlands, pp. 351–365, 2013.)

biological maker. They are commonly simple, µm-sized cavities left in rocks. These could be molds of cells found within mineral grains, or they could be tubular microbial borings resulting from an organism actively penetrating a solid substrate in search of nutrients or protection (Figure 8.2a). It is difficult to prove that such small and simple traces were formed by an organism but clues may come from the replication of cellular morphology by the trace, the presence of organic material within the trace, and the distribution of a suite of trace fossils (e.g., do they show preference for a particular rock or mineral that may have been rich in nutrients).

Structures known as ambient inclusion trails (AITs) are also microtubular in nature and can closely resemble tubular microbial borings (Figure 8.2b). However, they possess a specific set of characteristics that permit their differentiation. These include the following: the presence of a mineral crystal at the end of the microtube that appears to have

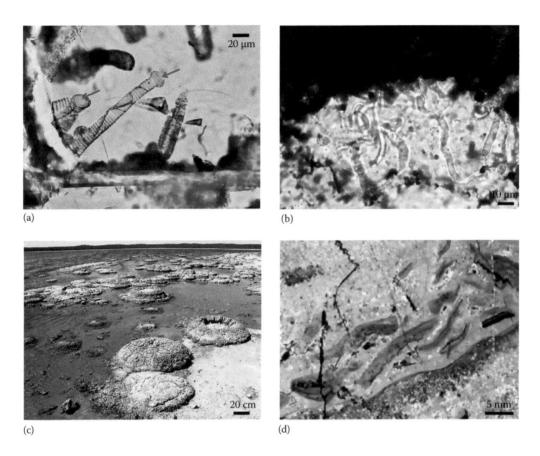

FIGURE 8.2 Examples of trace fossils from the rock record. (a) Tubular microbial borings into volcanic glass (two examples arrowed) from the 92 Ma Troodos Ophiolite, Cyprus. (Modified from Wacey, D. et al., *Chem. Geol.*, 363, 1, 2013a.) (b) AITs from the 1900 Ma Gunflint Formation, Canada. *Note*: terminal pyrite grain (arrowed) that has been propelled along one microtube. (c) Modern domal stromatolites from Lake Thetis, Western Australia. (Modified from Wacey, D., *Astrobiology*, 10, 381, 2010.) (d) Fragments of microbial mat that have been ripped up by currents from a sedimentary surface (rolled up example is arrowed). This is one type of MISS. Material is from the 3480 Ma Dresser Formation, Western Australia. (Courtesy of N. Noffke; Modified from Noffke, N. et al., *Astrobiology*, 13(12), 1103, 2013.)

been propelled along the tube; the presence of striations along the edge of the microtube caused by movement of the aforementioned mineral crystal; microtubes may crosscut one another, make sharp changes in direction, and form branches of different diameters, caused by fragmentation of the propelled crystal or interception of a second mineral grain; and an AIT is more than often polygonal in cross section, since this is the common shape of the propelled crystal. There is still some doubt about whether biology is needed to create AIT, so they are best classed as only possible trace fossils. A biological formation mechanism involves the degassing of decomposing organic material within an impermeable host sediment or rock material, with crystal movement caused by increased gas pressure and dissolution of the host rock by organic acids (Knoll and Barghoorn 1974).

Stromatolites are commonly cited as evidence for Earth's earliest life (Figure 8.2c). Modern stromatolites, such as those found in Shark Bay or Lake Thetis, Western Australia, are clearly formed by the interaction of biology with sedimentation and can be classed as trace fossils, following the definition of Awramik et al. (1976) "...an organo-sedimentary structure produced by sediment trapping, binding, and/or precipitation as a result of the growth and metabolic activity of microorganisms." In the ancient rock record, however, the biological nature of stromatolites is much less certain. Indeed, in early life studies, the term stromatolite should have a nongenetic definition "...an attached, laminated, lithified sedimentary growth structure, accretionary away from a point or limited surface of initiation" (Semikhatov et al. 1979) until the presence of biology is proven. Abiotic mimics of true biological stromatolites include soft sediment deformation, later structural deformation, and chemical precipitation of stromatolite-like crusts. Recently, stromatolites have even been produced in the laboratory without the aid of biology, using a simple spray paint mechanism (McLoughlin et al. 2008).

A further group of trace fossils are microbially induced sedimentary structures (MISS). MISS tend to occur in siliciclastic sediments (e.g., sandstones) where sediment deposition and preservation is modified by large communities of mat-forming microorganisms (Figure 8.2d). MISS may be macroscopic in nature; examples include sedimentary bedding planes with leveled ripple marks, wrinkles, microbial mat chips, mat curls, and shrinkage cracks. They may also be microscopic in nature, observed in petrographic thin sections; these include sponge pore fabrics, gas domes, fenestrae, sinoidal laminae, and oriented grains that appear to float in a carbon-rich mat matrix (see Gerdes et al. 2000 for examples of all these features). These are an intriguing group of structures, often with many types of MISS occurring in close proximity to one another. The oldest examples so far reported come from the 3200 Ma Moodies Group of South Africa (Noffke et al. 2006) and potentially the 3480 Ma Dresser Formation of Western Australia (Noffke et al. 2013).

Chemical fossils are traces of biological activity indicated by specific chemical signals left in rocks including isotopic variations in, for example, carbon, sulfur, nitrogen, or iron; distinctive ratios of elements; or molecular compounds that may be tied to a particular group of organisms. Although chemical fossils have no morphological expression, some may be more stable through geological time than the organism that created them, and thus some are the most common life signals in early Archean rocks. They are also highly debated, because it is difficult to unambiguously attribute a chemical signal to life in preference to some abiotic chemical reaction. The two most common chemical fossils in the early Archean are ^{12}C-enriched carbon isotope signals found in organic matter and ^{32}S-enriched sulfur isotope signals found both in sulfide minerals such as pyrite and, more rarely, also in organic matter. Other chemical fossils could include highly localized concentrations of biologically important elements such as nitrogen or phosphorus and trace elements such as zinc that play key roles in biological enzymes, but these will be equally as difficult to attribute unambiguously to life.

8.3 PROBLEM OF NONBIOLOGICAL ARTIFACTS

Body fossils, trace fossils, and chemical fossils can all be mimicked by nonbiological artifacts and this has led to cases of mistaken identity and great controversy in studies of early life on Earth. Microbial body fossils comprise simple shapes such as spheres, rods, and filaments that are often only a few μm in size. These are difficult to distinguish from some mineral crystals that grow with distinct spheroidal (e.g., some forms of silica) or filamentous (e.g., numerous silicate minerals) habits. This is especially difficult when carbon has been remobilized in a rock, for example, by a hydrothermal or metamorphic fluid, and redistributed along mineral grain boundaries. In such a case, a ring of carbon around a spheroidal silica grain will look almost identical to a true cell. Even more complex curved and helical filamentous microfossil-like objects have been synthesized in the laboratory without the need for biology (Garcia Ruiz et al. 2003). To reject such nonbiological mimics requires their study at the highest possible spatial resolution, often looking at the sub-μm scale to see how carbon is distributed in relation to mineral crystal boundaries and investigate the ultrastructural features of the candidate cell wall.

Likewise, microbial impressions and trace fossils can be mimicked by simple chemical and physical erosion and dissolution processes, creating cavities similar in size and shape to those left by microbes. Chemical fossils can be mimicked by nonbiological chemical reactions. For example, a light (^{12}C-enriched) carbon isotope signature may be caused by biological processing of carbon but may also be caused by specific hydrothermal reactions such as Fischer–Tropsch-type synthesis (McCollom and Seewald 2006).

A long list of *biogenicity criteria* has been developed over the past 30 years or so to give researchers the best possible chance of differentiating a true biological signal from a nonbiological artifact. These criteria are not infallible but close adherence should prevent most cases of mistaken identity. These criteria are summarized from Brasier et al. (2004, 2005), Buick (1990), Hofmann (2004), McLoughlin et al. (2007), Schopf and Walter (1983), Sugitani et al. (2007), and Wacey (2009):

a. Structures should exhibit biological morphology that can be related to extant cells, sheaths, traces of activity, or waste products. Ideally, life cycle variants should be identifiable (reproductive stages), comparable to that found in morphologically similar modern or fossil microorganisms.

b. More than a single step of biology-like processing should be evident. These steps may take the form of biominerals (e.g., some forms of pyrite), geochemical fractionations of isotopes (e.g., carbon and sulfur), specific organic compounds (e.g., hopanoid biomarkers), or distinctive elemental ratios.

c. Structures should occur within a geological context that is plausible for life, that is, at temperatures and pressures that modern organisms are known to survive.

d. Structures should fit within a plausible evolutionary context.

e. Structures should be abundant and ideally occur in a multicomponent assemblage.

f. Following from e, ideally, they should show colonial/community behavior.

g. Following from f, a preferred orientation indicating a role in the formation of biofabrics would be an additional bonus criterion.

h. Microfossils should ideally be composed of kerogenous carbon. However, if mineralized, this should be a result of microbially mediated precipitation. Later mineral replacement of carbonaceous material may also be permissible but then doubts upon their age will be raised.

i. Microfossils should be largely hollow. Cell walls and sheaths are by far the most likely parts of the microbe to be preserved; cellular constituents are rarely preserved in more modern examples.

j. Ideally, the microfossils should show some sort of cellular elaboration, for example, not just smooth cell walls, although this may be difficult to detect.

k. Microfossils should show taphonomic degradation, that is, collapse of cells, folding of films, and fracturing.

l. The object must exceed the minimum size for independently viable cells (~0.25 μm diameter). Note: This criterion is not applicable to fossil viruses.

m. Microfossils should be demonstrably dissimilar from potentially coexisting nonbiological organic bodies (e.g., spherulitic mineral coatings) and should occupy a restricted biological morphospace.

n. Evidence of extracellular polymeric substances surrounding the putative microfossils would be an added bonus criterion.

o. Trace fossils should show preferential exploitation of certain substrates or horizons, for example, those that are rich in trace metals utilized by microbial metabolisms or those that contain structural defects and weaknesses that facilitate microboring.

p. Trace fossils should show enrichments of biologically important elements, for example, carbon- and/or nitrogen-enriched linings or biomineral infillings.

q. Endolithic microborings should show preferred growth orientations; that is, they should penetrate from the outside of a grain and grow inwards and may also cluster on one side of the grain.

r. Trace fossils should be demonstrably dissimilar from coexisting nonbiological etch pits and cracks; that is, they should be circular to elliptical in cross section and be of a restricted range of diameters.

s. If trace fossils are branched, ideally, there should not be a change of diameter of the structure at the branching point.

t. The area immediately surrounding a trace fossil should ideally show depletion in biologically important elements.

8.4 PROBLEM OF POSTDEPOSITIONAL CONTAMINATION

Rocks containing the earliest signs of life have had 3–3.5 billion years to be modified and contaminated with more recent biological material. The three most common ways that this can occur are transport of biological material into rock fractures and pore spaces by later fluids, for example, during periods of metamorphism or even percolating modern groundwater; later microbes actively boring into rocks in search of nutrients or protection; and human error, whereby biological material is introduced during sample processing. Careful observation and adherence to the antiquity criteria outlined in the following is necessary to navigate the minefield of most postdepositional contaminants. These antiquity criteria have been developed by many researchers over the last three decades and are here summarized from Brasier et al. (2004, 2005), Buick (1990), Hofmann (2004), McLoughlin et al. (2007), Schopf and Walter (1983), Sugitani et al. (2007), Wacey (2009), and Westall and Folk (2003):

a. Structures must occur in rocks of known provenance; that is, detailed location information must be presented so that independent resampling is possible.

b. Structures must occur in rocks of demonstrable or established (Archean) age; that is, the host rock must be dated directly by radiometric techniques (Section 8.5.1) or the age of the rocks can be accurately inferred by correlation to nearby rocks that have been dated.

c. Structures must be indigenous to the primary fabric of the host rock; that is, they must be physically embedded within the rock, not products of sample collection or preparation. They should, therefore, be present in petrographic thin sections of the rock.

d. Structures must be syngenetic with the primary fabric of the host rock; that is, they must not have been introduced by ancient or modern postdepositional fluids.

e. Following from d, any structures found within metastable mineral phases, void filling cements, veins, or crosscutting fabrics must be viewed with extreme caution.

f. Structures should not occur in high-grade metamorphic rocks, because delicate organic structures rarely survive these extremes of pressure and/or temperature; the likelihood of nonbiological artifacts in such rocks is substantially increased.

g. The geological context of the host rock must be fully understood at a range of scales; that is, the host unit must show geographical extent and fit logically within the regional geological history.

h. Specific to microfossils—potential microfossils should not be significantly different in color from that of other particulate carbonaceous material in the remainder of the rock matrix. For example, brown *microfossils* in a largely black carbonaceous rock would immediately be suspicious.

i. There should be evidence for organosedimentary interaction, for example, sediment grains trapped or supported by fossils, coatings of distinctive composition or texture precipitated around the fossils, or perhaps alternating layers of prostrate and erect filaments in stromatolite-like sediments.

j. Specific to trace fossils—trace fossils should be concentrated in detrital grains or primary rock matrix, not around later conduits for fluid or microbial entry. They should be crosscut by later stage veins and fractures (if present) and should be filled with a mineral phase that is capable of surviving the history of burial and heating that the rock unit is known to have endured.

8.5 WHAT TECHNIQUES SHOULD BE USED TO SEARCH FOR EARTH'S EARLIEST LIFE?

A wide selection of techniques is required to study ancient rocks for signs of early life. These range from investigation at the kilometer to meter scale in order to establish the geological context of the rock down to investigation at the μm to nanometer (nm) scale to investigate the morphology and chemistry of a single candidate cell. Here, I summarize the most commonly applied techniques in this field.

8.5.1 Geological Mapping

The geological context of samples and their spatial relationships are of critical importance in the study of early life. The context for potential biological signals should be mapped and studied at a wide range of scales from kilometers to millimeters. Evidence for, and interpretation of, the context should be clearly separated. Any potential biological signals should be referable to a well-defined history within this context, to show if they are early or late, indigenous or exogenous. Plausibility of the context for early life can then be assessed.

As part of the geological mapping, *radiometric dating* may be employed to give an absolute age for the rock being studied. This relies on the radioactive decay of an unstable isotope (parent nuclide) to a stable form (daughter nuclide); the relative proportions of each of these isotopes in the rock, together with the experimentally determined *half-life* of the decay reaction can then be used to date the mineral in which these isotopes are found. For early Archean rocks, parent–daughter nuclide pairs must have a relatively long *half-life*; otherwise, all of the parent isotope would have decayed long ago. Common isotopic systems for dating early Archean rocks are uranium–lead (^{238}U–^{235}U–Pb) found in the mineral zircon ($ZrSiO_4$), potassium–argon (^{40}K–Ar) found in minerals such as feldspar and hornblende, and rubidium–strontium (^{87}Rb–Sr) found in several minerals within granites. The age of a given mineral is calculated by substituting the data acquired into the age equation

$$t = 1/\lambda \ln (1 + D/P)$$

where
t is the age of the sample
D is the number of atoms of the daughter nuclide in the sample
P is the number of atoms of the parent nuclide in the sample
λ is the decay constant of the parent isotope (inversely proportional to the *half-life*)

This age can be as accurate as ±2 Ma for a mineral that is 3000 Ma, using the U–Pb system, but may be as poor as ±30 Ma for a similarly aged mineral using the Rb–Sr system. Not all rock types can be dated in this way because they may not contain suitable minerals.

This is indeed the case for most sedimentary rocks that contain evidence for early life. In these cases, one must use geological relationships obtained from mapping to constrain the age of the candidate rock in between two rock strata that have known radiometric ages. This can vastly increase the error on the age of a candidate rock and many early Archean examples can only be dated to within ±50–100 Ma.

8.5.2 Optical Microscopy

The next logical step in the investigation of a candidate rock sample for signs of life is to prepare thin sections of the rock that can be examined using an optical microscope. Optical microscopy (also known as petrography) has been, and continues to be, the primary screening tool of the Precambrian paleontologist. Authentic Precambrian microfossils must be observed by optical microscopy in petrographic thin sections (Figure 8.3a). Optical microscopy shows the morphology of the candidate structure and is also an essential extension of geological mapping because it allows the context of any structure to be mapped on the mm to μm scale. Various types of equipment can be used to enhance a standard optical microscope to obtain the maximum information from a given sample using plane polarized transmitted light, cross polarized transmitted light, reflected light, cathodoluminescence, lasers, and various digital imaging and processing packages.

8.5.3 Laser Raman Microspectroscopy

Laser Raman (LR) uses laser excitation of a candidate sample to obtain molecular and structural data and is mainly used to identify minerals and to inform on the crystallinity and bonding of organic material (see Fries and Steele 2011 for details). It is a nondestructive technique meaning that it is often one of the first to be applied to a candidate sign of life after investigation by optical microscopy. LR requires no additional sample preparation with good results being obtained from standard petrographic thin sections. It operates at similar magnifications to optical microscopy so it can provide μm-scale information; indeed, LR systems are usually attached to a standard optical microscope body so that optical and Raman data can be superimposed on one another. Raman data come in two main types: point spectra and maps. With point spectra, the laser is targeted at a specific point within a mineral; when the chemical bonds of the mineral are excited by the laser, a spectrum of peaks is produced and the positions and intensities of these peaks differ from mineral to mineral. The unknown spectrum is compared to a database (e.g., RUFF database of Downs 2006) and minerals can be rapidly identified. The widths and intensities of Raman peaks can also alter depending on the crystallinity or order of a mineral phase; this feature has been used extensively to investigate the structure of carbon in ancient rocks and has been useful in decoding the relative age of carbon compared to the known age and metamorphic history of the host rock (e.g., Pasteris and Wopenka 2003; Tice et al. 2004). This often satisfies antiquity criteria for a candidate microfossil or other organic structure. However, LR cannot distinguish biological and nonbiological carbon. In mapping mode, Raman spectra are obtained from multiple points within a sample and combined to form maps with μm-scale spatial resolution (Figure 8.3b). These maps could show the spatial distribution of a certain mineral or of a specific Raman

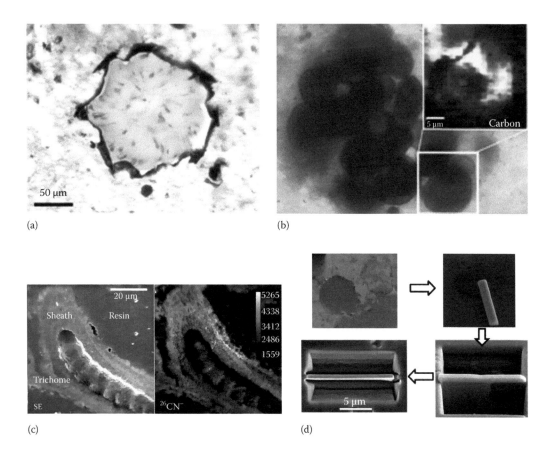

FIGURE 8.3 Examples of the techniques commonly applied to studies of early life on Earth. (a) Optical microscopy showing exquisite preservation of large, probable eukaryote cell in phosphate from the 1000 Ma Torridon Group, Scotland. (Photo courtesy of M. Brasier.) (b) LR microspectroscopy demonstrating the carbonaceous nature of a cell from the 1000 Ma Torridon Group. (Reproduced from Wacey, D., In situ, high spatial resolution techniques in the search for the origin of life, in: Seckbach, J., ed., *Genesis—In The Beginning. Cellular Origin, Life in Extreme Habitats and Astrobiology*, Vol. 22, Springer, Dordrecht, the Netherlands, pp. 391–411, 2012. With permission.) (c) NanoSIMS ion images demonstrating the high level of carbon and nitrogen within a modern stromatolite-building cyanobacterium from Lake Clifton, Western Australia. (Modified from Wacey, D. et al., *Geobiology*, 8, 403, 2010b.) (d) Series of images demonstrating the FIB sample preparation process. Features of interest are located in standard petrographic thin sections (top left, rounded cell-like feature). A protective platinum strip is deposited across the area of interest prior to FIB milling (top right). Trenches are milled either side of the area of interest and the user can check that the feature continues below the sample surface (bottom right). Using reduced beam currents, the wafer is gradually thinned and cleaned to ~100 nm (bottom left). The wafer can then be removed from the sample for subsequent TEM analysis while retaining the context of the feature of interest. (Reproduced from Wacey, D., In situ, high spatial resolution techniques in the search for the origin of life, in: Seckbach, J., ed., *Genesis—In The Beginning. Cellular Origin, Life in Extreme Habitats and Astrobiology*, Vol. 22, Springer, Dordrecht, the Netherlands, pp. 391–411, 2012. With permission.)

spectral peak (e.g., to show how the crystallinity of carbon changes over a given area). 3D data are also obtainable by creating maps with the laser focused at various depths within a thin section and then stacking them together using a 3D software package (e.g., McKeegan et al. 2007). Finally, Raman can be used to look at the orientation of minerals since many minerals produce spectral peaks that vary in intensity depending on their crystallographic orientation to the laser. This feature can be used, for example, to image the distribution of quartz crystallographic axes to see whether candidate biological material occurs between grain boundaries, is enclosed by entire grains, or occurs in cracks. Examples of the techniques commonly applied to studies of early life on Earth are shown in Figure 8.3.

8.5.4 Secondary Ion Mass Spectrometry

Secondary ion mass spectrometry (SIMS) can determine the elemental, isotopic, or molecular composition of a sample (see Ireland 1995; McKibben et al. 1998 for details). It is a destructive surface analysis technique whereby a sample is bombarded with ions, causing secondary ions from the sample surface to be ejected; these are then identified based upon their mass, using a mass spectrometer. *Large radius SIMS* instruments are used to obtain high-precision isotopic data from objects as small as about 10 μm in diameter. This may be employed to date minerals by measuring the amounts of, for example, U and Pb in a sample (see radiometric dating) or may be used to measure the proportions of stable isotopes such as ^{12}C versus ^{13}C or ^{32}S versus ^{34}S that may in turn point to particular types of biological metabolisms. These stable isotopes of biologically important elements are particularly useful in studies of early life and SIMS allows such data to be obtained directly from microfossils, carbonaceous laminations within microbial mats or stromatolites, and potential biominerals (e.g., pyrite). This is preferential to bulk analyses (e.g., from a powdered rock), which may suffer from contamination and/or isotopic homogenization during sample processing, preparation, and analysis. *NanoSIMS* instruments have even better spatial resolution than large radius SIMS and can obtain isotopic data from objects as small as about 2 μm; however, the precision of the data is much poorer. Hence, it is up to the individual to weigh up whether spatial resolution or precision is more important, before deciding which type of SIMS to use. NanoSIMS is also an excellent tool for chemical mapping because it is highly sensitive to elements that may be present only in very low concentrations, can measure up to seven elements at once, and has a spatial resolution of about 100 nm. This is especially useful for mapping biologically important elements such as nitrogen that might only be present in a few parts per million and only occur over a few hundred nm in ancient microfossils (Figure 8.3c).

8.5.5 Transmission Electron Microscopy

Transmission electron microscopy (TEM) is a versatile technique that can provide both images and chemical and structural information from a sample at the nm (and sometimes even atomic) scale. As the name suggests, beams of electron are passed through (transmitted) a sample and data come from the interactions of these electrons with features in the sample. Specific sample preparation is required for TEM; in order to extract the maximum possible information, TEM samples need to be ultrathin (~100 nm). The best way to achieve this is to use a *focused ion beam* (FIB) (Figure 8.3d). This procedure uses a highly focused beam of heavy gallium ions to sputter ions from the sample surface, essentially cutting into the sample with very high precision (see Wirth 2009 for details). Hence, small wafers, typically about 15 µm × 10 µm × 0.1 µm, can be cut directly out of thin sections that have already been characterized using optical microscopy or some other nondestructive technique. These are not only uniformly thin but they retain the context of the object of interest and eliminate the possibility of contamination. A wide range of TEM subtechniques can then be applied to these samples including regular imaging of morphology; high-resolution imaging (HRTEM) that shows the arrangement of atoms in sample, providing information on its crystallinity and lattice structure (Figure 8.4a); electron energy loss spectroscopy (EELS) to provide information on elemental composition, bonding of a given element, and oxidation state of a given element; maps of elemental distributions; and selected area electron diffraction (SAED) that allows the identification of different mineral phases (Figure 8.4b). All of this is done at extremely high spatial resolution, beyond that achieved with any other technique.

8.5.6 Scanning Electron Microscopy

Scanning electron microscopy (SEM) is a highly flexible surface analysis technique being able to analyze a range of samples including thin sections, crushed and powdered rock samples, single mineral crystals, and relatively large rock chips. Both images and compositional data can be attained at a wide range of magnifications (approximately 25× to 250,000×). Samples are scanned with a high-energy beam of electrons, and these interact with the atoms of the sample producing numerous secondary signals (secondary electrons, backscattered electrons, x-rays, and cathodoluminescence). With suitable detectors in place, each of these secondary signals reveals information about the sample. Secondary electrons give very high spatial resolution images of the sample surface, backscattered electrons give qualitative information on sample chemical composition, x-rays give both qualitative and quantitative chemical information, and cathodoluminescence can also be related to specific minerals. In early studies of Archean life, because SEM only images and detects surface features, contamination and artifacts from sample processing and preparation were significant issues.

The use of SEM to investigate early life has been reinvigorated by recent advances in the design of dual-beam FIB–SEM instruments. This now means that a sample can be manipulated and cut (milled) by a focused beam of ions (as in the TEM sample preparation outlined earlier) while simultaneously being imaged using an electron beam. Additional detectors can also be inserted to permit elemental analysis (EDS detector) or phase detection and crystallographic mapping (EBSD detector). FIB–SEM is therefore very useful for looking at what happens below the surface within a thin section or rock chip; a site-specific hole can quickly be milled and then imaged at nm-scale spatial resolution (Figure 8.4c). An exciting

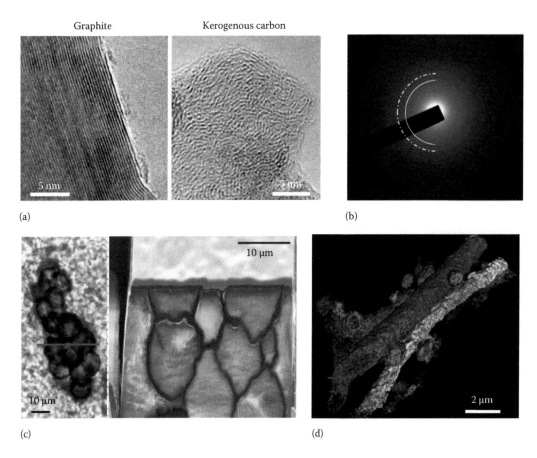

FIGURE 8.4 Examples of the techniques commonly applied to studies of early life on Earth. (a) High-resolution TEM (HRTEM) demonstrating the different patterns and atomic spacings exhibited by different forms of carbon. (Reproduced from Wacey, D., In situ, high spatial resolution techniques in the search for the origin of life, in: Seckbach, J., ed., *Genesis—In The Beginning. Cellular Origin, Life in Extreme Habitats and Astrobiology*, Vol. 22, Springer, Dordrecht, the Netherlands, pp. 391–411, 2012. With permission.) (b) SAED used to identify mineral phases. In this example from a 3400 Ma rock, we see a mixture of sharp rings (traced over in white) corresponding to the atomic spacings in the iron oxide mineral magnetite, plus a diffuse background indicating amorphous (noncrystalline) material. (Reproduced from Wacey, D., In situ, high spatial resolution techniques in the search for the origin of life, in: Seckbach, J., ed., *Genesis—In The Beginning. Cellular Origin, Life in Extreme Habitats and Astrobiology*, Vol. 22, Springer, Dordrecht, the Netherlands, pp. 391–411, 2012. With permission.) (c) FIB milling combined with SEM (FIB–SEM) showing the detailed morphology of a number of cells from the 1000 Ma Torridon Group, Scotland. Here, the red line shows the location of milling, perpendicular to the optical microscopy image (left), while the SEM image (right) shows additional cellular detail that is obscured in the optical image. (Modified from Wacey, D., In situ morphologic, elemental and isotopic analysis of Archean life, in: Dilek, Y. and Furnes, H., eds., *Evolution of Archean Crust and Early Life, Modern Approaches in Solid Earth Sciences*, Vol. 7, Springer, Dordrecht, the Netherlands, pp. 351–365, 2013.) (d) 3D visualization of FIB–SEM data showing a reconstruction of filamentous microbes being decomposed by tiny spheroidal heterotrophic microbes (orange) from the 1900 Ma Gunflint Formation, Canada. (Modified from Wacey, D. et al., *Proc. Natl. Acad. Sci. USA*, 110, 8020, 2013b.)

extension to this is the sequential milling of multiple slices through an object. After each slice is milled, an image is taken or a chemical map is produced, and each of these images or maps is stacked together to give a true 3D representation of the structure of interest (see Wacey et al. 2012; Wirth 2009 for details). The spatial steps between successive FIB slices can be set by the user and can be smaller than 50 nm so that even fine scale detail within microfossils can be captured (Figure 8.4d). It must be noted that this is the most destructive of all the techniques described here, since the structures of interest are actively milled away during the analysis. Hence, it is not particularly suitable for samples containing very rare or *type* microfossils.

8.5.7 Synchrotron Radiation Techniques

Synchrotron radiation provides a source of high-intensity x-rays that can be used for a suite of imaging and spectroscopy techniques at high spatial resolution and high sensitivity. *X-ray tomography* can create stacks of images at sub-μm-scale resolution that can be combined into 3D visualizations. The spatial resolution is slightly inferior to 3D-FIB–SEM but it has the advantage of being nondestructive so it could potentially be used on unique microfossil specimens. *X-ray absorption spectroscopy* allows the bonding, coordination, and oxidation states of an element to be investigated, much like EELS but with better energy resolution. This is particularly useful to investigate the bonding of carbon to see if any biological functional groups (e.g., phenol or carboxylic acid) are present. *X-ray fluorescence* provides semiquantitative elemental maps over flexible spatial scales (mm down to μm) and can be especially useful to obtain a relatively large-scale overview of the chemistry of a sample before isolating a smaller area for more detailed study. Finally, *scanning transmission x-ray microscopy* can be used to obtain both images and spectral (chemical) data at the nm scale.

8.6 WHERE DO WE LOOK FOR THE EARLIEST EXAMPLES OF LIFE ON EARTH?

Rocks from the earliest periods of Earth history are rare. Some have been uplifted into mountain belts and subsequently eroded, while others have been forced below the surface of the Earth by subduction. Of those that remain on the Earth's surface, many have never been habitable for life (e.g., granite formed in magma chambers), and many that were once habitable have been significantly compressed and heated by contact and regional metamorphism so that any previous signs of life have very likely been destroyed. Only two places on Earth are currently known to possess sedimentary and volcanic rocks of early Archean (>3000 million years) age that were habitable for life and have only experienced minimal postdepositional modification. These are the Pilbara Craton of Western Australia and the Kaapvaal Craton of South Africa. Some would argue that small regions of Greenland should also be included here, but these rocks have been deformed to a greater extent and no body or trace fossils have yet been described from them, their only signs of life being controversial chemical fossils.

The *Pilbara Craton* is composed of a mixture of ancient granites and greenstones (i.e., volcanic and sedimentary rocks that have been slightly altered by low-temperature greenschist

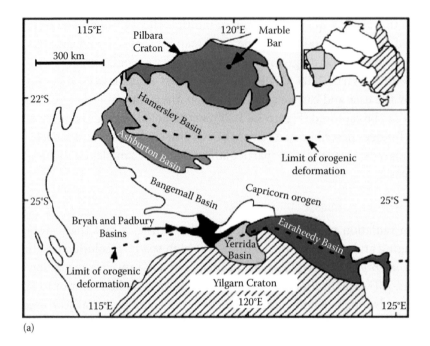

(a)

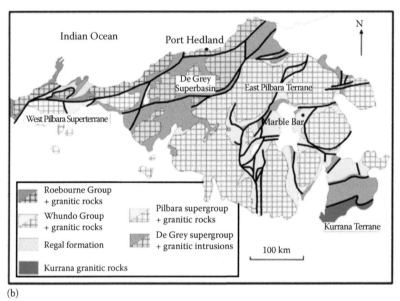

(b)

FIGURE 8.5 (a) Location of the Pilbara Craton, Western Australia. (Reproduced from Brasier, M.D. et al., *Precambrian Res.*, 140, 55, 2005. With permission.) (b) Location of the Pilbara Supergroup (pale green), home to the oldest claims for life in this region. (Reproduced from Wacey, D., *Early Life on Earth: A Practical Guide*, Springer, Amsterdam, the Netherlands, 285p., 2009. With permission.)

(c)

(d)

(e)

FIGURE 8.5 (continued)   (c–e) Typical rock types that host signs of life in the Pilbara Supergroup, including banded black and white chert (c), sandstone with occasional well-preserved black cores (d), and pillow basalt (e). (d: Modified from Wacey, D., et al., The 3426-3350 Ma Strelley Pool Formation in the East Strelley greenstone belt—A field and petrographic guide. *Geological Survey of Western Australia Record*, 10, 2010a.)

facies metamorphism). It is these greenstones that are viable for hosting life and the oldest examples occur in the Pilbara Supergroup, exposed in the East Pilbara Terrane (Figure 8.5; Van Kranendonk 2006). The Pilbara Supergroup is subdivided into four rock packages or groups. The oldest is the Warrawoona Group, deposited from 3515 to 3420 Ma. This consists mostly of dark, mafic volcanic rocks of the double-bar formation, table top formation, North Star Basalt, Mount Ada Basalt, and Apex Basalt. These are interspersed with thin chert horizons, light felsic volcanics and volcaniclastics of the 3515–3500 Ma Coucal Formation, the 3472–3465 Ma Duffer Formation, the ~3460 Ma *Apex Chert*, and the 3458–3427 Ma Panorama Formation. A notable and spatially restricted horizon is the ~3480 Ma Dresser Formation, consisting of bedded chert, sulfate, carbonate, and jasper, together with pillow basalt (Van Kranendonk et al. 2008).

The Warrawoona Group is separated from the overlying Kelly Group by a marker unit, the ca. 3400 Ma Strelley Pool Formation, composed mainly of sandstone and

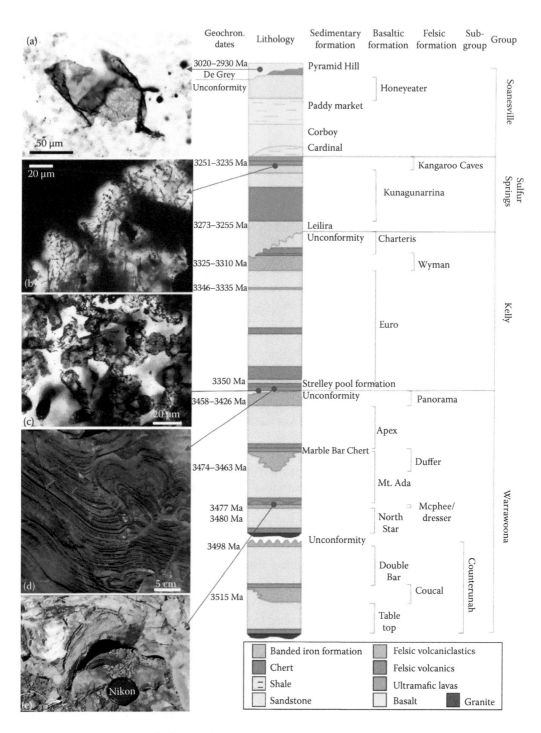

FIGURE 8.6 (see caption on facing page)

chert (Wacey et al. 2010a). The Kelly Group is mostly volcanic, save for a few thin chert horizons, containing the 3350–3325 Ma Euro Basalt and the 3325–3315 Ma Wyman Formations. Above this comes the 3270–3240 Ma Sulphur Springs Group, a mixture of sediments, volcanics, and one of the Earth's oldest massive sulfide deposits. Finally, the uppermost package of the Pilbara Supergroup is the Soanesville Group (~3200–3165 Ma), which is mostly sedimentary in nature (Cardinal, Corboy, and Paddy Market Formations) with minor basalt (Honeyeater Basalt) and banded iron formation (Pyramid Hill Formation). The >3000 Ma stratigraphy in the East Pilbara is completed by the lowermost sediments of the De Grey Supergroup, packaged together as the ~3020–2930 Ma Gorge Creek Group.

Signs of life have been reported in (from youngest to oldest; Figure 8.6) the 3000 Ma Farrel Quartzite of the Gorge Creek Group (Grey and Sugitani 2009; Oehler et al. 2009, 2010; Sugitani et al. 2007, 2009a,b), the 3240 Ma Kangaroo Caves Formation of the Sulphur Springs Group (Duck et al. 2007; Rasmussen 2000), the 3350 Ma Euro Basalt of the Kelly Group (Banerjee et al. 2007; Schopf and Packer 1987), the 3400 Ma Strelley Pool Formation (Allwood et al. 2006, 2007, 2009; Brasier et al. 2006; Hofmann et al. 1999; Lowe 1980; Sugitani et al. 2010, 2013; Van Kranendonk et al. 2003; Wacey 2010; Wacey et al. 2008, 2010a, 2011a,b, 2012), the 3450 Ma Panorama Formation of the Warrawoona Group (Westall et al. 2006b), the 3460 Ma Apex Basalt of the Warrawoona Group (Schopf 1992, 1993; Schopf and Packer 1987), the 3470 Ma Mount Ada Basalt (Awramik et al. 1983), and the 3480 Ma Dresser Formation of the Warrawoona Group (Dunlop et al. 1978; Glickson et al. 2008, 2010; Noffke et al. 2013; Philippot et al. 2007; Shen et al. 2001, 2009; Ueno et al. 2001, 2006a, 2008; Van Kranendonk 2006; Walter et al. 1980). There have also been claims for life in the slightly younger West Pilbara Terrane, notably in the 3200 Ma Dixon Island Formation (Kiyokawa et al. 2006) and the 3020 Ma Cleaverville Formation (Ueno et al. 2006a,b). These claims for life are summarized in Table 8.1, where the types of evidence are listed and an evaluation of the robustness of each claim is given based upon this evidence. The most robust examples of life in these rocks are illustrated in Figure 8.6.

The Barberton region of the *Kaapvaal Craton* in South Africa contains volcanic and sedimentary greenstone rocks up to about 3550 million years old (Figure 8.7). The oldest

FIGURE 8.6 (see figure on facing page) **Stratigraphy of rocks older than 3000 Ma in the Pilbara** region of Western Australia (Modified from Wacey, D., *Early Life on Earth: A Practical Guide*, Springer, Amsterdam, the Netherlands, 285p., 2009), plus selected example of signs of life from these rocks (a–e). (a) Folded carbonaceous biofilm from the 3000 Ma Farrel Quartzite. (Courtesy of K. Sugitani; Modified from Wacey, D., *Early Life on Earth: A Practical Guide*, Springer, Amsterdam, the Netherlands, 285p., 2009.) (b) Filamentous pyritized microfossils from the 3240 Ma Kangaroo Caves Formation. (Reproduced from Wacey, D., *Early Life on Earth: A Practical Guide*, Springer, Amsterdam, the Netherlands, 285p., 2009. With permission.) (c) Tubular sheath-like microfossils from the 3430 Ma Strelley Pool Formation. (d) Complex stromatolite from the 3430 Ma Strelley Pool Formation. (Modified from Wacey, D., *Early Life on Earth: A Practical Guide*, Springer, Amsterdam, the Netherlands, 285p., 2009.) (e) Simple domal stromatolite from the 3480 Ma Dresser Formation.

TABLE 8.1 Summary of Reported Evidence for Life in >3000 Ma Rocks in the Pilbara Region of Western Australia

Age (Ma)	Geological Unit	Morphological Evidence	Geochemical Evidence	Comments
~3000	Farrel Quartzite, Gorge Creek Group	*Microfossils*: spheroids, 2.5–80 μm in diameter; lens/spindle shapes up to 40 μm in length and 35 μm in width; threadlike filaments <1 μm diameter and up to 100s μm in length; potential carbonaceous biofilms (Oehler et al. 2009, 2010; Sugitani et al. 2007, 2009a,b).	*Bulk carbon isotopes*: $\delta^{13}C_{PDB}$ ca. –30‰ to –35‰ consistent with biology. *Laser Raman*: fossils are made of carbon and carbon structure is consistent with age and biology. *Elemental mapping*: C, N, and S found in microfossil walls.	Strong evidence for life.
~3020	Cleaverville Formation, West Pilbara	*Putative microfossils*: spheroids, 4–89 μm in diameter; solitary and paired, usually enclosed by outer envelopes; many have internal structures; possible postmortem degradation (Ueno et al. 2006a,b).	None except for carbonaceous composition.	Possible life; many rather larger than expected for cells.
~3200	Dixon Island Formation, West Pilbara	*Putative microbial mat and microfossils*: clumps and laminations of carbon; ~1 μm diameter filaments and spheres; 50–80 μm long rods (Kiyokawa et al. 2006).	*Bulk carbon isotopes*: $\delta^{13}C_{PDB}$ ca. –27‰ to –33‰ consistent with biology. *Raman spectra*: consistent with metamorphic grade of the rock.	Possible life but none of the morphologies are particularly distinctive of cells.
~3240	Kangaroo Caves Formation, Sulphur Springs Group	*Putative microfossils*: solid pyritic filaments 0.5–2 μm in diameter and up to 300 μm in length, often intertwined and densely clustered (Rasmussen 2000).	None.	Possible life; geochemistry needed to back up promising morphology.
		Putative microfossils: filaments and tubes <1 μm wide, often bundled in packages 1–5 μm in diameter and up to 100 μm in length; aggregates of tiny spheres <50–100 nm in diameter (Duck et al. 2007).	*Bulk carbon isotopes*: $\delta^{13}C_{PDB}$ ca. –27‰ to –34‰ consistent with biology.	Not seen in situ within the rock so exact context unknown; some are smaller than expected for cells.
~3350	Euro Basalt, Kelly Group	*Trace fossils*: microtubes 1–5 μm in diameter and up to 150 μm in length, often branched and segmented (Banerjee et al. 2007).	*Elemental mapping*: possible enrichments of C and N within the microtubes.	Possible life, but formation mechanism poorly understood.
		Putative microfossils: spheroids ca. 8 μm in diameter; spheroids ca. 21 μm in diameter sometimes enclosed by a potential sheath (Schopf and Packer 1987).	None.	Possible life, chemistry needed to back up morphology; debate over exact location of the sample.

~3400	Strelley Pool Formation	*Microfossils*: carbonaceous spheroids (ca. 5–25 μm in diameter), tubes (ca. 7–20 μm in diameter), and rare cellular envelopes (up to 80 μm in diameter), associated with micron-sized pyrite. Spheroids often cluster and organize into chains. Microfossils colonize sand grains and some show taphonomic degradation (Wacey et al. 2011a, 2012).	*In situ carbon isotopes*: $\delta^{13}C_{PDB}$ −33‰ to −46‰ consistent with biology. *In situ sulfur isotopes*: $\delta^{34}S_{V\text{-}CDT}$ range of 18‰ and $\Delta^{33}S$ of +1.43‰ to −1.65‰ indicating sulfur-processing metabolisms. *Elemental mapping*: nitrogen enrichment in microfossil walls. *Laser Raman*: carbon structure consistent with age and biology.	Strong evidence for life; the most comprehensive suite of evidence so far presented from the early Archean.
		Stromatolites: best examples are conical and columnar morphologies with trapped and reworked grains (Allwood et al. 2006, 2007, 2009; Van Kranendonk et al. 2003).	*Elemental distributions*: co-occurrence of C, N, and S in detrital carbonaceous grains; rare-earth element enrichment in carbonate laminae relative to chert laminae is consistent with younger microbial carbonates.	Probable life.
		Microfossils: lenticular/spindle shapes (long axis ca. 35–120 μm), large spheroids (up to ca. 100 μm diameter), clusters of smaller spheroids, wrinkled and folded carbonaceous films, and threadlike filaments (Sugitani et al. 2010, 2013).	*Bulk carbon isotopes*: $\delta^{13}C_{PDB}$ ca. −31‰ to −37‰ consistent with biology. *Laser Raman*: shows microfossils are carbonaceous and carbon has structure consistent with age and biology.	Probable life.
		Trace fossils: microborings (channels and spherical to elliptical surface pits, with near constant diameter of 1–5μm) often concentrated in clumps along one side of a detrital pyrite grain (Wacey et al. 2011b).	*Elemental mapping*: C and N enrichments in some microborings. *Laser Raman*: shows carbonaceous biofilms associated with some microborings; carbon has structure consistent with age and biology.	Probable life.
		Possible trace fossils: microtubes containing a pyrite grain at their termination—known as ambient inclusion trails (Wacey et al. 2008).	*In situ carbon isotopes*: mean $\delta^{13}C_{PDB}$ of ca. −26‰ from carbonaceous microtube linings. *Elemental mapping*: enrichments in C, N, P, and S in some microtubes.	Possible life, but formation mechanism poorly understood.

(continued)

TABLE 8.1 (continued) Summary of Reported Evidence for Life in >3000 Ma Rocks in the Pilbara Region of Western Australia

Age (Ma)	Geological Unit	Morphological Evidence	Geochemical Evidence	Comments
~3450	Panorama Formation, Warrawoona Group	*Microbial mat with microfossils*: spheroids (<1 μm in diameter), often clustered or in chains; filaments (<1 μm diameter and up to 40 μm in length); occasional rod shapes; films interpreted as fossilized extracellular polymeric substances (Westall et al. 2006b).	*Bulk carbon isotopes*: $\delta^{13}C_{PDB}$ ca. −26‰ to −30‰ consistent with biology. *Laser Raman*: spectra indicate carbon with a structure consistent with age and biology.	Probable life.
~3460	Apex Basalt, Warrawoona Group	*Possible microfossils*: filaments—11 *species* varying in average diameter from 0.5 to 16.5 and 28 to 89 μm in length (Schopf 1993). *BUT*: many are C-, J-, and L-shaped structures around spherulitic silica crystals; some follow ghost rhombic crystal outlines (Brasier et al. 2002).	*Bulk carbon isotopes*: $\delta^{13}C_{PDB}$ averages −27.7‰ consistent with biology. *Laser Raman*: spectra show objects are made of carbon and structure consistent with biology. *BUT*: carbon isotope and Raman signals are also consistent with abiotic carbonaceous artifacts; some may not be carbonaceous.	The most widely debated and controversial evidence for early life.
~3470	Mount Ada Basalt, Warrawoona Group	Putative filamentous and septate microfossil (Awramik et al. 1983)—but severe doubts over age and location of specimen.	None.	No map or GPS coordinates; attempts to recollect the material have failed.

~3480	Dresser Formation, Warrawoona Group	*Stromatolites*: coniform, nodular, domical, and wavy-laminated morphotypes (Van Kranendonk et al. 2008; Walter et al. 1980).	None.	Possible life but similar structures have been produced without biology.
		Putative microfossils: three types of filaments: (a) average 1 μm in diameter, some mutually interwoven and branched; (b) segmented, unbranched, average 8.7 μm in diameter; and (c) nonsegmented, tubular, average 12.6 μm in diameter (Ueno et al. 2001).	*Bulk and in situ carbon isotopes*: $\delta^{13}C_{PDB}$ ca. $-30‰$ to $-42‰$ consistent with biology.	Possible life but could also be nonbiological carbonaceous artifacts.
		Putative microfossils: entire *hollow cells* and fragments such as *cell walls* observed under the transmission electron microscope (Glickson et al. 2008—also described from Hooggenoeg Formation).	*Bulk carbon isotopes*: $\delta^{13}C_{PDB}$ ca. $-32‰$ to $-36‰$ consistent with biology.	Possible life but not observed *in situ* within rock thin sections.
		Putative microfossils: spheres and ellipses 0.2–7.2 μm in diameter, often in pairs, chains, or clusters (Dunlop et al. 1978).	None, except probable carbonaceous composition.	Probable nonbiological bitumen droplets (Buick 1990).
		Possible biominerals: microscopic pyrite (no biological morphology).	*Bulk and in situ sulfur isotopes*: $\delta^{34}S_{V\text{-}CDT}$ range of 22‰; $\Delta^{33}S$ of $+6‰$ to $-1.2‰$ consistent with sulfur-processing biological metabolisms (e.g., Shen et al. 2009; Ueno et al. 2008).	A strong isotopic claim for life but more still needs to be understood about the Archean sulfur cycle.
		Methane in fluid inclusions: no biological morphology.	*Carbon isotopes*: $\delta^{13}C_{PDB}$ $-36‰$ to $-56‰$ consistent with microbial methanogenesis (Ueno et al. 2006a,b).	Possible life, but this isotopic signature could also be produced without biology (Sherwood Lollar and McCollum 2006).

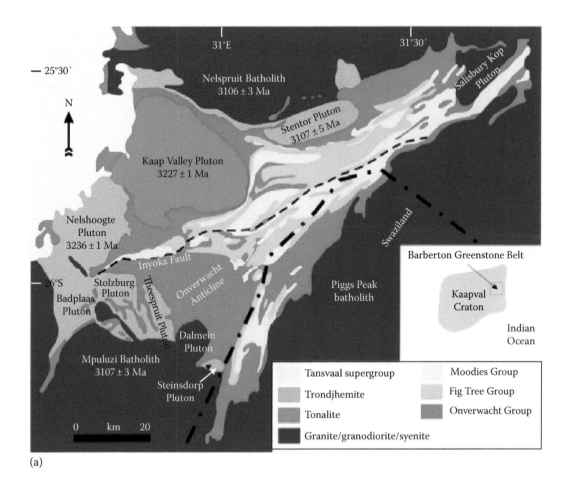

(a)

(b)

FIGURE 8.7 (a) Location of the Barberton greenstone belt in the Kaapvaal Craton of South Africa, with (b) a typical field photograph showing the terrain and outcrop in this area.

FIGURE 8.7 (continued) (c–e) Typical rock types that contain signs of life in the Barberton, including pillow basalt (c), black and white banded chert (d), and sandstone (e). (Reproduced from Wacey, D., *Early Life on Earth: A Practical Guide*, Springer, Amsterdam, the Netherlands, 285p., 2009. With permission.)

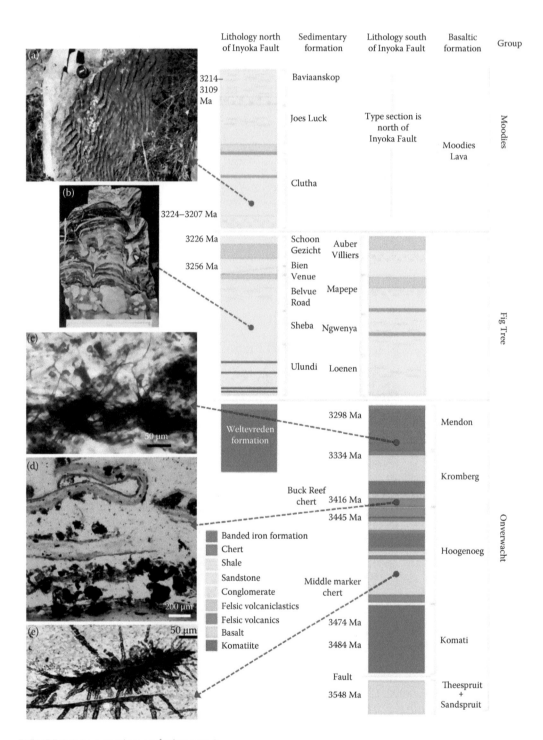

FIGURE 8.8 (see caption on facing page)

of these occurs in the Swaziland Supergroup, which itself is divided into three smaller rock packages. The lowermost Onverwacht Group (~3550–3300 Ma) is mostly volcanic with a handful of thin chert units. Above this comes the Fig Tree Group (~3260–3225 Ma) followed by the Moodies Group (~3220 Ma); both of these are dominated by clastic sedimentary rocks such as sandstone, siltstone, and shale. As with the Pilbara, these greenstone rocks are interspersed and intruded by large granite domes, >3000 Ma, but these are not habitable for life.

Signs of life have been reported in (from youngest to oldest; Figure 8.8) 3220 Ma sandstones from the lower part of the Moodies Group (Noffke et al. 2006), the 3245 Ma Sheba Formation of the Fig Tree Group (Byerly et al. 1986; Schopf and Barghoorn 1967), the 3260 Ma Swartkoppie Formation of the Fig Tree Group (Knoll and Barghoorn 1977), the 3400 Ma Kromberg Formation of the Onverwacht Group (Furnes et al. 2004; Tice and Lowe 2004; Walsh 1992; Walsh and Lowe 1985; Westall et al. 2001), and the 3450 Ma Hooggenoeg Formation of the Onverwacht Group (Banerjee et al. 2006; Engel et al. 1968; Furnes et al. 2004; McLoughlin et al. 2012; Walsh 1992; Walsh and Lowe 1985; Westall et al. 2001, 2006). These claims for life are summarized in Table 8.2, where the types of evidence are listed and an evaluation of the robustness of each claim is given based upon this evidence. The most robust examples of life from these rocks are illustrated in Figure 8.8. Table 8.3 summarizes claims for life from older 3700 to 3850 Ma rocks from Greenland; these comprise chemical fossils only and are particularly controversial.

FIGURE 8.8 (see figure on facing page)   Stratigraphy of rocks older than 3000 Ma in the Barberton region of South Africa (Modified from Wacey, D., *Early Life on Earth: A Practical Guide*, Springer, Amsterdam, the Netherlands, 285p., 2009), plus selected examples of signs of life from these rocks (a–e). (a) MISS (microbial mat wrinkles on sedimentary ripples) from the ~3220 Ma Moodies Group. (Courtesy of N. Noffke.) (b) Stromatolite from the ~3245 Ma Fig Tree Group. (Courtesy of G. Byerly; Modified from Wacey, D., *Early Life on Earth: A Practical Guide*, Springer, Amsterdam, the Netherlands, 285p., 2009.) (c) Filamentous microfossils from the ~3400 Ma Kromberg Formation. (Courtesy of M. Walsh; Modified from Wacey, D., *Early Life on Earth: A Practical Guide*, Springer, Amsterdam, the Netherlands, 285p., 2009.) (d) Ripped up portion of a microbial mat plus carbonaceous grains from the 3416 Ma Buck Reef chert. (Modified from Wacey, D., *Early Life on Earth: A Practical Guide*, Springer, Amsterdam, the Netherlands, 285p., 2009.) (e) Tubular microbial borings from the ~3450 Ma Hooggenoeg Formation. (Courtesy of N. McLoughlin.)

TABLE 8.2 Summary of Reported Evidence for Life in >3000 Ma Rocks in the Barberton Region of South Africa

Age (Ma)	Geological Unit	Morphological Evidence	Geochemical Evidence	Comments
~3220	Moodies Group	*Microbially influenced sedimentary structures*: wrinkle and roll-up structures; aligned grains in a matrix of carbonaceous laminae (Noffke et al. 2006).	*Bulk carbon isotopes*: $\delta^{13}C_{PDB}$ around −21‰ consistent with biology.	A promising biosignature indicating the presence of microbial mats at 3220 Ma.
~3245	Sheba Formation, Fig Tree Group	*Stromatolites*: laterally linked domes, pseudocolumns, and crinkly stratiform varieties (Byerly et al. 1986).	None.	Probable life, but nonbiological experiments can produce *stromatolites* with similar morphology (McLoughlin et al. 2008).
		Putative microfossils: spheroidal, average ~19 μm diameter; possible internal contents; thick walls with reticulate surface texture (Schopf and Barghoorn 1967).	None, except microfossil walls are carbonaceous.	Possible life, but no geological mapping or geochemistry reported. Could equally be nonbiological.
~3260	Swartkoppie Formation, Fig Tree Group	*Putative microfossils*: spheroidal 1–4 μm diameter; possible internal contents; possible binary division (Knoll and Barghoorn 1977).	None, except microfossil walls are carbonaceous.	Possible life; higher resolution study needed to discount nonbiological artifacts mimicking cell morphology.
~3400	Kromberg Formation, Onverwacht Group	*Microbial mat*: carbonaceous laminations with portions *ripped up* and/or folded over; some filaments preserved (Tice and Lowe 2004).	*Bulk carbon isotopes*: $\delta^{13}C_{PDB}$ −20‰ to −35‰ consistent with biology.	A strong claim for life.
		Carbonaceous grains: possible eroded fragments of a microbial mat (Tice and Lowe 2004).	*Bulk carbon isotopes*: $\delta^{13}C_{PDB}$ −20‰ to −35‰ consistent with biology.	Occur in same samples as the ripped up microbial mat, probably biological.
		Putative microfossils: a selection of <2 μm spheroidal and rodlike morphologies, often clustered or paired (Westall et al. 2001).	*Bulk carbon isotopes*: $\delta^{13}C_{PDB}$ −27‰ consistent with biology.	Possible life but morphology too simple to be uniquely biological. Only one bulk $\delta^{13}C$ value.
		Putative microfossils: tubular and solid carbonaceous and pyritic filaments up to 2.5 μm diameter and 200 μm long, plus large spindle shapes (Walsh 1992).	None.	Possible life; geochemistry and higher resolution morphology needed to discount simple mineral filaments and coatings.

Age (Ma)	Formation	Evidence	Geochemistry	Interpretation
~3450	Hooggenoeg Formation, Onverwacht Group	*Trace fossils*: segmented microtubes, ca. 4 μm in diameter and up to 200 μm in length, plus clustered spheres (Banerjee et al. 2006; Furnes et al. 2004; McLoughlin et al. 2012).	*In situ sulfur isotopes*: $\delta^{34}S_{V\text{-}CDT}$ −40‰ to −3‰ from pyrite in microtubes indicates sulfur-based metabolism. *Bulk carbon isotopes from carbonate*: $\delta^{13}C$ −16‰ to +4‰.	Probable life, but formation mechanism still poorly understood even in modern examples.
		Putative microfossils: solid filaments up to 2.6 μm diameter and 200 μm long (Walsh and Lowe 1985).	None.	Possible life; geochemistry and higher resolution morphology needed to discount mineral filaments.
		Putative microfossils: sausage-shaped, up to 3.8 μm long and 1.1 μm wide (Westall et al. 2001, 2006a).	None.	Unreliable; only seen in surface SEM images, not in thin sections.
		Putative microbial mat: crinkly carbonaceous laminations; some bundled, broken, or folded over; carbonaceous grains (Walsh 1992).	None.	Possible life, but nonbiological carbon could also organize itself in this manner.
		Microbial mat: comprising parallel and interwoven filaments; mat was flexible (Westall et al. 2006a).	*Bulk carbon isotopes*: $\delta^{13}C_{PDB}$ −23‰ consistent with biology.	Probable life; more geochemistry needed to make it compelling.
		Putative microfossils: spheres 5–105 μm in diameter (Engel et al. 1968).	None.	Nonbiological artifacts (Schopf and Walter 1983).

TABLE 8.3   Summary of Reported Evidence for Life in >3000 Ma Rocks from Greenland

Age (Ma)	Geological Unit	Morphological Evidence	Geochemical Evidence	Comments
~3700	Garbenschiefer Formation, Isua, Greenland	None	*Bulk carbon isotopes*: $\delta^{13}C_{PDB}$ averages −19‰ consistent with biological processing (Rosing 1999).	Carbon isotope signature is not uniquely biological.
~3700	An unnamed *banded iron formation*, Isua, Greenland	None	*In situ carbon isotopes*: $\delta^{13}C_{PDB}$ averages −30‰ consistent with biological processing (Mojzsis et al. 1996).	Carbon isotope signature is not uniquely biological; carbon may have formed via thermal-metamorphic reactions (e.g., Van Zuilen et al. 2002, 2003).
~3850	An unnamed *quartz–pyroxene rock*, Akilia Island, Greenland	None	*In situ carbon isotopes*: two sets of $\delta^{13}C_{PDB}$ data, one averages −37‰, the other averages −29‰, both consistent with biological processing (Mojzsis et al. 1996).	Carbon isotope signature is not uniquely biological; geological setting is very controversial, in terms of both the rocks' age and original sedimentary nature (e.g., Fedo and Whitehouse 2002).

GLOSSARY

Abiotic: Not derived from living organisms.

AIT: Ambient inclusion trail.

Archean: The period of time between 4 and 2.5 billion years ago.

Biomarker: A substance whose presence indicates the existence of living organisms (can be specific to a type of organism).

Biomineral: A mineral produced by the activity of living things.

Carboxyl group: An organic functional group consisting of a carbon atom double bonded to an oxygen atom and single bonded to an OH group.

DNA: Deoxyribonucleic acid; the molecule that encodes genetic instructions.

EBSD: Electron backscatter diffraction.

EDS: Energy dispersive spectroscopy.

EELS: Electron energy loss spectroscopy.

Endolith: An organism living/growing inside a rock.

Exogenous: Having an external origin.

Fischer–Tropsch synthesis: A reaction (or series of reactions) that converts carbon monoxide and hydrogen into liquid hydrocarbons.

FIB: Focused ion beam.

FIB–SEM: Focused ion beam milling combined with scanning electron microscopy.

Half-life: The time taken for half the radioactive nuclei in a given sample to undergo radioactive decay.

HRTEM: High-resolution transmission electron microscopy.

Indigenous: Native (to the rock in which it is found).

Isotope: Each of two or more forms of the same element that have the same number of protons but different numbers of neutrons and hence differ in relative atomic mass.

Kerogen: A mixture of high-molecular-weight polymers formed from the decomposition of living matter.

LR: Laser Raman.

Metamorphism: The change in form of a rock (i.e., change in minerals and/or texture).

MISS: Microbially induced sedimentary structure.

NanoSIMS: Nanometer-scale secondary ion mass spectrometry.

Peptidoglycan: Polymer consisting of sugars and amino acids.

Petrographic thin section: A thin slice of rock that can be studied with an optical microscope.

Phospholipids: A lipid bonded to two fatty acids and a phosphate group.

Prebiotic: Before the evolution of living things.

RNA: Ribonucleic acid.

SAED: Selected area electron diffraction.

SEM: Scanning electron microscopy.

SIMS: Secondary ion mass spectrometry.

Synchrotron: A cyclic particle accelerator producing a high-energy electron beam.

Syngenous/syngenetic: Formed at the same time as the surrounding rock.

Taphonomic: The conditions and processes by which organisms become fossilized.

TEM: Transmission electron microscopy.

Tomography: Imaging by sectioning through use of any penetrating wave.

REVIEW QUESTIONS

1. What conditions are needed for remarkable preservation of body fossils?

2. Summarize the biogenicity criteria for (a) body fossils and (b) trace fossils.

3. Summarize the antiquity criteria for (a) body fossils and (b) trace fossils.

4. What are the best techniques to analyze a unique or rare microfossil?

5. Choose two analysis techniques commonly applied to studies of early life on Earth and describe the information that can be obtained from them.

6. Why are habitable rocks from the earliest periods of Earth history so rare?

7. What are the typical types of rocks that host Earth's early life?

8. What are the most robust examples of life from Pilbara and Barberton rocks?

9. Why are signs of life from rocks in Greenland controversial?

10. Why is the study of the early fossil record important for astrobiology?

RECOMMENDED READING

Books

Awramik, S. M., Margulis, L., and Barghoorn, E. S., 1976. Evolutionary processes in the formation of stromatolites. In: Walter, M. R. (ed.), *Stromatolites*. Elsevier, Amsterdam, the Netherlands, pp. 149–162.

Glickson, M., Hickman, A. H., Duck, L. J., Golding, S. D., and Webb, R. E., 2010. Integration of observational and analytical methodologies to characterize organic matter in early Archean rocks: Distinguishing biological from abiotically synthesized carbonaceous matter structures. In: Golding, S. D. and Glickson, M. (eds.), *Earliest Life on Earth: Habitats, Environments and Methods of Detection*. Springer Science and Business Media B.V., the Netherlands.

Ireland, T. R., 1995. Ion microprobe mass spectrometry: Techniques and applications in cosmochemistry, geochemistry, and geochronology. In: Hyman, M. and Rowe, M. (eds.), *Advances in Analytical Geochemistry*. JAI Press, Greenwich, CT, pp. 1–118.

Schopf, J. W., 1992. Paleobiology of the Archean. In: Schopf, J. W. and Klein, C. (eds.), *The Proterozoic Biosphere, A Multidisciplinary Study*. Cambridge University Press, New York, pp. 25–39.

Schopf, J. W. and Walter, M. R., 1983. Archean microfossils: New evidence of ancient microbes. In: Schopf, J. W. (ed.), *Earth's Earliest Biosphere, Its Origin and Evolution*, Princeton University Press, Princeton, NJ, pp. 214–239.

Wacey, D., 2009. *Early Life on Earth: A Practical Guide*. Springer, Amsterdam, the Netherlands, 285p.

Wacey, D., 2012. In situ, high spatial resolution techniques in the search for the origin of life. In: Seckbach, J. (ed.), *Genesis—In The Beginning. Cellular Origin, Life in Extreme Habitats and Astrobiology*, Vol. 22. Springer, Dordrecht, the Netherlands, pp. 391–411.

Wacey, D., 2013. In situ morphologic, elemental and isotopic analysis of Archean life. In: Dilek, Y. and Furnes, H. (eds.), *Evolution of Archean Crust and Early Life, Modern Approaches in Solid Earth Sciences*, Vol. 7. Springer, Dordrecht, the Netherlands, pp. 351–365.

Journal Articles

Allwood, A. C., Grotzinger, J. P., Knoll, A. H. et al., 2009. Controls on development and diversity of early Archean stromatolites. *Proceedings of the National Academy of Sciences of the United States of America* 106: 9548–9555.

Allwood, A. C., Walter, M. R., Burch, I. W., and Kamber, B. S., 2007. 3.43 Billion-year-old stromatolite reef from the Pilbara Craton of Western Australia: Ecosystem-scale insights to early life on Earth. *Precambrian Research* 158: 198–227.

Allwood, A. C., Walter, M. R., Kamber, B. S., Marshall, C. P., and Burch, I. W., 2006. Stromatolite reef from the Early Archaean era of Australia. *Nature* 441: 714–718.

Awramik, S. M., Schopf, J. W., and Walter M. R., 1983. Filamentous fossil bacteria from the Archaean of Western Australia. *Precambrian Research* 20: 357–374.

Banerjee, N. R., Furnes, H., Muehlenbachs, K., Staudigel, H., and de Wit, M., 2006. Preservation of ~3.4–3.5 Ga microbial biomarkers in pillow lavas and hyaloclastites from the Barberton Greenstone Belt, South Africa. *Earth and Planetary Science Letters* 241: 707–722.

Banerjee, N. R., Simonetti, A., Furnes, H. et al., 2007. Direct dating of microbial ichnofossils. *Geology* 35: 487–490.

Brasier, M. D., Green, O. R., Jephcoat, A. P. et al., 2002. Questioning the evidence for Earth's oldest fossils. *Nature* 416: 76–81.

Brasier, M. D., Green, O. R., Lindsay, J. F., McLoughlin, N., Steele, A., and Stoakes, C., 2005. Critical testing of Earth's oldest putative fossil assemblage from the ~3.5 Ga Apex Chert, Chinaman Creek, Western Australia. *Precambrian Research* 140: 55–102.

Brasier, M. D., Green, O. R., and Mcloughlin, N., 2004. Characterization and critical testing of potential microfossils from the early Earth: The Apex 'microfossil debate' and its lessons for Mars sample return. *International Journal of Astrobiology* 3: 1–12.

Brasier, M. D., McLoughlin, N., and Wacey, D., 2006. A fresh look at the fossil evidence for early Archaean cellular life. *Philosophical Transactions of the Royal Society B* 361: 887–902.

Buick, R., 1990. Microfossil recognition in Archaean rocks: An appraisal of spheroids and filaments from 3500 M.Y old chert-barite at North Pole, Western Australia. *Palaios* 5: 441–459.

Byerly, G. R., Lowe, D. L., and Walsh, M. M., 1986. Stromatolites from the 3,300–3,500-Myr Swaziland Supergroup, Barberton Mountain Land, South Africa. *Nature* 319: 489–491.

Downs, R., 2006. The RRUFF project: An integrated study of the chemistry, crystallography, Raman and infrared spectroscopy of minerals. *Program and Abstracts of the 19th General Meeting of the International Mineralogical Association*, Kobe, Japan, O03–13.

Duck, L. J., Glikson, M., Golding, S. D., and Webb, R. E., 2007. Microbial remains and other carbonaceous forms from the 3.24 Ga Sulphur Springs black smoker deposit, Western Australia. *Precambrian Research* 154: 205–220.

Dunlop, J. S. R., Muir, M. D., Milne, V. A., and Groves, D. I., 1978. A new microfossil assemblage from the Archaean of Western Australia. *Nature* 274: 676–678.

Engel, A. E. J., Nagy, B., Nagy, L. A., Engel, C. G., Kremp, G. O. W., and Drew, C. M., 1968. Algal-like forms in Onverwacht Series, South Africa: Oldest recognised lifelike forms on Earth. *Science* 161: 1005–1008.

Fedo, C. M. and Whitehouse, M. J., 2002. Metasomatic origin of quartz–pyroxene rock, Akilia, Greenland, and its implications for Earth's earliest life. *Science* 296: 1448–1452.

Fries, M. and Steele, A., 2011. Raman spectroscopy and confocal Raman imaging in mineralogy and petrography. *Springer Series in Optical Sciences* 158: 111–135.

Furnes, H., Banerjee, N. R., Muehlenbachs, K., Staudigel, H., and de Wit, M., 2004. Early life recorded in Archean pillow lavas. *Science* 304: 578–581.

Garcia-Ruiz, J. M., Hyde, S. T., Carnerup, A. M., Christy, A. G., Van Kranendonk, M. J., and Welham, N. J., 2003. Self-assembled silica carbonate structures and detection of ancient microfossils. *Science* 302: 1194–1197.

Gerdes, G., Klenke, T., and Noffke, N., 2000. Microbial signatures in peritidal siliciclastic sediments: A catalogue. *Sedimentology* 47: 279–308.

Glickson, M., Duck, L. J., Golding, S. D. et al., 2008. Microbial remains in some earliest Earth rocks: Comparison with a potential modern analogue. *Precambrian Research* 164: 187–200.

Grey, K. and Sugitani, K., 2009. Palynology of Archean microfossils (c. 3.0 Ga) from the Mount Grant area, Pilbara Craton, Western Australia: Further evidence of biogenicity. *Precambrian Research* 173: 60–69.

Hofmann, H. J., 2004. Archean microfossils and abiomorphs. *Astrobiology* 4: 135–136.

Hofmann, H. J., Grey, K., Hickman, A. H., and Thorpe, R. I., 1999. Origin of 3.45 Ga coniform stromatolites in the Warrawoona Group, Western Australia. *Bulletin of the Geological Society of America* 111: 1256–1262.

Kiyokawa, S., Ito, T., Ikehara, M., and Kitajima, F., 2006. Middle Archean volcano-hydrothermal sequence: Bacterial microfossil-like bearing 3.2 Ga Dixon Island Formation, coastal Pilbara terrane, Australia. *Bulletin of the Geological Society of America* 118: 3–22.

Knoll, A. H. and Barghoorn, E. S., 1974. Ambient Pyrite in Precambrian Chert: New evidence and a theory. *Proceedings of the National Academy of Sciences of the United States of America* 71: 2329–2331.

Knoll, A. H. and Barghoorn, E. S., 1977. Archean microfossils showing cell division from the Swaziland System of South Africa. *Science* 198: 396–398.

Lowe, D. R., 1980. Stromatolites 3,400-Myr old from the Archean of Western Australia. *Nature* 284: 441–443.

McCollom, T. M. and Seewald, J. S., 2006. Carbon isotope composition of organic compounds produced by abiotic synthesis under hydrothermal conditions. *Earth and Planetary Science Letters* 243: 74–84.

McKeegan, K. D., Kudryavtsev, A. B., and Schopf, J. W., 2007. Raman and ion microscopic imagery of graphitic inclusions in apatite from older than 3830 Ma Akilia supracrustal rocks, west Greenland. *Geology* 35: 591–594.

McKibben, M. A., Shanks III, W. C., and Ridley, W. I., 1998. Applications of microanalytical techniques to understanding mineralizing processes. *SEG Reviews in Economic Geology* 7: 263.

McLoughlin, N., Brasier, M. D., Wacey, D., Green, O. R., and Perry, R. S., 2007. On biogenicity criteria for endolithic microborings on early Earth and beyond. *Astrobiology* 7: 10–26.

McLoughlin, N., Grosch, E. G., Kilburn, M. R., and Wacey, D., 2012. Sulfur isotope evidence for a Paleoarchean sub-seafloor biosphere, Barberton, South Africa. *Geology* 40: 1031–1034.

McLoughlin, N., Wilson, L. A., and Brasier, M. D., 2008. Growth of synthetic stromatolites and wrinkle structures in the absence of microbes—Implications for the early fossil record. *Geobiology* 6: 95–105.

Mojzsis, S. J., Arrenhius, G., McKeegan, K. D., Harrison, T. M., Nutman, A. P., and Friend, C. R. L., 1996. Evidence for life on Earth 3,800 million years ago. *Nature* 384: 55–59.

Noffke, N., Christian, D., Wacey, D., and Hazen, R. M., 2013. Microbially induced sedimentary structures recording an ancient ecosystem in the 3.48 billion-year-old Dresser Formation, Pilbara, Western Australia. *Astrobiology* 13(12): 1103–1124.

Noffke, N., Eriksson, K. A., Hazen, R. M., and Simpson, E. L., 2006. A new window into Early Archean life: Microbial mats in Earth's oldest siliciclastic tidal deposits (3.2 Ga Moodies Group, South Africa). *Geology* 34: 253–256.

Oehler, D. Z., Robert, F., Walter, M. R. et al., 2009. NanoSIMS: Insights to biogenicity and syngenicity of Archaean carbonaceous structures. *Precambrian Research* 173: 70–78.

Oehler, D. Z., Robert, F., Walter, M. R. et al., 2010. Diversity in the Archaean biosphere: New insights from NanoSIMS. *Astrobiology* 10: 413–424.

Pasteris, J. D. and Wopenka, B., 2003. Necessary, but not sufficient: Raman identification of disordered carbon as a signature of ancient life. *Astrobiology* 3: 727–738.

Philippot, P., Van Zuilen, M., Lepot, K., Thomazo, C., Farquhar, J., and Van Kranendonk, M. J., 2007. Early Archaean microorganisms preferred elemental sulfur, not sulfate. *Science* 317: 1534–1537.

Rasmussen, B., 2000. Filamentous microfossils in a 3235-million-year-old volcanogenic massive sulphide deposit. *Nature* 405: 676–679.

Rosing, M. T., 1999. ^{13}C depleted carbon microparticles in >3700-Ma sea-floor sedimentary rocks from West Greenland. *Science* 283: 674–676.

Schopf, J. W., 1993. Microfossils of the early Archaean Apex Chert: New evidence for the antiquity of life. *Science* 260: 640–646.

Schopf, J. W. and Barghoorn, E. S., 1967. Alga-like fossils from the early Precambrian of South Africa. *Science* 156: 508–512.

Schopf, J. W. and Packer, B. M., 1987. Early Archaean (3.3 billion to 3.5 billion-year-old) microfossils from Warawoona Group, Australia. *Science* 237: 70–73.

Semikhatov, M. A., Gebelein, C. D., Cloud, P., Awramik, S. M., and Benmore, W. C., 1979. Stromatolite morphogenesis: Progress and problems. *Canadian Journal of Earth Sciences* 19: 992–1015.

Shen, Y., Buick, R., and Canfield, D. E., 2001. Isotopic evidence for microbial sulphate reduction in the early Archaean era. *Nature* 410: 77–81.

Shen, Y., Farquhar, J., Masterson, A., Kaufman, A. J., and Buick, R., 2009. Evaluating the role of microbial sulfate reduction in the early Archean using quadruple isotope systematics. *Earth and Planetary Science Letters* 279: 383–391.

Sherwood Lollar, B. and McCollum, T. M., 2006. Biosignatures and abiotic constraints on early life. *Nature* 444: E18.

Sugitani, K., Grey, K., Allwood, A. et al., 2007. Diverse microstructures from Archaean chert from the Mount Goldsworthy–Mount Grant area, Pilbara Craton, Western Australia: Microfossils, dubiofossils, or pseudofossils? *Precambrian Research* 158: 228–262.

Sugitani, K., Grey, K., Naaoka, T., and Mimura, K., 2009a. Three-dimensional morphological and textural complexity of Archean putative microfossils from the northeastern Pilbara craton: Indications of biogenicity of large (>15 mm) spheroidal and spindle-like structures. *Astrobiology* 9: 603–615.

Sugitani, K., Grey, K., Nagaoka, T., Mimura, K., and Walter, M. R., 2009b. Taxonomy and biogenicity of Archaean spheroidal microfossils (ca. 3.0 Ga) from the Mount Goldsworthy–Mount Grant area in the northeastern Pilbara Craton, Western Australia. *Precambrian Research* 173: 50–59.

Sugitani, K., Lepot, K., Nagaoka, T. et al., 2010. Biogenicity of morphologically diverse carbonaceous microstructures from the ca. 3400 Ma Strelley Pool Formation, in the Pilbara Craton, Western Australia. *Astrobiology* 10: 899–920.

Sugitani, K., Mimura, K., Nagaoka, T. et al., 2013. Microfossil assemblage from the 3400 Ma Strelley Pool Formation in the Pilbara Craton, Western Australia: Results from a new locality. *Precambrian Research* 226: 59–74.

Tice, M. M., Bostick, B. C., and Lowe, D. R., 2004. Thermal history of the 3.5–3.2 Ga Onverwacht and Fig Tree Groups, Barberton greenstone belt, South Africa, inferred by Raman microspectroscopy of carbonaceous material. *Geology* 32: 37–40.

Tice, M. M. and Lowe, D. R., 2004. Photosynthetic microbial mats in the 3,416-Myr-old ocean. *Nature* 431: 549–552.

Ueno, Y., Isozaki, Y., and McNamara, K. J., 2006b. Coccoid-like microstructures in a 3.0 Ga chert from Western Australia. *International Geology Review* 48: 78–88.

Ueno, Y., Isozaki, Y., Yurimoto, H., and Maruyama, S., 2001. Carbon isotopic signatures of individual Archean microfossils(?) from Western Australia. *International Geology Review* 43: 196–212.

Ueno, Y., Ono, S., Rumble, D., and Maruyama, S., 2008. Quadruple sulfur isotope analysis of ca. 3.5 Ga Dresser Formation: New evidence for microbial sulfate reduction in the early Archean. *Geochimica et Cosmochimica Acta* 72: 5675–5691.

Ueno, Y., Yamada, K., Yoshida, N., Maruyama, S., and Isozaki, Y., 2006a. Evidence from fluid inclusions for microbial methanogenesis in the early Archaean era. *Nature* 440: 516–519.

Van Kranendonk, M. J., 2006. Volcanic degassing, hydrothermal circulation and the flourishing of early life on Earth: New evidence from the Warrawoona Group, Pilbara Craton, Western Australia. *Earth Science Reviews* 74: 197–240.

Van Kranendonk, M. J., Phillipot, P., Lepot, K., Bodorkos, S., and Pirajno, F., 2008. Geological setting of Earth's oldest fossils in the ca. 3.5 Ga Dresser Formation, Pilbara Craton, Western Australia. *Precambrian Research* 167: 93–124.

Van Kranendonk, M. J., Webb, G. E., and Kamber, B. S., 2003. Geological and trace element evidence for a marine sedimentary environment of deposition and biogenicity of 3.45 Ga stromatolitic carbonates in the Pilbara Craton, and support for a reducing Archean ocean. *Geobiology* 1: 91–108.

Van Zuilen, M. A., Lepland, A., and Arhenius, G., 2002. Reassessing the evidence for the earliest traces of life. *Nature* 418: 627–630.

Van Zuilen, M. A., Lepland, A., Teranes, J., Finarelli, J., Wahlen, M., and Arrhenius, G., 2003. Graphite and carbonates in the 3.8 Ga old Isua Supracrustal Belt, southern West Greenland. *Precambrian Research* 126: 331–348.

Wacey, D., 2010. Stromatolites in the ~3400 Ma Strelley Pool Formation, Western Australia: Examining biogenicity from the macro- to the nano-scale. *Astrobiology* 10: 381–395.

Wacey, D., Gleeson, D., and Kilburn, M. R., 2010b. Microbialite taphonomy and biogenicity: New insights from NanoSIMS. *Geobiology* 8: 403–416.

Wacey, D., Kilburn, M. R., McLoughlin, N., Parnell, J., Stoakes, C. A., and Brasier, M. D., 2008. Use of NanoSIMS to investigate early life on Earth: Ambient inclusion trails in a c. 3400 Ma sandstone. *Journal of the Geological Society of London* 165: 43–53.

Wacey, D., Kilburn, M. R., Saunders, M., Cliff, J., and Brasier, M. D., 2011a. Microfossils of sulfur metabolizing cells in ~3.4 billion year old rocks of Western Australia. *Nature Geoscience* 4: 698–702.

Wacey, D., McLoughlin, N., Kilburn, M. R. et al., 2013b. Nano-scale analysis reveals differential heterotrophic consumption in the ~1.9 Ga Gunflint Chert. *Proceedings of the National Academy of Sciences of the United States of America* 110: 8020–8024.

Wacey, D., McLoughlin, N., Saunders, M., and Kong, C., 2013a. The nano-scale anatomy of a complex carbon-lined microtube in volcanic glass from the ~92 Ma Troodos Ophiolite, Cyprus. *Chemical Geology* 363: 1–12.

Wacey, D., McLoughlin, N., Stoakes, C. A., Kilburn, M. R., Green, O. R., and Brasier, M. D., 2010a. The 3426–3350 Ma Strelley Pool Formation in the East Strelley greenstone belt—A field and petrographic guide. *Geological Survey of Western Australia Record* 10: 64p.

Wacey, D., Menon, S., Green, L. et al., 2012. Taphonomy of very ancient microfossils from the ~3400 Ma Strelley Pool Formation and ~1900 Ma Gunflint Formation: New insights using a focused ion beam. *Precambrian Research* 220–221: 234–250.

Wacey, D., Saunders, M., Brasier, M. D., and Kilburn, M. R., 2011b. Earliest microbially mediated pyrite oxidation in ~3.4 billion-year-old sediments. *Earth and Planetary Science Letters* 301: 393–402.

Walsh, M. M., 1992. Microfossils and possible microfossils from the early Archean Onverwacht Group, Barberton Mountain Land, South Africa. *Precambrian Research* 54: 271–293.

Walsh, M. M. and Lowe, D. L., 1985. Filamentous microfossils from the 3,500 Myr-old Onverwacht Group, Barberton Mountain Land, South Africa. *Nature* 314: 530–532.

Walter, M. R., Buick, R., and Dunlop, J. S. R., 1980. Stromatolites, 3,400–3,500 Myr old from the North Pole area, Western Australia. *Nature* 284: 443–445.

Westall, F., de Ronde, C. E. J., Southam, G. et al., 2006a. Implications of a 3.472–3.333 Gyr-old subaerial microbial mat from the Barberton greenstone belt, South Africa for the UV environmental conditions on the early Earth. *Philosophical Transactions of the Royal Society B* 361: 1857–1875.

Westall, F., de Vries, S. T., Nijman, W. et al., 2006b. The 3.446 Ga "Kitty's Gap Chert", an early Archean microbial ecosystem. *Geological Society of America Special Paper* 405: 105–131.

Westall, F., de Witt, M. J., Dann, J., van der Gaast, S., de Ronde, C. E. J., and Gerneke, D., 2001. Early Archean fossil bacteria and biofilms in hydrothermally-influenced sediments from the Barberton greenstone belt, South Africa. *Precambrian Research* 106: 93–116.

Westall, F. and Folk, R. L., 2003. Exogenous carbonaceous microstructures in Early Archaean cherts and BIFs from the Isua Greenstone belt: Implications for the search for life in ancient rocks. *Precambrian Research* 126: 313–330.

Wirth, R., 2009. Focused Ion Beam (FIB) combined with SEM and TEM: Advanced analytical tools for studies of chemical composition, microstructure and crystal structure in geomaterials on a nanometre scale. *Chemical Geology* 261: 217–229.

Prebiotic Chemistry

In Water and in the Solid State

Vera M. Kolb

CONTENTS

9.1 PREBIOTIC CHEMISTRY IN AQUEOUS MEDIA

Water is a preferred prebiotic reaction medium. One talks about prebiotic soup and pre-biotic oceans as common places where the organic reactions have occurred on prebiotic Earth. Prebiotic chemists have focused mostly on the organic molecules that are water soluble, such as small molecules, or those possessing hydrophilic groups, such as hydroxyl and carboxyl. This became a limiting factor in the prebiotic synthetic repertoire, since most organic compounds are either insoluble in water or are only poorly soluble. Other media that are commonly used as solvents in the regular (nonprebiotic) organic syntheses, such as aliphatic and aromatic hydrocarbons, halogenated hydrocarbons, ethers, and alcohols, were not available on the prebiotic Earth as pools of liquids. They could be found only in relatively small amounts as compared to the large oceans of water.

This has created a great problem for prebiotic chemistry.

9.1.1 Organic Reactions *on Water* and *in Water*

It has been shown recently that many organic materials that are not soluble in water are still capable of reacting in water, often at faster rates than in the organic solvents in which they are completely soluble. This has provided a new era in the study of the prebiotic reactions.

The old view that prebiotic reactions in water are hampered by the low solubility of the organic compounds in water is now being revised due to the discoveries of the reactions *on water*. These reactions occur in the heterogeneous system comprised of the organic compounds and water. Unexpectedly, such reactions are extremely efficient; they often give quantitative yields and are accelerated in the presence of water as compared to the organic solvents. These *on water* reactions are not the same as the *in water* reactions, which occur in solution, and are thus homogenous. We briefly address here the nomenclature issue. Reactions in aqueous medium may be described as *in water, in the presence of water,* and *on water*. These terms are sometimes used interchangeably, although they describe reactions that occur under quite different conditions. The *in water* reactions are those in which the reactants dissolve in water, giving a homogeneous solution. The traditional prebiotic chemistry that was done with the water-soluble substrates would fall into this category. The *on water* reactions are those that proceed in aqueous organic emulsions or suspensions. Often, all the starting materials for such reactions are prebiotically feasible, but since they are not water soluble, the reactions themselves were not examined in the past by the prebiotic chemists. Since such reactions were typically left out of the prebiotic chemists' synthetic repertoire, we focus on them in this chapter. Examples of the *on water* reactions include Diels–Alder, Claisen, Passerini, and Ugi reactions, among many others. Some of these reactions are multicomponent but give a single product. We survey a selected number of these *on water* reactions, which have potential prebiotic applications.

9.1.2 Diels–Alder Reaction *on Water*

Diels–Alder reaction is a cycloaddition reaction that is used for making carbon–carbon bonds and specifically six-membered cyclic compounds. The cycloaddition occurs between two simple types of organic compounds, a diene and a dienophile (a compound that *loves* dienes), which add to each other to provide a cyclic compound. The importance of this reaction for organic chemistry cannot be overstated. Its discovery led to a Nobel Prize for the reaction itself and its synthetic utility and another Nobel Prize for the elucidation of its reaction mechanism, which involves specific interactions of the molecular orbitals of diene and dienophile. In the past, this reaction has been typically performed in organic solvents. Therefore, its great synthetic utility was not introduced to prebiotic chemistry, despite the fact that both dienes and dienophiles were abundant on the prebiotic Earth.

In the 1980s, Breslow and coworkers found out that some Diels–Alder reactions, when performed in the aqueous suspensions (thus *on water*), unexpectedly gave improved reaction rates and selectivity. Continuing studies of this unusual reactivity by Breslow's and other research groups made the Diels–Alder reaction a representative and one of the best-known *on water* reactions. These studies have also opened the doors for inclusion of Diels–Alder reaction in the prebiotic chemists' synthetic toolkit.

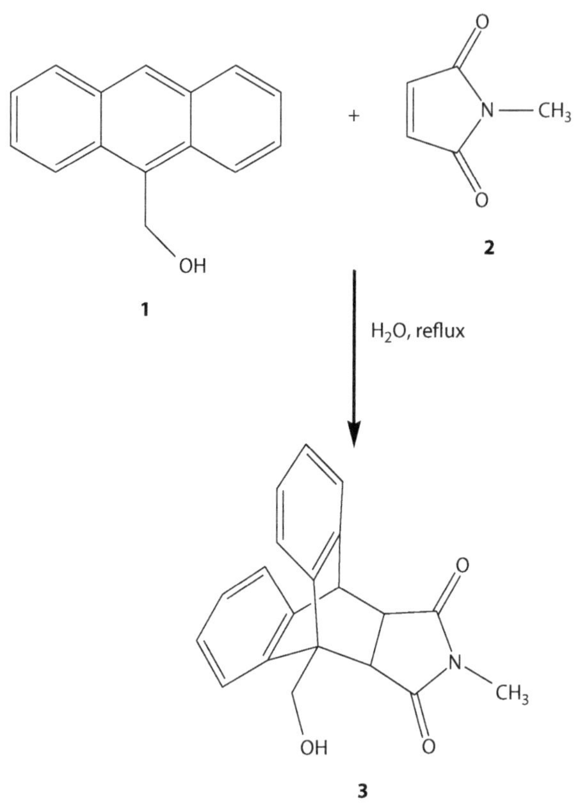

FIGURE 9.1   An example of a Diels–Alder reaction performed *on water*.

We show in Figure 9.1 a typical *on water* Diels–Alder reaction, between anthracene-9-carbinol (a diene) (**1**) and *N*-methylmaleimide (a dienophile) (**2**), which provides the Diels–Alder product (often called *adduct*) (**3**).

We focus on the following two critical questions about the mechanism of this *on water* reaction. How does this reaction occur when the reagents are not soluble in water? Why is the reaction accelerated in water as compared to the organic solvents? Before we explain the mechanism, let us introduce the reader into the strange world of the *on water* reactions as far as the experimental observations are concerned. When one mixes the starting materials shown in Figure 9.1, the water-insoluble anthracene-9-carbinol (a diene, **1**) and *N*-methylmaleimide (a dienophile, **2**) with water, one observes a distinct powder and water. The powder does not appear to dissolve. When one refluxes (heats at the boiling temperature) the mixture, one does not observe any changes; the powder looks the same. Yet, if one follows the reaction progress, for example, by the thin-layer chromatography, one sees a steady reaction progress towards formation of the Diels–Alder adduct **3**. After about 1 h of refluxing, the reaction is complete. When one filters the insoluble powder, one finds that it is a virtually pure product **3**!

This reaction success was initially explained by a hydrophobic effect in which the water-insoluble reactants **1** and **2** are pushed towards each other in an attempt to escape water. Their close proximity allows for an efficient molecular orbital overlap between them, and thus a faster and more selective reaction. The operation of the hydrophobic effect is supported by the influence various salts exert on the reaction. Depending on the structure of the added salts, the hydrophobic effect and the resulting reaction rates may increase (salting-out effect, by, e.g., LiCl) or may slightly decrease (salting-in, by, e.g., guanidinium chloride). Much work has been done on the additives that promote hydrophobic interaction. For example, the addition of mono- or disaccharides and mono- or polyhydroxyl alcohols increases the hydrophobic effect. This would have an immediate applicability to the prebiotic soup, in which a pure water as a medium is unrealistic, but the presence of salts, small polar molecules, or amino acids is expected.

9.1.3 Passerini Multicomponent Reaction *on Water*

We now examine another example of the *on water* reaction, the Passerini multicomponent reaction. Various multicomponent organic reactions are known to occur *on water*. They typically give products of high purity and in almost quantitative yields. Generally, they are accelerated by water. These reactions have been studied recently from both synthetic and mechanistic points of view, because many of them are important for the pharmaceutical industry. For example, the Passerini reaction is useful in producing numerous derivatives via a combinatorial chemistry approach. An example of the Passerini reaction is shown in Figure 9.2. In this figure, we see that the three components, benzaldehyde (**4**), *tert*-butyl isocyanide (**5**), and benzoic acid (**6**), react in water to give a single product, benzoic acid, *tert*-butylcarbamoyl-phenyl-methyl ester (**7**).

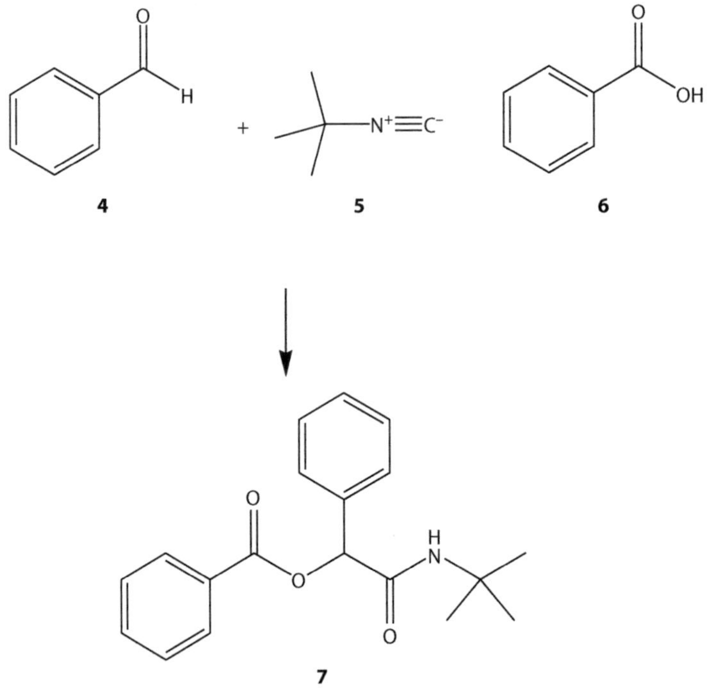

FIGURE 9.2   An example of a multicomponent Passerini reaction performed *on water*.

When the Passerini *on water* reaction that is shown in Figure 9.2 is performed in the laboratory, the observations may be misleading, just like in the case of the previously discussed *on water* Diels–Alder reaction. When the ingredients **4**, **5**, and **6** for the Passerini reaction are mixed together in water, one observes a white solid, which looks just like benzoic acid (**6**) and which does not seem to dissolve. Upon mixing, within a few minutes, the white solid never disappears, but the reaction is complete. If one removes the white solid from the aqueous medium and analyzes it, one finds that this solid is not the unreacted benzoic acid (**6**), as it would appear, but is the essentially pure product, namely, the *tert*-butylcarbamoyl-phenyl-methyl ester of benzoic acid (**7**). The speed of the reaction is remarkable.

All the starting materials of the Passerini reaction are feasible prebiotic compounds. By *feasible* we mean that these and similar compounds have either been found on the meteorites, such as Murchison, or could be reasonably easily synthesized by the known simulated prebiotic reactions. Since Passerini reaction can be performed *on water*, it has a great prebiotic potential. The resulting product could help build chemical diversity in the prebiotic soup.

9.2 IMPORTANCE OF MULTICOMPONENT REACTIONS IN PREBIOTIC CHEMISTRY

Not so long ago, it seemed that the field of prebiotic synthesis has exhausted all the imaginable synthetic pathways. The success in synthesizing various complex prebiotic molecules that are biorelevant was sporadic. The progress in expanding the basic synthetic menu was not steady. Two major synthetic difficulties interfered with the progress.

The first difficulty is that the prebiotic syntheses in the laboratory should not be assisted, or be minimally assisted, in order to mimic the primordial systems. For example, we are allowed to mix together all the chemicals that are needed for the synthesis of the desired product, but are not allowed to add them sequentially, even though the synthesis we are trying to model may occur in several distinct steps. In addition, we are not allowed to use the procedures that in regular syntheses assure good yield and purity of the product, such as the protection and deprotection of the reactive functional groups, filtering out the solids, separation of the reaction layers, extractions, and various purifications steps.

Organic chemists typically perform either a linear synthesis, shown in Scheme 9.1, or a convergent synthesis, which is presented in Scheme 9.2.

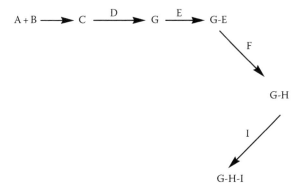

SCHEME 9.1   Linear synthesis of the target molecule G-H-I, achieved by a series of sequential steps.

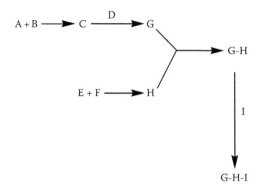

SCHEME 9.2 Convergent synthesis of the target molecule G-H-I, achieved by a combination of two building blocks that were synthesized separately.

According to Scheme 9.1, chemists would first combine A and B, with the required solvent and catalyst as needed, and would monitor the reaction progress by some standard analytical means (e.g., spectroscopically or via a thin-layer chromatography). After the reaction is done, chemists would purify C to remove any unreacted starting materials A and B and other possible impurities, such as the organic by-products and inorganic compounds. Chemists would also employ protecting groups as needed. Let us assume that C has several functional groups, as often is the case. If the reactant D would destroy some functional group in C, before it could productively react with C to give G, chemists would need to protect the sensitive group in C first. Then chemists would combine the protected compound with D to obtain G, which would still have the protecting group attached. The protecting group can be removed at this or at the later steps, as needed. None of these options are available in the laboratory simulations of the prebiotic reactions. All the reagents would have to be mixed at once.

Organic chemists employ also convergent synthesis, depicted in Scheme 9.2.

This type of synthesis is popular for making complex molecules. In convergent synthesis, two (or more) building blocks of the target molecule are synthesized separately and are then reacted together. Scheme 9.2 shows the convergent synthesis of G-H-I, a molecule whose linear synthesis was previously shown in Scheme 9.1. Prebiotic chemists sometimes utilize the convergent approach for making components of complex molecules, such as nucleic acids, by synthesizing separately their building blocks and then combining them in a subsequent step. Since prebiotic reactions are not supposed to be assisted, prebiotic chemists justify this approach by a possibility that the building blocks in some cases may be made at different locations and then be mixed by chance.

The prebiotic equivalent of Schemes 9.1 and 9.2 is presented in Scheme 9.3.

In Scheme 9.3, all the ingredients are mixed together and the desired product G-H-I is supposed to form with no intervention. Intractable mixtures often result, with only a small amount of the desired product being formed. The obtained mixture is often composed

$$A + B + D + E + F + I \longrightarrow G\text{-}H\text{-}I$$

SCHEME 9.3 Prebiotic equivalent of Schemes 9.1 and 9.2. All the chemicals are mixed together in one step.

mostly of the undesired by-products and side reactions. Scheme 9.3 dramatically illustrates difficulty of any complicated multistep reaction to occur prebiotically to give the desired product in a reasonable yield and purity.

The multicomponent reactions offer great advantages over the synthetic pathways that we have described in Schemes 9.1 and 9.2. For example, in the Passerini reaction (Figure 9.2), we have three components that react fast and give a single product. If the structures of the components are varied, numerous new compounds can be synthesized easily. This could be a great generator of diversity of chemicals in the prebiotic soup. All three components, the isocyanide, the aldehyde, and the carboxylic acid, are prebiotically feasible. New multicomponent reactions have been discovered based on isocyanide chemistry in the regular, nonprebiotic chemistry. This opened new possibilities for the prebiotic chemistry. Thus, without protection/deprotection, isolation, purification, and similar manipulations, the Passerini multicomponent reaction gives the product in one step, in virtually quantitative yield, and in high purity.

Recently, Sutherland and his coworkers have successfully applied multicomponent reactions of the Passerini type to various aspects of prebiotic synthesis of nucleotides and their components, which were previously notoriously difficult. In some applications, they have used the isocyanide chemistry that was modeled by the multicomponent Ugi reaction (related to the Passerini reaction), in which an isocyanide, an aldehyde, and ammonium chloride were reacted.

9.3 SOLVENTLESS AND SOLID-STATE REACTIONS IN PREBIOTIC CHEMISTRY

Myriad organic compounds have been found on meteorites that belong to the type of meteorites called carbonaceous chondrites. The most famous and most extensively studied such meteorite is named Murchison meteorite. Tens of thousands of different molecular compositions were revealed on the Murchison meteorite by the application of ultrahigh-resolution analytical methods that combine state-of-the-art mass spectrometry, liquid chromatography, and nuclear magnetic resonance. The examples of organic compounds found on the Murchison meteorite include aliphatic and aromatic hydrocarbons, carboxylic acids, hydroxyl carboxylic acids, dicarboxylic acids, carboxamides, amino acids, aldehydes, ketones, alcohols, amines, purines and pyrimidines, nitrogen heterocycles, sulfonic acids, and phosphonic acids, among many others. Meteorites are obtained from meteors, and the latter from asteroids. It has been proposed that the organic materials were brought to the early Earth on the meteorites. These chemicals were important for chemical evolution that led to life. A question arises how these chemicals were made under the extraterrestrial conditions. Chemistry on asteroids is considered important for the early chemical evolution. Water on asteroids was available only sporadically, causing so-called aqueous alteration (the change in composition of a rock, produced in response to interactions with H_2O-bearing ices, liquids, and vapors by chemical weathering). Although most organic compounds are not water soluble, they could react *on water*, as described in the previous sections. We have proposed an additional way for the reactions on meteorites to occur, without water or any solvent, as solventless or solid-state reactions.

9.3.1 How Solventless and Solid-State Reactions Occur

Any student of organic chemistry knows that if the melting point of the product is lowered, this typically indicates impurities. Students may recall the mixed-melting-point procedure, by which one can identify the unknown compounds. One prepares approximately 1:1 mixture of the unknown compound with the known compound, which one suspects the unknown could be. Then one takes the melting point of the mixture. If the melting point of the mixture is the same as that of the unknown, the unknown is identical to the known compound. If the melting point is substantially lower, the guess is incorrect.

The mixed-melting-point procedure is based on the phenomenon that the melting point of a pure compound becomes lower when a different compound is added to it. The two compounds act as impurities to each other. The incorporation of the impurities into the crystalline lattice of a pure compound breaks up its regular crystalline pattern and results in the lowering of the melting point.

As more impurities are added, the melting point becomes progressively lower. In some instances, the lowering of the melting point is sufficient to cause the mixture to melt. If so, chemical reactions could occur in the melted state and the need for the solvent is eliminated. Such reactions are known as *solventless*.

Another scenario is when the solid reactants are mixed, the melting point is lowered, but the mixture does not melt. Such a mixture, especially if well ground, may become reactive. Sometimes mild heating speeds up the reactions. Such reactions are known as solid–solid reactions. Other variations exist. For example, when a liquid and a solid component are mixed or ground together, the suspension becomes reactive. Sometimes a gas–solid reaction is successful. All these scenarios could be fruitful on asteroids.

Our main focus is on solventless and solid–solid reactions. The asteroids are presumably rich in the organic compounds, judging by the analysis of Murchison and similar carbonaceous chondrites. Some of these compounds may be reactive towards each other, but others may not be. However, the unreactive compounds could lower the melting points of the mixtures in which the reactive compounds are found, enabling their reactions to occur in the melted state. The temperature estimates on the asteroids include a range from 25°C to 100°C, which is consistent with aqueous alterations and is friendly to the organic reactions and preservation of the organic compounds. The heat to maintain this temperature range is provided by the radioactive processes or impacts.

9.3.2 Examples of Prebiotically Relevant Solventless and Solid-State Reactions

Literature describes many organic reactions that occur as solventless or in the solid state. These reactions often occur rapidly and are remarkably efficient. Examples include all major types of organic reactions, such as oxidations, reductions, eliminations, substitutions, condensations, cyclizations, rearrangements, formation of carbon–carbon, carbon–oxygen, and carbon–nitrogen bonds, among many others. Detailed experimental procedures for solvent-free reactions are compiled also in a recent book by Tanaka. Many such procedures fulfill the prebiotic requirements either completely or they could be made prebiotic rather easily, by introducing some minor modifications. Especially interesting is the formation of

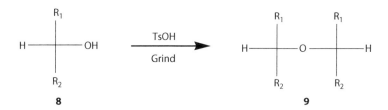

FIGURE 9.3   Solid-state formation of ethers from alcohols.

ethers in the solid state, from the solid mixture of the alcohols and in the presence of acid catalysts. Such a synthesis could be a prebiotically feasible way to produce various ethers. Such ethers could be used as membranes for the primitive life. While most contemporary life uses lipids, which are esters, some Archaea's membranes are ethers. Other interesting examples include syntheses of rather large molecules that are related to porphyrins. These and other examples are presented in the following, with the chemical equations and essential conditions. They illustrate a great potential for prebiotic chemistry.

In Figure 9.3, we show a solid-state reaction of alcohols (**8**) to form ethers (**9**) under acid catalysis of TsOH (*p*-toluenesulfonic acid).

Alcohols (**8**) are prebiotically feasible, but the corresponding ethers (**9**) are difficult to make prebiotically in the aqueous solution, since water needs to be driven off during the reaction. The solid-state etherification reaction provides a viable prebiotic path to the ethers. TsOH could be substituted by prebiotic acid catalysts, such as various acidic clays. Heating to remove water would be helpful.

In Figure 9.4, we show the solid-state oxidative coupling of a phenol, 2-naphthol (**10**), which is converted to [1,1′-binaphthalene]-2,2′-diol, known as BINOL (**11**) with $FeCl_3$ as a catalyst.

BINOL is formed as a racemic mixture of two enantiomers, which are shown in Figure 9.4. BINOL enantiomers can be readily separated and are stable towards racemization. This synthesis is prebiotically feasible and the chiral properties of BINOL may have a prebiotic catalytic potential.

Aldol-type reactions, in which carbon–carbon bonds are made, play an important role in chemistry and biology and are also of a great prebiotic importance. An example of aldol reaction in which basic alumina is used as a catalyst, under solvent-free conditions, is shown in Figure 9.5.

An aromatic aldehyde (**12**), which does not possess an α-H (hydrogen next to the aldehyde carbonyl), reacts with an aromatic ketone (**13**), which has an α-H (hydrogen next to the ketone carbonyl). The ketone **13** is able to make an α-carbanion (enolate) in a base-catalyzed reaction in which the α-H is removed. This α-carbanion reacts with the carbonyl of **12** in an addition reaction to eventually yield the product **14**. The aldehyde **12** cannot make an α-carbanion, since it does not have an α-H, and thus is unreactive with another aldehyde **12** or ketone **13** molecule. The α-carbanion of ketone **13** reacts only with the aldehyde **12** but does not react with other molecules of the ketone **13**, since the reaction is thermodynamically unfavorable. Therefore, the reaction is clean since there is no cross-reactivity and a single product is obtained. This reaction also illustrates that a belief

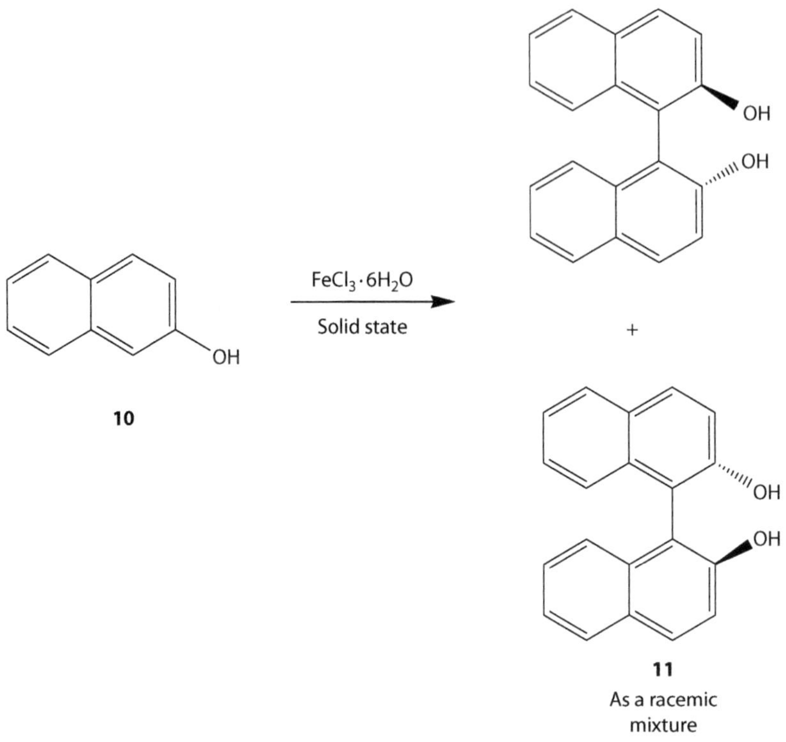

FIGURE 9.4 Solid-state oxidative coupling of phenols: conversion of 2-naphthol to BINOL.

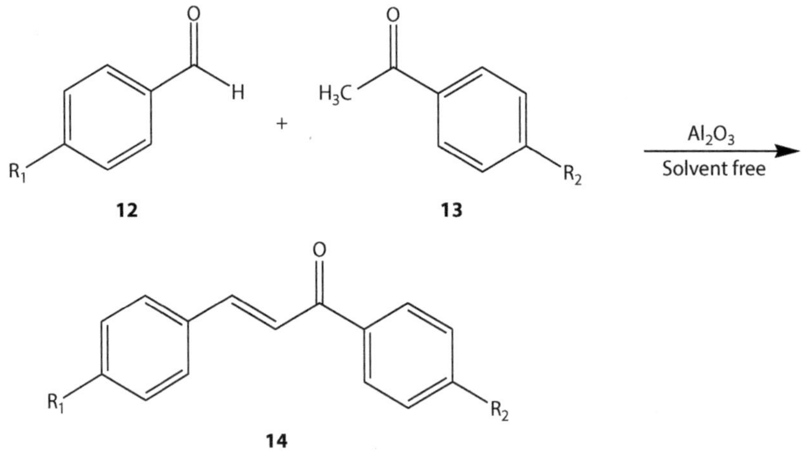

FIGURE 9.5 Solvent-free aldol condensation with basic alumina as a catalyst.

that prebiotically "everything reacts with everything" may be sometimes misleading. Single products may be obtained due to the intrinsic limitations in the reactivities of the starting materials, as in this example.

A solventless aldol reaction in which two solids melt upon grinding is shown in Figure 9.6. The aldehyde **15** reacts with the ketone **16** to provide a single product **17**, with the help of NaOH as a basic catalyst.

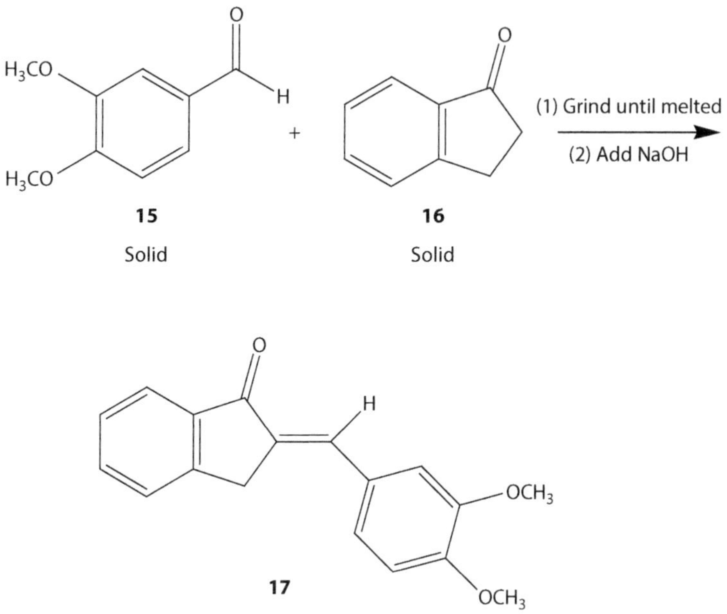

FIGURE 9.6   Solventless aldol condensation with NaOH as a catalyst.

The mechanism of the reaction is the same as in the previous case, shown in Figure 9.5. Again, a single product is obtained for the same reason. In this case, when the two solids, **15** and **16**, are mixed together, the melting point of the mixture is lowered sufficiently for melting to occur.

A rather spectacular synthesis is shown in Figure 9.7, in which a corrole, a porphyrin analogue, is formed. The reaction occurs under solventless conditions with Al_2O_3 as catalyst. Pyrrole (**18**) reacts with an aromatic aldehyde (**19**) to provide a corrole (**20**). Such a corrole could serve as a primitive enzyme, since it has a metal complexing site in the center of the molecule. A diversity of the structures of corroles can be achieved by varying the aromatic aldehyde (**19**) structure via introduction of different substituents into the aromatic ring.

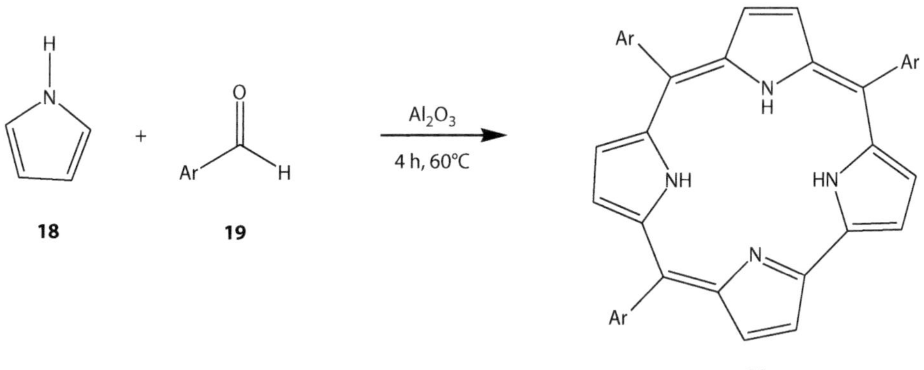

FIGURE 9.7   Solventless condensation of pyrrole and aromatic aldehyde to give a corrole.

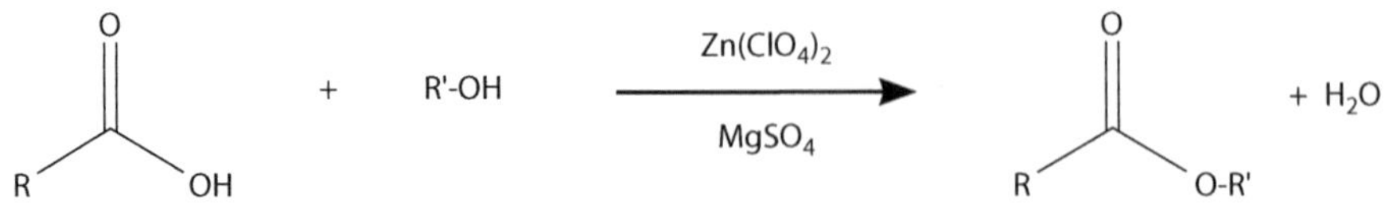

FIGURE 9.8   Solvent-free formation of esters from carboxylic acids and alcohols.

Esters can be prepared from carboxylic acids and alcohols, with strong acid as a catalyst, in a process named Fischer esterification. To obtain the ester product, the equilibrium needs to be shifted towards the products, which are the desired ester and undesired water. This is achieved in the organic laboratory typically by using conc. H_2SO_4 as a catalyst and a dehydrating agent, or by contraptions such as Dean–Stark trap, to remove water. Neither procedure can be worked out to be prebiotic. Although both carboxylic acids and alcohols are prebiotically feasible, the reaction is problematic in the aqueous solution, since water needs to be removed to shift the equilibrium towards the desired ester. The solvent-free esterification shown in Figure 9.8 provides a more feasible prebiotic alternative.

The carboxylic acid **21** reacts with the alcohol **22** to provide the ester **23**, without the solvent, with Lewis acid catalysts. The $MgSO_4$ salt could bind water to form a hydrate. The water from the hydrate could then be removed by heating. $MgSO_4$ could thus serve as a recyclable dehydrating agent. This appears to be prebiotically feasible.

The solid–solid esterification of cholesterol is definitely feasible for the prelife. The reaction is shown in Figure 9.9.

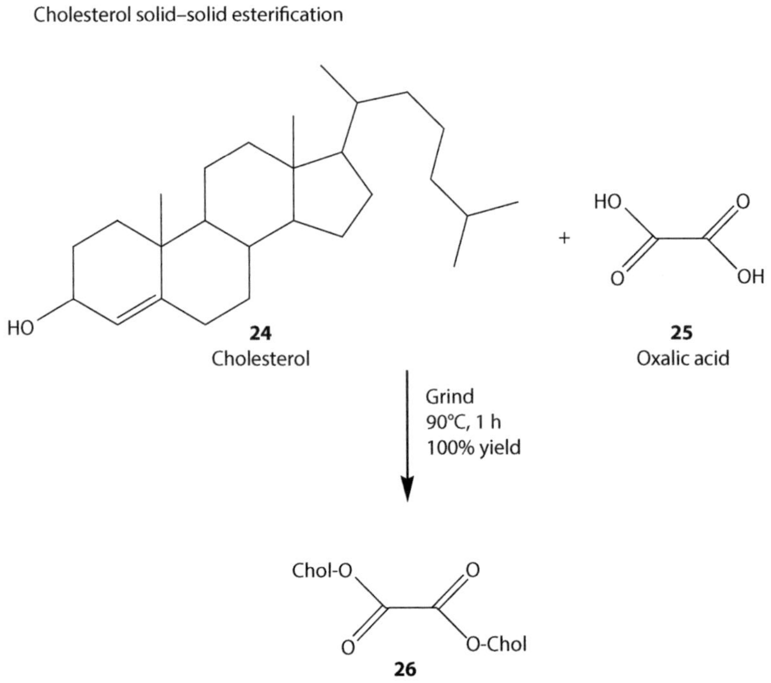

FIGURE 9.9   Solid-state esterification of cholesterol.

Cholesterol (**24**) reacts with oxalic acid (**25**) to give the corresponding ester (**26**), upon mixing and heating in a solid–solid reaction. The ester **26** could perhaps serve as a primitive lipid membrane in the primordial protocells.

It is often difficult to get a high yield of well-defined macromolecular materials in the aqueous solution. The example in Figure 9.10 dramatically illustrates success of the solid-state chemistry in creating such materials.

In this example, resorcinol (**27**) reacts with benzaldehyde or its derivatives (**28**), under the solvent-free conditions. Acid catalyst (TsOH) is used. A remarkable cyclocondensation occurs at room temperature. After only 1 h, a large cyclic molecule **29** is formed. It could perhaps serve as a primitive enzyme. The TsOH could probably be substituted with some acidic clay.

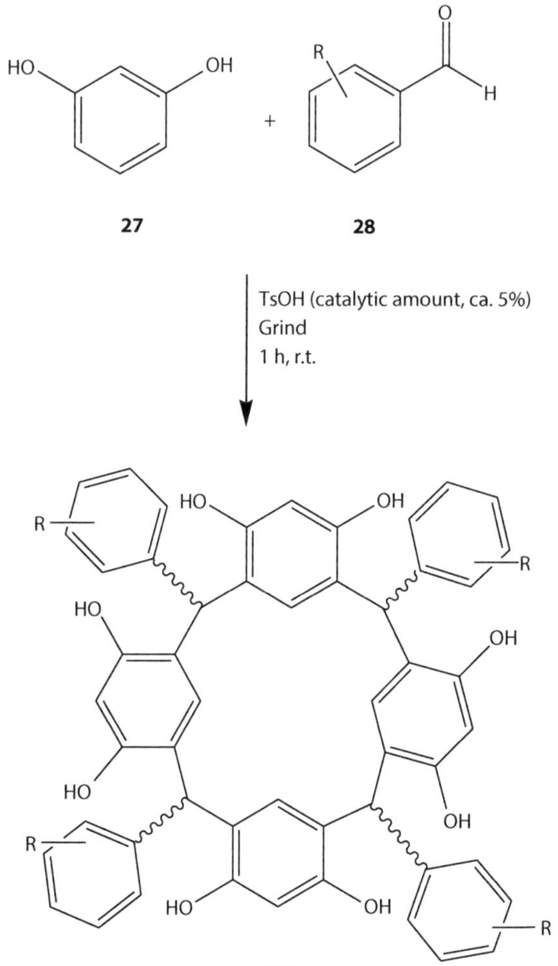

FIGURE 9.10   Resorcinol/aldehyde solvent-free cyclocondensation.

9.4 CONCLUSIONS

In this chapter, we have shown new synthetic avenues for prebiotic chemistry. These include *on water*, solventless, and solid–solid reactions. These reactions expand the repertoire of prebiotic chemistry in the prebiotic soup and in the oceans, as well as on the asteroids.

We have discussed Diels–Alder and Passerini multicomponent reactions, which provide rich chemistry starting from the prebiotically feasible starting materials. These starting materials are not water soluble, but the reactions occur smoothly, in a selective manner and high yields, in the aqueous suspensions (*on water*). We have also addressed the mechanism by which such *on water* reactions occur. There are many more known organic *on water* reactions, and they should be gradually explored for their prebiotic potential.

Solventless and solid–solid reactions are relevant to the chemistry on asteroids. We show selected examples of the key organic reactions, including formation of carbon–carbon and carbon–oxygen bonds, which occur with ease in the solventless media. High yields and clean products result. Especially important and relevant are esterifications, etherifications, aldol-type condensations, and preparations of the large cyclic molecules. The latter could serve as primitive enzymes. We have also discussed a general mechanism by which the solventless and solid–solid reactions occur. Again, many more such reactions are known within the general organic chemistry repertoire, and they should be explored for their prebiotic potential.

GLOSSARY

Aldol reaction: A reaction between aldehydes and ketones that can make enolate ions (carbanions next to the carbonyl) and that produces new C—C bonds.

Archaea: A domain (kingdom) of single-celled microorganisms that are of ancient origin; many live in the extreme environments.

Combinatorial chemistry: A synthetic approach by which a large number of compounds can be made in a single process.

Convergent synthesis: A synthetic method in which two (or more) building blocks of the target molecule are synthesized separately and are then reacted together.

Diels–Alder reaction: A cycloaddition reaction between a diene and dienophile. This reaction makes C—C bonds and cyclohexane rings.

Functional groups: Chemical groups that characterize chemical behavior of the organic molecules, such as alcohols, ethers, esters, carboxylic acids, amines, and amides, among many others.

Hydrophobic effect: A tendency of nonpolar molecules to aggregate in aqueous solution and exclude water molecules in the process.

In water reactions: Reactions between components that are soluble in water and that occur in water.

Linear synthesis: A synthetic method in which synthesis of the target molecule is achieved by a series of sequential steps.

Mechanism of reaction: A stepwise path from the reactants to the products; a step-by-step sequence.

Mixed melting point: Melting point of a mixture, typically a 1:1 ratio between an unknown compound and a known compound that is believed to be identical to the unknown.

***On water* reactions:** Reactions between components that are not soluble in water and yet react in water (aqueous suspension).

Passerini reaction: A multicomponent reaction, with three components, an aldehyde, a carboxylic acid, and an isocyanide.

Protecting groups: Chemical groups that modify a selected functional group in the molecule to make it resistant to selected reagents. The protecting groups should come off easily to regenerate the original group. This is called a deprotection process.

Reflux: Heating a solution at its boiling point. This ensures heating at a constant temperature. A typical setup includes a flask with the solution with the attached condenser that is placed in a vertical position. This ensures that the boiled liquid is not lost, but it condenses and returns to the flask.

Solid–solid reactions: Reactions in which the solids are mixed, but do not melt. The reaction then occurs in the solid state.

Solventless reactions: Reactions that occur without solvent, typically by grinding two solids that then act as impurities to each other, which results in the overall melting of the mixture. Then the melted medium acts in lieu of the solvent.

Spectroscopy methods: Methods used for identification and often quantification of mostly organic materials. Examples include infrared, nuclear magnetic resonance, and mass spectrometry, among others; please consult basic organic books for a more detailed coverage.

Thin-layer chromatography: A method of separation of compounds on a thin layer of adsorbent material, such as silica gel of aluminum oxide, which is deposited on a glass plate or a plastic or aluminum sheet (so-called TLC plate). The material is applied on the TLC plate and is moved on the adsorbent by an appropriate solvent. The positions of the spots that correspond to the various compounds from the mixture can be visualized often by the ultraviolet light lamp.

REVIEW QUESTIONS

1. Explain on an example of a Diels–Alder reaction how two components (a diene and a dienophile) that are not soluble in water still can react in a water suspension (*on water*) to give the cycloaddition product (Diels–Alder *adduct*).

2. Diels–Alder reactions are often faster when performed *on water* than in an organic solvent in which both starting components are freely soluble. Offer a reasonable explanation for this observation.

3. Why are multicomponent reactions important for prebiotic chemistry? Provide more than one reason for their importance.

4. What is the difference between *in water* and *on water* reactions? Why are *on water* reactions important in prebiotic chemistry?

5. Give an example of a solid-state method for making ethers. Why is making ethers prebiotically important? What are the possible uses of ethers in primitive protocells?

6. Simulated prebiotic reactions are typically done in water. What other organic and inorganic ingredients should be added to water to make the experiments more reflective of the composition of the prebiotic soup?

7. Research the papers by Sutherland and his coworkers about prebiotic synthesis of nucleotides and their components, which are listed under the recommended readings. What have they achieved that others were not able to?

8. Explain the mixed-melting-point principle and how it provides a foundation for the solventless reactions.

9. Consider the following situation on an asteroid. Two solid compounds could react with each other chemically (thus their reactivities are appropriately matched), but their mixed melting point is very high. What options would exist on the asteroid to lower their mixed melting point?

10. Examine the experiments in your organic laboratory textbook. Look for the experiments that are labeled *green*. Some of the green (environmentally friendly) experiments are performed in water. Which products do such green experiments give? Do you think that such products would be of a prebiotic value? Use your knowledge of a general biochemistry as a guide and assume that some of the biochemical compounds were originally assimilated by the primitive protocells from the prebiotic environment.

RECOMMENDED READING

Books

Carey, F. A. and Sundberg, R. J. 2001. *Advanced Organic Chemistry, Part B: Reactions and Synthesis*, 4th edn. New York: Kluwer/Plenum Publisher.

Kolb, V. M. 2012. Application of the organic "on water" reactions to prebiotic chemistry. In *Instruments, Methods, and Missions for Astrobiology XV*, R. B. Hoover, G. V. Levin, and A. Y. Rozanov, eds., *Proceedings of SPIE*, Bellingham, Washington: SPIE (The International Society for Optics and Photonics), Vol. 8521, 85210D, pp. 1–6.

Lindström, U. M., ed. 2007. *Organic Reactions in Water: Principles, Strategies and Applications*. Oxford, U.K.: Blackwell Publishing.

Mason, S. F. 1991. *Chemical Evolution, Origins of the Elements, Molecules and Living Systems*. Oxford, U.K.: Oxford University Press.

Miller, S. L. and Orgel, L. E. 1974. *The Origins of Life on Earth*. Englewood Cliffs, NJ: Prentice Hall, Inc.

Shaw, A. M. 2006. *Astrochemistry, From Astronomy to Astrobiology*. New York: Wiley, especially Chapter 6: Meteorite and comet chemistry, pp. 157–192, and Chapter 8: Prebiotic chemistry, pp. 225–255.

Solomons, G. and Fryhle, C. 2013. *Organic Chemistry*, 11th edn. New York: Wiley.

Tanaka, K. 2009. *Solvent-Free Organic Synthesis*, 2nd edn. Weinheim, Germany: Wiley-VCH.

Zhang, W. and Cue, B. W., eds. 2012. *Green Techniques for Organic Synthesis and Medicinal Chemistry.* Chichester, U.K.: Wiley. (Please observe that "green chemistry books" contain valuable information about organic reactions in the aqueous medium, since water is a "green" solvent, namely environmentally friendly.)

Journals

Breslow, R. 1991. Hydrophobic effects on simple organic reactions in water. *Acc. Chem. Res.* 24: 159–164.

Breslow, R. 2006. The hydrophobic effect in reaction mechanism studies and in catalysis by artificial enzymes. *J. Phys. Org. Chem.* 19: 813–822.

Chanda, A. and Fokin, V. V. 2009. Organic synthesis 'on water'. *Chem. Rev.* 109: 725–748.

Hooper, M. H. and DeBoef, B. 2009. A green multicomponent reaction for the organic chemistry laboratory, the aqueous Passerini reaction. *J. Chem. Ed.* 86: 1077–1079.

Kaupp, G. 2005. Organic solid-state reactions with 100% yield. *Top. Curr. Chem.* 254: 95–183.

Klijn, J. E. and Engberts, J. B. F. N. 2005. Fast reactions 'on water'. *Nature* 435: 746–747.

Kolb, V. M. 2010. On the applicability of the green chemistry principles to sustainability of organic matter on asteroids. *Sustainability* 2: 1624–1631.

Kolb, V. M. 2012. On the applicability of solventless and solid state reactions to the meteoritic chemistry. *Int. J. Astrobiol.* 11: 43–50.

Kumar, A. 2001. Salt effects on Diels–Alder reaction kinetics. *Chem. Rev.* 101: 1–19.

Lindström, U. M. 2002. Stereoselective organic reactions in water. *Chem. Rev.* 102: 2751–2772.

Manna, A. and Kumar, A. 2013. Why does water accelerate organic reactions under heterogeneous condition? *J. Phys. Chem.* 117: 2446–2454.

Mullen, L. B. and Sutherland, J. D. 2007. Simultaneous nucleotide activation and synthesis of amino acid amides by a potentially prebiotic multi-component reaction. *Angew. Chem., Int. Ed.* 46: 8063–8066.

Narayan, S., Muldoon, J., Finn, M. G., Fokin, V. V., Kolb, H. C., and Sharpless, K. B. 2005. "On water": Unique reactivity of organic compounds in aqueous suspension. *Angew. Chem., Int. Ed.* 44: 3275–3279.

Otto, S. and Engberts, J. B. F. N. 2000. Diels–Alder reactions in water. *Pure Appl. Chem.* 72: 1365–1372.

Pirrung, M. C. and Sarma, K. D. 2004. Multicomponent reactions are accelerated in water. *J. Am. Chem. Soc.* 126: 444–445.

Pirrung, M. C., Sarma, K. D., and Wang, J. 2008. Hydrophobicity and mixing effects on select heterogeneous, water-accelerated synthetic reactions. *J. Org. Chem.* 73: 8723–8730.

Powner, M. W. and Sutherland, J. D. 2011. Prebiotic chemistry: A new modus operandi. *Philos. Trans. R. Soc. B* 366: 2870–2877.

Powner, M. W., Sutherland, J. D., and Szostak, J. W. 2010. Chemoselective multicomponent one-pot assembly of purine precursors in water. *J. Am. Chem. Soc.* 132: 16677–16688.

Rideout, D. C. and Breslow, R. 1980. Hydrophobic acceleration of Diels–Alder reaction. *J. Am. Chem. Soc.* 102: 7816–7817.

Rothenberg, G., Downie, A. P., Raston, C. L., and Scott, J. L. 2001. Understanding solid/solid organic reactions. *J. Am. Chem. Soc.* 123: 8701–8708.

Sarma, D., Pawar, S. S., Deshpander, S. S., and Kumar, A. 2006. Hydrophobic effects in a simple Diels–Alder reaction in water. *Tetrahedron Lett.* 47: 3957–3958.

Schmitt-Kopplin, P., Gabelica, Z., Gougeon, R. D., Fekete, A., Kanawati, B., Harir, M., Gebefuegi, I., Eckel, G., and Hertkorn, N. 2010. High molecular diversity of extraterrestrial organic matter in Murchison meteorite revealed 40 years after its fall. *Proc. Natl. Acad. Sci. USA* 107: 2763–2768.

Shapiro, N. and Vigalok, A. 2008. Highly efficient organic reactions "on water", "in water", and both. *Angew. Chem.* 120: 2891–2894.

Sutherland, J. D., Mullen, L. B., and Buchet, F. F. 2008. Potentially prebiotic Passerini-type reactions of phosphates. *Synlett* 14: 2161–2163.

Tanaka, K. and Toda, F. 2000. Solvent-free organic syntheses. *Chem. Rev.* 100: 1025–1074.

Thomas, J. M. 1979. Organic reactions in the solid state: Accident and design. *Pure Appl. Chem.* 51: 1065–1082.

Toda, F. 1995. Solid state organic chemistry: Efficient reactions, remarkable yields, and stereoselectivity. *Acc. Chem. Res.* 28: 480–486.

Toda, F., Takumi, H., and Akehi, M. 1990. Efficient solid-state reactions of alcohols: Dehydration, rearrangement, and substitution. *J. Chem. Soc. Chem. Commun.* 1270–1271.

CHAPTER 10

Encapsulation of Organic Materials in Protocells

Erica A. Frankel, Daniel C. Dewey, and Christine D. Keating

CONTENTS

10.1 INTRODUCTION

Compartmentalization was important in the origins of life. This chapter will outline general physical and chemical principles that could lead to compartmentalization of organic matter on the prebiotic Earth. Possible types of compartments and mechanisms for concentrating organic matter within them will be discussed. Special emphases will be placed on amphiphile self-assemblies and the phenomenon of aqueous phase separation in polymer-containing solutions as means of forming prebiotic reactors.

Thus far, in earlier chapters, we have learned about probable conditions on the early Earth and processes that could lead to the formation of organic molecules, including polymeric precursors to the molecules of life. In this chapter, we examine physical mechanisms by which this organic material could have become concentrated together and encapsulated within a precursor of living cells. A characteristic of living organisms is that they have a clear boundary between inside and outside, or self and nonself. They are distinct from their surroundings. This chapter will introduce several classes of compartments that would likely have been present on the early Earth that could have assisted in the chemical synthesis and concentration of organic molecules (Koonin, 2007). Once sufficient quantities of organic molecules were present, additional types of compartmentalization would have become possible. We will learn about two main classes of self-assembled microcompartments that form spontaneously from nonliving organic matter: amphiphile vesicles and polymer-rich aqueous phase droplets. Both types of organic compartments can provide a semipermeable boundary while encapsulating internal compositions distinct from the external milieu (Rasmussen et al., 2003). Such structures could have served as important microreactors during the chemical evolution of nonliving matter and as templates for the formation of the first cellular life.

10.2 IMPORTANCE OF COMPARTMENTS

Compartmentalization was likely essential on the early Earth to facilitate the chemical reactions that led to the appearance of organic macromolecules and would have further enabled noncovalent interactions and molecular self-assembly (Chen and Silver, 2012). Many chemical reactions are concentration dependent. For example, the rate of amino acid polymerization to form polypeptides depends upon the concentration of amino acid precursors. Prior to cellular life, special sites such as mineral surfaces, compartments within porous rock, salty liquid pools left behind when water freezes or evaporates, or aerosol droplets could have served as sites for molecule concentration and reaction (Koonin, 2007). This is thought to be crucial because the earliest organic molecules would have been very dilute, and hence mechanisms for their selection and concentration for reaction were needed. Various types of microenvironments on the early Earth could have supplied necessary reagents or conditions to facilitate the formation of macromolecules such as peptides or nucleic acids. These include thermal vents, air–ocean, and ocean–rock interfaces, which remain an important resource for many elements and organic compounds today.

Once synthesized, the prebiotic organic materials would need to be further compartmentalized into discrete units that could serve as early, nonliving progenitors of the first cellular life. These precellular assemblies, termed protocells, are thought to have self-assembled from organic molecules and to have facilitated the conversion of the organic matter into functional macromolecules such as catalytically active nucleic acids and proteins (Chen and Silver, 2012). Protocellular compartments are generally considered to be characterized by a semipermeable boundary and to capture one or more additional aspects of living cells; ultimately, these structures must have developed mechanisms to grow and divide while replicating their internal informational and metabolic molecules (Luisi et al., 2006).

Compartmentalization of organic materials into protocells could serve several important functions. First, protocells could concentrate the otherwise dilute progenitors of biomolecules to enable their reactions (e.g., polymerization, addition of functional groups) and interactions (e.g., binding, self-assembly, competition). This could be accomplished selectively, such that only some molecules were concentrated into the compartments and allowed to interact (by selective interactions or semipermeable boundaries), while others remained in the dilute external media. Separate compartments could have harbored dissimilar collections of molecules, resulting in different reaction outcomes (Branciamore et al., 2009). The interior of protocells would have had a distinct microenvironment that could facilitate certain processes, while preventing others, to protect against adverse reactions and conditions occurring in the surroundings.

Despite clear benefits of compartmentalization for selectively increasing local molecular concentrations, protection from adverse reactions, and providing favorable microenvironments, not all means nor all consequences of compartmentalization will necessarily be beneficial. For example, the permeability of the protocellular compartment must be limited such that molecules of interest can be retained while still enabling precursors to enter and unnecessary or waste molecules to depart (Morse and Mackenzie, 1998). Modern cells are contained within a proteolipid bilayer, in which sophisticated methods of passive and active transport are used to move ions and molecules in and out of the cell, generally involving membrane proteins (Chen and Silver, 2012). New materials are added to the membrane by biosynthesis, allowing it to expand as part of the cell's growth and division cycle. For early probiotic life to grow and divide, there would need to be some corresponding means of producing new molecular area and of controlling transport in the absence of membrane transport proteins. Even for much simpler compartments such as molecular layers on mineral clays, one must be alert for possible challenges such as the greater binding affinity that can arise with increased molecular length, which could make it difficult to release polymers after their formation.

10.3 COMPARTMENTS WERE PRESENT PRIOR TO THE APPEARANCE OF ORGANIC MATTER

The geophysics and properties of the prebiotic ocean, atmosphere, and rock formations are thought to have set the stage for the conversion of simple organic molecules into complex biopolymers, as shown in Figure 10.1.

Within these macrocosms existed a diverse selection of potential microenvironments, able to catalyze prebiotic chemistry. Solid–liquid and liquid–gas interfaces could have promoted the emergence of complex structures necessary for life (Vaughan and Lloyd, 2012). By preconcentrating certain rare organic molecules together, reactions were able to occur forming new, functional biopolymers. Material at these interfaces could have ultimately formed one of the organic microcompartments described in Sections 10.4 and 10.5; such structures may have been initially stabilized by interactions with structures such as mineral surfaces. In this section, we introduce several general classes of compartments that could have been present even before organic matter was prevalent on the early Earth and that may have facilitated the formation and concentration of the organic matter.

10.3.1 Concentration by Water Removal

In the warm pond originally envisioned by Darwin, repeated cycles of water evaporation and condensation would have provided a means of concentrating initially dilute organic matter (Darwin, 1870). One can envision several scenarios in which aqueous mixtures of organic molecules become concentrated, perhaps all the way to dryness, by loss of water to the atmosphere, and are rehydrated at a later time. This process could have facilitated reactions between the organic materials, which would otherwise be too dilute to interact with each other. The cyclic nature of concentration/dilution processes could give an added advantage to molecules in this type of an environment, by allowing reversible binding interactions between different sets of molecules that might have binding affinities too high under one set of conditions (i.e., cannot dissociate) and too unstable under another (i.e., cannot associate).

A related scenario could occur when water that contains organic solutes freezes. When ice forms from solutions that contain dissolved salts and organic molecules, these solutes can be excluded from the crystal and become concentrated. When particular salt concentrations are reached, an ice/solute mixture, the eutectic, forms with the lowest complete melting point of any combination of the same ingredients. At particular salt concentrations, a eutectic forms with the lowest complete melting point of any combination of the same ingredients. The eutectic can remain liquid well below the freezing point of water. The combination of various salts and metal ions present could have led to a eutectic composition within oceanic environments. The resulting salty, organic-rich eutectic mixtures could serve as microreactors in much of the same way that a partially evaporated *warm pond* could. Scientists have used the ice eutectic premise to demonstrate the compartmentalization of RNA and other cofactors for RNA replication reactions. A liquid pool remaining after sample freezing permitted RNA replication despite low overall concentrations of precursor molecules by colocalizing them within the eutectic liquid (Attwater et al., 2010). Although cold temperatures ($\sim-7°C$) are not generally associated with fast reaction rates, they could prove favorable in some cases due to the reduced risk of RNA degradation. While this particular example dealt with complex

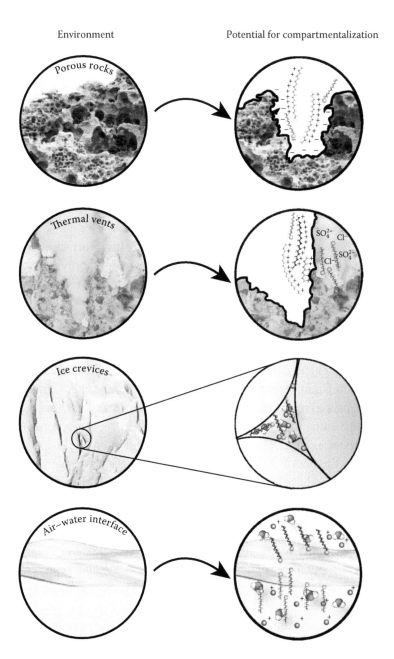

FIGURE 10.1 Regions on the Earth have the potential to form compartments of organic matter. Porous rock is an example of a solid–liquid interface, where charged solids have the ability to attract and compartmentalize organic matter. Thermal vents also have the potential to act as compartments and can display a wide range of gradients in temperature and ion concentration. Ice crevices act as a compartment through the removal of water upon freezing, leaving organic matter concentration in a small pool. Interfaces between water and air can compartmentalize matter through aerosol droplets.

functional macromolecules, it seems plausible that simpler reactions could have benefited from this manner of compartmentalization as well.

10.3.2 Solid–Liquid Interfaces

The solid–liquid interface provides a different chemical and physical environment from bulk fluid. Films of macromolecules can form and persist at such interfaces due to interactions with the mineral surface and may experience locally different pH or ion concentrations (Vaughan and Lloyd, 2012). Adsorption of molecules at mineral surfaces thus provides a means of selection and concentration. For example, clay mineral surfaces are able to facilitate polymerization of amino acids or nucleotides to form small, protein-like polypeptides or RNA nucleotides, respectively (Deamer, 1997). In addition, the surface of minerals may have acted as catalysts for redox transformation and respiration of oxygen and nitrate species, which precellular assemblies may have used for an electron source. Crevices and pores in rocks and clays provide particularly high surface area microenvironments where material required for the production of prebiotic polymers or amphiphiles could have been concentrated and assembled into compartments (Knauth, 2005).

10.3.3 Liquid–Gas Interfaces

Liquid–gas interfaces such as between planetary ocean and atmosphere also present potential routes to reaction preorganization that could have facilitated the emergence of simple organic molecules. In particular, aerosols close to the water–air interface on ancient Earth could have been significant. Aerosol particles, which are composed of either liquid or solid suspensions in air, have a large surface area due to their small size, contributing to the adsorption of organic films along their surface. Depending on their makeup, aerosols can undergo photochemical processes, nucleation and growth, coagulation and division, evaporation, and deposition into the ocean. All of these processes represent feasible ways in which biochemical processes could have been catalyzed (Griffith et al., 2012; Griffith and Vaida, 2012).

10.3.4 Environmental Gradients

Interfaces occurring in the environment, such as those discussed in the previous sections, are not the only locations where functional molecules could have emerged. Gradients in ionic strength, temperature, pH, and dissolved materials also represent special environments where reactions could preferentially occur. Modern cells need physicochemical ion gradients to live, grow, and divide. Gradients in specific ions and/or overall ionic strength may also have been important on the prebiotic Earth (Spitzer and Poolman, 2009). Temperature gradients could also have been important in molecule formation and association. The rates of many chemical reactions are temperature dependent, hence, diffusion between regions of higher and lower temperature, for example, could have provided a primitive sort of *annealing* of inter- and intramolecular binding. Gradients of pH would also have been important since the pH controls the charge on many organic molecules. For example, a carboxylic acid with a pKa of 4.5 will exist primarily in its protonated form below pH 4.5 (i.e., neutral) and primarily in its deprotonated state above this pH (i.e., negatively charged). The pKa represents the point at which half the molecules are protonated and half are deprotonated. This can greatly alter electrostatic interactions between molecules that have functional groups that can be protonated. Lastly, dissolved materials such as metal ions have important functions for the catalysis of functional molecules. In modern cells, enzymes harness metal ions for general acid–base catalysis, so it seems plausible that metal ions may have served the same function for early molecules and protocells. Flux of metal ions would increase catalytic activity, such as bond cleavage, to make new molecules (Spitzer and Poolman, 2009). It should be noted that the different types of physical microenvironments described in Sections 10.3.1 through 10.3.3 could also have been subjected to gradients in composition and physical properties. For example, porous rocks near hydrothermal vents could have experienced particularly favorable conditions for the formation of organic material and its ultimate assembly to form protocells. Under simulated hydrothermal vent conditions, pyruvate has been synthesized from alkyl thiols and carbon monoxide at high temperature (250°C) in the presence of metal sulfides, all reagents that were likely to be found in hydrothermal vents (Novikov and Copley, 2013).

10.4 AMPHIPHILIC MOLECULES FORM BILAYER COMPARTMENTS

Once organic materials started to accumulate, it is likely that their chemical and physical properties led to their association with each other. Of particular interest are molecular amphiphiles. Biological membranes and vesicles are compartments formed from a bilayer of such molecules. The structure of an amphiphilic molecule consists of a hydrophilic head group that carries charge(s) or highly polar bonds as well as a nonpolar chain.

These groups, like those shown in Figure 10.2, can interact with water through hydrogen bonding and dipole–dipole interactions. The other part of the molecule is hydrophobic, typically composed of carbon–hydrogen chains or rings and cannot hydrogen bond with water. The amphiphilic molecules can assemble together to form a bilayer in aqueous solutions because only one side interacts favorably with solvent. Vesicles and micelles formed from such bilayers are attractive as early compartments because they form spontaneously under biologically relevant conditions from molecules thought to be plausible in the prebiotic environment.

Inside the bilayer, van der Waals forces determine the stability. The longer the hydrophobic chain is and the fewer double bonds it has, the more likely the layer will hold itself together in a solid-like form. Short chains, and double bonds (which can cause bending in the chain), lead to less intermolecular attraction and less ordering of the chains, resulting in a more fluid layer at a given temperature. The temperature at which the transition from solid-like to fluid layers occurs is the melting transition temperature, or T_M of the membrane. Lipids above and below the T_M will behave similar to oil and butter at room temperature, respectively.

Amphiphiles will hydrate spontaneously in aqueous solution to form several structures as shown in Figure 10.2. Micelles form when the hydrophobic tails assemble together in the center to form small, oil-filled spheres. When bilayers form into a compartment with an aqueous interior, it is a vesicle. Vesicles can vary greatly in size, have multiple layers, and contain other vesicles, as described in Figure 10.2. Vesicles have many similarities to modern cell membranes and can be made from the same molecules experimentally. A modern cell membrane contains many proteins that expand its functionality beyond that of just an amphiphilic bilayer. However, lipid membranes are crucial to the modern cell's compartmentalization, and therefore, vesicles must be strongly considered for early life compartmentalization as well.

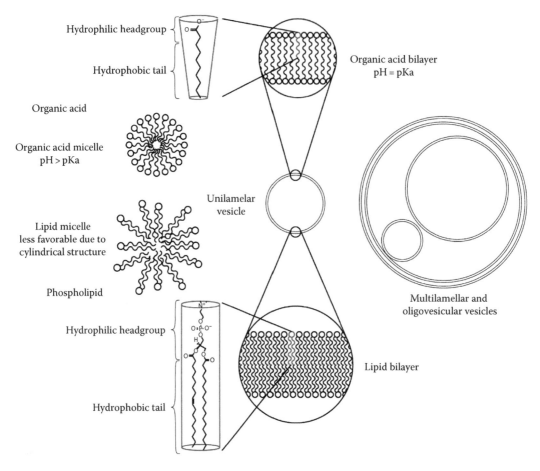

FIGURE 10.2 Self-assembled structures are formed by amphiphiles in aqueous conditions. The top portion of the figure shows the structures formed by organic acids that may have existed on the early Earth. These structures are sensitive to pH, which changes the molecular packing, and in turn the membrane curvature. The modern phospholipids in the bottom portion organize into layers that stack with less sensitivity, largely due to the second tail group. However, this change makes high-curvature structures such as micelles unfavorable. All representations are cross-sectional (2-D) representations of 3-D structures that are spherical in nature.

10.4.1 Difference between Modern and Early Earth Amphiphile Membranes Can Be Derived from Molecular Structure

Modern lipids tend to form bilayer sheets rather than micelles. The phospholipid bilayer sheets form at a low concentration of component phospholipids due to the low water solubility of these molecules and have a low permeability to most solutes. By reviewing their molecular structure, the cause of these properties can be elucidated. A modern phospholipid (like those shown in Figure 10.2) has two hydrophobic tails and a relatively large head group possessing both a negatively charged phosphate moiety and generally another positively charged functional group (most typically an amine to form phosphatidylcholine) to be hydrophilic. The zwitterionic head group near neutral pH prevents like-charged groups in close proximity from destabilizing the membrane. The result of two tails and a large head group is a nearly cylindrical shape that arranges side by side to form planar structures. The long hydrophobic chains, low solubility, and relatively large size of the molecules, as compared to other amphiphiles, lead to higher membrane stability that is beneficial to complex, more independent modern cells.

Modern lipid bilayers present a problem for the early Earth environment in that the membrane is too stable and would limit solute entry/egress by diffusion unless pores or channels were incorporated. Modern phospholipids were unlikely for the first membranes, because of low probability of the chemical reactions that produce them in high yield. So what amphiphilic molecules were available? From studies of the early solar system and Earth, such as data from the Miller–Urey experiment, Murchison meteorite, and Fischer–Tropsch reactions, many amphiphiles are probable but would vary in abundance. Those likely to be abundant include short-chain carboxylic acids, but long-chain acids, sulfonic acids, phosphonic acids, alcohols, and phospholipids may have also been present (Botta and Bada, 2002; Walde, 2006). It has also been demonstrated that the aromatic hydrocarbons, which comprised the bulk of extraterrestrial organic material, can be oxidized to form additional soluble acids (Huang et al., 2007).

Of all proposed prebiotic amphiphiles, the most attention has been given to the soluble organic acids. These molecules are not zwitterions and tend to be smaller and more soluble, like those shown in Figure 10.3. The following discusses the effect of these changes on membrane formation and function in respect to early life.

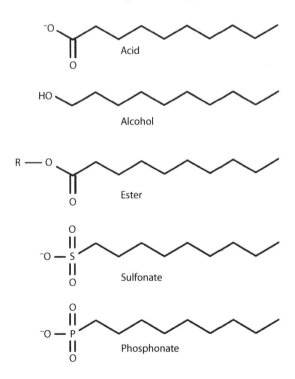

FIGURE 10.3   These are structures of potential early Earth amphiphiles. The names refer to the polar functional group on each molecule. These molecules have been found in meteorite samples and in experiments that mimic early Earth conditions. The carbon chains could vary in size, with shorter chains in higher abundance. (From Botta, O. and Bada, J., *Surv. Geophys.*, 23, 411, 2002. doi: 10.1023/A:1020139302770; Walde, P. *Orig. Life Evol. Biosph.*, 36, 109, 2006. doi: 10.1007/s11084-005-9004-3.)

10.4.2 Assembly of Amphiphiles into Compartmental Structures Is Thermodynamically Favorable

Amphiphilic compartment self-assembly starts with solute dissolution. For a solute to dissolve, the entropy gain from mixing must be greater than the interaction free energy of the separate solute and solvent. When the molecular properties (typically bond polarity) are similar, the change in interaction free energy is small, and molecules are surrounded by and remain in solution. This is the case for the head group of the amphiphile. For the tail group of the amphiphile, the difference is large, and a separate phase is formed that has been described by Guida (2010):

$$\mu(N) = \varepsilon_N + \frac{k_B T}{N} \ln\left(\frac{X_N}{N}\right) \tag{10.1}$$

where
N is the number of phase molecules forming an aggregate
μ is the chemical potential of a molecule within the aggregate
ε_N is the mean interaction free energy of the molecule with itself
X_N is the volume fraction of solute present in the aggregate
k_B is the Boltzmann constant
T is the time

From Equation 10.1, the implication is that for any molecule forming an aggregate (not just the tail group of an amphiphile),

$$X_N = N\left(X_1 \exp\left[\frac{(\varepsilon_1 - \varepsilon_N)}{k_B T} \right] \right)^N \tag{10.2}$$

where
X_1 is the mole fraction of solvent
ε_1 refers to the mean interaction free energy of the hydrophobic solute with solvent molecules

For molecules to associate (i.e., $N > 1$), $\varepsilon_1 - \varepsilon_N > 0$ at relevant concentrations. Since an amphiphile will have a thermodynamically solvated head group, the aggregated tail group will take on dimensionality in relation to the volume it occupies in the aggregate. To account for the variety of hydrated amphiphile structures,

$$\varepsilon_N = \varepsilon_\infty + \frac{\alpha k_B T}{N^{1/d}} \tag{10.3}$$

where
d is the dimensionality
ε_∞ is the interaction free energy of the aggregate bulk
α is a constant to describe the border interaction

For this equation, $d = 1$ refers to cylindrical micelles (line), $d = 2$ refers to bilayers (plane), and $d = 3$ refers to an emulsion droplet or micelle (sphere). Combining Equations 10.2 and 10.3 yields

$$X_N = N\left[X_1 e^\alpha\right]^N \exp\left(-\alpha N^{(d-1)/d}\right) \tag{10.4}$$

This equation will effectively predict emergence of amphiphilic structures (formation of bilayers or micelles), given a high enough concentration is reached but has difficulties describing bilayer curvature. Refinements to include curvature can be made by using a more advanced term for bulk interaction than ε_∞ that accounts for the strain of curvature. The lowest concentration at which structures form for a given amphiphile is the critical micelle concentration (CMC), sometimes referred to as the critical bilayer concentration (CBC). Greater details are discussed by Guida (2010).

10.4.3 Early Earth Amphiphiles Have Increased Water Solubility and Membrane Permeability

The smaller, single-tailed structures of early Earth amphiphiles lead to greater water solubility than modern amphiphiles. Solubility is typically expressed in terms of the CMC, above which molecules begin to self-assemble. For modern phospholipids, the CMC is generally under 1 mM phospholipid and can be as low as 1 nM for longer chain tail groups with 16–18 carbons (Avanti, 2013). In comparison, for octanoic acid, the smallest fatty acid that will self-assemble into such structures, the CMC is 130 mM (Thomas and Rana, 2007). In certain natural waters, amphiphilic molecules can have a concentration of 120 μg/L (Fatoki and Vernon, 1989). The greater solubility of fatty acids as compared to phospholipids results in a different form of membrane hydration. Individual fatty acid molecules equilibrate between the membrane and solution, frequently escaping and rejoining the same or other membranes from solution (Monnard and Deamer, 2002). As a result, the membranes formed by early Earth amphiphiles are more dynamic than modern phospholipid membranes. Fatty acid vesicles and micelles change composition by fatty acid molecular exchange and can adjust their shape in response to these changes for optimized cumulative membrane curvature.

The vesicular structures that formed from early Earth amphiphiles are composition and pH dependent. While each modern lipid represents one large, cylindrical block, early Earth amphiphiles tend toward conical structure, depicted in Figure 10.4, depending on head group deprotonation. At high pH, most acids would be deprotonated, leading to conical units and micelles owing to repulsion between adjacent head groups (Monnard and Deamer, 2002). As pH decreases, the positive charges balancing the membrane will weaken negative head group repulsion to effectively reduce head group size and encourage cylindrical structure. Bilayers and vesicles form near the pKa of the carboxylic acid membrane, which is higher than the pKa of the acid free in solution. For acidic amphiphiles, low pH leads to mostly protonated head groups and reduced solubility and aggregation. When pH is changed, the dynamic nature of these structures is apparent (Budin and Szostak, 2010).

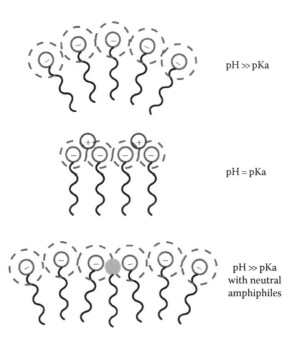

pH ≫ pKa

pH = pKa

pH ≫ pKa
with neutral
amphiphiles

FIGURE 10.4 The curvature of membranes changes based on the head group charge interactions. Since pH determines the ratio of deprotonated to protonated head groups, it has a direct relation to membrane curvature. Micelles form at pH higher than the pKa of the acid membrane because no positive charge is present to reduce membrane repulsion. When the pH is equal to the pKa of the membrane, then bilayers can form, because the lower occupied volume of the head groups leads to less curvature. When ~10% of a neutral amphiphile such as an alcohol is added, bilayers can form above the pKa of the membrane because this increases separation between the charges.

The molecules in solution that were deposited in the most curvature-stable structures persist, while those less-favorable structures redissolve more rapidly. Given time, structures with unfavorable curvature are dissolved, and the molecules are assembled in preferred structures, minimizing free energy.

Due to the proximity of charges in the membrane, the organic acids are less likely to deprotonate when they are in the membrane as compared to free in solution. Therefore, the acid dissociation constant is evaluated as an aspect of the cumulative membrane instead of the individual substituents. The pKa of pure amphiphilic acid membranes is basic, such as 8.5 for oleic acid, but actual early Earth membranes would be likely to possess a variety of head groups and chain lengths (Monnard and Deamer, 2003). These would shift the transition points for structure conversions. For example, neutral groups such as alcohols and esters would increase distances between like charges, reducing curvature induced by head group charge repulsion (Mansy, 2009). When adding 10% 1-nonanol to nonanoic acid, vesicles could be formed up to pH 11 (Monnard and Deamer, 2002). Complementary molecules also lead to a change in solubility. Here, the dilution of membrane charge results in less force driving the membrane apart, and therefore a lower CMC. The 1-nonanol and nonanoic acid mixture mentioned earlier changes CMC from over 85 to 20 mM (Guida, 2010). Therefore, the additional neutral amphiphiles make the membrane more robust by reducing the CMC and expanding the vesicular pH range.

10.4.4 Mechanisms of Solute Transport and Loading in Early Vesicles

The probability that a single vesicle would form holding enough reactants for competitive growth and reproduction is low. Selective compartmentalization would be necessary to contain active molecules while providing opportunity for the entry of fresh reactants and removal of waste. While the modern cell has a membrane that prevents diffusion of ionic species and uses complex pores and channels created by membrane proteins for selective diffusion and transport, compartments created by early Earth amphiphiles meet this requirement by forming a membrane that is more permeable than modern amphiphiles. Early Earth amphiphiles are favorable in this regard. The weaker associations and solubility lead to higher permeability via membrane defects. For ionic solutes, the change in diffusion can reach 100-fold for a decrease in chain length of 18–14 carbon (Guida, 2010). These early Earth compartments have been demonstrated to retain macromolecules, while remaining permeable to small molecules. Over time, this could lead to the macromolecular crowding present in modern cells.

The higher instability of the membrane can also be observed via another property of bilayer membranes, the flip-flop frequency of the bilayer molecules. By flip-flop frequency, we refer to the rate at which an amphiphilic molecule is reoriented between one bilayer leaflet and the opposite leaflet. Flip-flop results in different pH equilibration properties for modern vesicles and those of early Earth amphiphiles. The loosely associated organic acids can transfer a protonated or deprotonated state from exterior to interior when they flip-flop. An example of this is shown in Figure 10.5. This equilibration occurs in seconds for organic acid vesicles and takes hours for phospholipid vesicles (Chen and Szostak, 2004).

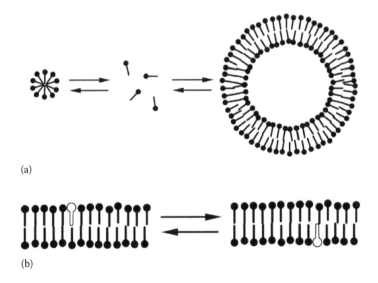

(a)

(b)

FIGURE 10.5 Organic acids are highly mobile relative to phospholipids. In part (a), the aqueous solubility leads to greater interstructure transfer. In (b), the smaller molecules have greater intrastructure transfer. They can both diffuse laterally through the membrane at a higher rate and flip between leaflets with greater frequency. (From Mansy, S.S., *Int. J. Mol. Sci.*, 10, 835, 2009. With permission.)

While compartments formed from smaller amphiphiles would be ion permeable, impermeable ones could also play a role. Ion-impermeable membranes would still provide selectivity through their interaction with charged species. First, membranes would better shield the contents from unfavorable outside conditions via the same properties that let them retain specific interior conditions.

Selective charge permeability can also cause concentration of molecules to create unique environments within greater surroundings. For example, a high- or low-pH vesicle in neutral media may have neutral species pass through the membrane. However, once inside the vesicle, deprotonation or protonation occurs due to a locally different pH. This both forms membrane-impermeable charged species and drives continued uptake of the uncharged species by Le Chatelier's principle. The result is concentration of charged species inside the vesicle by depletion of the surrounding solution (Walde, 2006). This property is beneficial for early life in creating diverse reaction conditions in close proximity and allowing reactions to occur that would not occur in the external medium.

Charged solutes including nucleic acid derivatives such as imidazole-activated nucleotides and adenosine diphosphate have been shown to permeate at a low rate through organic acid membranes. Increased permeability of these specific solutes is important because if they make it to the interior, then larger nucleic acids inside the vesicle are now able to undergo further growth by polymerization (Murtas et al., 2007).

The initial encapsulation is also important in considering vesicular contents. In one example, a fluorescent reporter protein produced only when all parts of the transcription and translation biochemical machinery were present together, which requires ~80 macromolecules, was used to report if the entire pathway was encapsulated (Pereira de Souza et al., 2011). While the probability of all parts being present to function in one vesicle is small enough to be unexpected, the presence of the protein could be observed reliably (though infrequently) in samples (Murtas et al., 2007). One explanation for this result is that the molecules in the pathway associate with each other in solution prior to encapsulation. This would reduce the number of effective parts and greatly increase the probability of full encapsulation of the pathway. Binding of molecules with each other or with the amphiphilic molecules of the membrane prior to vesicle formation provides scenarios for increased encapsulation efficiency that could also be possible on the early Earth.

10.4.5 The Bilayer Itself Is Also a Compartment

Besides creating internal aqueous compartments, the oily interior of bilayers and micelles is a compartment of its own. The membrane's oily interior partitions other hydrophobic and amphiphilic molecules from solution. This compartmentalization has a similar effect to that of the aqueous compartment. Besides the host of modern membrane-mediated reactions with proteins, reactions such as amino acid polymerization have been demonstrated (Walde, 2006). In this case, the membrane structure hosts molecules that interact with the aqueous surroundings. It is also possible for the oily interior to be a reaction vessel in itself.

Micelles and membranes can increase reactant concentration and keep reacting with molecules in close proximity. In this way, micelles have exhibited enzyme-like behavior by compartmentalization. The compartment is enzyme-like in that it is a molecular assembly with an active volume to promote chemical reaction and is not consumed by the reaction. These structures are referred to as lipozymes. An example of this includes the enantioselective reaction where the ketone acid hydrolysis of a substrate is inhibited by sodium dodecyl sulfate micelles, while nitrosation of the enol is accelerated (Walde, 2006). Since a key aspect of compartmentalization is the increase in local concentration, these findings are significant in that a vesicle can both confine oil-soluble compounds to increase reaction rate and at the same time confine water-soluble compounds in its interior.

RNA self-cleavage has also been observed to be catalyzed by micelles in the case of hammerhead ribozymes (Riepe et al., 1999). In this case, the need for metal ions was replaced by NH_4^+ in combination with neutral or zwitterionic micelles. Modern surfactants were used, such as Triton X-100, but the proof-of-principle remains. The effect of the surfactants in solution was independent of head group or molecule size, but a direct relation between CMC and self-cleavage was observed when monitoring the reaction by gel electrophoresis. While the specific interaction of the RNA with the micelle (surface vs. interior) was not determined, micellar catalysis of the self-cleavage reaction was observed. In the upcoming section, membrane-creating lipozymes are used to form autocatalytic self-replicating structures.

Peptides can also play a role in the formation of, reactions within, and transport through bilayer compartments. Amphiphilic peptides can be made by using different R groups (side chains branching from the amino acid core groups) to create hydrophobic and hydrophilic portions (Zhang, 2012). The amino acids used to create these amphiphilic peptides need be no more complex than glycine, alanine, and aspartic acid, which were all likely to be present on the early Earth. As an example, the hydrogen and methyl side groups glycine (G) and alanine (A), respectively, contribute to a tail because less charge and hydrogen bonding is available. An acid side group (aspartic acid, D) gives charge and has been demonstrated as a head group. These peptides could assemble into nanotubes and nanovesicles and mix with other amphiphiles in bilayers. Depending on the number of amino acids in each portion, the peptide can change in scale to resemble other amphiphiles, for example, the structure GGGGDD is 2.4 nm long and resembles a phospholipid in its molecular shape and packing preferences.

Along with adding to the composition of the membrane, peptides can also play an active role, as in the example of polycondensation mentioned earlier. Peptides can interact with molecules in solution with reactive groups. They also can act enzymatically with active sites, as covered extensively in investigations into modern membrane proteins, and they can form a bilayer with function as a whole, as shown in Figure 10.6. For example, bilayers of only the peptide KLVFFAE have been demonstrated to bind the fluorescent dye Congo red, as evidenced by Förster resonance energy transfer from covalently attached Rhodamine 110 (Childers et al., 2009; Pohorille et al., 2003). Bilayer binding sites are important for early life compartmentalization because they are another way for molecules to be concentrated from dilute solution. The other compartmental role that peptides can play is in the transfer of solutes through the bilayer via the existence of beta-barrels (Childers et al., 2009). Beta-barrels are peptide sheets that form through a multitude of peptide bonds and that have curled around to form a cylinder. Beta-barrels have been observed to incorporate in the bilayer and form a pore that spans the depth of the layer.

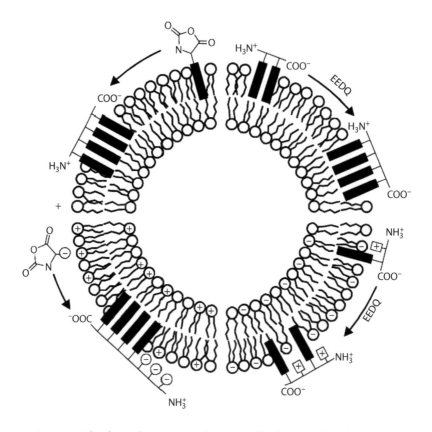

FIGURE 10.6   Amino acid polycondensation can be assisted by functional molecules in vesicular membranes. (From Walde, P., *Orig. Life Evol. Biosph.*, 36(2), 109, 2006. With permission.) On the left, amino acids both membrane bound (top) and electrostatically associated (bottom) were used and were initiated by activated (*N*-carboxyanhydride) amino acids. On the right, the hydrophobic condensing agent 2-ethoxy-1-ethoxycarbonyl-1,2-dihydroquinoline (EEDQ) was used to condense two different membrane-bound amino acids. (From Walde, P., *Orig. Life Evol. Biosph.*, 36(2), 109, 2006. With permission.)

Therefore, early development of peptides in membranes becomes a solution to the predicament of transport in and out of the vesicular compartment.

10.4.6 Amphiphilic Compartments Can Grow and Divide

As mentioned in the previous section, early Earth amphiphiles equilibrate between positions in the membrane and free in solution. This equilibrium permits growth of more stable membranes by depletion of less stable surrounding structures. Destabilization can occur due to changes in curvature from pH change or insertion of other amphiphilic molecules. A striking example is the addition of modern lipids to fatty acid membranes. When phospholipids were added at 10% of the overall membrane composition, the vesicles became much more stable and better retain the organic acids. As a result, the membrane grows rapidly in the presence of other organic acid structures that lack modern lipids. The rapid growth leads to an excess membrane relative to volume and the vesicle loses spherical morphology. The long, tubular structures present a smaller barrier to membrane closure and division can occur (Budin and Szostak, 2010).

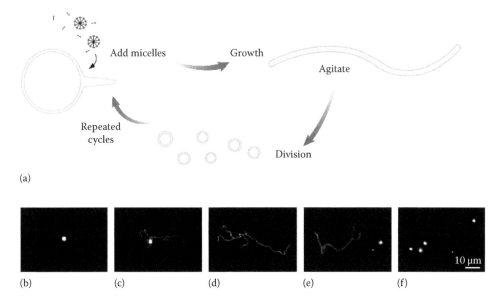

FIGURE 10.7 Micelle-induced growth and agitation-induced division of fatty acid vesicles. (a) Schematic of growth by assimilation micellar amphiphiles to create vesicles with excess surface area, followed by agitation and division. (b–e) Microscope images of an initial ~4 μm fatty acid vesicle and subsequent growth as micelles are assimilated over ~30 min. (f) After agitation, the excess membrane present after growth is now distributed between several vesicles. The vesicles are visible due to encapsulated dye, which remains trapped throughout the process. (Adapted from Zhu, T.F. and Szostak, J.W., *J. Am. Chem. Soc.*, 131(15), 5705, 2009. With permission.)

This phenomenon is relevant to early life in a possible route to individual, competing cell-like compartments (Chen et al., 2005). Structures that could stabilize amphiphiles, that is, lead to greater residence times for individual amphiphilic molecules in the membrane, would be able to grow at the expense of less stable membranes or other amphiphile assemblies in the surrounding environment, as seen in Figure 10.7. They would grow until division and produce more lipids to continue the process.

Other methods of creating excess membrane have been demonstrated to cause budding and division in modern lipids. Osmotic shock by adding external solutes to draw water volume out of vesicles is one example. The reduced volume from lost water leads to excess membrane, budding, and eventual division (Giant Vesicles, 1999). These examples demonstrate that excess membrane and division are a general property of amphiphilic vesicles that can occur from physical processes. Membrane can also be created through reaction to form amphiphiles. In the case of acyl-CoA and lysophosphatidylcholine, both are single-chain amphiphiles that form micelles when dissolved. The micelles react via acyltransferase (a membrane protein) to form double-chained phosphatidylcholine lipids (Walde, 2006). The result is depletion of micelles and the growth of bilayer vesicles. This autocatalysis of membrane can lead to growth and reproduction via vesicle division. Autocatalytic reproduction is analogous to the self-generated reproduction of life and is a precursor to natural selection and evolution.

10.4.7 Practical Concerns and Ongoing Investigation of Amphiphilic Compartments

The primary concerns with the suggestion of amphiphilic compartments in the development of early life stem from the availability of amphiphiles and transport through the membrane. While many potential sources of amphiphiles exist, few of them suggest an abundance of amphiphilic molecules without restricting the presence to a localized environment. Similarly, the high levels of metallic ions present in the early ocean present a further problem. Many multivalent cations cause amphiphiles to aggregate. This is also a problem for modern lipids, which is countered by binding the ions with other molecules to keep the multivalent ion concentration low. Decomposition is another threat to early Earth amphiphiles, which can undergo hydrolysis, photochemical reaction, and pyrolysis (Deamer, 1997). To address these problems, investigations into further sources of amphiphiles and studies that focus on simple, stable amphiphiles are preferred (such as acids and peptides). Better exploration of early Earth environments may also help to address this issue, as the degradation mechanisms present in one environment may be absent in another. Extended research into the details of the early Earth has already yielded potential solutions for the other issue of membrane permeability. As discussed earlier, by employing amphiphiles more appropriate to the early Earth, permeability was found to be greater than with initial studies that focused on modern phospholipids. The discovery of amino acids, understanding their role as amphiphilic peptides, and possibility of pore formation also greatly increases the chance for transport as well as providing another amphiphile source. Therefore, while initially daunting, the continued study of amphiphiles as early compartments continues to make progress as a probable early contributor in the origin of life.

Experimental models for protocells are an active area of current research. Scientists have coupled the growth and division of fatty acid vesicles with uptake and polymerization of nucleotides (Hanczyc et al., 2003), such that the daughter vesicles that result from division events *inherit* nucleic acids. An example of this is featured in Figure 10.8.

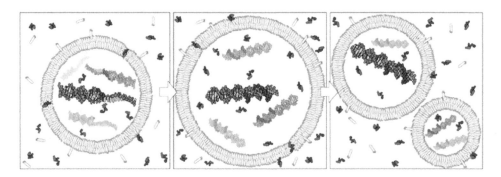

FIGURE 10.8 Growth of the protocell membrane results from the incorporation of environmentally supplied amphiphiles, whereas division may be driven by intrinsic or extrinsic physical forces. Externally supplied activated nucleotides permeate across the protocell membrane and act as substrates for the nonenzymatic copying of internal templates. Complete template replication followed by random segregation of the replicated genetic material leads to the formation of daughter protocells. (From Mansy, S.S. et al., *Nature*, 454(7200), 122, 2008. With permission.)

Models based on modern lipids and purified biochemicals such as enzymes, DNA, and RNA have also been explored (Kurihara et al., 2011; Luisi and Stano, 2011). Transcription and translation processes can be performed inside lipid vesicles and can be coupled to vesicle growth and division.

Increasingly complex and lifelike assemblies are being produced, which not only help us to understand how the first cells may have formed but also point to physical and chemical mechanisms still at play in modern cells. The self-assembly behavior of fatty acids, lipids, and other related molecules is crucially important in all current living organisms. The membranes of today's cells must have been preceded by simpler versions that could potentially have appeared early in prebiotic evolution and may have played important roles in the development of living from nonliving matter.

10.5 MACROMOLECULAR SOLUTIONS CAN UNDERGO PHASE SEPARATION TO GENERATE SIMPLE COMPARTMENTS

A second class of organic compartment can be expected to have appeared once polymers become prevalent on the early Earth: macromolecule-rich aqueous phase droplets. Solutions of macromolecules can undergo several types of phase separation, leading to coexisting aqueous volumes of differing composition that could serve as compartments. Driving forces for demixing can be enthalpic and/or entropic and can involve solvent molecules and counterions in addition to the polymers themselves. This can be expressed in terms of a standard thermodynamic equation relating the change in a system's free energy (ΔG) to changes in enthalpy (ΔH) and entropy (ΔS); T indicates temperature (Ronca and Russell, 1985):

$$\Delta G_{\text{separation}} = \Delta H_{\text{separation}} - T\Delta S_{\text{separation}} \tag{10.5}$$

Phase separation will occur spontaneously when the Gibbs free energy of the process is favorable (i.e., when ΔG is negative). This is favored by negative changes in enthalpy, which could arise, for example, when repulsive chemical interactions present between solutes can be relieved by a phase transition that results in their separation into different phases. Often, however, entropy drives phase separation (Ronca and Russell, 1985). For example, when oppositely charged polyelectrolytes combine to form a polymer-rich phase, the favorable change in free energy that allows phase separation to occur results from a large entropic gain associated with release of counterions.

Liquid–liquid demixing leads to regions of differing composition between which solutes can partition based on their chemical activity in the phases. In fact, aqueous two-phase systems (ATPSs) formed from polymers such as poly(ethylene glycol) (PEG) and dextran (a polyglucose) are commonly used as an analytical separation technique. Many solutes will accumulate in one phase of a multiphase system. Solute partitioning depends on many factors such as the chemical structure, hydrophobicity, size, and shape of the solute and the polymer content, ionic strength, and pH of the liquid phases. The equilibrium distribution of a solute molecule of interest in a two-phase system is described by the partitioning coefficient, K:

$$K = \frac{C_{\text{phaseA}}}{C_{\text{phaseB}}} \tag{10.6}$$

where C_{phaseA} and C_{phaseB} indicate the concentration of solute in each phase. By convention, the less dense phase that appears on top in bulk samples is used in the numerator. Solute size is an important determinant of the difference in its chemical activity between the phases, and hence the degree of partitioning between them. This is because as the size of a molecule increases, it has a greater area of contact with the phase. We can therefore anticipate that larger organic molecules would have accumulated preferentially into polymer-rich aqueous phase droplets, becoming locally concentrated (Rito-Palomares, 2004).

Figure 10.9 shows the strong length dependence for partitioning of RNA molecules between two aqueous phases formed in a solution of polyethylene glycol and dextran.

Several general classes of aqueous phase separation can occur, depending on the type of polymers present and solution characteristics such as ionic strength. Examples of those classes are found in Figure 10.10.

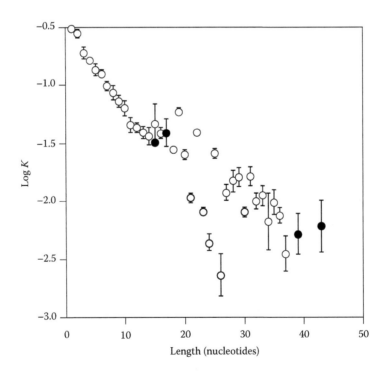

FIGURE 10.9 Partitioning of RNA strands in ATPS as a function of length (in nucleotides). This figure exhibits that longer polymers partition more favorably than polymers of smaller chain lengths. (Adapted from Strulson, C.A. et al., *Nat. Chem.*, 4(11), 941, 2012. With permission.)

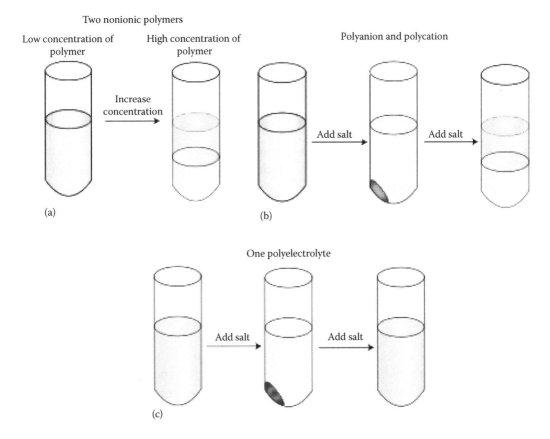

FIGURE 10.10 Types of aqueous phase separation. (a) Two nonionic polymers will phase separate if present above a certain weight percent. (b) Two polyelectrolytes bearing opposite charges will phase separate into complex coacervates. Addition of more salt will cause the coacervates to dissociate, forming a bottom and top later, enriched in each of the polymers. (c) One polyelectrolyte will form simple coacervates in the presence of a small counter ion or other molecule that lowers the overall solubility of the polymer. Both the middle and bottom panel show the coacervate phase after centrifugation. The coacervates are drawn to the bottom of the tube due to their increased density and weight.

When polyelectrolytes are present, coacervation can occur to produce a small polymer-rich phase—termed the coacervate phase—and a much larger dilute phase. Mixtures of uncharged polymers can also form multiple phases, generally each phase is enriched in a different polymer, and the phases may or may not be similar in size. Macromolecule concentrations required to achieve phase separation are dependent in part on the molecular weight (MW) of the polymers; higher MW polymers more readily phase separate. Phase separation is common in mixtures of aqueous macromolecules and does not require specific macromolecular structure or functional groups (although polymer chemistry impacts the properties of the resulting phases). The generality of this phenomenon suggests the possibility that phase separation may have acted as a vehicle for development of prebiotic and protocellular evolution. Indeed, more than 50 years ago, Oparin hypothesized that coacervate droplets were the basis of the origin of life (Oparin, 1957). Since then, scientists have learned a great deal about roles for nucleic acids and lipids in modern life and their likely roles in the predecessors of modern cells, and the coacervate theory of life has fallen out of favor (Lazcano, 2010). Nonetheless, the physical chemistry of macromolecules indicates that these phase droplets were likely present on the early Earth prior to the earliest cellular life and may have facilitated its development. Moreover, recent findings indicate that today's cells contain aqueous phase microcompartments within their cytoplasm and nucleoplasm (Brangwynne, 2011).

Like the surfactant amphiphiles discussed in the previous section, aqueous phase separation in polymer-containing solutions can be evaluated in the laboratory as model systems relevant to early compartments. By changing the amount of polymer and their corresponding mass ratios, compartments of very small volume can be synthesized, mimicking early compartments. The following subsections introduce several major classes of aqueous phase separation and discuss the potential of these systems for understanding early and modern cellular life.

10.5.1 Oppositely Charged Polymers Phase Separate into Complex Coacervates

Two polyelectrolytes bearing opposite charges in solution will exhibit a tendency to associate in solution and can phase separate into liquid droplets known as complex coacervates. Complex coacervates present a feasible method to form highly concentrated phases of macromolecules in the presence of a very dilute equilibrium (continuous) liquid. Given that early macromolecules may have been charged, this perhaps presents the best simulator of evolutionary conditions due to the relatively small amount of organic material required. Because the formation mechanism is based on electrostatics rather than any specific chemical functionality of the polyelectrolyte, coacervates can form from a wide range of polymeric molecules and are not limited to two species. When interacting with one another, these polyelectrolytes can be considered as new colloidal entities and are virtually uncharged and distinct from the equilibrium liquid. Increasing ionic strength diminishes the interaction. Moreover, this type of phase separation can occur at a wide range of temperatures, which could have been beneficial in different primordial ocean conditions, such as near hydrothermal vents (de Souza-Barros and Vieyra, 2007).

A common theory to describe complex coacervation behavior is the Voorn–Overbeek model, which characterizes the thermodynamics and entropy associated with mixing and the electrostatic interactions between charged polymers (Veis, 2011). The model treats the polyions as essentially random chains in both the dilute and concentrated (coacervate) phases, with the chain elements distributed randomly in the model lattice within both phases. One can imagine small ions being released into solution and having more degrees of freedom when the two oppositely charged polyelectrolytes become associated in a very dense, concentrated phase. Release of the counterions increases the overall entropy of the system.

Several factors, including pH and salt concentration, contribute to the characteristics of a complex coacervate aqueous phase and should be considered when considering a system as a mimic for prebiotic compartmentalization (de Kruif et al., 2004). Aggregation of the oppositely charged polyelectrolytes is also possible in these types of solutions. Aggregates precipitate as solids and form when attractive interactions between the polyelectrolytes are very strong, for example, in solutions with very low ionic strength. These structures have

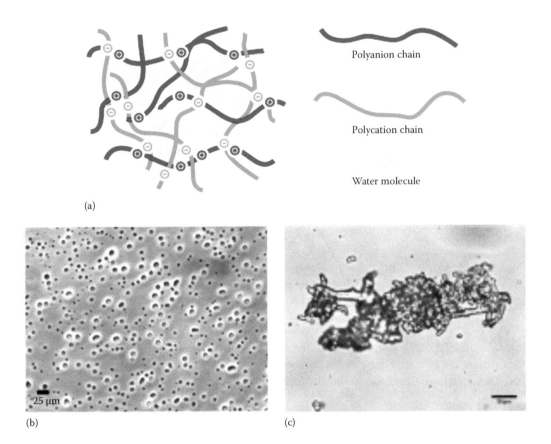

(a)

(b)   (c)

FIGURE 10.11 Examples of regions along the phase diagram where precipitates and coacervates form. (a) Schematic of complex coacervate process. (b) Examples of coacervates made from an aqueous solution of poly(L-lysine hydrochloride) (PLys) and poly(L-glutamic acid sodium salt) (PGlu). (c) Example of an aggregate of PLys and PGA for comparison. (From Chollakup, R. et al., *Macromolecules*, 46(6), 2376, 2013. With permission; Priftis, D. et al., *Langmuir*, 28(23), 8721, 2012. With permission.)

low water content and cannot act as compartments for molecules to diffuse in and out and as such are less interesting as prebiotic reactors.

Figure 10.11 shows optical microscope images of aggregates and coacervates. The coacervates are spherical, liquid droplets with high polymer concentration that can serve as compartments into which solutes can diffuse. A precipitate will more likely form than a liquid coacervate phase if one of the polyions is a strong polyelectrolyte, for example, polymers with sulfate or quaternary amine charged groups. Strong polyelectrolytes are always charged in solution, in contrast to weak polyelectrolyte, such as polymers with carboxylic acid or primary amines, which will carry partial charges that depend on pH. Coacervate formation is maximal where the ratio of polyelectrolytes in the mixture produces complexes of neutral charge. Even when one polyelectrolyte is present in excess, the complexes formed are only moderately charged. Complexes between weakly associating polyelectrolytes can be dissociated by low concentrations of salt (~10 mM), while other pairs of polyelectrolytes are stable beyond 1 M salt. The concentration of salt in solution

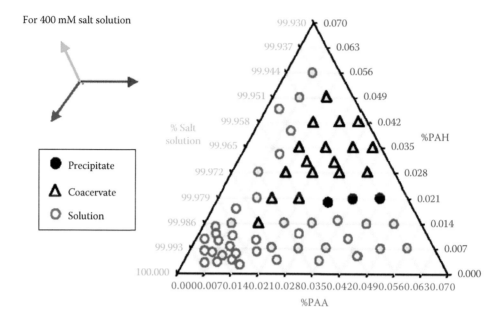

FIGURE 10.12   Ternary phase diagrams of PAA and PAH. Total polyelectrolyte concentration was held constant at 0.05% wt. Figure shows phase behavior in an aqueous solution containing 400 mM NaCl. (Adapted from Chollakup, R. et al., *Macromolecules*, 46(6), 2376, 2013. With permission.)

is approximately equal in both the equilibrium and coacervate phases. Temperature has almost no effect on the phase boundary where coacervates form. Many scientists have studied the phase separation of biologically relevant proteins and polysaccharides to better understand the dynamics involved.

Figure 10.12 displays the phase diagram for two oppositely charged polyelectrolytes in 400 mM NaCl. Complex coacervates can form at very low weight percents of polymer in the overall solution. For example, in solutions of poly(acrylic acid) (PAA) and poly(allylamine) (PAH), phase separation can occur at total polymer concentrations of 0.05% wt.

10.5.2 Neutral or Charged Polymers Will Phase Separate into Simple Coacervates in the Presence of Salt or Other Species That Decrease Their Solubility

Coacervation can also happen in the presence of only one type of polymer. Here, a polyelectrolyte or neutral polymer associates strongly with a counterion or another solute that decreases the overall solubility of the macromolecule. For example, a polycarboxylate could form such a coacervate when mixed with a metal cation that was chelated by the carboxylic acid groups or a neutral polymer could form one in the presence of an alcohol (Menger and Sykes, 1998). Under certain conditions, the concentrated phase is a gel-like substance, such as in the case of a gelatin system (Mohanty and Bohidar, 2005). Coacervates can be redissolved by reducing the interactions between their major components. For example, by increasing ionic strength, the charge screening length is decreased and attraction between a polyelectrolyte and counterion can be diminished. Phase separation occurring from the interaction between one charged polyelectrolyte and high concentrations of salt or solute is referred to as a simple coacervate. Simple coacervates can be made from a variety of poly-electrolytes, such as a protein, polysaccharide, or other types of charged polymers, and can be formed at polymer weights along the order of 0.001% (de Kruif et al., 2004; Shimokawa et al., 2013). Decreasing the solubility can be achieved in different manners, including adding an alcohol, such as methanol or ethanol, or a small ion, such as a monovalent or divalent salt (de Kruif et al., 2004). The higher the MW of the polyelectrolyte, the lower the concentration required to achieve phase separation. Moreover, the types of salts used influence whether two phases form or not. Salts of polyvinyl sulfonic acids are able to form two phases in the presence of NaCl, KCl, and RbCl, but not LiCl, CsCl, and NH_4Cl (Eisenberg, 1977). A similar system includes dextran sulfate, which forms two phases upon addition of KCl and CsCl, but not with LiCl, NaCl, or NH_4Cl. In terms of early Earth conditions, the abundance of sodium (about two times as much as present day at 20 g/L) could have likely provided the ionic strength required to induce phase separation in the prebiotic oceanic conditions (Knauth, 2005).

Electrostatics, as mentioned in the previous section, plays a critical role in the formation of a coacervate phase. To help overcome electrostatic barriers, small counter ions such as Na^+ and Cl^- are used to screen the overall charge of the polymer, allowing it to separate into a microdroplet. Too much salt will decrease the Debye length, the external electric field length from a charged species. Decreasing the Debye length can cause the macromolecule to remain dissolved in solution or precipitate out. Precipitation occurs when the macro-molecule has an increased affinity for itself, leading to aggregation into solid particles. The concentration of salt required for phase separation changes with charge valency, MW, and overall charge on the polymer. Thus, an anionic polymer with a charge on every monomer unit would require more ionic strength than a polymer with a charge dispersed along different monomers throughout the polymer chain.

Simple coacervates could have appeared alongside or before complex coacervates, depending on the prevalence of polyelectrolytes and other ions and molecules on the early Earth. Consolidation of polymers, such as proteins, polysaccharides, and nucleic acids into small volumes, could have been accomplished by either type of coacervate compartment.

10.5.3 Incompatibility of Two Nonionic Polymers Leads to Separation into Two Aqueous Phases

The absence of polyelectrolytes does not necessarily prevent aqueous phase separation from occurring. For systems consisting of two nonionic polymers in a common solvent, incompatibility between the polymers will lead to the separation into two phases. Above a certain concentration, most aqueous solutions of two or more structurally distinct, uncharged polymers will give rise to a multiphase solution. Water remains the primary component in all of the phases, but each polymer is enriched in one of the phases, causing a solution with two distinct layers (Dobry and Boyer-Kawenoki, 1947). One of the most widely used and well-studied examples is the PEG and dextran system, which has a top phase enriched in PEG and a bottom phase enriched in dextran. This system has been widely used due to the fact that both polymers are relatively inexpensive, require moderate concentrations, separate rapidly, are compatible with biomolecules for separations, and can be buffered to different pH values.

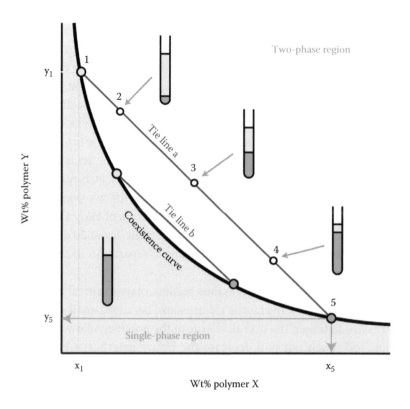

FIGURE 10.13 Typical phase diagram for two neutral polymers. Above a certain weight percent of each polymer, phase separation occurs. Tie lines can be drawn between points along the coexistence curve. Tie lines give valuable information about the volumes and composition of the phases present. (From Keating, C.D., *Acc. Chem. Res.*, 45, 2114, 2012. doi: 10.1021/ar200294y.)

ATPSs, such as the one composed of PEG and dextran, can be made at multiple combinations of polymer concentrations. Phase diagrams are often used to characterize the separation of two polymers as a function of concentration.

Figure 10.13 exhibits a generic phase diagram for two immiscible polymers. At low concentrations of polymer, the solution exists as one phase and exhibits complete miscibility, with phase separation only occurring at higher concentrations. Specifically for a PEG 6 kDa/dextran 30 kDa system, the two polymers would exist as one phase if both were present at 5% w/v in water. However, if the concentration of both polymers were then doubled to 10% w/v, phase separation would occur. Separating these phase regions is the coexistence curve, or binodal. Closer to this region, the system is more sensitive to alterations. An example of a composition near the binodal for the PEG 6 kDa/dextran 30 kDa system is at 5% w/v PEG and 9% w/v dextran. Concentrations close to this point will be sensitive to temperature or other changes in environment. Tie lines, as shown in the figure, can be made by determining the composition of both polymers in each phase; tie lines become shorter with reduced polymer concentrations due to greater similarity between the phases. Determining the volumes of each phase with such tie lines can be a useful tool for partitioning studies, where activity of a chemical reaction may be enhanced by decreasing the volume of the phase enriched with an enzyme or analyte. Experiments

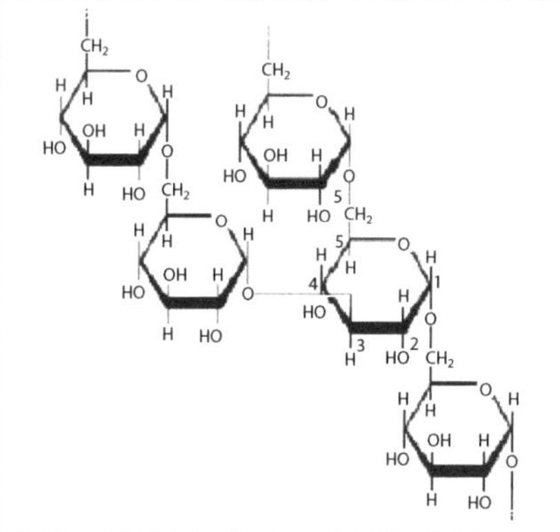

Dextran Type	Molecular Weight (Da)	Concentration (w/v)
Dextran 5	3,400	20%
Dextran 17	30,000	5%
Dextran 37	179,000	<2%
Dextran 48	460,000	<1%
Dextran 68	2,200,000	<0.5%

FIGURE 10.14   MW dependence of dextran polymer on phase separation with PEG. As the MW of the dextran increases, the concentration required to initiate phase separation decreases. (From Albertsson, P.Å., *Partition of Cell Particles and Macromolecules; Distribution and Fractionation of Cells, Mitochondria, Chloroplasts, Viruses, Proteins, Nucleic Acids, and Antigen–Antibody Complexes in Aqueous Polymer Two-Phase Systems*, Wiley-Interscience, New York, 1971.)

can be tailored so that volume can be excluded just by tuning the mass ratios between the two immiscible polymers.

Typically phase separation can occur with moderate overall amounts of polymer, approximately less than 10% wt/wt, but is influenced by the MWs of the polymeric components. The higher the MW of the polymers, the lower the concentration required for phase separation. A specific example of this includes methylcellulose, which has the ability to make two phases at low concentrations of total polymer (approximately 0.01% wt). Moreover, it is important to note the differences in the MWs between the two polymers. Increased differences between the MWs of the two polymers give rise to phase diagrams that are more asymmetrical in shape, such as seen in Figure 10.14 (Asenjo and Andrews, 2011). By changing the MW of just one of the polymers, it is possible to greatly alter the concentration required to achieve phase separation.

Figures 10.14 and 10.15 show the concentration of different dextran polymers required to achieve phase separation using a 7.5% w/v PEG 6 kDa concentration (Albertsson, 1971).

Phase separation is not limited to 2 polymers; even 10 different polymers can be used to make a 10-phase system (Albertsson, 1971). Even though two-phase systems have been discussed in this section, the same properties still apply for aqueous multiphase systems and they can be made using the same techniques. A range of polymers, when present in solution at required concentrations, will phase separate based on their density, MWs, and polydispersity as shown in Table 10.1. Substitution along the polymer backbone, like the addition of sulfate groups, can be sufficient to induce separation. Phase separation of aqueous polymers suggests a straightforward experimental design that can be a representative sample of the compartmentalization of biomolecules that may have occurred during the origin of life through the ages toward more modern cells. It is plausible to consider that some form of coacervation would have appeared on the early Earth prior to the type of phase separation that occurs between uncharged polymers as described in this section,

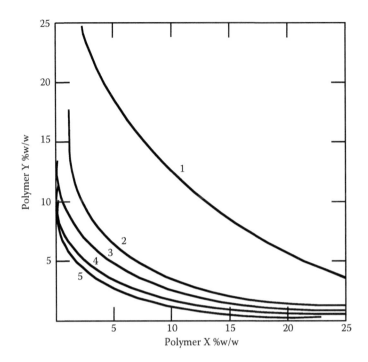

FIGURE 10.15 Effect of polymer MW on phase behavior of dextran–polyethylene glycol–water systems. All systems have the same polyethylene glycol mass (PEG 6 kDa) and the different size dextran polymers: dextran 5 kDa (1), dextran 17 kDa (2), dextran 37 kDa (3), dextran 48 kDa (4), and dextran 68 kDa (5).

just due to the generally higher concentrations needed for phase separation in nonionic polymer solutions. Nonetheless, as the amount of organic matter available increased over time, this type of phase separation would have coexisted with coacervation and provided additional types of compartments with varying chemical and physical properties.

10.5.4 Reactions Can Be Facilitated by Compartmentalization in Phase Droplets

Compartmentalization of molecules presents an attractive way to facilitate reactions. There have been several examples where reactions of biomolecules, including proteins and RNA, have demonstrated rate increases due to crowding and colocalization of otherwise dilute materials. Albertsson was one of the first to demonstrate the fundamentals of protein partitioning in aqueous phase systems. To target and collect certain subcellular proteins over others into one phase, he conjugated a biospecific ligand onto one of the polymers, increasing the partitioning of one specific protein into one of the phases. Instead of being diffuse throughout the entire solution, the protein was concentrated primarily into one small phase volume of the liquid (Albertsson, 1971). Colocalizing the protein therefore increased the chances of enzymatic reaction with the ligand of interest. Even in the absence of specific binding interactions with the polymers, large differences in solute concentration can be maintained by partitioning in an ATPS and can lead to improved reaction kinetics. Unfortunately, although a great deal of research has been performed to understand and

TABLE 10.1   Examples of Representative Molecules Able to Participate in Aqueous Phase Separation and Their Corresponding Structures

Name	Structure	Description
Poly(allylamine)		Polycation pKa 9.7
Poly(diallyl dimethylamine) chloride		Polycation pKa ≈ 6.4
Poly(acrylic acid)		Polyanion pKa 4.2
Poly(styrenesulfonate)		Polyanion pKa ≈ 1
Poly(2-acrylamido-2-methylpropanesulfonate)		Polyanion pKa ≈ 0.36

(continued)

TABLE 10.1 (continued)   Examples of Representative Molecules Able to Participate in Aqueous Phase
Separation and Their Corresponding Structures

Name	Structure	Description
Adenosine triphosphate		Polyanion pKa ≈ 4, 6.5
Poly(ethylene glycol)		Neutral
Dextran		Neutral
Poly(propylene glycol)		Neutral
Ficoll		Neutral
Citrate		Anionic pKa 6.4

TABLE 10.1 (continued)   Examples of Representative Molecules Able to Participate in Aqueous Phase
Separation and Their Corresponding Structures

Name	Structure	Description
Phosphate		Anionic pKa 2.1
Sulfate		Anionic pKa 1.99

control solute partitioning in ATPS, much less is known about the effect of partitioning in ATPS on reaction rates and outcomes. A few examples will be described here.

The rate of reactions using catalytic RNA, also known as ribozymes, has been shown to increase significantly upon localization into one phase of an ATPS. RNA partitioned into the dextran-rich phase of a PEG/dextran ATPS, achieving an approximately 3000-fold concentration excess in these phase compartments as compared to the external PEG-rich phase. By varying the volume ratios between the two phases, the local RNA concentration within the droplets was increased and consequently the reaction rate was enhanced. Figure 10.16 shows that as the volume ratio between PEG and dextran increases, the overall activity of the RNA is enhanced upward of 50-fold.

Reactions using protein enzymes have also been performed in aqueous phase systems and have been shown to provide control over the location and rate of the reaction. For example, the protein urease accumulates in the dextran-rich phase of a PEG/dextran ATPS, where it catalyzes urea hydrolysis and has been coupled to local formation of $CaCO_3$ mineral restricted to just this phase.

Coacervate phase compartments have been less thoroughly investigated with respect to partitioning than neutral polymer ATPS, because the latter are more amenable to bioseparations. Nonetheless, some work has been done to explore partitioning in these systems. Researchers have demonstrated complexation of nucleic acid precursors, such as adenosine triphosphate (ATP), with a polycation to form robust droplets, that can be used to sequester a variety of chemical species (Koga et al., 2011). Among those, sequestered species include proteins, nucleic acids, nanoparticles, and small dye molecules (Williams et al., 2012).

Figure 10.17 displays the coacervate's ability to sequester fluorescent dyes and other small molecules using ATP and oligolysines. Additionally, enhanced yields have been observed for reactions catalyzed by protein enzymes when performed in coacervate droplets under either low or high ionic strength (Crosby et al., 2012). Droplets in this case were also composed of ATP nucleotides as well as a polymeric quaternary amine. Although this work was

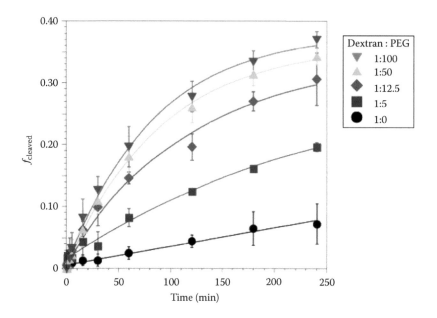

FIGURE 10.16 Phase volume ratios $(V_D:V_P)$ are 1:0 (filled black circles), 1:5 (blue squares), 1:12.5 (red diamonds), 1:50 (blue triangles), and 1:100 (inverted green triangles) in 10 mM Mg^{2+} experiments that included 2 pmol of the E strand. These data show that RNA catalysis is enhanced by decreasing the compartment size. (From Strulson, C.A. et al., *Nat. Chem.*, 4(11), 941, 2012. With permission.)

done with complex, modern-day protein enzymes, the same principles can be expected to apply for simpler functional molecules, and compartmentalization in coacervate droplets presents a plausible mechanism for reaction enhancement in the prebiotic milieu.

10.6 CONCLUDING THOUGHTS

The early Earth likely had a large number of different types of microenvironments in which particular chemistries may have been aided by special chemical and/or physical conditions. Once sufficient quantities of organic materials had been formed, new types of organic compartments became possible, based on interactions between molecules. Much of the form and function of both vesicle assemblies and aqueous phase droplets can be understood from the chemical structures of their component molecules. Comparing modern lipids and early Earth amphiphiles illustrates how different chemical properties control membrane stability and function. The simple amphiphiles such as fatty acids, which may have been present on early Earth, could in fact have been better suited for the development of early life than the phospholipid molecules used in modern membranes. Moreover, given the prevalence of phase separation in aqueous macromolecule solutions, it is reasonable to anticipate that the various polyelectrolytes and uncharged polymers formed by abiotic chemistry on the early Earth would have led to polymer-rich phase droplets of one form or another. The very low concentrations of total polymer needed to form coacervates suggest that these polyelectrolyte-rich compartments may have been the most likely.

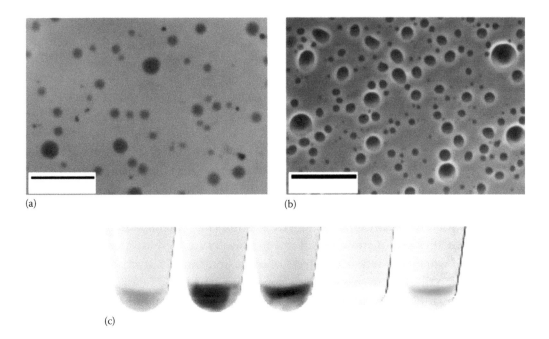

FIGURE 10.17 Optical micrographs of droplets prepared at pH 8 in water from 20 mM solutions of peptides and ATP: oligolysine–ATP (1–5 kDa, 5–24 monomer units; bright field). (a, b) Two examples of droplets made from poly(lysine) and ATP. The droplets were stained with methylene blue. Scale bars for panel (a) and (b) are 20 and 50 μm, respectively. (c) Optical images of poly(lysine) and ATP droplets where [PLys] = [ATP] = 25 mM in the presence of various solutes and nanoparticles. From left to right; CMDex–Co$_3$O$_4$ (pH 5), protoporphyrin IX, copper chlorophyllin, fluorescein, and methylene blue. Coloration within the PLys/ATP coacervate layer is indicative of preferential sequestration into the membrane-free microcompartments. (Adapted from Koga, S. et al., *Nat. Chem.*, 3(9), 720, 2011. With permission.)

Structures formed by self-assembly of amphiphiles or phase separation could supply favorable microenvironments needed for diverse chemical reactions and subsequent chemical evolution to occur. Such compartments can accumulate reactants within their aqueous interiors, control pH by limiting transfer of charged species, and confine molecules by size and/or chemistry (e.g., charge, hydrophobicity). Additionally, crowding in two dimensions on the surface of a membrane or in three dimensions within a polymer-rich phase can influence the chemical activity of molecules in these regions, potentially inducing folding, association, or other types of volume-reducing transformations. These self-assembled compartments meet many of the demands required for the complex and diverse reactions of life to progress and even exhibit some of the fundamental process that are normally associated with living matter, like growth and division.

It is important to note that these general classes of compartments, whether phase separated or membrane bound, did not necessarily act alone. We can certainly imagine that multiple types of compartments coexisted and that multiple physical effects could have combined. The organic microcompartments described in this chapter would have appeared in the context of an environment that already had several types of inorganic microcompartments such as liquid–solid interfaces, and it is reasonable to hypothesize that they would have interacted. For example, the surfaces of minerals could have served to bind and preconcentrate amphiphiles, making it possible for them to self-assemble into vesicles. Similarly, polymers could have become concentrated in rock crevices, ice eutectics, or other types of physical compartments prior to undergoing phase separation. Inorganic ions leaching from rocks could have also aided phase separation.

Organic compartments that formed in close association with existing inorganic microenvironments such as clay mineral particles could benefit from preconcentration of functional polymers such as RNA. Researchers have demonstrated that RNA molecules added to montmorillonite clay particles, and entire clay particles, can be encapsulated within fatty acid vesicles formed from fatty acid micelles. The resulting structures, which thus can contain catalytic and informational materials (i.e., clay and RNA), were further demonstrated to be able to grow by adding fatty acids from solution and to divide when exposed to the stress of extrusion through a small pore (Hanczyc et al., 2003). Finally, the two classes of organic compartments discussed in this chapter are not incompatible and may have interacted in the prebiotic milieu (Keighron and Keating, 2011). Amphiphilic compartments such as micelles or vesicles may have partitioned into polymer-rich phase droplets and/or assembled at their interfaces. Phospholipid vesicles have been demonstrated to both partition in, and accumulate at interfaces of, ATPS formed from neutral polymers (Albertsson, 1971). Researchers have further demonstrated encapsulation of sufficient polymer concentrations during lipid self-assembly to support phase separation in the interior volume of the vesicle (Keating, 2012).

ACKNOWLEDGMENT

The authors thank the NASA Exobiology program, grant number NNX13AI01G, for financial support.

GLOSSARY

Amphiphile: A molecule consisting of a hydrophilic (polar, water-soluble) group attached to a hydrophobic (nonpolar, water-insoluble) group that is typically a hydrocarbon chain.
Coacervate: Polyelectrolyte-rich aqueous droplets.
Compartmentalization: Creation of areas of different composition than surrounding areas. Compartments can be formed by, for example, barriers to diffusion or environments for which solutes have different affinity.
Debye length: The measure of a molecules net electrostatic effect in solution and the length over which those electrostatic effects persist. It is the length of the persistence charge from one molecule felt by an adjacent molecule.
Electrostatic repulsion: The unfavorable interaction between two species of like charges.
Intermolecular forces: Forces of attraction or repulsion between atoms, molecules, or ions. Intermolecular forces can be weaker than the force felt to keep a molecule

together, known as intramolecular forces. Intermolecular forces include van der Waals, dipole–dipole, and hydrogen bonding.

Lipid bilayer: Phospholipids organized in two layers wherein their hydrophobic tails are projecting inward while their polar head groups are projecting outward.

Micelle: A submicroscopic aggregate of amphipathic molecules in water, with the nonpolar portions in the interior and the polar portions at the exterior surface, exposed to water.

Phase diagram: A diagram representing the limits of stability of the various phases in a chemical system at equilibrium, with respect to variables such as composition and temperature.

pKa: The negative log of the acid dissociation constant. It represents the pH at which the molecule is half protonated and half deprotonated.

Protocells: Endogenously ordered spherical compartments containing nucleic acids, lipids, and/or polypeptides proposed as the stepping-stone for the origin of life.

Protonation: The addition of a hydrogen ion (H^+) to an atom, molecule, or ion, such that a charge of one is added for each hydrogen atom added.

Vesicle: A hollow sac formed by a surrounding amphiphilic bilayer that contains aqueous solution and/or other structures.

Zwitterion: A molecule with an overall neutral charge but contains both a negative and positive electrical charge. Zwitterions may also be referred to as dipolar molecules.

REVIEW QUESTIONS

1. Why would compartments be *advantageous* for the early development of prebiotic life? Are there potential *disadvantages* can arise from compartmentalization?

2. Name two types of organic compartments that may have been important on the early Earth. What kinds of processes might these environments have facilitated?

3. Explain why fatty acids are amphiphilic molecules. Why does this make them a favorable molecule for the construction of vesicles and micelles?

4. Compare fatty acid amphiphiles with phospholipids. Explain how each type of molecule forms self-assemblies, and how the properties of these assemblies differ.

5. What does selective compartmentalization mean? How does this relate to the permeability of vesicular structures? Explain several examples where vesicles can become permeable to certain molecules and how this can occur.

6. Why do you think multivalent ions have an aggregating effect on amphiphiles? How do modern cells combat this issue? What factors would amphiphiles have to overcome to stabilize vesicle formation in an environment rich in metal cations (e.g., Ca^{2+}, Fe^{3+}, Mg^{2+}), such as the ocean?

7. Describe three main types of phase separation that can occur in aqueous solutions when macromolecules are present. What types of polymeric molecules do they require? How do the compartments formed from the different types compare?

8. In what region(s) of a phase diagram would you expect ATPSs to be most sensitive to changes in temperature, pH, and salt? At what composition would you expect ATPS to be most robust in changing microenvironments?

9. The ionic strength of a solution has a significant impact on phase separation. Explain the concept of ionic strength and what happens when salt interacts with a charged polymer. Draw pictures to help explain. (*Hint*: what happens to the interaction between the polyelectrolytes as a function of ionic strength?)

10. In your own words, describe the thermodynamic process behind phase separation of two uncharged polymers. What are the entropic and enthalpic factors that go into phase separation? Use the common equation for free energy to help explain.

11. Explain why vesicles form with some molecules and aqueous two systems phase occur with others. In what other way are vesicles and aqueous phases different?

12. List the similarities and differences between aqueous phases and vesicles. What processes do you think each of these structures could have facilitated on the early Earth? Can you think of processes that are not compatible with these types of early compartments?

13. What aspects of living cells do you think are most important for consideration in generating protocells either on the early Earth or in the laboratory? Explain.

BIBLIOGRAPHY

Books

Albertsson, P. Å. (1971). *Partition of Cell Particles and Macromolecules; Distribution and Fractionation of Cells, Mitochondria, Chloroplasts, Viruses, Proteins, Nucleic Acids, and Antigen–Antibody Complexes in Aqueous Polymer Two-Phase Systems*. New York: Wiley-Interscience.

Keighron, J. and Keating, C. (2011). Towards a minimal cytoplasm. In P. L. Luisi and P. Stano (eds.), *The Minimal Cell* (pp. 3–30). Dordrecht, the Netherlands: Springer.

Luisi, P. L. and Stano, P. (2011). *The Minimal Cell*. Dordrecht, the Netherlands: Springer.

Luisi, P. L. and Walde, P. (1999). *Giant Vesicles: Perspectives in Supramolecular Chemistry*. Vol. 6. New York: John Wiley & Sons, Inc.

Oparin, A. I. (1957). *The Origin of Life on the Earth*. New York: Academic Press.

Vaughan, D. J. and Lloyd, J. R. (2012). Mineral–organic–microbe interfacial chemistry. In *Fundamentals of Geobiology*, A. H. Knoll, D. E. Canfield, and K. O. Konhauser (eds.), (pp. 131–149). John Wiley & Sons, Ltd.

Journals

Asenjo, J. A. and Andrews, B. A. (2011). Aqueous two-phase systems for protein separation: A perspective. *Journal of Chromatography A, 1218*(49), 8826–8835. doi: http://dx.doi.org/10.1016/j.chroma.2011.06.051.

Attwater, J., Wochner, A., Pinheiro, V. B., Coulson, A., and Holliger, P. (2010). Ice as a protocellular medium for RNA replication. *Nature Communications, 1*, 76. doi: http://www.nature.com/ncomms/journal/v1/n6/suppinfo/ncomms1076_S1.html.

Avanti. (2013). Critical Micelle Concentrations (CMCs). Retrieved October 15, 2013, from http://avantilipids.com/index.php?option=com_content&view=article&id=1703&Itemid=422.

Botta, O. and Bada, J. (2002). Extraterrestrial organic compounds in meteorites. *Surveys in Geophysics, 23*(5), 411–467. doi: 10.1023/A:1020139302770.

Branciamore, S., Gallori, E., Szathmáry, E., and Czárán, T. (2009). The origin of life: Chemical evolution of a metabolic system in a mineral honeycomb? *Journal of Molecular Evolution, 69*(5), 458–469. doi: 10.1007/s00239-009-9278-6.

Brangwynne, C. P. (2011). Soft active aggregates: Mechanics, dynamics and self-assembly of liquid-like intracellular protein bodies. *Soft Matter, 7*(7), 3052–3059. doi: 10.1039/C0SM00981D.

Budin, I. and Szostak, J. W. (2010). Expanding roles for diverse physical phenomena during the origin of life. *Annual Review of Biophysics, 39*(1), 245–263. doi: 10.1146/annurev.biophys.050708.133753.

Chen, A. H. and Silver, P. A. (2012). Designing biological compartmentalization. *Trends in Cell Biology, 22*(12), 662–670. doi: http://dx.doi.org/10.1016/j.tcb.2012.07.002.

Chen, I. A., Salehi-Ashtiani, K., and Szostak, J. W. (2005). RNA catalysis in model protocell vesicles. *Journal of the American Chemical Society, 127*(38), 13213–13219. doi: 10.1021/ja051784p.

Chen, I. A. and Szostak, J. W. (2004). Membrane growth can generate a transmembrane pH gradient in fatty acid vesicles. *Proceedings of the National Academy of Sciences of the United States of America, 101*(21), 7965–7970. doi: 10.1073/pnas.0308045101.

Childers, W. S., Ni, R., Mehta, A. K., and Lynn, D. G. (2009). Peptide membranes in chemical evolution. *Current Opinion in Chemical Biology, 13*(5–6), 652–659. doi: http://dx.doi.org/10.1016/j.cbpa.2009.09.027.

Chollakup, R., Beck, J. B., Dirnberger, K., Tirrell, M., and Eisenbach, C. D. (2013). Polyelectrolyte molecular weight and salt effects on the phase behavior and coacervation of aqueous solutions of poly(acrylic acid) sodium salt and poly(allylamine) hydrochloride. *Macromolecules, 46*(6), 2376–2390. doi: 10.1021/ma202172q.

Crosby, J., Treadwell, T., Hammerton, M., Vasilakis, K., Crump, M. P., Williams, D. S., and Mann, S. (2012). Stabilization and enhanced reactivity of actinorhodin polyketide synthase minimal complex in polymer–nucleotide coacervate droplets. *Chemical Communications, 48*(97), 11832–11834. doi: 10.1039/C2CC36533B.

Darwin, C. (1870). Letter to J.D. Hooker. *Darwin Correspondence Database*, from http://www.darwinproject.ac.uk/entry-7471 (accessed November 7, 2013).

de Kruif, C. G., Weinbreck, F., and de Vries, R. (2004). Complex coacervation of proteins and anionic polysaccharides. *Current Opinion in Colloid & Interface Science, 9*(5), 340–349. doi: http://dx.doi.org/10.1016/j.cocis.2004.09.006.

de Souza-Barros, F. and Vieyra, A. (2007). Mineral interface in extreme habitats: A niche for primitive molecular evolution for the appearance of different forms of life on Earth. *Comparative Biochemistry and Physiology Part C: Toxicology & Pharmacology, 146*(1–2), 10–21. doi: http://dx.doi.org/10.1016/j.cbpc.2006.12.018.

Deamer, D. W. (1997). The first living systems: A bioenergetic perspective. *Microbiology and Molecular Biology Reviews, 61*(2), 239–261.

Dobry, A. and Boyer-Kawenoki, F. (1947). Phase separation in polymer solution. *Journal of Polymer Science, 2*(1), 90–100. doi: 10.1002/pol.1947.120020111.

Eisenberg, H. (1977). Polyelectrolytes, thirty years later. *Biophysical Chemistry, 7*(1), 3–13. doi: http://dx.doi.org/10.1016/0301-4622(77)87010-5.

Fatoki, O. S. and Vernon, F. (1989). Determination of free fatty-acid content of polluted and unpolluted waters. *Water Research, 23*(1), 123–125. doi: http://dx.doi.org/10.1016/0043-1354(89)90070-5.

Griffith, E. C., Tuck, A. F., and Vaida, V. (2012). Ocean–atmosphere interactions in the emergence of complexity in simple chemical systems. *Accounts of Chemical Research, 45*(12), 2106–2113. doi: 10.1021/ar300027q.

Griffith, E. C. and Vaida, V. (2012). In situ observation of peptide bond formation at the water–air interface. *Proceedings of the National Academy of Sciences, 109*(39), 15697–15701. doi: 10.1073/pnas.1210029109.

Guida, V. (2010). Thermodynamics and kinetics of vesicles formation processes. *Advances in Colloid and Interface Science, 161*(1–2), 77–88. doi: http://dx.doi.org/10.1016/j.cis.2009.11.004.

Hanczyc, M. M., Fujikawa, S. M., and Szostak, J. W. (2003). Experimental models of primitive cellular compartments: Encapsulation, growth, and division. *Science, 302*(5645), 618–622. doi: 10.1126/science.1089904.

Huang, Y., Alexandre, M. R., and Wang, Y. (2007). Structure and isotopic ratios of aliphatic side chains in the insoluble organic matter of the Murchison carbonaceous chondrite. *Earth and Planetary Science Letters, 259*(3–4), 517–525. doi: http://dx.doi.org/10.1016/j.epsl.2007.05.012.

Keating, C. D. (2012). Aqueous phase separation as a possible route to compartmentalization of biological molecules. *Accounts of Chemical Research, 45*(12), 2114–2124. doi: 10.1021/ar200294y.

Knauth, L. P. (2005). Temperature and salinity history of the Precambrian ocean: Implications for the course of microbial evolution. *Palaeogeography, Palaeoclimatology, Palaeoecology, 219*(1–2), 53–69. doi: http://dx.doi.org/10.1016/j.palaeo.2004.10.014.

Koga, S., Williams, D. S., Perriman, A. W., and Mann, S. (2011). Peptide–nucleotide microdroplets as a step towards a membrane-free protocell model. *Nature Chemistry, 3*(9), 720–724. doi: http://www.nature.com/nchem/journal/v3/n9/abs/nchem.1110.html-supplementary-information.

Koonin, E. V. (2007). An RNA-making reactor for the origin of life. *Proceedings of the National Academy of Sciences, 104*(22), 9105–9106. doi: 10.1073/pnas.0702699104.

Kurihara, K., Tamura, M., Shohda, K.-i., Toyota, T., Suzuki, K., and Sugawara, T. (2011). Self-reproduction of supramolecular giant vesicles combined with the amplification of encapsulated DNA. *Nature Chemistry, 3*(10), 775–781. doi: http://www.nature.com/nchem/journal/v3/n10/abs/nchem.1127.html-supplementary-information.

Lazcano, A. (2010). Historical development of origins research. *Cold Spring Harbor Perspectives in Biology, 2*(11), a002089. doi: 10.1101/cshperspect.a002089.

Luisi, P., Ferri, F., and Stano, P. (2006). Approaches to semi-synthetic minimal cells: A review. *Naturwissenschaften, 93*(1), 1–13. doi: 10.1007/s00114-005-0056-z.

Mansy, S. S. (2009). Model protocells form single-chain lipids. *International Journal of Molecular Sciences, 10*(3), 835–843. doi: 10.3390/ijms10030835.

Mansy, S. S., Schrum, J. P., Krishnamurthy, M., Tobe, S., Treco, D. A., and Szostak, J. W. (2008). Template-directed synthesis of a genetic polymer in a model protocell. *Nature, 454*(7200), 122–125. doi: http://www.nature.com/nature/journal/v454/n7200/suppinfo/nature07018_S1.html.

Menger, F. M. and Sykes, B. M. (1998). Anatomy of a coacervate. *Langmuir, 14*(15), 4131–4137. doi: 10.1021/la980208m.

Mohanty, B. and Bohidar, H. B. (2005). Microscopic structure of gelatin coacervates. *International Journal of Biological Macromolecules, 36*(1–2), 39–46. doi: http://dx.doi.org/10.1016/j.ijbiomac.2005.03.012.

Monnard, P.-A. and Deamer, D. W. (2002). Membrane self-assembly processes: Steps toward the first cellular life. *The Anatomical Record, 268*(3), 196–207. doi: 10.1002/ar.10154.

Monnard, P.-A. and Deamer, D. W. (2003). Preparation of vesicles from nonphospholipid amphiphiles. *Methods in Enzymology, 372*, 133–151. doi: 10.1016/S0076-6879(03)72008-4.

Morse, J. W. and Mackenzie, F. T. (1998). Hadean ocean carbonate geochemistry. *Aquatic Geochemistry, 4*(3/4), 301–319. doi: 10.1023/A:1009632230875.

Murtas, G., Kuruma, Y., Bianchini, P., Diaspro, A., and Luisi, P. L. (2007). Protein synthesis in liposomes with a minimal set of enzymes. *Biochemical and Biophysical Research Communications, 363*(1), 12–17.

Novikov, Y. and Copley, S. D. (2013). Reactivity landscape of pyruvate under simulated hydrothermal vent conditions. *Proceedings of the National Academy of Sciences, 110*(33), 13283–13288. doi: 10.1073/pnas.1304923110.

Pereira de Souza, T., Steiniger, F., Stano, P., and Fahr, A. (2011). Spontaneous crowding of ribosomes and proteins inside vesicles: a possible mechanism for the origin of cell metabolism. *Chembiochem: A European Journal of Chemical Biology, 12*(15), 2325–2330. doi: 10.1002/cbic.201100306.

Pohorille, A., Wilson, M., and Chipot, C. (2003). Membrane peptides and their role in proto-biological evolution. *Origins of Life and Evolution of the Biosphere*, 33(2), 173–197. doi: 10.1023/A:1024627726231.

Priftis, D., Farina, R., and Tirrell, M. (2012). Interfacial energy of polypeptide complex coacervates measured via capillary adhesion. *Langmuir*, 28(23), 8721–8729. doi: 10.1021/la300769d.

Rasmussen, S., Chen, L., Nilsson, M., and Abe, S. (2003). Bridging nonliving and living matter. *Artificial Life*, 9(3), 269–316. doi: 10.1162/106454603322392479.

Riepe, A., Beier, H., and Gross, H. J. (1999). Enhancement of RNA self-cleavage by micellar catalysis. *FEBS Letters*, 457(2), 193–199. doi: http://dx.doi.org/10.1016/S0014-5793(99)01038-8.

Rito-Palomares, M. (2004). Practical application of aqueous two-phase partition to process development for the recovery of biological products. *Journal of Chromatography B*, 807(1), 3–11. doi: http://dx.doi.org/10.1016/j.jchromb.2004.01.008.

Ronca, G. and Russell, T. P. (1985). Thermodynamics of phase separation in polymer mixtures. *Macromolecules*, 18(4), 665–670. doi: 10.1021/ma00146a015.

Shimokawa, K.-i., Saegusa, K., Wada, Y., and Ishii, F. (2013). Physicochemical properties and controlled drug release of microcapsules prepared by simple coacervation. *Colloids and Surfaces B: Biointerfaces*, 104, 1–4. doi: http://dx.doi.org/10.1016/j.colsurfb.2012.11.036.

Spitzer, J. and Poolman, B. (2009). The role of biomacromolecular crowding, ionic strength, and physicochemical gradients in the complexities of life's emergence. *Microbiology and Molecular Biology Reviews*, 73(2), 371–388.

Strulson, C. A., Molden, R. C., Keating, C. D., and Bevilacqua, P. C. (2012). RNA catalysis through compartmentalization. *Nature Chemistry*, 4(11), 941–946.

Thomas, J. and Rana, F. R. (2007). The influence of environmental conditions, lipid composition, and phase behavior on the origin of cell membranes. *Origins of Life and Evolution of Biospheres*, 37(3), 267–285. doi: 10.1007/s11084-007-9065-6.

Veis, A. (2011). A review of the early development of the thermodynamics of the complex coacervation phase separation. *Advances in Colloid and Interface Science*, 167(1–2), 2–11. doi: http://dx.doi.org/10.1016/j.cis.2011.01.007.

Walde, P. (2006). Surfactant assemblies and their various possible roles for the origin(s) of life. *Origins of Life and Evolution of Biospheres*, 36(2), 109–150. doi: 10.1007/s11084-005-9004-3.

Williams, D. S., Koga, S., Hak, C. R. C., Majrekar, A., Patil, A. J., Perriman, A. W., and Mann, S. (2012). Polymer/nucleotide droplets as bio-inspired functional micro-compartments. *Soft Matter*, 8(22), 6004–6014. doi: 10.1039/C2SM25184A.

Zhang, S. (2012). Lipid-like self-assembling peptides. *Accounts of Chemical Research*, 45(12), 2142–2150. doi: 10.1021/ar300034v.

Zhu, T. F. and Szostak, J. W. (2009). Coupled growth and division of model protocell membranes. *Journal of the American Chemical Society*, 131(15), 5705–5713. doi: 10.1021/ja900919c.

CHAPTER 11

Role of Phosphorus in Prebiotic Chemistry

Matthew Pasek

CONTENTS

11.1 ELEMENTS OF LIFE

Life is characterized by a set of self-sustaining chemical reactions involving carbon, hydrogen, oxygen, nitrogen, phosphorus, sulfur, and an assortment of salts and metals such as potassium and iron. These elements are not found in equal parts in life, but instead follow distributions that can be termed the stoichiometry of life. In chemistry, stoichiometry describes the relative ratios of elements or molecules in chemical systems. Biological stoichiometry relates to the modern (present day) and ancient (over the last 4 billion years) environmental availability of these elements and to their importance to biochemical processes.

The stoichiometry of several materials relevant to life is given in Table 11.1. In this table, you should see a few interesting features. For one, all elements are normalized to one phosphorus atom. This is done so as to keep most of the numbers whole and to allow for easy comparison between environments. Next, the chemistry of the environments listed (cosmos, ocean, bulk silicate earth) changes significantly. This is due to the fact that we live on a rocky planet, not a gas planet such as Jupiter, and hence the more volatile elements, such as nitrogen and hydrogen, are less abundant in rocks (represented by the bulk silicate earth) than they are in the cosmos. Additionally, four biochemical compositions are listed.

TABLE 11.1   Stoichiometry (on a per Atom Basis) of the Cosmos, as Defined by Elemental Abundance Patterns in the Sun and in Meteorites, of the Oceans, and of the Bulk Silicate Earth (Earth's Crust, Mantle, and Hydrosphere), as well as Biological Portions Described in the Text

	Cosmic	Oceans	Bulk Silicate Earth	Bulk Bacteria	P-Lipid	RNA	Metabolic Core
H	2.8×10^6	4.9×10^7	21	203	90	10	15.4
O	1,400	2.5×10^7	10,800	71.7	8	7	9
C	680	974	4	116	46	9.5	9.5
N	230	16	0.06	15.5	1	3.75	2
S	43	12,400	3.3	0.19	0	0	0.05
P	1	1	1	1	1	1	1

The *whole cell* composition refers to the dried-down material of a cell and shows the composition of the stuff left after all water is removed. Water typically makes up about 70% of the mass of a cell. The next three columns give the composition of the cellular membrane—the material that separates cells from the outside world—as well as the composition of the average RNA molecule and the composition of the metabolome. The metabolome is the set of molecules, excluding enzymes, which are used to make and break organic compounds to make new ones by cells.

What then might be the composition of life near its start? Is the composition that we see today reflective of its composition in antiquity? We do indeed have some reason to suspect that life's current composition is fairly ancient. This idea has come about due to studies of how elements likely behaved on the early Earth and how these behaviors compare to modern biological composition. If the modern day elemental composition of the ocean is compared to the modern day composition of cells, there is a decent correlation between the two (see schematic Figure 11.1), though there is significant scatter. However, since life has been modifying the geochemistry of the Earth for the last 3.5 billion years at least, using the modern ocean composition does not present an accurate picture of the environment in which cells arose. The early Earth, lacking oxygen, was influenced by a reducing environment, rich in free electrons, as evidenced by the presence of pyrite and uranitite minerals on the early Earth's surface. Both of these minerals react away in an oxidizing environment. If the composition of the ocean is changed to account for a reducing environment, the relationship between modern cells and the ancient ocean significantly improves (Figure 11.1). The composition of living cells reflects in part the environment in which those cells evolved and has not changed too much in the last few billion years. For more discussion on early geochemistry relationships with stoichiometry (from which Figure 11.1 is derived), see Byrne (2002).

Although the current stoichiometry is similar to that of the ancient ocean, it is probably unlikely that this stoichiometry describes the composition of the chemical mixture at the origin of life. Additionally, it is unlikely that this stoichiometry is required for life elsewhere in the universe. A comparison of the stoichiometry of life and the stoichiometry of the environment reveals that the element phosphorus, P, is often a limiting reagent, especially compared to its relative importance in biology. As a result, we should pose the question, *Why phosphate?*

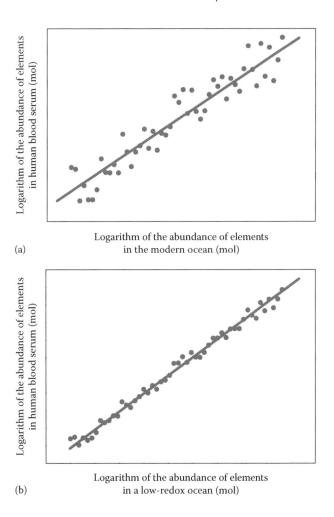

FIGURE 11.1   Schematic of the elemental abundances of human blood serum (used as a proxy for general biochemical systems), compared to the elemental abundances of two oceans, with (a) showing the modern ocean and (b) showing a reducing ocean.

11.1.1 Why Phosphorus and Phosphate?

The element phosphorus generally will bond to oxygen under typical conditions at the Earth's surface at a ratio of four oxygen atoms for every phosphorus atom. This arrangement is tetrahedral in shape with phosphorus at the center and forms the phosphate tetrahedron. Most phosphorus on the Earth's surface is in phosphate, and this molecular arrangement describes the *speciation* of phosphorus. There are a number of other arrangements of P possible, including surrounding P with three hydrogen atoms to form the poisonous gas phosphine (PH_3), binding it to two oxygen atoms and two hydrogen atoms to form hypophosphite ($H_2PO_2^-$), or binding it to three oxygen atoms and an organic radical to form a phosphonate (RPO_3^{2-}). There are other potential phosphorus species, though the phosphate is generally the most stable (Figure 11.2).

Why is phosphate used so extensively by life? Phosphorus, as phosphate, has a few key characteristics that make it ideal for metabolic processes and for the construction of biomolecules.

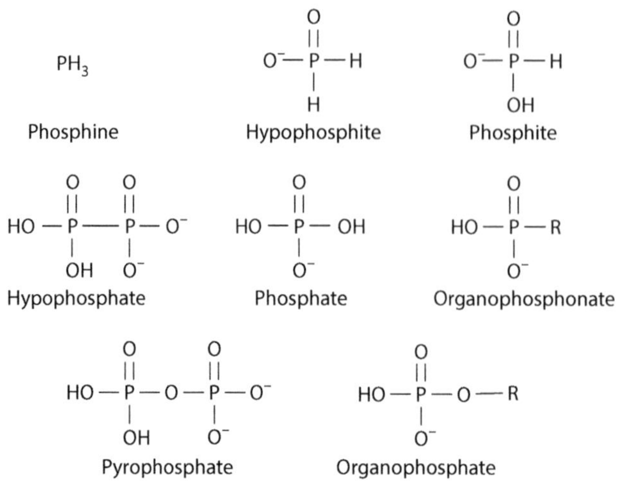

FIGURE 11.2 Phosphorus speciation diagrams with names. Species are shown ionized, as they would be at pH 6.

Phosphate resides primarily in four reservoirs in cells: as free phosphate within cellular plasma, in membranes as phospholipids, in the nucleic acids, and as metabolic molecules including adenosine triphosphate (ATP). At its most basic level, free phosphate, specifically $H_2PO_4^-$ and HPO_4^{2-}, is highly useful as a pH buffer. The pK_A of $H_2PO_4^-$ is about 7, meaning that at equal concentrations of $H_2PO_4^-$ and HPO_4^{2-}, a fluid will have a pH of 7, thus ensuring a static chemical system with respect to acidity.

Free phosphate is not bound to organic compounds and is a solute. Modern cells have a dissolved phosphate concentration of about 10^{-4} M. Most of the phosphate in cells is actually bound through covalent bonds to organic molecules, where it serves several important roles. A significant amount of phosphate is in phospholipids. Phospholipids are lipids (compounds with both hydrophobic and hydrophilic ends that self-assemble) that make up cell membranes and separate the cell from the outside world. These molecules consist of a phosphate group bound to a glycerol, which is in turn bound to two fatty acids (Figure 11.3). The phosphate group in phospholipids is strongly hydrophilic, whereas the fats are strongly hydrophobic. As a result, these two ends of the molecule act to create lipid bilayers, a sort of oily separation of one set of material from the environment.

An extremely important role for phosphorus in biology is in the backbone of the nucleic acids DNA and RNA. Both DNA and RNA consist of three main parts: nucleobases that serve as the letters of the genetic code (e.g., A, G, C, T, or U); a sugar, either ribose (RNA) or deoxyribose (DNA); and a phosphate. The phosphate binds the sugars together to make the backbone of the nucleic acid, which in turn holds the nucleobases together to make a double helix (DNA) or single helix (RNA). As the phosphate does so, it has a few major effects. For one, phosphate is capable of forming bridges between sugar atoms through two oxygen atoms while maintaining a charge. The resulting chemistry is a large, charged, polymer that stores genetic information. Having a large, charged polymer ensures that the polymer cannot leak out through a cell membrane. Cell membranes are quite permeable to hydrophobic

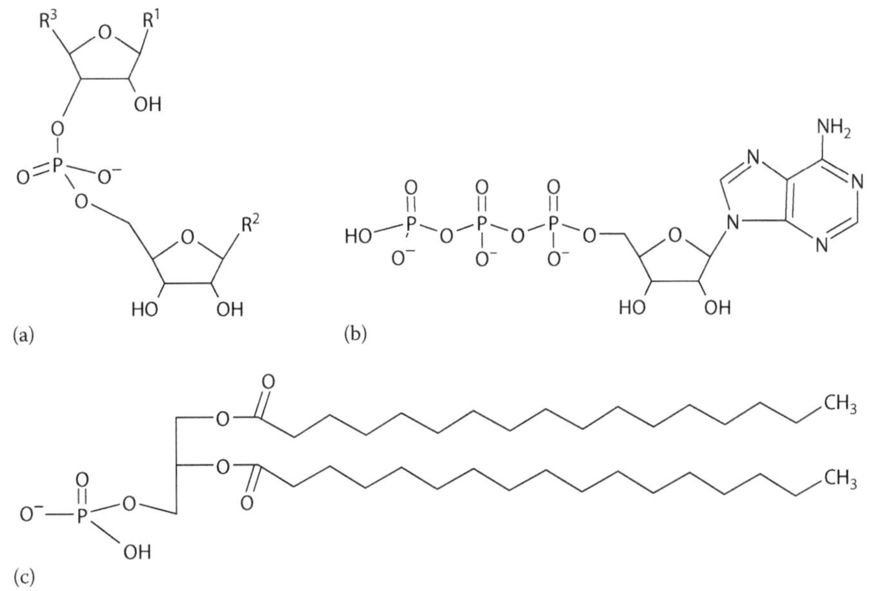

FIGURE 11.3 Phosphorus biomolecules acting in (a) replication (RNA), (b) metabolism (ATP), and (c) cellular membranes (phospholipid). Ionizations are appropriate for pH 6. R^1 and R^2 are nucleobases such as adenine or guanine, and R^3 is the continuing repetition of this structure.

materials (such as organic polymers) but not to charged molecules. The charged phosphate keeps the nucleic acid in place. Additionally, it does so without increasing the chirality of the molecules (see Chapters 4 and 5).

The addition of phosphate to an organic molecule has other effects as well. For one, the addition of the strongly charged phosphate group increases the solubility of the molecule. Since most cells are made of water, this allows organics to be shuffled about within a cell. Additionally, since phosphate molecules are strongly attracted to divalent cations such as Mg^{2+}, the addition of a phosphate group to a molecule allows it to be directed to an enzyme site by attracting the phosphate group with a bound Mg^{2+} atom. For instance, Mg–phosphate bonds determine the shape of large RNA molecules (see Bowman et al. 2012 for more details).

The final, and probably the most useful, features of phosphate to life are its energetic characteristics. Phosphate is capable of bonding to another phosphate anion to make a polyphosphate (Figure 11.3), of which the major one occurring in biologic systems is ATP. The formation of polyphosphates is accompanied by a loss of water and a gain of energy. Polyphosphates are critical materials that shuffle cellular energy about: when a cell is at rest, it burns sugars to make CO_2 and ATP. When the cell needs to do work, it uses its stores of ATP to affect its environment and change its chemistry. For more information on the role of phosphorus in biochemistry, see Westheimer (1987).

Though phosphate is important in life due to its unique chemistry, other alternatives may play similar roles in life elsewhere. One such molecule that has been suggested to serve as an alternative to phosphate in life is arsenate. The arsenate anion consists of an arsenic atom surrounded by four oxygen atoms and is chemically quite similar to phosphate: it has similar K_As, is capable of forming bridging bonds with no increase in chirality, and increases the solubility of organic molecules when bonding. For more information on the arsenic-based backbone idea, see Wolfe-Simon et al. (2011).

Indeed, the concept of element periodicity allows for a range of unusual substitutions in possible astrobiological (or science fiction!) alternatives to terrestrial life. Element periodicity is the concept that elements within the same column of the periodic table have very similar chemical behaviors. An extreme version of this view is an organism consisting of silicon-based molecules living in a sea of H_2S (an analog to H_2O) with arsenate backbones (an analog for phosphate) for its silicon-based DNA (an analog to carbon-based organic chemistry). Please refer to Chapter 7 for the role of silicon in life.

Although possible, such a scenario is unlikely, as element periodicity only goes so far. For one, the cosmic abundances of the elements limit the chemistry of life to the first three rows of the periodic table. The elements occurring in the lower rows are just too rare to be likely major constituents of life. Additionally, although the chemistry of elements within a period is similar, it is not identical. Arsenate participates in redox chemistry much more readily than does phosphate, with redox changes from +5 to +3 occurring under only mildly reducing conditions. Thus, the stability of arsenate as a global constituent is unlikely. Finally, the fundamental chemical behavior of elements within the same period can be quite difficult. Although arsenate can bond to sugars just like phosphate, it also breaks off of sugars at a much faster rate (thousands of times faster). Thus, the long-term persistence of arsenate biomolecules, at least on planets like the Earth, is unlikely.

11.2 THERMODYNAMICS OF PHOSPHORYLATION

The formation of an organophosphate may have been a critical step in the development of life on Earth. An organophosphate molecule is a molecule with a phosphate bound to it through an oxygen atom (Figure 11.3). The synthesis of an organophosphate is difficult and does not occur spontaneously. The reasons why this reaction is so heavily impeded are due to the fundamentals of chemical thermodynamics. Understanding the thermodynamics of phosphorylation requires an understanding of chemical thermodynamic equations.

A thermodynamic equation is composed of two terms: ΔH and ΔS. The ΔH term is the change in *enthalpy* of a chemical reaction. This has a very obvious relationship to the physical world as this is the heat of a reaction. If a chemical reaction releases or absorbs heat as it proceeds, this is directly linked to the enthalpy of the reaction. If a reaction releases heat, it is called *exothermic*. An example would be the combustion of gasoline. An *endothermic* reaction is the opposite: it absorbs heat to proceed. The evaporation of isopropanol (rubbing alcohol) on your skin is a good example of this process.

The ΔS term is the change in entropy of a chemical reaction. Although entropy is commonly referred to as *disorder*, these two terms are not equivalent. Effectively, entropy refers to the ability of a system to explore its physical environment. For instance, if a gas is confined to a vial, and if the vial's volume is doubled, the gas can now explore twice as much space and it has increased in entropy. All chemical processes increase the entropy of the universe on some level, though many do so with a smaller loss of entropy elsewhere.

The crystallization of salt from water decreases the entropy of the salt (as it goes from freely moving ions to a confined mineral), but this occurs as a result of evaporation of water (which increases entropy).

These two terms taken together give the Gibbs free energy of a reaction:

$$\Delta G \text{ (rxn)} = \Delta H \text{ (rxn)} - T\Delta S \text{ (rxn)} \tag{11.1}$$

In general, we care primarily about the ΔG (rxn) at standard state (i.e., 1 atmosphere of pressure and 298 K). Under these conditions, ΔG (rxn) is equal to ΔG^0 (rxn).

The thermodynamic value most useful for determining compositions in geochemical fluids is the ΔG value or the Gibbs free energy of a reaction. Thermodynamics uses two equations to do this:

$$\Delta G^0 \text{ (Reaction)} = \Delta G^0 \text{ (Products)} - \Delta G^0 \text{ (Reactants)} \tag{11.2}$$

$$\Delta G^0 \text{ (Reaction)} = - RT \ln K \quad \text{or} \quad e^{\frac{-\Delta G^0 \text{ (Reaction)}}{RT}} = K \tag{11.3}$$

Note that R is the universal gas law constant (8.314 J/mol K, or 0.001987 kcal/mol K), and T is the temperature in Kelvin (K). At standard state, T is equal to 298 K, and the pressure is 1 atmosphere.

The ΔG^0 values of individual species are compiled as tables of data. These ΔG^0 values are all referenced to the ΔG^0 values of their standard elemental state at a given temperature. In other words, at 298 K, all compounds bearing oxygen are referenced to the thermodynamic properties of O_2 (g). To make it easy, the ΔG^0 of O_2 (g) at 298 K is 0 kcal or 0 kJ or 0 J. Compounds that are composed of multiple elements, for instance, Mg_2SiO_4, are referenced to elemental Mg metal, elemental Si, and O_2 gas at 298 K.

The ΔG value of a chemical reaction determines the direction of a chemical reaction. If the ΔG (rxn) is greater than 0, then the reaction will move toward reactants; if it is less than 0, then it will move toward products; and if it is 0 (or close), then it is at equilibrium and will not change much at all. The value of ΔG (rxn) also tells you how much energy can be extracted from a reaction or whether energy needs to be added to the reaction to push it one direction or another.

In most cases, we use thermodynamics to solve for the ratio of products to the reactants, known as the K value. Using Equation 11.3 requires a bit of knowledge on what the K represents. For the generic reaction,

$$aA + bB \rightarrow cC + dD \tag{11.4}$$

The K value of this reaction is equal to

$$K = \frac{[C]^c[D]^d}{[A]^a[B]^b} \tag{11.5}$$

Note that each of the coefficients in front of the compounds reacting becomes exponents in the K. The value used for [C] is the *activity* of C. The activity of C is unitless and represents a reactive characteristic of the element. In many cases, the activity is set equal to the molarity of C (the moles of C per liter of solution). This is accurate for most dilute solutions.

There are some other curiosities associated with activity. Most solids are defined in this system as having an activity of 1. In other words, solids do not participate in K calculations (as a ratio multiplied or divided by 1 is equal to itself). Naturally, this is not always true, but serves as a useful approximation. Additionally, the activity of water will usually have a value of 1 as well, so, like solids, it drops out of the K calculation.

11.2.1 Nonequilibrium Systems

For the most part, these calculations are used for equilibrium systems. For a nonequilibrium system, which is one that has not completely reached the lowest energy, Equation 11.3 is modified to

$$\Delta G = \Delta G^0 + RT \ln Q \tag{11.6}$$

where
   ΔG^0 is the Gibbs free energy at equilibrium
   Q is the reaction quotient at the disequilibrium state

Note that the ΔG being calculated does not need to be at a standard state and is not necessarily at equilibrium. Indeed, at equilibrium, the ΔG value is equal to 0, and there is no chemical energy free to be extracted from the reaction. Q is equivalent to K, but has the values that are measured at a given point in the reaction placed into the activities instead of the equilibrium values.

11.2.2 Thermodynamics of Phosphorylation

A phosphorylation reaction is a reaction that generates an organophosphate compound. Organophosphate compounds are the primary biologic phosphates used by life and include DNA, RNA, and the metabolic molecule ATP. The phosphorylation of an organic compound is generally not spontaneous and usually requires energy. For instance, the phosphorylation of glucose

$$C_6H_{12}O_6 + H_2PO_4^- \rightarrow C_6H_{12}O_6PO_3^- + H_2O \tag{11.7}$$

has a ΔG^0 at 298 K of +13.8 kJ/mol if the phosphate attaches to the 6′ carbon. We can turn this into a K:

$$K = e^{\frac{-\Delta G^0 \text{(Reaction)}}{RT}} = \frac{[C_6H_{12}O_6PO_3^-]}{[C_6H_{12}O_6][H_2PO_4^-]} = 0.0038 \tag{11.8}$$

This means that the ratio of the concentration of glucose-6-phosphate to the product of the concentrations of glucose and phosphate is only 0.0038. The activity of water in this reaction is set to 1.

Chemists often think in terms of yield: the amount of product produced divided by the amount of product expected if the reactants reacted completely. We do not know the maximum concentration of phosphate or glucose in the early Earth (prebiotic) water, but if the concentration is similar to that of today, we might expect to see concentrations on the order of 10^{-6} M. At these concentrations, you would expect to see 3.8×10^{-15} M of glucose phosphate, an exceedingly low concentration. Glucose is actually one of the easier compounds to phosphorylate, and most phosphorylation reactions have much lower yields. If phosphates were critical for the development of life on Earth, then how did they arise spontaneously?

We can use the K to determine ways of increasing the yield. One of the simplest ways of increasing yield is to increase the concentrations of reactants. Indeed, if the glucose and phosphate concentrations are raised to 1 mol/L, you would expect 3.8×10^{-3} M of glucose-6-phosphate. In biology, cells increase concentrations locally by binding molecules to enzymes that increase their local activity and simultaneously increase yield. Alternatively, if the glucose-6-phosphate is removed from the system (e.g., it may be shuffled to another part of the cell or used in another chemical reaction), then more glucose-6-phosphate can be made as its activity has gone down again. Still, another way to increase the yield of phosphorylation is to remove water. Since water is a product of this reaction, drying the system decreases the activity of water (from 1 to less than 1), enhancing the yield of glucose-6-phosphate. Water can be removed by heating (boiling it off), freezing (turning it to ice), or by changing the solvent to an organic solvent such as formamide ($HCONH_2$).

Two other ways that may provide increased reactivity and enhance phosphorylation yields are to use reactive compounds to activate the organic compounds or to use more reactive phosphate compounds. In this way, the thermodynamic barrier for phosphorylation is overcome by adding energy in the form of a more reactive reagent. One class of these compounds are called *condensing agents*, so name because they cause a *condensation* reaction, or a loss of water. Condensing agents can be formed through natural atmospheric phenomena and include the compounds cyanate, OCN^-, and cyanamide, $NCNH_2$. Alternatively, use of reactive phosphates, such as polyphosphates, can overcome these energetic barriers and promote phosphorylation.

Prebiotic chemists have attempted to form phosphorylated organic compounds using all of these methods. Indeed, several organophosphates have been made by increasing the concentration of phosphate, heating solutions to evaporate water and thus decreasing water activity, adding condensing agents, and adding polyphosphates (further information on prebiotic phosphorylation methods can be found in Pasek and Kee [2011]). Unfortunately, a significant problem has come about through these reactions: very few accurately represent a geologic system that might actually use both naturally occurring phosphorus minerals and organics and subject them to reasonable geochemical processes (heating, UV light, lightning) to generate organophosphates. As a result, the phosphorylation of organics under plausible geologic or cosmic conditions has presented a conundrum that has been colloquially termed *the phosphate problem*.

11.3 PHOSPHORUS MINERALS

The *phosphate problem* is a problem because the geochemistry of phosphate is different from all of the other major biogenic elements. For one, there is no significant volatile phase for phosphorus. Oxygen and hydrogen form water, carbon forms methane and CO_2, nitrogen forms ammonia, and sulfur can form H_2S. Phosphorus, however, is almost uniformly in phosphate minerals on the Earth's surface. As a result, the cycling of phosphate on Earth is slow, as it is tied to the rock cycle, which is the slow building of mountains, weathering of these mountains to sediment, and burying the sediment by plate tectonics, eventually uplifting them again to form new mountains. As you can figure, this process is slow. The other elements move around the Earth rapidly, since they have volatile phases. Additionally, this suggests that the phosphorus used by early life also had to come from a mineral source, since most phosphorus is bound up in rocks.

There are at least 100 distinct phosphate minerals on the surface of the Earth. Have there always been this many phosphate minerals on the Earth's surface? We can convincingly argue that the actual phosphate mineral inventory of the early Earth was much lower on the basis of a few key features. For one, many of the modern phosphates are biological in origin. Some of these minerals are found in guano, or as biominerals, or as microbial precipitates. Other phosphate minerals use oxidized metals, such as copper(II) or iron(III) as cations. Since the early Earth was less oxidizing than the present-day Earth, these minerals are unlikely. Still, other phosphate minerals are generated as a result of extensive plate tectonics or hydrothermal processing and probably took 2 or 3 billion years to accumulate enough to form distinct deposits of these minerals. Indeed, the phosphate inventory of the early Earth was probably limited to under 20 unique phosphate minerals. For a perspective on what other minerals were present on the early Earth, see Hazen et al. (2008).

The phosphate mineral inventory of the early Earth was likely small and probably consisted of a few robust phosphate minerals. These phosphate minerals include a number of calcium phosphates as calcium is a very strong scavenger of phosphate, as well as a few rare earth phosphates. Both phosphate and rare earth elements are poorly soluble in the average silicate mineral, and as many rocks crystallize, rare earth elements, calcium, and phosphate bond together to form phosphate minerals. These minerals include a mineral named apatite, which has a formula of $Ca_5(PO_4)_3$ (F, OH, Cl) and is near and dear to us vertebrates, as it makes up our bone (Figure 11.4).

The primordial phosphate minerals were robust, which means they were pretty stable. Since these minerals were stable, they are poorly soluble. Indeed, if apatite is added to pure water, you would expect to see less than 10^{-6} M of phosphate in the water resulting from apatite dissolution. The other primordial minerals provided similarly little phosphate. Since the phosphorylation of organics is dependent on a high activity for phosphate, the low concentration of phosphate resulting from the dissolution of primordial phosphate minerals implies that these minerals may not have been a great source of phosphate for the development of these key biomolecules.

One potential alternative to phosphate minerals that may have been important on the early Earth is phosphorus from meteorites. Meteorites bear phosphorus in a form not

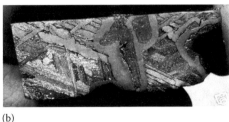

(a)   (b)

FIGURE 11.4   (a) Apatite forms translucent crystals. (b) Schreibersite is a metallic mineral (slightly different texture in the middle of meteorite).

usually encountered on Earth, in minerals called phosphides. One phosphide, the mineral schreibersite, $(Fe,Ni)_3P$, is especially common in meteorites. When this mineral reacts with water, it releases phosphate and phosphite (HPO_3^{2-}) anions that are not impeded by low solubility as these anions do not bond to iron very well at neutral pH. As a result, the activity of phosphate in a body of water with meteoritic phosphorus in it can be quite high. Although this meteoritic alternative to phosphates may seem far-fetched, recent studies have shown that the phosphorus in some of the oldest sedimentary rocks on Earth may have a significant meteoritic origin, with meteoritic phosphorus being 100 times more abundant in oceans than phosphate on the early Earth. If interested, please see Pasek (2008) and Pasek et al. (2013) for more details on the meteoritic origin of prebiotic phosphorus.

11.4 CONCLUSIONS

Phosphorus is an important element in biology today due to several key features, including charge, bonding structures, and energy. The spontaneous formation of phosphorylated biomolecules is difficult as the most facile routes to phosphorylated biomolecules require conditions that are not geochemically plausible. This problem comes about as a result of the unique thermodynamic features of phosphorylated biomolecules. Pathways leading to organophosphates under plausible geochemical conditions may have to invoke meteorites, redox chemistry, or production of polyphosphates.

GLOSSARY

Activity: A chemical term describing the reactive concentration of a compound in a solution. If the solution is ideal, then the ratio of the activity to concentration of the solute is 1. If the solution is not ideal, perhaps by being too salty, then this ratio deviates from 1.

Apatite: A set of calcium phosphate minerals that include hydroxyapatite $Ca_5(PO_4)_3OH$, fluorapatite $Ca_5(PO_4)_3OH$, and chlorapatite $Ca_5(PO_4)_3OH$. Most phosphate is trapped in these minerals on the Earth's surface, and these minerals also make up a significant portion of the minerals in meteorites.

Bulk silicate earth: The part of the Earth that includes the mantle, hydrosphere, and crust, excluding the metallic core.

Condensation reaction: A reaction that must release water to occur.

Enthalpy: The change in heat of a chemical reaction.

Entropy: The change in physical arrangement of a chemical reaction, often in terms of order and disorder, with disorder being a higher entropy state than order.

Gibbs free energy: A measure of the free energy of a chemical reaction, which considers both enthalpy and entropy changes taking place. The Gibbs free energy determines whether a reaction will be spontaneous or not. It is named after the nineteenth-century physical chemist Willard Gibbs.

Hydrophilic: A material or molecule that is attracted to water, due to similarity in molecular bonds.

Hydrophobic: A material or molecule that is repelled by water.

Metabolome: The set of molecules, excluding enzymes, which are used to make and break organic compounds to make new ones by cells.

Meteorite: A naturally occurring rock from space that has landed on the surface of the Earth.

Orthophosphate: Dissolved inorganic phosphate in water, with composition that ranges from H_3PO_4 at low pH to PO_4^{3-} at high pH.

Phosphate problem: A problem in prebiotic chemistry that describes the difficulty of the spontaneous phosphorylation of organic compounds by minerals likely present on the early Earth's surface.

Phosphorylation: A chemical reaction that adds a phosphate group to an organic compound reducing the early Earth.

Reduced early Earth: A hypothesis on the chemical environment of the early Earth that proposes the Earth's surface was electron rich or reducing. This would be accompanied by reducing gases in the atmosphere such as H_2 and CH_4 and differences in the speciation of several elements such as Fe and U.

Schreibersite: A meteoritic mineral rarely found on Earth with formula $(Fe,Ni)_3P$. This mineral may have acted as the primordial source of phosphate for phosphorylation reactions on the early Earth.

Stoichiometry: A branch of chemistry describing the relative molar ratios of reactants and products in a chemical system.

REVIEW QUESTIONS

1. If you wanted to assemble an RNA molecule from seawater, what ratio of seawater to RNA molecules might you expect? If the amount of phosphate in seawater is 10^{-7} M, how much water would you need to build 1 mol of RNA molecules?

2. Which of the following is NOT a change that occurs to an organic molecule when phosphate is bound to it?

   a. Increasing its charge

   b. Increasing its solubility

   c. Increasing its volatility

   d. Allowing the molecule to bind to divalent cations, such as Mg^{2+}

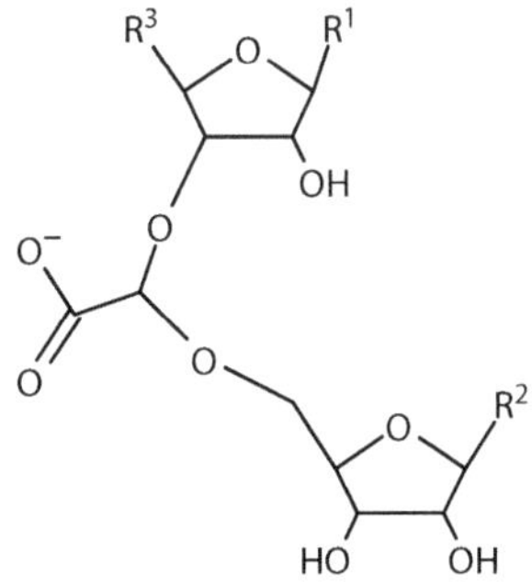

FIGURE 11.5   Schematic of glyoxylate acting in RNA replication. R^1 and R^2 are nucleobases such as adenine or guanine, and R^3 is the continuing repetition of this structure.

3. Certain plants and algae make sulfolipids instead of phospholipids, where a sulfur-bearing organic replaces the phosphate group (see, for instance, Van Mooy et al. 2006). Why might this be?

4. One potential alternative to phosphate in prebiotic assembly is the molecule glyoxylate, which bonds spontaneously to nucleosides to form polymers (e.g., the ΔG is less than 0). What issues might arise from the spontaneous assembly of polymers using glyoxylate (Figure 11.5)?

   For more on the glyoxylate hypothesis, see Bean et al. (2006).

5. The phosphorylation reaction may work better if polyphosphates are used in place of phosphate. One polyphosphate that is relatively easy to synthesize under plausibly prebiotic conditions is pyrophosphate, or $H_2P_2O_7^{2-}$. If pyrophosphate spontaneously hydrolyzes with a free energy of -19.2 kJ/mol,

$$H_2P_2O_7^{2-} + H_2O \rightarrow 2H_2PO_4^-$$

   And if the phosphorylation reaction for glucose requires $+13.8$ kJ/mol, what would the free energy be of the reaction

$$C_6H_{12}O_6 + H_2P_2O_7^{2-} \rightarrow C_6H_{12}O_6PO_3^- + H_2PO_4^-?$$

   Is this reaction spontaneous?

6. Using the thermodynamic properties of several substances in the following table, estimate the solubility of the mineral apatite (in mg/L). Assume the dissolution reaction is

$$Ca_5(PO_4)_3F \rightarrow 5Ca^{2+} + 3PO_4^{3-} + F^-$$

Species	ΔG^0 (kJ/mol)
$Ca_5(PO_4)_3F$	-6475.72
Ca^{2+}	-553.676
PO_4^{3-}	-1026.1
F^-	-277.884

REFERENCES

Bean, H. D., F. A. L. Anet, I. R. Gould, and N. V. Hud. Glyoxylate as a backbone linkage for a prebiotic ancestor of RNA. *Origins of Life and Evolution of Biospheres* 36, 1 (2006): 39–63.

Bowman, J. C., T. K. Lenz, N. V. Hud, and L. D. Williams. Cations in charge: Magnesium ions in RNA folding and catalysis. *Current Opinion in Structural Biology* 22, 3 (2012): 262–272.

Byrne, R. H. Inorganic speciation of dissolved elements in seawater: The influence of pH on concentration ratios. *Geochemical Transactions* 2, 2 (2002): 11.

Hazen, R. M., D. Papineau, W. Bleeker, R. T. Downs, J. M. Ferry, T. J. McCoy, D. A. Sverjensky, and H. Yang. Mineral evolution. *American Mineralogist* 93, 11–12 (2008): 1693–1720.

Pasek, M. A. Rethinking early Earth phosphorus geochemistry. *Proceedings of the National Academy of Sciences* 105, 3 (2008): 853–858.

Pasek, M. A., J. P. Harnmeijer, R. Buick, M. Gull, and Z. Atlas. Evidence for reactive reduced phosphorus species in the early Archean ocean. *Proceedings of the National Academy of Sciences* 110, 25 (2013): 10089–10094.

Pasek, M. A. and T. P. Kee. On the origin of phosphorylated biomolecules. In *Origins of Life: The Primal Self-Organization*, pp. 57–84. Springer, Berlin, Germany, 2011.

Van Mooy, B. A. S., G. Rocap, H. F. Fredricks, C. T. Evans, and A. H. Devol. Sulfolipids dramatically decrease phosphorus demand by picocyanobacteria in oligotrophic marine environments. *Proceedings of the National Academy of Sciences* 103, 23 (2006): 8607–8612.

Westheimer, F. H. Why nature chose phosphates. *Science* 235, 4793 (1987): 1173–1178.

Wolfe-Simon, F., J. S. Blum, T. R. Kulp, G. W. Gordon, S. E. Hoeft, J. Pett-Ridge, J. F. Stolz et al. A bacterium that can grow by using arsenic instead of phosphorus. *Science* 332, 6034 (2011): 1163–1166.

CHAPTER 12

Cold and Dry Limits of Life

Christopher P. McKay, Alfonso F. Davila, and Henry J. Sun

CONTENTS

12.1 INTRODUCTION

Life as we know it exists within a set of boundary conditions defined by temperature, pH, water activity, etc. This imagined multidimensional space is composed of many specific environments, each of which is occupied by an assemblage of well-adapted organisms. To the extent that organisms from one environment may not survive in another, all these environments might be considered extreme. Astrobiologists are interested in delineating the boundary conditions of life, conditions beyond which long-term existence is no longer possible. Such absolute extreme environments provide a basis for assessing the limits of habitability on Earth and elsewhere.

Mars has been at the center of exobiological explorations. This is partly because of easy accessibility and partly because conditions on early Mars were wetter and possibly conducive to the origins of life. Today, however, Mars is an extreme cold desert. If life originated there, some organisms may have adapted to the drastically changing planet and survived, perhaps even to the present. It is in this context that scientists have been exploring the Antarctic Dry Valleys, one of the coldest and driest places on Earth, for the last 60 years and more recently the Atacama, one of the driest. Much is known about the organisms that survive there. Sufficient information indicates that in Antarctica, the cold limit is approached but may not be crossed. In the Atacama, the dry limit is reached in the hyperarid core.

12.2 ALL EXTREMES ARE NOT EQUAL

For many environmental conditions, the limits of habitability are usually sharply defined. With regard to temperature, the upper limit appears to be about 125°C (Takai et al. 2008), perhaps due to an unfavorable change in the dielectric properties of water. There are organisms well adapted to grow and reproduce up to this temperature, and many require the high temperature for growth. A similar pattern is observed for low and high pH and salinity. Obligate alkaliphiles, for instance, can grow at pH 11. They possess H^+/Na^+ antiporters that pump proton in and help maintain a neutral cytoplasmic milieu. Neutral pH environments are lethal to these specialized adapted organisms.

However, environments that are very cold or very dry are quite different in that we do not find organisms successfully adapted to them. Although there are reports of microorganisms growing down to −15°C (Mykytczuk et al. 2013), in general, microorganisms found in permafrost environments are similar to their temperate counterparts (Vorobyova et al. 1997, Gilichinsky et al. 2007). For example, growth rates in whole samples of Siberian permafrost decreased exponentially with temperature and were highest at the highest temperature measured (+5°C) (Rivkina et al. 2000), despite the environment from which these samples originated never experience temperatures above freezing. The community tolerates the cold but is not optimized for the cold. As noted by Friedmann (1994), there is a reduction in bacterial numbers with advancing permafrost age that points to a gradual elimination of organisms over time. Permafrost microbiota is therefore better described as survivors. In that respect, it is noteworthy that permafrost microbiota survives by depleting nonrenewable resources, and the community is bound to an end point when its resources are depleted (Friedmann 1994). Cavicchioli (2006) discusses the question of cold adaptation in a review of cold-adapted Archaea. He suggests that organisms that can grow in cold environments should be labeled cold adapted even if the temperature at which they grow fastest and their upper growth temperature limit can be well above the temperature of their environment. This would change the terminology but leave intact the observation that the survival in the extreme cold is a situation of compromise, that is, the organisms prefer much higher temperatures.

The reasons for the apparent lack of successful adaptations in extremely cold environments are still speculative. The survival strategy for life in these environments appears to be rooted in the ability to survive freezing with little damage and rapidly switch on metabolism when clement conditions become available. Even during these brief, clement periods, growth is extraordinarily slow. This imposes obvious constraints on the capacity of organisms to adapt, since evolutionary processes would be accordingly slow (Friedman 1994). It is also possible that investment in complex biochemical adaptations to grow at low temperature or water activity may not be cost effective.

Life in the Atacama Desert follows a similar pattern. Across a rainfall gradient from wetter Copiopo to drier Yungay, there is a sharp reduction in soil taxonomic diversity due to increasing aridity (Crits-Christoph et al. 2013). Soil microbiota in the hyperarid core also may be considered survivors, as less desiccation-tolerant species are eliminated. Although occasional, small-volume rain events occur, growth during the wet periods may be limited. Like in permafrost, the long-term survival in hyperarid soil is constrained by a finite, nonrenewable energy resource, and therefore sooner or later, the community would die.

Endolithic cyanobacteria living within salt rocks (halite) are the only photosynthetic life-form in the arid core of the Atacama Desert (Wierzchos et al. 2006). The deliquescence of halite condenses vapor from the atmosphere when relative humidity (RH) is above 75%, and the condensate is retained in small pore spaces. In the Antarctic, sandstone can melt snow and retain moisture during the summer (Friedmann et al. 1987, 1993, Sun 2013). The fact that in both environments endolithic organisms are the sole survivor suggests that this is the form of life to search for on Mars. For a recent review of endolithic communities, see Wierzchos et al. (2013).

Cold and dry environments are unusual when compared to other extremes in two additional ways; the organisms best able to survive these challenges are not prokaryotes and there is no lower limit to survival. Many organisms can survive to absolute zero for both temperature and water activity. The types of organisms that can withstand extreme cold and dry warrant some elaboration. From an environmental perspective, it is clear that there are many eukaryotes that can survive cold conditions as well as any prokaryote (e.g., Onofri et al. 2004), and many eukaryotes survive dry conditions better (e.g., Palmer and Friedmann 1990). Indeed, the organisms that live at the lowest value of water activity and spoil food at low water activity are molds, yeasts, and fungi (e.g., Scott 1957).

12.3 DRY LIMIT OF LIFE ON EARTH

In the context of the dry limit, nonrain sources of water, including fog and dew, are also important. The role of fog and rain in the hypolithic habitat has been studied in the Namib Desert, where there is a clear and gradual transition from fog-dominated moisture sources on the West Coast to rain-dominated moisture sources on the interior 200 km or so inland. Along this transect, hypolithic cyanobacteria growing below translucent stones (e.g., Pointing and Belnap 2012, Wierzchos et al. 2013) are equally present in the foggy and rainy ends of the transect (Stomeo et al. 2013, Warren-Rhodes et al. 2013). Azúa-Bustos et al. (2011) also reported on hypolithic cyanobacteria supported mainly by fog in the coastal range of the Atacama Desert.

In the hyperarid core of the Atacama Desert, exemplified by the Yungay region, fog and dew are not enough to create wet conditions below translucent stones, and hypolithic cyanobacteria are not found (McKay et al. 2003, Warren-Rhodes et al. 2006). Here, the water activity (RH) of the atmosphere and surface soils has a daily average between 0.2 and 0.4 throughout the year, well below the lower limit for life (McKay et al. 2003). However, there are occasional periods when morning RH rises to high levels, even causing condensation. The absence of hypolithic cyanobacteria in the hyperarid core at first seemed to indicate that photosynthesis was not occurring in this environment. However, Wierzchos et al. (2006) discovered endolithic cyanobacteria living just below the surface of submeter-scale halite (NaCl) nodules in the Yungay region. Because of the persistent low water activity and lack of rain, the halite is not washed away and is stable over geological time. Saturated brine is generated inside the halite by the combined action of deliquescence at atmospheric RH values above 0.7 and the matrix potential of the small pore spaces in the rock (Davila et al. 2008, Wierzchos et al. 2012). These sources of water are sufficient to sustain photosynthetic activity inside the halite (Davila et al. 2013), possibly the last habitable niche before the dry limit of life. Figure 12.1 shows visually the three habitable niches in extremely dry environments: endoliths in halite from the Atacama, typical desert hypolith from the Mojave Desert, and the sandstone endoliths from the Antarctic Dry Valleys.

An interesting aspect of life in dry environments is the apparent link between radiation resistance and dehydration resistance. Such a link has clear implications for Mars where water is scarce and radiation is high. It has been established for over three decades that desiccation resistance and radiation resistance are correlated. The level of radiation tolerated by desiccation-resistant organisms does not occur in their natural environments. The nature of the link between desiccation and radiation resistance remains enigmatic (Mattimore and Battista 1996, Fredrickson et al. 2004, 2008).

(a)

(b)

(c)

FIGURE 12.1 Photosynthesis in the driest environments on Earth occurs inside rocks. Panel (a), a green layer of cyanobacteria living just below the surface of halite rocks in the dry core of the Atacama Desert. (From Wierzchos, J. et al., *Astrobiology*, 6(3), 415, 2006.) Panel (b), a typical desert hypolith shown in an inverted samples of red-coated, carbonate translucent rocks from the Mojave Desert showing green biofilm of cyanobacteria that live beneath the rock. Panel (c), lichen forming a green and black layer inside sandstone from the Dry Valleys of Antarctica. (From Friedmann, E.I., *Science*, 215(4536), 1045, 1982.) The scale bar in all images is 1 cm. (Images (a) and (c) are courtesy of J. Wierzchos and E.I. Friedmann, respectively.)

Cellular damage during radiation has been attributed to the formation of reactive oxygen species, particularly hydroxyl and peroxyl radicals (Nauser et al. 2005), and Daly et al. (2007) established a link between oxidative damage and intracellular Mn^{2+}/Fe^{2+} ratios. Mn is a radical scavenger. Iron catalyzes radical formation. Indeed, cells containing high Mn are more resistant to radiation than cells with lower Mn. Bacteria with more intracellular iron are more sensitive to ionizing radiation than those with lower iron levels. A key insight from studies of radiation tolerance in *Deinococcus radiodurans* is that Mn^{2+} plays a key role in protecting the proteins needed for repair of radiation damage (Frederickson et al. 2008, Daly 2009). Conversely, the presence of Fe^{2+} causes the production of $HO^{\cdot}$ radicals through Fenton reactions: $H_2O_2 + Fe^{2+} \rightarrow Fe^{3+} + OH^- + HO^{\cdot}$ (Daly 2009). One might therefore expect that environments in which organisms must grow in arid conditions— both hot and cold—would be the natural breeding grounds of radiation-tolerant bacteria.

12.4 COLD LIMIT OF LIFE ON EARTH

If sufficient water is available, then temperature becomes the most serious constraint to life. In cold environments, water is available as ice but this poses two problems for use by biology. First and obviously, ice is a solid and therefore cannot function as a solution medium within cells. Second, ice is dry—its water activity is less than unity. The activity of any pure substance is always unity and ice is no exception. The activity, a_i, of ice is unity but not the activity with respect to liquid water, a_w. (Water activity, a_w, is defined as the partial pressure of water divided by the saturation pressure of pure water. For vapor, it is equivalent to the RH; for saline solutions, it is approximately equivalent to the mole fraction of water in the solution.) Figure 12.2 shows a plot of liquid water activity versus temperature. Lines for

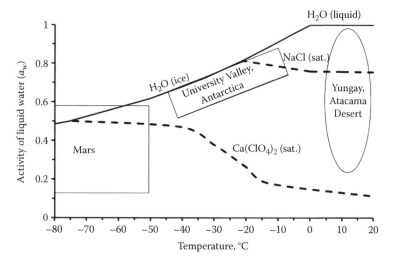

FIGURE 12.2   Water activity and temperature showing values for liquid water and ice. Shown are the water activities of NaCl (the dominant salt on Earth) and $Ca(ClO_4)_2$ (the dominant salt on Mars). Environmental regions for Mars, the upper elevation Dry Valleys of Antarctica, and the Yungay region of the Atacama are shown. (Data for $Ca(ClO_4)_2$ are from Nuding, D.L. et al., Deliquescence of calcium perchlorate: An investigation of stable aqueous solutions relevant to Mars, Paper No. 2584, *44th Lunar and Planetary Science Conference*, The Woodlands, TX.)

liquid and solid (ice) are shown. There are two mechanisms that can maintain liquid water at temperatures below 0°C: salts that lower the freezing point and films of liquid water that form at the boundary between ice and air or between ice and mineral grains. The suppression of freezing by salt solutions is well known. This is shown in Figure 12.2 in which the water activity for saturated solutions is shown as a function of temperature for two salts, NaCl—the dominant form of Cl on Earth—and $Ca(ClO_4)_2$ thought to be a primary form of the perchlorate salt that is the dominant form of Cl on Mars. For both salts, the water activity of the saturated solution is always less than the value for water at the same temperature. From a biological perspective, the advantage of the saline solution is that it remains liquid below the freezing point of pure water. When the water activity of the saturated saline solution intersects the water activity for pure, this defines the eutectic point—the lowest temperature at which the saturated solution remains liquid. For NaCl, this is −22.1°C and for $Ca(ClO_4)_2$, this is about −70°C. It is an interesting fact, possibly a coincidence, that the curve for NaCl nearly defines the lower limit of a_w for microbial life at higher temperatures and the lowest temperature for which metabolism seems possible. As a rule of thumb, on Earth, if environmental temperature and water activity are such that saturated NaCl is liquid, then growth is possible. Perhaps, this reflects the adaptation of life on Earth to the dominant salt present on Earth. If so, a speculative application of this logic to Mars might suggest that life on that planet would have evolved to use the dominant salt there $Ca(ClO_4)_2$; thus, if environmental temperature and water activity on Mars are such that saturated $Ca(ClO_4)_2$ is liquid, then growth of Martian life is possible. As shown in Figure 12.2, this speculation would suggest life at much lower temperature and lower a_w than is possible on Earth.

Even in the absence of salt, liquid water can be present below 0°C due to surface effects. At the interface between water and air or between water and mineral grains, the edge of ice is liquid. The thickness of this layer of unfrozen water decreases from about 15 nm at −1.5°C to about 5 nm at −10°C, to less than a molecular layer for temperatures below −40°C. The exact size depends on the temperature and curvature (size) of the particles (e.g., Anderson 1967, Ostroumov and Siegert 1996, Wettlaufer 2001). This subfreezing surface melt is what makes ice slippery (Rosenberg 2005).

In permafrost at low temperatures, it appears that the diffusion of nutrients, not the low temperature, is the main limitation for life. This was suggested by Rivkina et al. (2000) based on experiments with Siberian permafrost and was supported theoretically by Rohde and Price (2007). Price (2009) presents a comprehensive assessment of limits to the survival of microorganisms frozen in ice and permafrost over geological time. He concludes that metabolism at low temperatures can be adequate to repair damage owing to spontaneous DNA depurination and amino acid racemization, as both rates drop rapidly with temperature. He suggests that on Earth (or Mars), microbial survival in frozen ice and soil is most likely due to α-particle radiation damage from U and Th. His analysis suggests that for a very long survival (longer than a million years), dormant spore formers may be unable to compete, and only cells that are able to repair at low temperatures, albeit at a low rate, will remain viable. There is a laboratory work that suggests a DNA double strand repair at −15°C (Dieser et al. 2013), and Price (2009) argues on theoretical grounds that genetic and macromolecule repair will continue to even lower temperatures. The lower limit to this sort of frozen-zombie-like metabolism is not determined, but any such activity will decrease exponentially with temperature following an Arrhenius law (Price 2009). However, the factors that thermally degrade biological molecules will also decrease similarly with temperature. The exception to this trend is ionizing radiation, which does not decrease with temperature and thus provides a lower limit on survival when the repair mechanisms become slower than the radiation damage. Smith and McKay (2005) estimate that for Martian permafrost, this may be many millions of years. But they also point out that while radiation might cause sufficient damage to frozen microorganisms to kill them, it would not destroy all their biomolecules. Therefore, organisms frozen, and dead, in Martian permafrost could be used for biochemical and genetic analysis.

Gilichinsky et al. (2005) describe the biodiversity of the indigenous microbial community in the sodium chloride water brines (cryopegs) left behind 100–120,000 years ago in what is now permafrost after the Arctic Ocean regression. Cryopegs remain liquid at the permafrost temperature of −10°C and are thus a stable, permanently subzero, saline liquid water environment. From these cryopegs, anaerobic and aerobic, sporeless and spore-forming, halotolerant and halophilic, psychrophilic and psychrotrophic bacteria, mycelial fungi, and yeast were isolated and their activity was detected below 0°C (Gilichinsky et al. 2005).

It is interesting to note that studies of permafrost find numerous organisms and subfreezing activity when the permafrost is organic rich—as is typically the case in the Arctic. However, recent studies of the permafrost in University Valley at high elevation in the Dry Valleys of Antarctica do not find a corresponding level of organisms or activity in the organic-poor permafrost present there even when the temperature regimes are comparable (J. Goordial, work in progress).

12.5 CONCLUSION

The use of Earth analogs to guide the search for life on Mars continues. Over the past years, focus has intensified on extremely dry and extremely cold environments in the Atacama and the Antarctic, respectively. Much has been learned: key conclusions are as follows:

1. Liquid states can be maintained in permafrost down to −20°C due to salt and the physics of thin surface films.

2. Many diverse microorganisms can survive exposure to arbitrarily cold temperatures, can maintain viability over geological times down to −40°C and possibly lower, can metabolize to −20°C, and can grow and reproduce down to −15°C.

3. Microorganisms, and microbial communities, that dominate in cold environments are not specifically adapted to the low temperatures and grow better at higher temperatures.

4. In the most extreme of both cold and arid environments, endolithic organisms are the principal survivors using the physics of the rock to obtain and hold water.

Key outstanding questions that remain are as follows:

1. What is the role of nutrient status in supporting life in cold environments? Organic-rich permafrost in the Arctic contains a wider diversity of life than the organic-poor soils of the upper elevations in the Antarctic.

2. Why is the most radiation-resistant bacterium *D. radiodurans* selected in the Yungay region of the Atacama?

3. What is the lowest temperature at which repair mechanisms can maintain viability against crustal levels of radiation in permafrost?

4. On Mars, does $Ca(ClO_4)_2$ play an important role in biology? And set the limits to temperature and water activity?

5. Are the remains of endolithic microbial communities present on Mars?

ACKNOWLEDGMENTS

This chapter is dedicated to the memory of Imre Friedmann who recruited all three of us to the study of extreme environments. We acknowledge funding from the NASA Astrobiology Science and Technology for Exploring Planets (ASTEP) and Exobiology programs.

GLOSSARY

Amino acid racemization: Is the process in which amino acids of one handedness are converted to the other handedness.

Antiporters: Are membrane proteins that play a major role in pH and Na(+) homeostasis of cells.

Archaea: Constitute a domain or kingdom of single-celled microorganisms.

Arrhenius law: A reaction rate law in which the rate drops exponentially with temperature.

Atacama Desert: Is a coastal desert in the countries of Chile and Peru in South America, covering a 1000 km strip of land on the Pacific coast, west of the Andes mountains. It is the driest hot desert in the world.

Cryopeg: Is a layer or lens of subsurface liquid brine in permafrost and hence is always below 0°C.

Cyanobacteria: Is a phylum of bacteria that obtain their energy through photosynthesis.

Deliquescence: The process by which a substance absorbs moisture from the atmosphere until it dissolves in the absorbed water and forms a solution.

DNA depurination: Is the loss of a purine base from the nucleotide.

Endolithic cyanobacteria: Are cyanobacteria that live below the surface of porous rocks.

Eutectic point: Is the temperature that represents the lowest temperature at which a mixture remains liquid.

Halotolerant and halophilic: Microorganisms that can grow in (hyper)saline environments, but only *halophiles* specifically require salt.

Hypolithic: Refers to cyanobacteria living under translucent stones.

Microbiota: A population of microorganisms.

Obligate alkaliphiles: Are organisms that require high pH to survive.

Permafrost: Is soil at or below the freezing point of water 0°C (32°F) for 2 or more years.

Psychrophilic and psychrotrophic: Microorganisms that have the ability to grow at 0°C. Psychrotrophic microorganisms have a maximum temperature for growth above 20°C and are widespread in natural environments and in foods. Psychrophilic microorganisms have a maximum temperature for growth at 20°C or below and are restricted to permanently cold habitats.

Prokaryotes: Are a group of organisms whose cells lack a membrane-bound nucleus and comprise the domains Bacteria and Archaea.

Water activity a_w: Is defined as the partial pressure of water divided by the saturation pressure of pure water. For vapor, it is equivalent to the RH; for saline solutions, it is approximately equivalent to the mole fraction of water in the solution.

REVIEW QUESTIONS

1. What is permafrost? Does life exist in permafrost? If so, which kind of life?

2. What are endolithic organisms? In which kinds of environments are they found?

3. What are cryopegs? How old are they? Which kinds of organisms live there?

4. Describe the difference between halophilic and halotolerant organisms and psychrophilic and psychrotrophic.

5. In which ways are the conditions on the Atacama Desert similar to those on Mars?

6. Sodium chloride is well represented in biology on Earth. Would this be the case for life on Mars as well? What are the properties of the salts that matter for life?

7. Discuss the cold and dry limits of life. Give examples of microorganisms that live at such limits.

REFERENCES

Anderson, D.M. (1967). The interface between ice and silicate surfaces. *Journal of Colloid and Interface Science*, 25(2): 174–191.

Azúa-Bustos, A., González-Silva, C., Mancilla, R.A., Salas, L., Gómez-Silva, B., McKay, C.P., Vicuña, R. (2011). Hypolithic cyanobacteria supported mainly by fog in the coastal range of the Atacama Desert. *Microbial Ecology*, 61(3): 568–581.

Cavicchioli, R. (2006). Cold-adapted Archaea. *Nature Reviews Microbiology*, 4(5): 331–343.

Crits-Christoph, A., Robinson, C.K., Barnum, T., Frickle, W.F., Davila, A.F., Jedynak, B., McKay, C.P., DiRuggiero, J. (2013). Colonization patterns of soil microbial communities in the Atacama Desert. *Microbiome*, 1: 28.

Daly, M.J. (2009). A new perspective on radiation resistance based on *D. radiodurans*. *Nature Reviews Microbiology*, 7: 237–245.

Daly, M.J., Gaidamakova, E.K., Matrosova, V.Y., Vasilenko, A., Zhai, M., Leapman, R.D. et al. (2007). Protein oxidation implicated as the primary determinant of bacterial radioresistance. *PLoS Biology*, 5: 769–779.

Davila, A.F., Gomez-Silva, B., de los Rios, A., Ascaso, C., Olivares, H., McKay, C.P., Wierzchos, J. (2008). Facilitation of endolithic microbial survival in the hyper-arid core of the Atacama Desert by mineral deliquescence. *Journal of Geophysical Research: Biogeosciences, 113*, G01028, doi: 10.1029/2007JG000561.

Davila, A.F., Hawes, I., Ascaso, C., Wierzchos, J. (2013). Salt deliquescence drives photosynthesis in the hyperarid Atacama Desert. *Environmental Microbiology Reports, 4*: 583–587.

Dieser, M., Battista, J.R., Christner, B.C. (2013). DNA double-strand break repair at −15°C. *Applied and Environmental Microbiology, 79*(24): 7662–7668.

Fredrickson, J.K., Li, S.M.W., Gaidamakova, E.K., Matrosova, V.Y., Zhai, M., Sulloway, H.M. et al. (2008). Protein oxidation: Key to bacterial desiccation resistance? *The ISME Journal, 2*: 393–403.

Fredrickson, J.K., Zachara, J.M., Balkwill, D.L., Kennedy, D., Li, S.M.W., Kostandarithes, H.M. et al. (2004). Geomicrobiology of high-level nuclear waste-contaminated vadose sediments at the Hanford Site, Washington State. *Applied and Environmental Microbiology, 70*: 4230–4241.

Friedmann, E.I. (1982). Endolithic microorganisms in the Antarctic cold desert. *Science, 215*(4536): 1045–1053.

Friedmann, E.I. (1994). Permafrost as microbial habitat. In D.A. Gilichinsky (ed.), *Viable Microorganisms in Permafrost* (pp. 21–26). Russian Academy of Sciences, Pushchino, Russia.

Friedmann, E.I., Kappen, L., Meyer, M.A., Nienow, J.A. (1993). Long-term productivity in the crypto-endolithic microbial community of the Ross Desert, Antarctica. *Microbial Ecology, 25*(1): 51–69.

Friedmann, E.I., McKay, C.P., Nienow, J.A. (1987). The cryptoendolithic microbial environment in the Ross Desert of Antarctica: Satellite-transmitted continuous nanoclimate data, 1984 to 1986. *Polar Biology, 7*: 273–287.

Gilichinsky, D., Rivkina, E., Bakermans, C., Shcherbakova, V., Petrovskaya, L., Ozerskaya, S. et al. (2005). Biodiversity of cryopegs in permafrost. *FEMS Microbiology Ecology, 53*(1): 117–128.

Gilichinsky, D.A., Wilson, G.S., Friedmann, E.I., McKay, C.P., Sletten, R.S., Rivkina, E.M. et al. (2007). Microbial populations in Antarctic permafrost: Biodiversity, state, age, and implication for astrobiology. *Astrobiology, 7*(2): 275–311.

Mattimore, V., Battista, J.R. (1996). Radioresistance of *Deinococcus radiodurans*: Functions necessary to survive ionizing radiation are also necessary to survive prolonged desiccation. *Journal of Bacteriology, 178*(3): 633–637.

McKay, C.P., Friedmann, E.I., Gómez-Silva, B., Cáceres-Villanueva, L., Andersen, D.T., Landheim, R. (2003). Temperature and moisture conditions for life in the extreme arid region of the Atacama Desert: Four years of observations including the El Niño of 1997–1998. *Astrobiology, 3*(2): 393–406.

Mykytczuk, N.C., Foote, S.J., Omelon, C.R., Southam, G., Greer, C.W., Whyte, L.G. (2013). Bacterial growth at −15°C; molecular insights from the permafrost bacterium *Planococcus halocryophilus* Or1. *The ISME Journal, 7*(6): 1211–1226.

Nauser, T., Koppenol, W.H., Gebicki, J.M. (2005). The kinetics of oxidation of GSH by protein radicals. *Biochemistry Journal, 392*: 693–701.

Nuding, D.L., Gough, R.V., Chevrier, V.F., Tolbert, M.A. (2013). Deliquescence of calcium perchlorate: An investigation of stable aqueous solutions relevant to Mars. Paper No. 2584, presented at *44th Lunar and Planetary Science Conference*, The Woodlands, TX.

Onofri, S., Selbmann, L., Zucconi, L., Pagano, S. (2004). Antarctic microfungi as models for exobiology. *Planetary and Space Science, 52*(1): 229–237.

Ostroumov, V.E., Siegert, C. (1996). Exobiological aspects of mass transfer in microzones of permafrost deposits. *Advances in Space Research, 18*(12): 79–86.

Palmer Jr., R.J., Friedmann, E.I. (1990). Water relations and photosynthesis in the cryptoendolithic microbial habitat of hot and cold deserts. *Microbial Ecology, 19*(1): 111–118.

Pointing, S.B., Belnap, J. (2012). Microbial colonization and controls in dryland systems. *Nature Reviews Microbiology, 10*(8): 551–562.

Price, P.B. (2009). Microbial genesis, life and death in glacial ice. *Canadian Journal of Microbiology*, *55*(1): 1–11.

Rivkina, E.M., Friedmann, E.I., McKay, C.P., Gilichinsky, D.A. (2000). Metabolic activity of permafrost bacteria below the freezing point. *Applied and Environmental Microbiology*, *66*(8): 3230–3233.

Rohde, R.A., Price, P.B. (2007). Diffusion-controlled metabolism for long-term survival of single isolated microorganisms trapped within ice crystals. *PNAS*, *104*(42): 16592–16597.

Rosenberg, R. (2005). Why is ice slippery? *Physics Today*, *58*: 50.

Scott, W.J. (1957). Water relations of food spoilage microorganisms. *Advances in Food Research*, *7*: 83–127.

Smith, H.D., McKay, C.P. (2005). Drilling in ancient permafrost on Mars for evidence of a second genesis of life. *Planetary and Space Science*, *53*(12): 1302–1308.

Sun, H.J. (2013). Endolithic microbial life in extreme cold climate: Snow is required, but perhaps less is more. *Biology*, *2*(2): 693–701.

Takai, K., Nakamura, K., Toki, T., Tsunogai, U., Miyazaki, M., Miyazaki, J., Hirayama, H., Nakagawa, S., Nunoura, T., Horikoshi, K. (2008). Cell proliferation at 122°C and isotopically heavy CH_4 production by a hyperthermophilic methanogen under high-pressure cultivation. *PNAS*, *105*(31): 10949–10954.

Vorobyova, E., Soina, V., Gorlenko, M., Minkovskaya, N., Zalinova, N., Mamukelashvili, A., Gilichinsky, D., Rivkina, E., Vishnivetskaya, T. (1997). The deep cold biosphere: Facts and hypothesis. *FEMS Microbiology Reviews*, *20*: 277–290.

Warren-Rhodes, K.A., McKay, C.P., Boyle, L.N., Kiekebusch E.M., Wing, M.R., Cowan, D.A. et al. (2013). Physical ecology of hypolithic communities in the central Namib Desert: The role of fog, rain, rock habitat and light. *Journal of Geophysical Research: Biogeosciences*, *118*: 1451–1460, doi: 10.1002/jgrg.20117.

Warren-Rhodes, K.A., Rhodes, K.L., Pointing, S.B., Ewing, S.A., Lacap, D.C., Gómez-Silva, B. et al. (2006). Hypolithic cyanobacteria, dry limit of photosynthesis, and microbial ecology in the hyperarid Atacama Desert. *Microbial Ecology*, *52*(3): 389–398.

Wettlaufer, J.S. (2001). Dynamics of ice surfaces. *Interface Science*, *9*(1–2): 117–129.

Wierzchos, J., Ascaso, C., McKay, C.P. (2006). Endolithic cyanobacteria in halite rocks from the hyperarid core of the Atacama Desert. *Astrobiology*, *6*(3): 415–422.

Wierzchos, J., Davila, A.F., Sánchez-Almazo, I.M., Hajnos, M., Swieboda R., Ascaso, C. (2012). Novel water source for endolithic life in the Atacama Desert's hyper-arid core. *Biogeoscience Journal*, *9*: 2275–2286.

Wierzchos, J., de los Ríos, A., Ascaso, C. (2013). Microorganisms in desert rocks: The edge of life on Earth. *International Microbiology*, *15*(4): 171–181.

CHAPTER 13

Microorganisms in Space

Gerda Horneck and Ralf Moeller

CONTENTS

13.1 INTRODUCTION

Life on our planet Earth has evolved under the protective blanket of our atmosphere and the shield of the geomagnetic field, which hold off most of the hostile parameters of outer space thereby providing a clement climate for our biosphere. It is a thin veneer surrounding our globe, ranging from about 30 km below the Earth's surface to an altitude of maximum 100 km. Using meteorological rockets, fungi and pigmented bacteria were isolated from as high as 77 km—the highest altitude microbes have been detected so far. With the advent of spaceflight, opportunities arose to send microorganisms and other living beings to outer space—in order to study their responses to this unique environment. During human space missions, microorganisms are inevitable companions of the astronauts that populate the spacecraft and may pose health hazards to the astronauts. Their responses to spaceflight need to be known to safeguard astronauts' health.

13.2 SPACE AS TEST BED FOR STUDIES ON MICROORGANISMS IN SPACE

13.2.1 Environment of Outer Space

Once sent to space, living beings are confronted to an extremely hostile environment not experienced on Earth, characterized by an intense radiation field of solar and galactic origin, a high vacuum, and extreme temperatures (Table 13.1). This environment or selected parameters of it are the test bed for astrobiological investigations, thereby exposing chemical or biological systems to selected parameters of outer space or defined combinations of them.

TABLE 13.1 Parameters of Outer Space of Relevance for Astrobiological Experiments

Space Parameter	Interplanetary Space	Low Earth Orbit	Simulation Facility
Space vacuum			
Pressure (Pa)	10^{-14}	10^{-7}–10^{-4}	10^{-7}–10^{-4}
Residual gas (part/cm^{-3})	1	10^4–10^5 H	Different values[a]
		10^4–10^6 He	
		10^3–10^6 N	
		10^3–10^7 O	
Solar electromagnetic radiation			
Irradiance (W/m^2)	Different values[b]	1,360	Different values[c]
Spectral range (nm)	Continuum	Continuum	Different spectra[c]
Cosmic ionizing radiation			
Dose (Gy/year)	≤0.1[d]	400–10,000[e]	Wide range[c]
Temperature (K)	>4[b]	Wide range[b]	Wide range[c]
Microgravity (*g*)	<10^{-6}	10^{-3}–10^{-6}	0–1000

[a] Depending on pumping system and requirements of the experimenter.
[b] Varying with orientation and distance to the Sun.
[c] Depending on the radiation source and filter system.
[d] Varying with shielding, highest values at mass shielding of 0.15 g/cm^2.
[e] Varying with altitude and shielding, highest values at high altitudes and shield of 0.15 g/cm^2.

13.2.1.1 Radiation Field in Space

In the interplanetary space, the ionizing radiation field consists mainly of two components: the solar cosmic radiation (SCR) and the galactic cosmic radiation (GCR). In the vicinity of the Earth, a third radiation component is present: the radiation trapped by the Earth's magnetosphere, the so-called van Allen belts. Space radiation has played a decisive role in processes of chemical evolution, for example, in the interstellar medium, in comets, and in planetary atmospheres. Its biological effects need to be known, when assessing the risks for astronauts during long-term exploratory missions and when assessing the likelihood of Panspermia, that is, the transport of life forms between planets of our Solar System.

It is important to note that radiation in space is a mixed radiation, composed of particles of different masses, charges, and energies. This scenario cannot be simulated by any facility on ground. The different components of the space radiation field are described in the following.

One of the major radiation sources in our Solar System is the Sun itself. A stream of charged particles, mainly electrons and protons (hydrogen nuclei), is steadily ejected from the upper atmosphere of the Sun. They are called the solar wind. Their energy is relatively low, about 1 keV (kilo electron volt). The solar wind creates the heliosphere, which surrounds our Solar System. It produces a magnetic field that partially protects the planets in our Solar System from galactic cosmic rays (GCR) impinging from the outside. In addition to the continuous solar wind, there are sudden eruptions of high-energy particles. These solar particle events (SPE) are composed primarily of protons with a minor component of helium nuclei (alpha particles) and an even smaller part of heavy ions and electrons. Their energies can reach high values up to GeV.

The second component, the GCR, is a continuous, that is, chronic, component of the radiation in space. GCR originate in cataclysmic astronomical events in our Galaxy, such as supernova explosions. GCR consist of 98% baryons and 2% electrons. The baryonic component is composed of 85% protons, with the remainder being alpha particles (14%) and heavier nuclei (about 1%). This latter component comprises the so-called HZE particles (particles of High charge Z and high Energy). Their composition is very similar to that of the elements of the Universe. They can reach very high energies, up to energies higher than 1000 GeV. However, they are orders of magnitude less frequent than the SCR particles and those of the radiation belts. In our Solar System, the flux of the lower-energy part of GCR, that is, of energies below 10,000 MeV, is modulated by the Sun's magnetic field: it is reduced at solar maximum and increased at solar minimum.

Solar wind particles are trapped by the Earth's magnetic field, thereby forming the radiation belts: the van Allen belts (named after the discoverer of the belts). Two belts of radiation are formed comprising electrons and protons and some heavier particles trapped in closed orbits around the Earth. Electrons reach energies of up to 7 MeV and protons up to about 200 MeV. The energy of trapped heavy ions is less than 50 MeV.

The surface of the Earth is largely spared from this cosmic radiation due to the deflecting effect of the Earth's magnetic field and the huge shield of 1000 g/m² provided by the atmosphere. As a consequence, the terrestrial average annual effective dose rate due to cosmic rays amounts to 0.30 mGy, which is more than 100 times lower than that experienced in interplanetary space (Table 13.1). In low Earth orbit (LEO), that is, beyond our atmosphere, electrons and protons of the van Allen belts are found in addition to the GCR and SCR. Radiation dose rates in the range of 100–800 mGy/year are received.

Our Sun is also the source of electromagnetic radiation. The spectrum of solar electromagnetic radiation spans several orders of magnitude, from short-wavelength X-rays (<0.01 nm) to radio frequencies (several m) (Table 13.1). At 1 astronomical unit (AU), that is, the mean distance of the Earth from the Sun, the solar irradiance equals to 1366 W/m², the solar constant. Astrobiology is especially interested in the solar extraterrestrial ultraviolet (UV) radiation at wavelengths (λ < 290 nm) that do not reach the surface of the Earth due to the UV screen of the stratospheric ozone layer.

13.2.1.2 Space Vacuum

In the interplanetary space, vacuum reaches pressures down to 10^{-14} Pa (Pascal), whereas in LEO where most space experiments have been performed, pressures of 10^{-7}–10^{-4} Pa prevail (Table 13.1). The major constituents of this environment are molecular oxygen and nitrogen as well as highly reactive oxygen and nitrogen atoms. In the vicinity of a spacecraft, the pressure increases and varies depending upon the degree of out-gassing from the spacecraft. If the pressure reaches values below the vapor pressure of a certain material, the material's surface atoms or molecules vaporize. Vacuum desiccation is the main process affecting biological samples exposed to space vacuum.

13.2.1.3 Extreme Temperatures

The temperature of a body in space depends on its position with respect to the Sun and other orbiting bodies as well as on its surface, size, mass, and albedo. In LEO, the following energy sources are present: solar radiation (1366 W/m²), the Earth's albedo (480 W/m²), and terrestrial radiation (230 W/m²). During a typical 90 min orbit, a spacecraft experiences 60 min of exposure to the Sun and 30 min of darkness, when it moves into the Earth's shadow. As a consequence, in LEO, the temperature of a body can oscillate within 90 min between extremely high and extremely low values (Table 13.1).

13.2.2 Environmental Conditions of Spaceflight

Within the spacecraft, the inhabitants are protected from most of those hostile parameters of space by containment within a space capsule, that is, a pressurized module, and an efficient life support system (LSS). In this case, mainly microgravity and radiation are the parameters of interest or concern, respectively.

13.2.2.1 Microgravity

Gravity is a permanently acting force resulting from the attraction of two or more bodies of certain masses at a certain distance. It is one of the four fundamental interactions of nature and can never be switched off. Gravity is responsible for various phenomena observed on Earth and throughout the Universe, such as coalescence of matter to form galaxies, stars, and planets in the Universe; it maintains planets in stellar orbit and the Moon in Earth orbit. The gravity of a planet gives the acceleration that it imparts to objects on or near its surface. It is dependent on the mass of the planet and is on average 9.81 m/s^2 for the Earth, 3.71 m/s^2 for Mars, and 1.62 m/s^2 for the Moon.

Although the gravitational attraction decreases with distance from a planet, this effect is never sufficient to reach a gravity-free environment by going into Earth orbit. The observed reduction of Earth's gravity during spaceflight is gained by the centrifugal force experienced in a circular orbit around the Earth that compensates Earth's gravitational attraction. This condition has been termed *microgravity* and can reach values of 10^{-3}–$10^{-6}g$, with g being the gravity of the Earth.

13.2.2.2 Cosmic Radiation

Inside the spacecraft, the cosmic radiation field is modified by interactions with shielding material provided by the spacecraft wall, by equipment installed inside, and by stored consumables (e.g., water, fuel). Secondary radiation, both charged and uncharged, is also created by those interactions. The shielding of the walls of the International Space Station (ISS) has an equivalent thickness of 10 g/cm^2 of aluminum (compared to 1000 g/cm^2 of the Earth atmosphere and 5–16 g/cm^2 of the Mars atmosphere). The radiation doses inside a spacecraft are difficult to predict, because they vary depending on orbit parameters, flight data, and spacecraft shielding. During the Russian space mission MIR (51.5°, 400 km altitude), a total dose equivalent rate of 640 μSv/day was measured, and during the Spacelab mission D2 (28.5°, 296 km), a total dose equivalent rate of 192 μSv/day was measured. For comparison, the dose equivalent rates for the ISS (51.5°, 400 km) are about 20 mSv/month and at the surface of the Earth about 3 mSv/year.

13.2.2.3 Closed Cabin Environment

Human exploration of extreme and remote areas, such as space, polar regions, or deep sea, requires the provision of confined habitats to protect the crew from the otherwise hostile outside. Such confined habitats have restrictions on waste disposal, water, and fresh air supply as well as personal hygiene and inevitably generate a particular community of microorganisms within the habitat. The main source of those microbial populations is the crew themselves (skin, upper respiratory tract, mouth, and gastrointestinal tract); but environmental microorganisms were also identified. The airborne bacterial and fungal contamination levels were monitored during most space missions. During the occupation of the MIR space station (1986–2001), 95% of the air samples contained less than 500 bacterial colony-forming units (CFUs)/m^3. Occasional increases were due to human exercise. A similar result was obtained for the ISS (1998–2005). These data are below the maximal concentration of microorganisms (10,000 CFU/100 cm^2 for bacteria and 100 CFU/100 cm^2 for fungi) allowed in water/air and on surfaces of the ISS. *Staphylococcus* and *Bacillus* spp. were found to be the dominant bacterial species in the air aboard the ISS.

In nature, the majority of bacteria form surface-associated microbial communities known as biofilms, which often exhibit increased resistance to environmental stress, antibiotics, and host defense systems. Abundant biofilms were also found in the MIR space station. They were responsible for increased corrosion and a blocked water purification system.

13.3 MICROORGANISMS UNDER SPACEFLIGHT CONDITIONS

13.3.1 Responses to Microgravity

A key concern for human spaceflight, especially for long-term exploratory missions, is how microgravity and other aspects of spaceflight affect bacterial growth, physiology, and virulence. Several in-flight studies have reported that the microgravity environment encountered during spaceflight altered bacterial growth and physiology, leading in certain cases to an increase in final cell density, antibiotic resistance, and virulence. While the exact mechanisms of adaptation of microorganisms to microgravity have not yet been fully determined, it is suggested that a microbial cell perceives changes in gravity at its surface and transduces the resulting signals to its interior. This phenomenon of transduction of a mechanical force into biological responses, which has been observed also with other physical and mechanical forces, for example, changes in osmotic gradients and fluid shear, seems to play a remarkable role in microbial gene expression, physiology, and pathogenesis. Hence, the proposed gravity-driven cascade of events can be summed up as (1) starting with an altered physical force acting on the cell and its environment upon exposure to microgravity (the *gravity trigger*), resulting in (2) reduced extracellular transfer of nutrients and metabolic by-products moving toward and away from the cell, which consequently (3) exposes the cell to a modified chemical environment, the sum of which ultimately gives rise to (4) an observed biological response that differs from what occurs under normal $1 \times g$ conditions.

13.3.2 Responses to Radiation

Radiation interacts with biological matter primarily through ionization and excitation of the electrons of its atoms and molecules. There are two alternative ways of radiation damage to biological key substances, such as proteins, nucleic acids (RNA and DNA), and lipids:

- Direct radiation effects occur through direct energy absorption in those molecules.

- Indirect radiation effects occur through interactions of those molecules with radicals that are produced by radiation in other molecules, such as water.

The DNA is the most sensitive biological radiation target. The type of its damage is dependent on the deposited energy. Because of their highly localized energy deposition, cosmic ray HZE particles mainly induce DNA double-strand breaks and even multiple damage, so-called complex or bulk damage. If not efficiently repaired, they may lead to mutations or even cell death.

Whereas the biological damage caused by those HZE particles can be very high, it is difficult to localize it because of the very low flux of those particles, for example, 8 or 28 Fe ions/cm^2 day, depending on the solar activity (data shown for solar maximum or minimum, respectively). To understand the ways by which single particles of cosmic radiation interact with biological systems, it is necessary to precisely localize the trajectory of an HZE particle relative to the biological object and to correlate the physical data of the particle relative to the observed biological effects along its path. Such effect particle correlations were accomplished by the Biostack method, that is, the use of visual track detectors that were sandwiched between layers of biological objects, such as bacterial spores, plant seeds, shrimp cysts, or insect eggs (Figure 13.1a).

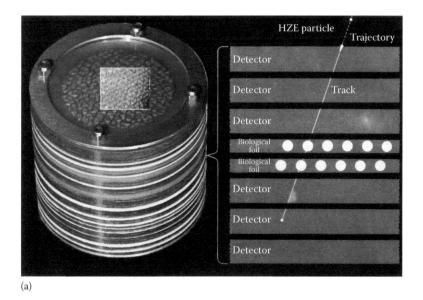

(a)

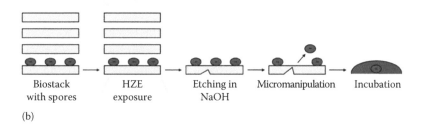

Biostack with spores → HZE exposure → Etching in NaOH → Micromanipulation → Incubation

(b)

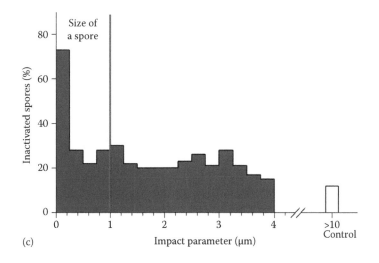

(c)

FIGURE 13.1   (a) Biostack concept to determine the biological effects of individual HZE particles of GCR, (b) analysis of *Bacillus subtilis* spores after spaceflight within the Biostack experiment, and (c) fraction of inactivated spores as a function of the distance from the particle track (impact parameter); results obtained from the Biostack experiment flown on the Apollo–Soyuz Test mission. (Modified from Horneck, G. and Möller, R., *BioSpektrum*, 17, 262, 2011., With permission.)

Biostack experiments were flown on several space missions, for example, Apollo 16 and 17, Apollo–Soyuz Test Project, Spacelab 1 and D2, and the European EURECA mission. This Biostack concept allows

- To localize each HZE particle's trajectory in relation to the biological specimens

- To investigate the responses of each biological individual hit separately, in regard to its radiation effects

- To measure the impact parameter, which is the distance between the particle track and the sensitive target

- To determine the physical parameters (charge [Z], energy [E], and linear energy transfer [LET])

- To correlate the biological effect with each HZE particle parameters

The responses of a single microbial cell to the passage of a single HZE particle of cosmic radiation were studied on spores of the bacterium *Bacillus subtilis*, which have a mean diameter of about 1 µm. Special microscopic techniques were required to identify the spores hit by an HZE particle (Figure 13.1a). First, the track detector that carried the spore layer was etched under the microscope in a way that the *spore side* was protected. Then the spores in the *hit area* were removed by micromanipulation and deposited in a microincubation chamber where their germination, growth, and formation of a micro-colony were microscopically monitored. Finally, the fraction of inactivated spores was determined as a function of the distance from the particle's track. Those spores that were directly hit by an HZE particle showed a high frequency of inactivation (Figure 13.1b). Also, spores that were located farther away from the HZE particle track (about 1–3 µm) were inactivated to a certain extent. This far-reaching effect of HZE particles that has now been observed in several other biological systems needs also to be considered when assessing the radiation risks for astronauts during long-term exploratory missions.

13.3.3 Combined Effects of Radiation and Microgravity

Besides the unique radiation field in space, the cessation of the gravity stimulus to which life on Earth has adapted is another important source for biological effects that are inevitably present during spaceflight. In response to microgravity, several essential cellular functions are affected (see Section 13.3.1). Different methods have been used to investigate the potential interaction of microgravity and radiation. Examples are the use of a 1*g* reference centrifuge as onboard control. Using the Biostack method (see Section 13.3.1) in combination with a 1*g* reference centrifuge permits to differentiate between the individual effects of cosmic radiation and microgravity and the interactions of both parameters of spaceflight. Another method to study the combined effects of microgravity and radiation is to expose biological systems to known quantities and qualities of radiation from an enclosed source while in orbit or on ground directly before launch.

In early space experiments—by use of the Biostack concept in combination with a 1*g* centrifuge—a synergistic effect of microgravity and radiation on embryonic development was demonstrated, such as reduced hatching rate, body anomalies, and increased mortality during embryogenesis of the stick insect *Carausius morosus*. Chromosome translocations and thorax deformations were observed in larvae and adults of the fruit fly *Drosophila melanogaster* after exposure of early development stages to [85]Sr γ-rays during spaceflight, which were not observed in the irradiated concurrent 1*g* controls. However, so far, synergistic interactions of microgravity and radiation were not observed in studies with microorganisms. When the kinetics of several repair pathways were examined in microorganisms that were irradiated prior to the space mission, the repair of radiation-induced DNA damage occurred in the microgravity environment at the same efficiency as in the 1*g* centrifuge or on ground. This was studied in *Escherichia coli* cells for the rejoining of radiation-induced DNA strand breaks and the induction of DNA repair processes and in *Bacillus subtilis* cells of different repair capacity for the efficiency of DNA repair. Because microgravity and radiation act continuously during spaceflight, further studies are required to understand the adaptation of the spacecraft microflora to those spaceflight conditions, in order to safeguard human health during long-term exploratory missions.

13.3.4 Adaptation of the Spacecraft Microflora to Spaceflight

This unique environment encountered in space habitats (i.e., close quarters, limited space, and recirculation of air) promotes the transmission of the normal flora and potentially opportunistic and pathogenic bacteria and viruses. In an orbiting spacecraft, airborne microorganisms as well as dust particles do not settle due to the absence of gravity; thermodiffusion or electrostatic forces gain in importance. This results in a more persistent (bio)aerosol and higher microbial contamination levels in cabin air that require continuous active removal of the aerosols from the air. To keep the microbial levels below the thresholds (see Section 13.2.2.3), the air in the ISS is continuously filtered.

The resident flora of the crew is by far the largest microbial reservoir on board. This microbial population is further shaped in both diversity and mass by the unique combination of environmental factors (e.g., restricted hygienic practices, confinement, microgravity, and radiation). Changes in the composition of the intestinal, oral, and nasal bacterial microflora have been noted already after short spaceflights. Two weeks into the confinement imposed by spaceflight led to a significant reduction in the number of bacterial species that could be isolated from the intestinal tract as well as an interchange of intestinal bacteria between crew. In the nasal flora of Soviet cosmonauts, the number of nonpathogenic bacteria was reduced and the number of opportunistic pathogens increased. Astronauts aboard the Russian space station MIR indicated that a significant number of episodes of microbial infection occurred, including conjunctivitis and acute respiratory and dental infections.

To make matters worse, the efficacy of antibiotics is reduced in microgravity, and microbial mutation rates increase dramatically (see Section 13.3.1). In addition, the opportunities for bacteria to establish foci of infection are enhanced due to the negative impact of space travel on immune functions. These factors will impinge on the capacity to treat effectively infections that will doubtless arise during long-term exploratory missions.

13.4 MICROORGANISMS IN OUTER SPACE

13.4.1 Likelihood of Panspermia

13.4.1.1 Original Panspermia Hypothesis

Are the planets in our Solar System isolated bodies, or is it possible for some resistant life forms to migrate from one planet to another? Of course, with current days' space technology, planetary space probes can provide the vehicle for microorganisms to travel to other planets and Moons of our Solar System (this, however, is subjected to strict planetary protection guidelines, which will be dealt with in Section 13.4.3).

Panspermia, the hypothesis on the transport of life within our Solar System and beyond, was first scientifically formulated in 1903 by the Nobel laureate Svante Arrhenius. He was impressed by the many dust particles that teem in space between the planets. From this observation, he followed that the radiation pressure of the Sun can drive very small particles of 1–1.5 μm in diameter away from the Sun into the interplanetary space and even out of the Solar System. As suitable candidates for this transport, he proposed bacterial spores: they are small enough to be transported through space by the radiation pressure of the Sun thereby seeding life from one planet to another.

When this Panspermia hypothesis was first published, it received severe criticism, such as (1) Panspermia shunts aside the question of the origin of life; (2) Panspermia cannot be experimentally tested; and (3) bacterial spores would not survive in the harsh environment of space, especially space vacuum and radiation. So, the Panspermia hypothesis was more or less forgotten.

13.4.1.2 Experimental Test of the Panspermia Hypothesis

With the advent of space technology, it became possible to send microorganism into space and to test their responses to the parameters of outer space, that is, to test the Panspermia hypothesis experimentally. This test was performed on spores of *Bacillus subtilis* that were exposed to selected parameters of outer space within the European exposure facility *Biopan* on board of a Russian satellite (Figure 13.2a). These spores are known to be very resistant to a variety of physical and chemical stresses, such as heat, intense radiation, desiccation, and also chemicals. Biopan was a box equipped with a lid and a bottom part, both accommodating biological test samples. During launch and reentry, Biopan was hermetically closed. Once in orbit, Biopan was opened to allow access of outer space to the samples. In the experiment *Survival* (Figure 13.2b), spores of *B. subtilis* in dry monolayers were exposed to the space environment, including the full spectrum of solar UV radiation, space vacuum, and GCR, as well as to selected parameters of space. After several space missions, each lasting for 10–15 days, it was clearly shown that nearly 100% of the spores survived in space, if they were shielded from solar UV radiation. However, in

(a)   (b)

FIGURE 13.2   Biopan mission of the ESA. (a) Russian satellite Foton with the European exposure facility Biopan and (b) Biopan facility opened with the experiment *Survival* in the lid. Each compartment accommodated 10^8 bacterial spores.

samples that were Sun-exposed, the survival was reduced by 4–7 orders of magnitude. From these relatively short space missions, it became already clear that isolated spores will not survive in outer space for extended periods of time, if they are exposed to solar extraterrestrial UV radiation. These results falsified the original Panspermia hypothesis that suggested a viable transport of single spores through space, driven by the radiation pressure of the Sun.

13.4.1.3 Lithopanspermia

The idea of Panspermia received new support after the detection of a group of meteorites that have originated from the Moon or from Mars. They prove that natural transport of material between the terrestrial planets has frequently occurred in the past. How could these rocks be expelled from the Moon or from Mars? The best explanation is that they were ejected from their parental body following the impact of a huge comet or meteoroid that led to the ejection of a large amount of material. In these highly energetic events, most of the material would be heated up to melting temperatures of the rocks. However, it was found that some of the Martian meteorites were not heated up very much, by less than 100°C. The explanation is the formation of a spall zone at the outer upper regions of the impact crater. In this zone, direct and reflected shock waves are interfering, which leads to lowered shock pressures and temperatures of the expelled rocks. This new impact–ejection-driven scenario has caused a revisiting of Panspermia, which has now been termed *Lithopanspermia* (Panspermia by aid of rocks).

13.4.1.4 Experimental Tests of Lithopanspermia

It is well known that on Earth, many rocks are inhabited by microorganisms, either in their interior (endolithic microbial communities) or on the surface (epilithic communities, e.g., lichens). The question arises whether such rock-colonizing microorganisms can survive (1) the impact and ejection process of Lithopanspermia, (2) a long-lasting journey through space, and finally, (3) the capture of its host rock by another planet followed by entry and landing on this planet.

The first step of Lithopanspermia, that is, the ejection process, was studied in laboratory shock recovery experiments within pressure ranges observed in Martian meteorites (5–50 GPa) using dry layers of microorganisms (spores of *Bacillus subtilis*, cells of the endolithic cyanobacterium *Chroococcidiopsis* sp., and thalli and ascocarps of the lichen *Xanthoria elegans*) sandwiched between gabbro disks (Martian analogue rock). From the results, a vital launch window for the ejection and transport of rock-colonizing microorganisms from a Mars-like planet was inferred (shock pressures in the range of 5–40 GPa for the bacterial spores and the lichens and 5–10 GPa for the cyanobacteria).

Such ejecta may wander irregularly for a long time within the Solar System before reaching another planet. This means that the microorganisms within their host rocks are confronted with the hostile environment of outer space for extended periods of time (see Section 13.2.1). To test this second step of Lithopanspermia, that is, the survival of rock-colonizing microorganisms in outer space, the European Space Agency (ESA) has provided the Biopan facility on board of the recovery satellite Foton for short-term space missions (Figure 13.2a,b) and the EXPOSE facility attached to the ISS (Figure 13.3)

FIGURE 13.3   EXPOSE facility of the ESA to be attached at the outside of the ISS.

for long-term missions of more than 1 year. The Biopan results showed that rock-colonizing lichens, for example, *Rhizocarpon geographicum* and *X. elegans*, still attached to their natural rock habitat, and the vagrant lichen *Aspicilia fruticulosa* (now renamed *Circinaria gyrosa*) was extremely resistant to outer space conditions, including the full spectrum of solar extraterrestrial electromagnetic radiation. This high resistance of the lichens to space that was confirmed by experiments on the 1.5 year lasting EXPOSE-E mission appears to be due to their symbiotic nature and protection by their upper pigmented layer, the cortex. In contrast, rock- or halite-inhabiting bacteria as well as bacterial spores of *B. subtilis* were severely damaged by exposure to solar extraterrestrial UV radiation. Action spectroscopy in space (during the EURECA mission of the ESA) identified DNA as the sensitive target for solar UV radiation. Mutations to rifampicin resistance (Rif^R) that were recovered from *B. subtilis* spores after 1.5 years of exposure to outer space (EXPOSE-E mission) all showed a C to T transition and were all localized to one hotspot: H482Y. In contrast, mutations isolated from the parallel ground control experiment showed a much wider mutagenic spectrum. From the data obtained in space, one can conclude that solar UV radiation is the most lethal component of outer space. They demonstrate also the unique mutagenic power of space. Therefore, microorganisms (spores) can only survive wandering through space if they are shielded against solar UV, for example, by rock material. Finally, GCR will become essential, if long wandering periods (millions of years) are considered. It has been calculated from space (e.g., Biostack; see Section 13.3.2) and laboratory experiments (e.g., at heavy ion accelerators) that for very long travel times (e.g., millions of years), the Lithopanspermia scenario requires at least 1 m of shielding by rock material in order to protect microorganisms in their center.

13.4.2 Habitability of Mars

The quest for life on Mars has received increased attention since the development of current space exploration programs (see Chapter 14). Mars with a mean distance to the Sun of 1.52 AU is located at the outer border of the habitable zone encircling the Sun, which is estimated under the premise of the presence of liquid water on the planet's surface at some time during its 4.5 billion years' lasting history. Compared to our planet Earth, the surface of Mars is less protected against solar electromagnetic radiation, GCR, and SCR. This is due to its thin atmosphere (600 Pa, 95% CO_2), the lack of an ozone screen, and the absence of an intrinsic magnetic field. As a consequence, the surface of Mars is exposed (1) to solar electromagnetic radiation at wavelengths $\lambda > 200$ nm, (2) to the chronic component of cosmic rays (GCR and solar wind), as well as (3) to the sporadic component (SPE), which all reach the Martian surface nearly unfiltered. This radiation environment needs to be taken into consideration when assessing the habitability of Mars but also when sending humans to Mars.

To test the habitability of Mars, bacterial spores (*Bacillus subtilis* and *Bacillus pumilus*) were subjected for 1.5 years to simulated Martian surface conditions during the EXPOSE-E mission of the ESA (Figure 13.3). The sealed exposure compartments allowed access of cosmic radiation, and they contained a simulated Martian atmosphere of 95% CO_2 at 600 Pa, whereas cutoff filters provided a Martian UV radiation climate ($\lambda > 200$ nm). Survival and mutagenesis were determined in these *stay on Mars* spores after retrieval. It was clearly shown that the Martian UV radiation spectrum ($\lambda > 200$ nm) was the most deleterious factor applied; in some samples, only a few survivors were recovered from spores exposed in monolayers. Spores in multilayers survived better by several orders of magnitude. All other environmental parameters encountered by the *stay on Mars* spores did little harm to the spores, which showed about 50% survival or more (if shielded from solar UV). It was shown that the survival of spores on Mars depends largely on the degree of shadowing, such as within multilayers or clumps, or hidden in cracks, than on the specific resistance of the bacterial strain under investigation. However, in all *stay on Mars* spores, a high rate of induced mutations was found, as tested for mutations to Rif^R, in those exposed to simulated Martian UV radiation ($\lambda > 100$ nm) as well as in those shielded from insolation. The data show both (1) the high chance of survival of spores on Mars, if protected against solar irradiation, and (2) the unique mutagenic power of Martian surface conditions as a consequence of DNA injuries induced by solar UV radiation, cosmic radiation, and atmospheric conditions.

13.4.3 Planetary Protection Needs

In 1967, shortly after spaceflight activities had started, the United Nations developed the Outer Space Treaty. This is the *Treaty on Principles Governing the Activities of States in the Exploration and Use of Outer Space, including the Moon and Other Celestial Bodies*, which has been signed and ratified by practically all spacefaring nations. This treaty is the foundation on which recommendations and guidelines of *planetary protection* are based. The current planetary protection guidelines have been developed and are controlled by the Committee on Space Research (COSPAR).

The rationale for planetary protection is the supposition that life may not be restricted to the Earth but that it could emerge at a certain stage of either cosmic or planetary evolution, if the right environmental requirements are given. Therefore, although the Earth is the only planet known to harbor life, there might be other habitable or even inhabited celestial bodies in our Solar System or beyond. With space exploration, we possess the tools to reach the candidate Moons and planets in our Solar System and to search *in situ* for signatures of indigenous life, either of fossils (extinct life) or of extant life forms. However, there would be the danger that the introduction—by means of orbiters, entry probes, or landing vehicles—and possible proliferation of terrestrial life forms on other planets could entirely destroy the opportunity to examine the target planets or Moons—and possible indigenous life forms—in their pristine conditions.

Planetary protection guidelines take care that the planet or Moon and its environment being explored is protected from terrestrial biological contamination; this process is called *forward contamination prevention.*

On the other side, the Earth and its biosphere and the human population need to be protected from potential hazards posed by extraterrestrial matter carried on a spacecraft returning from another celestial body; this process is called *backward contamination prevention.*

Based on the mission–target combination, different measures are required—up to complete sterilization of the landing device.

Because bacterial spores are highly resistant to a variety of environmental extremes, including the harsh environment of outer space, spore-forming microbes are of particular concern in the context of planetary protection. Assuming they are capable of coping with the different environmental attacks imposed on them during the transfer to, for example, Mars, spores may pose a serious hazard to the *in situ* life detection experiments and to the efforts to maintain the surface of the target planet in pristine condition. Therefore, the resistance of bacterial spores has been used as an international standard for planetary protection purposes.

In order to test the hardiness of bacterial spores during a hypothetical trip to Mars, spores of *Bacillus subtilis* mounted as dried layers on spacecraft-qualified aluminum coupons were exposed to selected conditions of outer space during the EXPOSE-E mission on board of the ISS. After 1.5 years of such a simulated journey to Mars, these *trip to Mars* samples were analyzed after retrieval. If shielded from solar UV radiation, about 50% of these *dark* flight spores of *B. subtilis* in multilayers survived the 559-day exposure to the combined action of cosmic radiation, space vacuum, and temperature fluctuations. If, in addition, solar electromagnetic radiation at wavelengths $\lambda \geq 110$ nm was experienced by the bacterial spores, they were further inactivated by about 2–3 orders of magnitude. The data clearly showed that microorganisms (spores) as *blind passengers* on a planetary mission would largely survive the journey if located inside of the satellite or in cracks, thereby shielded from solar UV radiation. These data confute the sometimes voiced argument of space as a huge sterilization medium, which would render preflight sterilization requirements unnecessary.

13.5 SUPPORTIVE GROUND-BASED STUDIES

Because flight opportunities are generally rare, various ground-based devices have been designed to simulate certain aspects of outer space or spaceflight. A deeper insight in gravity-induced biological phenomena has been obtained by ground-based studies under conditions of hypergravity (by the use of centrifuges) or artificial (simulated) weightlessness (by the use of 2-D and 3-D clinostats/random positioning machines and magnetic levitation). Planetary and space simulation facilities aimed at mimicking extraterrestrial conditions, for example, outer space (vacuum, temperature, radiation), other planetary surfaces (atmospheric composition and pressure, temperature fluctuations, radiation), or Moons (gas mixture, pressure, low temperature) are valuable instruments in the preparation of flight experiments (selection of suitable biological or chemical test systems and testing of flight hardware).

13.6 OUTLOOK

In view of the increasing public and political interest in space exploration at a global scale, microbiologists will face important tasks in order to realize this ambitious endeavor. In the next decades, human spaceflight to Mars, the next reachable planetary body of interest, will become reality and astronauts are likely to spend at least 2–3 years away from Earth. Time spent in such extreme environments will result in a diminution of immune status and profound changes in the human bacterial microflora. Various factors associated with the spaceflight environment have been shown to potentially compromise the immune system of astronauts, increase microbial proliferation and microflora exchange, alter virulence, and decrease antibiotic effectiveness. To safeguard astronauts' health during those exploratory missions, studies are needed to effectively eliminate or mitigate those adverse effects. Microorganisms will also play an increasing role as active participants in regenerative LSSs, including waste and water recycling. Another important task for space exploration is the development of efficient planetary protection guidelines and modes of their supervision, because the uncontrolled introduction of terrestrial microorganisms on another planet might destroy the opportunity to investigate that planet in its pristine conditions and might jeopardize further search-for-life experiments on that celestial body.

GLOSSARY

Apollo: NASA lunar missions.
AU: Astronomical unit (1 AU = the mean distance of the Earth from the Sun).
Biostack: A device to study the biological effects of HZE particles: visual track detectors are sandwiched between layers of biological objects.
CFU: Colony-forming unit.
COSPAR: Committee on Space Research.
DNA: Deoxyribonucleic acid.
EURECA: European Retrievable Carrier.
GCR: Galactic cosmic radiation.
HZE particles: Particles of High charge Z and high Energy.
ISS: International Space Station.
LEO: Low Earth orbit.
LET: Linear energy transfer.
MIR: Russian space station.
Rif^R: Rifampicin resistance.
RNA: Ribonucleic acid.
SCR: Solar cosmic radiation.
SPE: Solar particle events.

REVIEW QUESTIONS

1. What is the most deleterious parameter of space with regard to microorganisms exposed to it?

2. What are the effects of the UV radiation (extraterrestrial and terrestrial) on living systems? Describe the physical, chemical, and biological interactions and the spectral dependence of the effects.

3. What is the difference between the Panspermia hypothesis and the Lithopanspermia scenario? Describe the different steps of Lithopanspermia.

4. Describe the interaction between the Earth magnetosphere and the incoming GCR and the radiation originating in the Sun.

5. Why is Mars thought to be a possible place to search for life?

6. Give reasons, scientific and ethical ones, that request protection of the planets of our Solar System and justify them.

RECOMMENDED READING

Books or Book Chapters

Brinckmann, E. (ed.), *Biology in Space and Life on Earth*, Wiley-VCH, Weinheim, Germany, 2007.

Clancy, P., A. Brack, and G. Horneck, *Looking for Life, Searching the Solar System*, Cambridge University Press, Cambridge, U.K., 2005.

Gilles, C. and K. Slenzka (eds.), *Fundamentals of Space Biology*, Kluwer Academic Publishers, Dordrecht, the Netherlands, 2006.

Horneck, G., C. Baumstark-Khan, and F. Rainer, Radiation biology, in *Fundamentals of Space Biology*, Clément, G. and K. Slenzka (eds.), pp. 292–335, Kluwer Academic Publishers, Dordrecht, the Netherlands, 2006.

Horneck, G. and P. Rettberg (eds.), *Complete Course in Astrobiology*, Wiley-VCH, Berlin, Germany, 2007.

Articles in Journals

de la Torre, R., L.G. Sancho, G. Horneck et al. (2010) Survival of lichens and bacteria exposed to outer space conditions—Results of the Lithopanspermia experiments. *Icarus*, 208: 735–748.

Demets, R. (2012) Darwin's contribution to the development of the Panspermia theory. *Astrobiology*, 12: 946–950.

Horneck, G., D.M. Klaus, and R.L. Mancinelli (2010) Space microbiology. *Microbiol. Mol. Biol. Rev.*, 74: 121–156.

Horneck, G. and R. Möller (2011) Mikroben im All—Aufgaben für die Mikrobiologie. *BioSpektrum*, 17: 260–265. Springer-Verlag GmbH Springer Spektrum.

Horneck, G., D. Stöffler, S. Ott et al. (2008) Microbial rock inhabitants survive impact and ejection from host planet: First phase of Lithopanspermia experimentally tested. *Astrobiology*, 8: 17–44.

Horneck, G. and M. Zell (Guest editors) (2012) Special collection on EXPOSE-E mission. *Astrobiology*, 12(5): 373–528.

Mileikowsky, C., F. Cucinotta, J.W. Wilson et al. (2000) Natural transfer of viable microbes in space, part 1: From Mars to Earth and Earth to Mars. *Icarus*, 145: 391–427.

Moeller, R., G. Reitz, T. Berger, R. Okayasu, W.L. Nicholson, and G. Horneck (2010) Astrobiological aspects of the mutagenesis of cosmic radiation on bacterial spores. *Astrobiology*, 10: 509–521.

Nicholson, W.L., N. Munakata, G. Horneck, H.J. Melosh, and P. Setlow (2000) Resistance of *Bacillus* endospores to extreme terrestrial and extraterrestrial environments. *Microbiol. Mol. Biol. Rev.*, 64: 548–572.

Nicholson, W.L., A.C. Schuerger, and P. Setlow (2005) The solar UV environment and bacterial spore UV resistance: Considerations for Earth-to-Mars transport by natural processes and by human spaceflight. *Mutat. Res.*, 571: 249–264.

Nickerson, C.A., C.M. Ott, J.W. Wilson, R. Ramamurthy, and D.L. Pierson (2004) Microbial responses to microgravity and other low-shear environments. *Microbiol. Mol. Biol. Rev.*, 68: 345–361.

Novikova, N.D. (2004) Review of the knowledge of microbial contamination of the Russian manned spacecraft. *Microb. Ecol.*, 47: 127–132.

Novikova, N.D., P. De Boever, S. Poddubko et al. (2006) Survey of environmental biocontamination on board the International Space Station. *Res. Microbiol.*, 157: 5–12.

Valtonen, M., P. Nurmi, J.-Q. Zheng et al. (2009) Natural transfer of viable microbes in space from planets in the extra-solar systems to a planet in our solar system and vice-versa. *Astrophys. J.*, 690: 210–215.

CHAPTER 14

Search for Life on Mars

An Astrogeological Approach

Jesús Martínez-Frías

CONTENTS

14.1 INTRODUCTION

The detection of extraterrestrial life would be one of the most amazing discoveries of humankind. The unequivocal confirmation of the existence of life beyond Earth would shake up our view of humanity's place in the cosmos with exciting scientific and social–cultural implications. It would be, without any doubt, an exceptional success in our understanding of nature and the universe. One of the main features of astrobiology is its multidisciplinarity (in fact transdisciplinarity) (Soffen, 1997; Cockell, 2001; Lunine, 2005). The Earth and planetary geosciences and astrogeological studies are proving to be of great help if not crucial in facing the challenging and fascinating questions of whether or not Earth's life is unique and the links between life on Earth and the potential life on other planets. The NASA's Astrobiology Roadmap (Des Marais et al., 2008) contains three relevant questions: (1) How does life begin and evolve? (2) Does life exist elsewhere in the universe? (3) And what is the future of life on Earth and beyond? To address these questions, which are extremely complex, a crosscutting approach is required. We start from an

FIGURE 14.1   Eugene Shoemaker (1928–1997), renowned both as a geologist and an astronomer and a member of the Board of Directors of The Spaceguard Foundation. He is considered *the father of astrogeology*. The Near Earth Asteroid Rendezvous space probe was renamed *NEAR Shoemaker* in his honor. (Photo courtesy of USGS.)

institutional point of view and address the foundation of planetary geology (also known as astrogeology), which took place in 1961 by Eugene Shoemaker within the United States Geological Survey (USGS). This is shown in Figure 14.1.

Since then, planetary geology has been providing essential scientific information for addressing the aforementioned questions. These comprise (Martínez-Frías, 2009) the following, among other issues:

1. The study of the mineralogical and geochemical characteristics of asteroidal, lunar, and Martian meteorites.

2. The characterization of impact events and craterization processes.

3. The application of principles and techniques of geology and geosciences in planetary missions to the study of solid bodies.

4. The classification and comparative analysis of *terrestrial analogs*.

5. Theoretical modeling and experimental simulation in special planetary chambers.

6. Planetary protection issues and the consideration of geoeducational and geoethical approaches. Mars exploration and research represents an excellent example for illustrating how to move forward and for the importance of considering this astrogeological perspective.

14.2 PLANETARY GEOLOGY (OR ASTROGEOLOGY)

Science and scientific disciplines are in permanent evolution. This dynamic factor is decisive in itself as it determines the progress and adaptability of contents as well as the way of focusing new research studies. This was also the case, more than 40 years ago, for geology evolving toward planetary geology or astrogeology.

14.2.1 Definition and General Overview

Although a standard definition of planetary geology (or astrogeology) does not exist, probably the best characterization of this geoscientific subdiscipline came from the Arizona State University, one of the most significant institutions with the world's highest tradition in this field. It is the place where the term astrogeology was proposed for the first time by Eugene Shoemaker (Figure 14.1). Planetary geology aims to study, at many different scales, the origin, evolution, and distribution of the condensed matter in the universe in the form of planets, satellites, asteroids, comets, and particles of different sizes and genesis. It involves the incorporation of results from spacecraft data analyses, laboratory simulations of various planetary processes, and field studies of features on Earth analogous to extraterrestrial features. This definition mostly agrees with the scientific guidelines of the USGS Astrogeology Research Program (Martínez-Frías and Hochberg, 2007). At present, there is a great variety of astrogeological aspects that connect this discipline with astrobiology. However, in the beginning, the main goals for its launching and development were very different. In a way, they were associated with the issues of habitability. The Astrogeology Research Program (Schaber, 2005) started out as the Astrogeologic Studies Group at the USGS Center in Menlo Park, CA. Later on, in 1962, it was moved to Flagstaff, Arizona. In accordance with Beattie (2001) and ABOR-NAU (2010),

> this is the real moment that The Astrogeology Research Program started, when USGS and NASA scientists transformed the northern Arizona landscape into a re-creation of the Moon... Using the cinder cones and craters scattered around northern Arizona as models, USGS and NASA scientists, including Eugene Shoemaker, taught astronauts about geologic features and lunar formations... The training gave them the skills essential for the first successful manned missions to the Moon.

The majority of studies and topics of research covered by planetary geology fall into the United Nations Educational, Scientific and Cultural Organization (UNESCO) field (code 25) "Earth and Space Sciences," and mainly geologists and geoscientists are the researchers involved in this type of geological activities (Martínez-Frías and Hochberg, 2007). However, planetary geology only appears, misplaced and poorly represented as a

UNESCO system subdiscipline, in the field of "Astronomy and Astrophysics" (code 21), with the code 2104.4—"Planetary Geology"—whereas somewhat surprisingly, it does not appear in the field of "Earth and Space Sciences" (code 25). This obliges both geologists and geoscientists (mainly geochemists, geophysicists, and geobiologists) working in planetary sciences, to incorporate their scientific studies either in the field of "Astronomy and Astrophysics" or within the rather ambiguous box of "Other" (code 2512.99), in the field of "Earth and Space Sciences." In short, the current UNESCO classification does not match the actual reality of planetary geology.

But, what is the real scientific scenario telling us about this? Does everything related to space fall within the disciplinary framework of "Astronomy and Astrophysics"? Is Mars exploration and research an astronomic/astrophysical subject? A detailed search in the Scientific Thomson Reuters' Web of Science (WoS) database for papers published in the period 1900–2013 yields 361 records for articles including the keywords "Mars and Physics" (123 in the WoS category of "Astronomy and Astrophysics," 71 in "Geoscience Multidisciplinary," and 50 in "Meteorology and Atmospheric Sciences"). If we combine the keywords "Mars and Geology," the results are 426 records (146 for the WoS category of "Geochemistry and Geophysics," 138 for "Astronomy and Astrophysics," and 84 for "Geoscience Multidisciplinary"). The number of records increases for the keywords "Mars and Chemistry" (893 publications; 271 in the WoS category of "Geochemistry and Geophysics," 267 in "Astronomy and Astrophysics," and 144 in "Geosciences Multidisciplinary"). The combination of *Mars* and *Biology* yields only 280 records. Despite the relative youth of astrobiology, there are already 447 records for "Mars and Astrobiology" in the WoS database (280 for the category of "Astronomy and Astrophysics," 139 for "Geosciences Multidisciplinary," and 124 for "Biology").

Thus, independently from this systematic anomaly in the UNESCO classification, in the real world, time scientific and technological research and exploration are what they are, and they are putting things in place. Most lunar and Mars missions are not purely astronomic or astrophysical endeavors but are real astrogeological missions in the original and foundational sense of Shoemaker.

14.3 PLANETARY GEOLOGY (OR ASTROGEOLOGY) IN THE NASA ASTROBIOLOGY INSTITUTE ROADMAP

As previously defined, the NASA's Astrobiology Roadmap (Des Marais et al., 2008) faces three basic questions (Section 14.1). These complex questions are incorporated into a series of 10 science goals and 17 more specific science objectives and are stressing 4 principles that are essential to the operation of the astrobiology program. Planetary geology studies are implicit in practically all of them. In short, the goals are the following:

- Goal 1: Understand how life arose on Earth.

- Goal 2: Determine the general principles governing the organization of matter into living systems.

- Goal 3: Explore how life evolves on the molecular, organism, and ecosystem levels.

- Goal 4: Determine how the terrestrial biosphere has coevolved with the Earth.

- Goal 5: Establish limits for life in environments that provide analogs for conditions on other worlds.

- Goal 6: Determine what makes a planet habitable and how common these worlds are in the universe.

- Goal 7: Determine how to recognize the signature of life on other worlds.

- Goal 8: Determine whether there is (or once was) life elsewhere in our solar system, particularly on Mars and Europa.

- Goal 9: Determine how ecosystems respond to environmental change on timescales relevant to human life on Earth.

- Goal 10: Understand the response of terrestrial life to conditions in space or on other planets.

Goals 1–4 match with the first question, as it follows. If we want to understand the first question, namely, how life arose on Earth, there are many geological and astrogeological aspects that are clearly involved: (a) to carry out mineralogical and geochemical studies of meteorites and how their inorganic and organic compounds (and even impacts per se) played a role in the early Earth processes (Wallis and Wickramasinghe, 1995; Melosh, 2003; Cockell, 2006; Pontefracti et al., 2012), (b) to evaluate the degasification processes of the Earth's mantle (Ozima and Zahnle, 1993) and how active were the interactions of the pristine atmosphere–lithosphere, (c) to determine which geological environments (hydrothermal, deep sea, shallow marine, lacustrine, etc.) were the most favorable for the origin of life (Rotchschild and Mancinelli, 2001), (d) to evaluate if some specific volcanic types (e.g., komatiites) (Nna-Mvondo and Martínez-Frías, 2007) were more suitable for life origin, and (e) to understand how are organic compounds and minerals/rocks interrelated (considering both physical and chemical mineral processes), among others (Novoselov et al., 2013). All these geological issues have also to be taken into account in the framework of the origin of life, its main principles ruling the organization of matter, the evolution of organisms in the different ecosystems, and the course of action of biogeological coevolution.

The second question about the possible existence of life in the universe is covered by goals 5–8. Regarding the establishment of the limits for life environments that can be used as analogs for planetary exploration, geology is providing the most significant markers from the study of different terrestrial settings. It appears appropriate here to define the concept of geomarker and its relation to biomarker. A fundamental problem related to the paleoenvironmental study of early Earth rocks and minerals and astrobiological exploration of planetary materials (e.g., asteroidal and Mars meteorites, Martian rocks) is not only recognizing and quantifying carbon-related compounds that may be present but also differentiating those molecules formed abiotically from those generated by extinct or extant life. Only through the combination of biomarkers and geomarkers (Martínez-Frías et al., 2007) will we be able to understand the global framework. Biological markers or *biomarkers* are

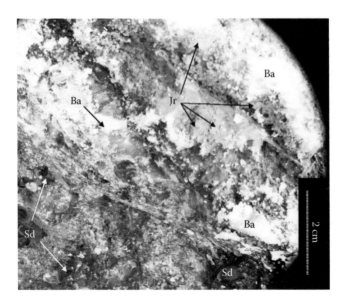

FIGURE 14.2   Jarosite patches from the Jaroso area (Jaroso Hydrothermal System), Sierra Almagrera (Spain). Various paragenetic stages are represented in the sample. Jr, jarosite; Sd, siderite; Ba, barite.

molecular fossils and are defined as "complex organic compounds, which originated from formerly living organisms and which are composed of carbon, hydrogen, and other elements. Abiotic organic compounds are not biomarkers per se because they do not originate from biosynthesis" (Simoneit, 2004). So, a particular mineral (e.g., magnetite, jarosite) (see Figure 14.2), a biologically related isotopic signature, and the presence of oxygen or water in a planet's atmosphere are not biomarkers, and the term is currently misused in many articles including geomicrobiology and astrobiology and also in exoplanetary research.

A definition of the term geomarker was proposed by Martínez-Frías et al. (2007). Geomarkers can be defined as any geological (i.e., mineralogical, geochemical, metallogenetic, sedimentological, petrological, tectonic) feature or set of features, which can be used as proxy indicators of the physical, chemical, and/or biological characteristics of the environment in which they occur, and/or of the processes that formed them. The investigation of Earth analogs, mainly connecting chemistry, geomicrobiology, and mineralogy, is being very successful to be optimistic about the robustness of life by the incorporation in the payloads of planetary missions of the science grounds connecting geo- and biomarkers from different terrestrial environments.

Likewise, in order to determine what makes a planet habitable and how common these worlds are in the universe, planetary geology is also proving to be of great importance. Intrinsically linked to the site in which the planet occupies in relation to the Sun (habitability zone) and the own geological features of our planet, many works have also stressed the plausible essential connection between the potential existence of life (or at least favorable habitability conditions) and the geological vitality of planetary bodies. There are several indicators. For instance, the geological study of the craterization index of planets, as shown in Figure 14.3, is serving as guide to date planetary surfaces (and to

FIGURE 14.3 Saturn's moon, Phoebe. Small planetary bodies do not have energy to reshape their surface. The image was taken from around 19,000 miles out from this 137 mile wide moon. (Courtesy of NASA/ESA/JPL/SSI.)

determine their internal energy that makes them capable of remodeling their surface) (Melosh, 1989; Holsapple, 1993; Collins et al., 2013).

Similarly, the existence of planetary geodynamic activity (plate tectonics, volcanism, hydrothermalism, etc.) is also important (considering a terrestrial approach). In particular, plate tectonics has been defined as a global catalyzer able to originate, modify, destroy, and rebuild sites for life to rise in and evolve in different environmental scenarios and at different temporal scales (Miller and Bada, 1988; Nisbet and Sleep, 2001; Condie, 2005). These factors are very important and need to be connected to goals 7 and 8. In order to find out how to recognize the signature of life on other worlds, we need first to be able to do it here, on our own planet. In addition, we need to define the signature of life criteria to proceed with the relevant exploration of Mars, Europa, Titan, or beyond, perhaps in other planetary systems. Finally, regarding the question about the life's future on Earth and beyond, NASA proposes goals 9 and 10. Goal 9 refers to the capability of response of ecosystems to environmental changes and to the need for understanding the response of terrestrial life to conditions in space or on other planets. Goal 10 is to understand the response of terrestrial life to space conditions. In both cases, biogeosciences are vital for considering not only by bio- and geointeractions at different levels but also by assessing the robustness of life under the harsh conditions in space of other planets, by well-designed experiments in low orbit and in planetary chambers.

Principle 1 refers to the multidisciplinary approach of astrobiology. In this sense, astrogeology could be considered as a redundancy, as it is also a multidisciplinary field in the framework of Earth and planetary sciences, covering topics from meteorites to planets; geological, mineralogical, geochemical, and metallogenetic studies of rock outcrops at different scales; atmosphere–lithosphere interactions; and planetary surface and underground processes, among others.

Principle 2 is related to planetary stewardship. This principle is well focused on biological contamination and ethical issues surrounding the export of terrestrial life to other planets. Recently, it has also been applied for consideration of the significance of geoethics (Martinez-Frías et al., 2011) and the need of having an appropriate astrogeological expertise in any bio- and geolinks associated with planetary protection issues (Rummel et al., 2002) (regarding scientific guidelines, protocols, and methodologies).

Principle 3 has a more societal character in connection with the search for extraterrestrial life and the adaptation of life to live on other worlds. Here, it is important to note how planetary geology is being incorporated to the classical search for the extraterrestrial intelligence (SETI) search issues, in the framework of the global concept of habitability, considering different planetary environments and geoethical approaches and also carrying out selected geobiological experiments and planetary environment simulations to test the robustness of life in extreme environments.

Finally, principle 4 refers to the significance of education and public outreach. In this context, planetary geoeducation is intrinsically linked to it, and one of the main lessons learned from astrogeology is the need for an appropriate combination of the biological and geological approaches to understand the coevolution of our planet.

14.4 SOME SELECTED DATA REGARDING PLANETARY GEOLOGY SUBJECTS

The study of extraterrestrial matter (mainly the mineralogical and geochemical investigation of meteorites and their impact events) and Mars exploration and research are two hot topics in the framework of astrobiology-related planetary geology studies. Meteorites (in particular undifferentiated meteorites) are extremely important to establish the features of the pristine matter, which contributed to the origin of our planet (and also to the origin of life). Likewise, meteorite impacts have marked the geobiological evolution of our planet (modifying the geological and atmospheric environments at different levels and scales and taking part, at least partially, in the extinction of some species). Mars is probably considered the hottest topic not only as the first astrobiological target but also as a fundamental planet to do comparative planetology and to understand the origin and evolution of terrestrial planets. In regard to these topics, a detailed search in the Scientific Thomson Reuters' WoS database for papers published in the period 1900–2013 shows that a combination of the terms *meteorites* and *astrobiology* is reflected in 166 records. The maximum value is assigned to the WoS category of "Astronomy and Astrophysics" (69%). The geoscientific WoS categories of "Geoscience Multidisciplinary" and "Geochemistry Geophysics" include 64 articles, which correspond to 39% of the whole contributions. The main three journals are *Astrobiology* (18 records), *Icarus* (14 records), and *Astrophysical Journal* (9 records); the geological topics are mainly included in the *Astrobiology* articles. If the keyword combination involves the term *meteorites* and *life*, the WoS database indicates more than 1000 records (1032 records). The WoS category of "Astronomy and Astrophysics" is again on the top position, with 374 records (36%), and the categories of "Geochemistry Geophysics" and "Geosciences Multidisciplinary" display a slight proportional decrease in the number of records (386 records; 37%). In this case, the top three journals are *Astrobiology* (62 records), *Meteoritics and Planetary Science* (58 records), and *Geochimica and Cosmochimica Acta* (50 records). The geoscientific topics are much more diversified and are distributed in these three journals. As previously stated, the combination of the terms *Mars* and *astrobiology* yields more than 400 records (447 records). Planetary geology studies are well represented in the WoS category of "Geoscience Multidisciplinary" in the second position, with 139 records, after the category of "Astronomy and Astrophysics" (280 records). The two most significant journals are *Astrobiology* (64 records) and *Icarus* (50 records). Three most active countries reporting on both subjects are the United States (298 records), the United Kingdom (94 records), and Spain (47 records). The combination of the keywords *Mars* and life indicates a significantly higher number of results (2913 records) and a different distribution in which planetary geology also appears represented in the second position in the context of the WoS category of "Geosciences Multidisciplinary" (577 records), after the first position that is occupied by "Astronomy and Astrophysics" (986 records). In this case, the two most significant journals are *Astrobiology* (208 records) and *Planetary and Space Sciences* (132 records), and the three most active countries are the United States (1453 records), the United Kingdom (337 records), and France (241 records). Finally, the combination of the terms *Mars* and *analog* yields more than 1000 records (1147 records), in which the planetary geology studies are represented in the second and third positions by the WoS categories of "Geochemistry Geophysics" and "Geosciences Multidisciplinary" with 251 and 239 records, respectively. Regarding the scientific journals, the two most important are *Journal of Geophysical Research Planets* (137 records) and *Icarus* (126 records), in which the United States (703 records) and the United Kingdom (136 records) represent the most active countries.

14.5 MARS' ASTROBIOLOGY: MAIN ASTROGEOLOGICAL SOURCES

There are three main sources of information to learn about and investigate Mars' astrogeology: (1) Martian meteorites, (2) Mars missions, and (3) Mars analogs. Scientific and technological studies combining these sources are providing us with a general panorama of the red planet and are helping us in advancing the knowledge about Mars and are also enabling us to foresee future exploration and research steps. We should not neglect or underestimate the significance of planetary (Mars) simulation chambers, which are new tools for emulating different types of planetary processes.

14.5.1 Martian Meteorites

In the same way that it is possible to differentiate terrestrial rocks from meteorites, there are also mineralogical and geochemical (mainly isotopic) criteria to determine if meteorites come from asteroids, the Moon, or Mars. Here, a very brief overview about the Martian meteorites is presented. Further and much more detailed information can be found in Bogard and Johnson (1983), Bogard et al. (1984), Smith et al. (1984), Treiman (1995), Nyquist et al. (2001), Beck et al. (2005), and Treiman (2005), among others.

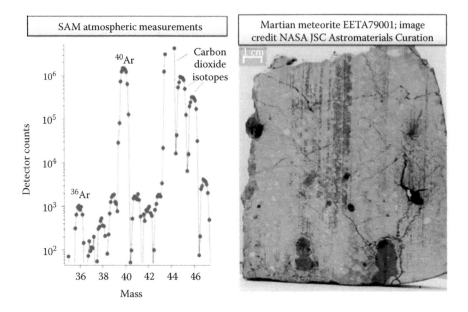

FIGURE 14.4   This meteorite, which originally weighed nearly 8 kg, provided the first strong proof that meteorites could come from Mars. EETA79001 is an achondrite meteorite, a basalt lava rock nearly identical to many Earth rocks.

The geochemical study, by the early 1980s, of various trapped gases in the EETA79001 meteorite (see Figure 14.4) provided evidence that they looked like those in the Martian atmosphere as analyzed by Viking.

Treiman et al. (2000) carried out an in-depth study of a group of Martian meteorites, so-called shergottites, nakhlites, and chassignites (SNCs), and stated that "there seems little likelihood that the SNCs are not from Mars. If they were from another planetary body, it would have to be substantially identical to Mars as it now is understood." In accordance with the Meteoritical Bulletin, there are 46,546 valid meteorite names and 10,723 provisional names. Of them, only 125 specimens are classified as Martian meteorites (October 8, 2013). Annual expeditions to Antarctica since 1975 by Japanese, the United States, and more recently Chinese government–sponsored teams have resulted in a steady increase in the number of Martian specimens. Nowadays, we know very well that Martian (SNC) group of meteorites are significantly different from most other meteorite types.

Shergottites are named after the Shergotty achondrite, which fell in Sherghati (India) in 1865. They represent the most abundant type of Martian meteorites. Shergottites are igneous rocks of volcanic or plutonic origin, and they look like the classical terrestrial material. They appear to have crystallized as recently as 180 million years ago, and they typically display evidence of severe shock metamorphism. In general, the plagioclase in shergottites has been transformed into maskelynite, a glass that is formed when plagioclase is subjected to shock pressures of at least 30 GPa. Based upon their mineral compositions, the shergottites are further (and broadly) subdivided into two distinct subgroups: the basaltic subgroup and the lherzolitic subgroup.

Nakhlites are named after the Nakhla achondrite, which fell in El-Nakhla, Alexandria, Egypt, in 1911. In accordance with Treiman (2005), nakhlites basically consist of euhedral to subhedral crystals of augite and olivine (to 1 cm long) in fine-grained mesostases. They formed in flows or shallow intrusions of basaltic magma on Mars about 1.3 billion years ago.

Chassignites are named after the Chassigny meteorite, which fell at Chassigny, Haute-Marne, France, in 1815. Basically, they consist of about 90% Fe-rich olivine, minor clinopyroxene, plagioclase, chromite, melt inclusions, and other accessory minerals and phases. There has been only one other chassignite recovered, named Northwest Africa (NWA) 2737. NWA 2737 was found in Morocco or Western Sahara in August 2000. Beck et al. (2005) indicated that its "mineralogy, major and trace element chemistry as well as oxygen isotopes revealed an unambiguous Martian origin and strong affinities with Chassigny."

There exists another group of Martian meteorite that is represented by the famous specimen Allan Hills 84001: an orthopyroxenite (OPX Martian meteorite) (see Figure 14.5).

ALH84001 received much attention from the astrobiological point of view, as it was the meteorite causing all of the excitement about life possibly existing on Mars (McKay et al., 1996). This meteorite was the first meteorite found by the Antarctic team in 1984,

FIGURE 14.5   ALH84001 is a cumulate rock consisting of 97% coarse-grained, Mg-rich orthopyroxene, with small amounts of plagioclase, chromite, and carbonate. It was initially classified as a diogenite; however, the presence of oxidized Fe in chromite led to its reclassification as a Martian meteorite. This is the meteorite causing all of the excitement about life possibly existing on Mars. (Courtesy of NASA.)

and it was initially identified as a diogenite, a singular type of achondrite meteorite. It was not until October 1993 that David Mittlefehldt (Astromaterials Research Office, NASA/Johnson Space Center) detected the error and properly identified it as an SNC. Recently (January 3, 2013), NASA reported that meteorite NWA 7034 (nicknamed "Black Beauty") was determined to be from Mars and found to contain 10 times the amount of water as any other Mars meteorites found on Earth. The rock quickly attracted special attention. This was well explained and synthesized by Amy Tikkanen:

> a research team was assembled with McKay as its leader. The study, which took more than two years, revealed several peculiarities. First was the presence of polycyclic aromatic hydrocarbons (PAHs). While these organic compounds are common-place, found throughout the solar system, the PAHs in the meteorite were unusual in appearance, resembling the type that result from the decay of organic matter. The presence of the molecules within the rock and their absence on its surface ruled out Earth contamination. The team also discovered carbonate globules, which are closely associated with bacteria found on Earth. Moreover, iron sulfides and magnetite were present. These compounds, which are so small that one billion of them can fit on the head of a pin, do not usually coexist. Certain bacteria, however, synthesize them simultaneously. These discoveries, published by McKay and his co-workers in the journal Science, indicated the possibility of ancient life on Mars. While the news generated a flurry of debate, McKay stressed that the findings were not definitive proof and that further research was planned.
>
> TIKKANEN (2010)

There is a paragraph in *The Telegraph* obituary of Prof. David Mackay that describes very well the significance of the McKay study about ALH84001 and astrobiology. It says:

> While final resolution of the dispute will probably have to await Nasa's "Mars Sample Return" project in the late 2020s, McKay's announcement in 1996 gave a welcome boost to the agency's plans for missions to Mars. It also prompted the establishment of an Astrobiology Institute which investigates signs of life in extreme environments – in space and on Earth.
>
> *THE TELEGRAPH* (2013)

14.5.2 Mars Missions

Due to the space limitations, we cannot describe here in detail all Mars missions connected with planetary geology and astrobiological issues. Mars missions (NASA, 2013) are numerous, and they range from the first flyby in 1960 that ended in failure (*Korabl4*, USSR) to the successful and currently active Mars Science Laboratory (MSL) (United States). Some specific missions are listed here. *Flyby missions* include Mariner 3–4 and Mariner 6–7. *Orbital missions* (Orbiters) include Mariner 8–9, Viking 1–2, Mars Observer, Mars Global Surveyor, Mars Climate Orbiter, 2001 Mars Odyssey, Mars Express, Mars Reconnaissance

Orbiter, and the planned NASA 2016 ExoMars Orbiter. *Landers and Rovers* include Viking 1–2, Pathfinder, Polar Lander/Deep Space 2, Mars Exploration Rovers, Phoenix, MSL, and planned NASA 2018 ExoMars Rover and 2020 Mission Plans. In addition, future plans include also missions *airplanes and balloons*, *subsurface explorers*, and *sample returns*.

In the same way as in the previous chapter, it appears appropriate to highlight some historical aspects for their crucial and emblematic significance linked to the search for life on Mars and astrogeological subjects (mainly with reference to the geochemical characterization of the Martian regolith). Of all the Mars missions, the two Viking spacecrafts, shown in Figure 14.6, were probably the most decisively related to biology (astrobiology).

Each of Viking spacecrafts carried four types of biological experiments to the surface of Mars. They were the first Mars landers to perform biogeochemical tests to search for biomarkers on Mars. The following four experiments were (1) gas chromatograph–mass spectrometer, (2) gas exchange, (3) labeled release (LR), and (4) pyrolytic release. The LR experiment is probably the one that provided the most guarantee for astrobiology. Martian regolith was inoculated with a drop of very dilute aqueous nutrient solution. The nutrients (seven molecules that were Miller–Urey products, such as some amino acids that can be formed under prebiotic conditions from simple gases and energy sources such as an electrical spark) were tagged with radioactive ^{14}C. The air above the soil was monitored for the evolution of $^{14}CO_2$ gas as evidence that possible microorganisms in the Martian soil had successfully metabolized the nutrients (Klein et al., 1976; Brown et al., 1978). According to NASA, "the only organic chemicals identified when the Viking landers heated samples of Martian soil were chloromethane and dichloromethane—chlorine

FIGURE 14.6   Viking 1 lander. It carried instruments to achieve the primary scientific objectives of the lander mission: to study the biology, chemical composition (organic and inorganic), meteorology, seismology, magnetic properties, appearance, and physical properties of the Martian surface and atmosphere. (Courtesy of NASA.)

compounds interpreted at the time as likely contaminants from cleaning fluids." Further information can be found in NASA-SP4212.

Later on, on August 2008, the Phoenix lander detected perchlorate (Hecht et al., 2009), a strong oxidizer when heated above 200°C. Perchlorates can destroy organics when heated and generate chloromethane and dichloromethane as by-products. These are the same chlorine compounds detected by both Viking landers. Given that perchlorate would have broken down any Martian organics, the question regarding the organic compounds is still wide open, something which, in fact, had already been highlighted before (Navarro-Gonzalez et al., 2006; Biemman, 2007). Very recently, the rover Curiosity's instruments (see Figure 14.7) found water, sulfur, and chlorine-containing compounds, including chlorinated methane gas.

According to the authors who analyzed the results of the Curiosity mission so far (Glavin et al., 2013), the origin of the chlorine species is certainly from Mars; however, the source of organics needed for the production of chlorinated organics is still unknown. Likewise, the presence of chloromethanes is tentatively thought to be a signature of perchlorates in the Martian soil.

This amazing and still controversial astrobiological subject links the first Viking experiments to the most recent Curiosity results. There are additional benefits of these and other Mars missions, since they have provided us with a very good scientific panorama about Mars and its general geological evolution. This is particularly important with regard to its past wet environment and water-related geomorphological features (fluvial and deltaic systems, lakes, gullies, etc.). Water-related minerals (gypsum, jarosite, clay minerals, etc.)

FIGURE 14.7   MSL (Curiosity) rover. The image was taken on May 26, 2011, in the Spacecraft Assembly Facility at NASA's Jet Propulsion Laboratory in Pasadena, California. Curiosity has a mass of 899 kg including 80 kg of scientific instruments. The rover is 2.9 m long by 2.7 m wide by 2.2 m high. The main scientific goals of the MSL mission are to help determine whether Mars could ever have supported life, as well as determining the role of water, and to study the climate and geology of Mars. (Courtesy of NASA.)

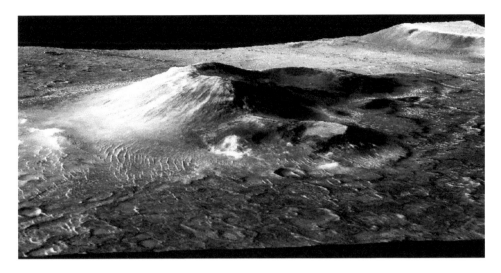

FIGURE 14.8 Various studies indicate that hydrothermal processes and mineralization could occur on Mars. This picture shows a volcanic cone in the Nili Patera caldera. Light-toned patches on the closest flank of the cone, and the entire field of light-toned material on the left of the cone, are hydrothermal deposits. The cone is about 5 km in diameter at the base. (Courtesy of NASA/JPL-Caltech/MSSS/JHU-APL/Brown Univ.)

are being currently used as geomarkers to understand the Martian paleoenvironments. This is shown in Figure 14.8 for Mars water search.

The geological, mineralogical, and geochemical characterization and study of these paleoenvironments and their habitability conditions have also benefited from the multi-disciplinary study of Mars analogs.

14.6 MARS ANALOGS

So far, we do not know of a place on Earth that is truly like Mars. Nevertheless, it is possible to identify sites on our planet where environmental conditions (geology, tectonics, mineralogy, hydrothermalism, geochemistry, etc.) approximate, in some specific ways, those possibly encountered on Mars at present or earlier in that planet's history. These environmental conditions are also privileged areas for testing new instrumentation and crucial for understanding planetary habitability issues. According to Farmer (2004), when defining a site-selection strategy for exploration of putative Martian fossil records, a key factor is contemporaneous chemical precipitation, or mineralization. On Earth, geological environments where microorganisms are often preserved in this way include, among others, (1) mineralizing systems (subaerial, subaqueous, and shallow subsurface hydrothermal systems and cold springs of alkaline lakes), (2) saline/alkaline environments of arid marine shorelines (sabkhas) or terminal (evaporative) lake basins, (3) duricrusts and subsoil hardpan environments formed by the selective leaching and reprecipitation of minerals within soil profiles, and (4) periglacial environments' ground ice or permafrost (frozen soils), which have captured and cryopreserved microorganisms and associated organic materials. There are many emblematic sites on Earth that have demonstrated to be extremely useful as

FIGURE 14.9 General view of an area rich in oxides and sulfates at the Río Tinto Mars analog (SW Spain). Various astrobiological studies were performed at this area in the context of the Mars Astrobiology Research and Technology Experiment (MARTE) project. (From Stoker, C.R. et al., *Astrobiology*, 8, 921, 2005.)

FIGURE 14.10 Outcrop rich in jarosite at the Jaroso Mars analog. The Jaroso ravine is the world-type locality of jarosite. The *Jaroso Hydrothermal System (JHS)* is an extremely interesting late-volcanic episode, and the *Jaroso Ravine*, located in Sierra Almagrera (Almeria province), is the best site (approximately 2 km × 4.5 km) where the hydrothermalism and alteration, associated with the JHS, have attained the maximum surface expression. (From Martínez-Frías, J. et al., *Earth Planet Space*, 56, 5, 2004.)

FIGURE 14.11   Los Azulejos area (note the blue and green outcrops), Parque de las Cañadas, Tenerife, Canary Islands. This area has been recently proposed as an excellent Mars analog for studying hydrothermalism and alteration processes (Lalla et al. 2013) and for testing the Raman instrument related to the ExoMars mission.

Mars analogs. They are found in Antarctica, Chile (e.g., Atacama), Iceland (e.g., Hvalfjordur and Berufjordur), Spain (Río Tinto, Jaroso, Tenerife) (see Figures 14.9 through 14.11), Australia (e.g., Pilbara), etc. All these Mars analogs (and many others, Fletcher, 2011) will be crucial to define biomarkers and geomarkers for Mars exploration.

14.7 MARTIAN HISTORICAL PERIODS AND THE FINAL REMARKS

Thanks to all the data obtained from these three main information sources about Mars (Martian meteorites, Mars missions, and Mars analogs) we have a general geological panorama of the planet (see Figure 14.12), which allows us to identify target areas for astrobiological exploration and research from a temporal perspective.

This perspective is crucial. Although it is well known that liquid water is not stable at the surface under today's atmospheric conditions (e.g., Ingersoll, 1970; Hecht, 2002), there is significant evidence that Mars once had a thicker atmosphere, that liquid water may have been much more abundant on the surface and in the subsurface earlier in Martian history, that it has at least sporadically flowed on the Martian surface, and that it may even still be present in the subsurface today (see Figures 14.13 and 14.14).

In a general planetary framework, we know that Mars is mainly characterized by its hemispheric dichotomy, the Tharsis and Elysium volcanic provinces, the existence of large impact basins, its equatorial canyon system, chaotic terrain and outflow channels, and its ice caps. Martian geochronology has been defined as (1) pre-Noachian (about 4.5 billion year ago), (2) Noachian (between 4.1 and 3.7 billion years ago), (3) Hesperian (between 3.7 and 3.0 billion Gya), and (4) Amazonian (from 3.0 Gya to present). The periods Noachian

FIGURE 14.12   Mars geology map. Composited with Mars Orbiter Laser Altimeter (MOLA) data. (Courtesy of NASA.)

FIGURE 14.13   Mars streambed conglomerate compared to example on Earth. This image was acquired by NASA's Curiosity rover on the surface of Mars. It shows an outcrop of conglomerate and some pebble-size weathering debris. In accordance with the scientific study (Williams et al., 2013), the round pebbles are too large to have been moved and shaped by wind; thus, they had to have been transported a significant distance by water. (Courtesy of NASA/JPL-Caltech/MSSS.)

and Hesperian (from 4.1 to 3.0 Ga) were very important in connection with the existence of liquid water. An alternative Martian timescale based on the predominant type of mineral alteration was recently proposed: Phyllocian (named after phyllosilicates or clay minerals that characterize the era), which lasted from the formation of the planet until around the early Noachian, Theiikian (named after sulfurous in Greek, for the sulfate minerals that

FIGURE 14.14 Outcrop at the *Sheepbed* locality, taken by NASA's Curiosity Mars rover with its right Mast Camera (Mastcam), shows show well-defined veins filled with whitish minerals, interpreted as calcium sulfate. (Courtesy of NASA/JPL-Caltech/MSSS.)

were formed), which lasted until about 3.5 Gya, and Siderikan (named for iron in Greek), for the iron oxides that formed, which lasted from 3.5 Gya until the present. As stated by Head (2007),

> The major dynamic forces shaping the surfaces, crusts, and lithospheres of planets are represented by geological processes which are linked to interaction with the atmosphere (e.g., aeolian, polar), with the hydrosphere (e.g., fluvial, lacustrine), with the cryosphere (e.g., glacial and periglacial), or with the crust, lithosphere, and interior (e.g., tectonism and volcanism). Interaction with the planetary external environment also occurs, as in the case of impact cratering processes.

Thus, planetary geology or astrogeology provides an essential scientific approach for any study about the origin of the evolution of Mars and its habitability conditions.

GLOSSARY

Abiotic: Not associated with or derived from living organisms.

Craterization: Alteration of the surface of a planet by the impact of cosmic objects.

Diogenite: A type of achondrite for which the parent body is believed to be the asteroid 2 Vesta.

EETA79001: Martian meteorite (shergottite).

Euhedral: Euhedral is a textural term used to describe crystals that are well formed with sharp, easily recognized faces.

Geomicrobiology: It concerns the role of microbe and microbial processes in geological and geochemical processes and vice versa.

Maskelynite: A type of naturally occurring glass, originated by the vitrification of plagioclase by shock melting in meteorites and meteorite impacts.

Planetary habitability: Planetary habitability is the capability of a planet or a natural satellite to develop and sustain life.

Planetary protection: A set of guidance in the design of a space mission to prevent biological contamination of the target celestial body and the Earth.

Plate tectonics: Scientific theory that describes the large-scale motions of the Earth's lithosphere.

Plutonic rock: Formed by slow solidification of magma deep within the Earth and crystalline throughout.

Regolith: A layer of loose, heterogeneous material that overlies the solid rock on the Earth, Moon, a planet, and some asteroids.

Shock metamorphism: Shock metamorphism is the term used to describe irreversible modifications in rocks and minerals caused by the passage of shock waves.

Undifferentiated meteorites: Primordial matter that has remained nearly unchanged for the last 4.5 billion years.

REVIEW QUESTIONS

1. Can you mention three main issues for which planetary geology provides essential information?

2. Who is considered the *father* of astrogeology and when and where did it start?

3. Is magnetite a biomarker?

4. Could you cite three examples of planetary geomarkers?

5. Is it possible to perform research on planetary geology of Mars while working on Earth?

6. Which are the main types of Martian meteorites?

7. In which principles of the NASA Astrobiology Institute (NAI) Roadmap would geo-ethics be specifically included?

8. Can you define three examples of Mars analogs?

9. Why would plate tectonics be important for astrobiology?

10. Are perchlorates important in the search for life on Mars and why?

REFERENCES

ABOR-NAU. 2010. *Days of Archives. A Real Life Soap Opera. Apollo Lunar Training.* Northern Arizona University, Flagstaff, AZ. http://library.nau.edu/speccoll/exhibits/daysofarchives/lunar.html (accessed September 3, 2013).

Beattie, D.A. 2001. *Taking Science to the Moon: Lunar Experiments and the Apollo Program.* Johns Hopkins University Press, Baltimore, MD, pp. 58–77.

Beck, P., Barrat, J.-A., Gillet, Ph. et al. 2005. The Diderot meteorite: The second chassignite. In *36th Annual Lunar and Planetary Science Conference*, League City, TX, Abstract No. 1326.

Biemann, K. 2007. On the ability of the Viking gas chromatograph–mass spectrometer to detect organic matter. *Proceedings of the National Academy of Sciences of the United States of America*, 104(25): 10310–10313.

Bogard, D.D. and Johnson, P. 1983. Martian gases in an Antarctic meteorite? *Science*, 221: 651–654.

Bogard, D.D., Nyquist, L.E., and Johnson, P. 1984. Noble gas contents of shergottites and implications for the Martian origin of SNC meteorites. *Geochimica et Cosmochimica Acta*, 48: 1723–1739.

Brown, F.S., Adelson, H.E., Chapman, M.C. et al. 1978. The biology instrument for the Viking Mars mission. *Review of Scientific Instruments*, 49(2): 139–182.

Cockell, C.S. 2001. Astrobiology and the ethics of new science. *Interdisciplinary Science Reviews*, 26(2): 90–96.

Cockell, C.S. 2006. The origin and emergence of life under impact bombardment. *Philosophical Transactions of the Royal Society of London B: Biological Sciences*, 361(1474): 1845–1856.

Collins, G.S., Melosh, H.J., and Osinski, G.R. 2013. The impact-cratering process. *Elements*, 8(1): 25–30.

Condie, K.C. 2005. *Earth as an Evolving Planetary System*. Elsevier Academic Press, Burlington, MA.

Des Marais, D., Nuth III, J.A., Allamandola, L.J. et al. 2008. The NASA astrobiology roadmap. *Astrobiology*, 8(4): 1–16.

Farmer, J. 2004. Targeting sites for future astrobiological missions to Mars. In: *Second Conference on Early Mars*, LPI, Jackson Hole, WY, Abstract No. 8088.

Fletcher, L.E. 2011. Mars on Earth: Discovery and characterization. First Year Report (electronic document). University of Oxford, Oriel College. http://www2.physics.ox.ac.uk/sites/default/files/2012-03-08/1_fletcher_pdf_71751.pdf (accessed July 12, 2013).

Glavin, D.P., Freissinet, C., and Miller, K.E. 2013. Evidence for perchlorates and the origin of chlorinated hydrocarbons detected by SAM at the Rocknest aeolian deposit in Gale Crater. *Journal of Geophysical Research Planets*, 118(10): 1955–1973. doi: 10.1002/jgre.20144.

Hecht, M.H. 2002. Metastability of liquid water on Mars. *Icarus* 156, 373–386.

Hecht, M.H., Kounaves, S.P., Quinn, R.C. et al. 2009. Detection of perchlorate and the soluble chemistry of Martian soil at the phoenix lander site. *Science*, 325(5936): 64–67.

Holsapple, K.A. 1993. The scaling of impact processes in planetary sciences. *Annual Review of Earth and Planetary Sciences*, 21: 333–373.

Klein, H.P., Lederberg, J., Rich, A. et al. 1976. The Viking mission search for life on Mars. *Nature*, 262(5563): 24–27.

Lalla, E., Lopez-Reyes, G., Rull, F. et al. 2013. Raman analysis of basaltic samples from Tenerife island (Cañadas, Azulejos and historical eruptions) with the ExoMars RLS instrument. In *44th Lunar and Planetary Science Conference*, The Woodlands, TX, Abstract No. 2403.

Lunine, J. 2005. *Astrobiology: A Multi-Disciplinary Approach*. Benjamin-Cummings Publishing Company, Subs of Addison Wesley Longman, Inc., San Francisco, CA.

Martínez-Frías, J. 2009. El geólogo planetario o astrogeólogo. In *La profesión de geólogo*, Ilustre Colegio Oficial de Geólogos (ed.). ICOG, CYAN, Proyectos y Producciones Editoriales, S.A., Madrid, pp. 201–219 (in Spanish).

Martínez-Frías, J., González, J.L., and Rull, F. 2011. Geoethics and deontology: From fundamentals to applications in planetary protection. *Episodes*, 34(4): 257–262.

Martínez-Frías, J. and Hochberg, D. 2007. Classifying science and technology: Two problems with the UNESCO system. *Interdisciplinary Science Reviews*, 32(4): 315–319.

Martinez-Frias, J., Lázaro, E., and Esteve-Núñez, A. 2007. Geomarkers versus biomarkers: Paleoenvironmental and astrobiological significance. *AMBIO: A Journal of the Human Environment*, 36(5): 425–426.

Martínez-Frías, J., Lunar, R., Rodríguez-Losada, J.A. et al. 2004. The volcanism-related multistage hydrothermal system of El Jaroso (SE Spain): Implications for the exploration of Mars. *Earth, Planets and Space*, 56: 5–8.

McKay, D.S., Gibson Jr., E.K., Thomas-Keprta, K.L. et al. 1996. Search for past life on Mars: Possible relic biogenic activity in Martian meteorite ALH84001. *Science*, 273(5277): 924–930.

Melosh, H.J. 1989. *Impact Cratering: A Geologic Process*. Oxford University Press, New York, pp. 126–162.

Melosh, H.J. 2003. Exchange of meteorites (and life?) between stellar systems. *Astrobiology*, 3(1): 207–215.

Miller, S.L. and Bada, J.L. 1988. Submarine hot spring and the origin of life. *Nature*, 334: 609–610.

NASA. 2013. Missions to Mars. http://www.nasa.gov/mission_pages/mars/missions/index.html.

NASA-SP4212. On Mars: Exploration of the red planet: 1958–1978 (electronic document). http://history.nasa.gov/SP-4212/contents.html.

Navarro-González, R., Navarro, K.F., de la Rosa, J. et al. 2006. The limitations on organic detection in Mars-like soils by thermal volatilization–gas chromatography–MS and their implications for the Viking results. *Proceedings of the National Academy of Sciences*, 103(44): 16089–16094.

Nisbet, E.G. and Sleep, N.H. 2001. The habitat and nature of early life. *Nature*, 409: 1083–1091.

Nna-Mvondo, D. and Martínez-Frías, J. 2007. Review komatiites: From Earth's geological settings to planetary and astrobiological contexts. *Earth, Moon, and Planets*, 100(3–4): 157–179.

Novoselov, A.A., Serrano, P., Forancelli Pacheco, M.L.A. et al. 2013. From cytoplasm to environment: The inorganic ingredients for the origin of life. *Astrobiology*, 13(3): 294–302.

Nyquist, L.E., Bogard, D.D., Shih, C.-Y. et al. 2001. Ages and geological histories of Martian meteorites. In *Chronology and Evolution of Mars*, Kallenbach, R., Geiss, J., Hartmann, W.K. (eds.), Vol. 6. Kluwer Academic Publishers, Dordrecht, the Netherlands, pp. 105–164.

Ozima, M. and Zahnle, K. 1993. Mantle degassing and atmospheric evolution: Noble gas view. *Geochemical Journal*, 27: 185–200.

Pontefracti, A., Osinski, G.R., Lindgren, P. et al. 2012. The effects of meteorite impacts on the availability of bioessential elements for endolithic organisms. *Meteoritics and Planetary Science*, 47(10): 1681–1691.

Rothschild, L.J. and Mancinelli, R.L. 2001. Life in extreme environments. *Nature*, 409: 1092–1101.

Rummel, J.D., Stabekis, P.D. Devincenzi, D.L. et al. 2002. COSPAR's planetary protection policy: A consolidated draft. *Advances in Space Research*, 30(6): 1567–1571.

Schaber, G.G. 2005. The U.S. Geological Survey, Branch of Astrogeology—A chronology of activities from conception through the end of project Apollo (1960–1973). US Geological Survey Open-File Report, US Department of the Interior, US Geological Survey, pp. 1–341.

Simoneit, B.R.T. 2004. Biomarkers (molecular fossils) as geochemical indicators of life. *Advances in Space Research*, 33(8): 1255–1261.

Smith, M.R. et al. 1984. Petrogenesis of the SNC (Shergottites, Nakhlites, Chassignites) meteorites: Implications for their origin from a large dynamic planet, possibly Mars. In *Proceedings of the 14th Lunar and Planetary Science Conference, Part 2. Journal of Geophysical Research*, 89(Supplement): B612–B630.

Soffen, G.A. 1997. Astrobiology from exobiology: Viking and the current Mars probes. *Acta Astronautica*, 41: 609–611.

Stoker, C.R., Cannon, H., Dunagan, S. et al. 2005. The 2005 MARTE robotic drilling experiment in Rio Tinto Spain: Objectives, approach, and results of a simulated mission to search for life in the Martian subsurface. *Astrobiology*, 8(5): 921–945.

The Telegraph. 2013. http://www.telegraph.co.uk/news/obituaries/9900905/David-McKay.html.

Tikkanen, A. 2010. McKay, David Stewart. http://universalium.academic.ru/252574/McKay_David_Stewart.

Treiman, A.H. 1995. SNC: Multiple source areas for Martian meteorites. *Journal of Geophysical Research*, 100: 5329–5340.

Treiman, A.H. 2005. The Nakhlite meteorites: Augite-rich igneous rocks from Mars. *Chemie der Erde*, 65: 203–270.

Treiman, A.H., Gleason, J.D., and Bogard, D.D. 2000. The SNC meteorites are from Mars. *Planetary and Space Science*, 48(12–14): 1213–1230.

Williams, M.E., Grotzinger, J.P., Dietrich, W.E. et al. 2013. Martian fluvial conglomerates at Gale Crater. *Science*, 340(6136): 1068–1072.

Wallis, M.K. and Wickramasinghe, N.C. 1995. Role of major terrestrial cratering events in dispersing life in the solar system. *Earth and Planetary Science Letters*, 130: 69–73.

CHAPTER 15

Elusive Definition of Life

A Survey of Main Ideas

Radu Popa

CONTENTS

15.1 INTRODUCTION

Reasons for defining life are wide ranging. At the pragmatic level, we cannot leave blank pages in a dictionary, especially on such important subject. In academia, the learning process requires conclusive definition on the path to understand living systems. In astrobiology, we seek *out of this world* life forms and a definition that is at the same time open minded, scientifically accurate, and experimentally verifiable. Technical (i.e., unambiguous and quantitative) definitions of life are needed to evaluate progress in the field of artificial life. Some realities (such as energy dissipative storms, GAIA, socioeconomical systems, the Internet, science, spirituality, and art) show properties of living systems. Some of them are evolving toward life, while others may be alive already. A clear life definition would help analyze them.

No definition of life ever became widely accepted. The type of education, preexisting prejudices about life, spirituality, and the intelligence of a person shape the definition he or she eventually accepts and determines which aspects of life make more sense and ought to be included in a definition. Explaining life and its origin is rich in paradoxes and far from definitive (Bedau, 1998). From a theoretical and technical standpoint, life may only exist in systems that are dynamic and very complex (Prigogine and Stengers, 1984; Kauffman, 1993). Thus, asking that a definition for life should be at the same time brief, popular, and scientifically accurate is wrong. Most dictionary-like abbreviated definitions or colloquial definitions of life are too truncated or too simplistic to be acceptable. They may serve but a segment of the public and certainly not the specialist. An acceptable definition for life cannot be constrained for size or simplistic on purpose. It will be in the form of a description of the essence and fundamental properties of life and will be tailored to a specific purpose and target audience. It is also possible that a definition of life is not something that a person with inadequate academic training can be briefed about in a rush, that is, "Life is not something that can be told about, it may have to be taught."

This chapter analyzes opinions and achievements in the history of life definitions and historical ways of explaining life and gives criteria for producing a life definition. It also distinguishes between *life* (in general) and *living entities* (in particular), as well as between *life at the individual level* and *life at the collective level*. Universal features of life are analyzed that are necessary and sufficient for producing technical and non-Earth-centric definitions. Poetic, insubstantial, or deliberately truncated statements such as "Life is a symphony of dynamic processes," "Life is creation," "It's alive if it can die," "I will know whether it is alive when I see it," and "Life is the ability to communicate" are not analyzed here.

15.2 CHRONOLOGICAL COLLECTION OF DIVERSE LIFE DEFINITIONS

The list presented in the succeeding text is not comprehensive nor gives sufficient credit to various authors. This collection emphasizes main trends for defining life and the diversity of life definitions. Bibliographic references for these definitions can be found in Popa (2004). More comprehensive lists of life definitions can be found in Pályi et al. (2002) and Koh and Ling (2013).

- Life is a power, force, or property of a special and peculiar kind, temporarily influencing matter and its ordinary force, but entirely different from, and in no way correlated with, any of these. (Beale, 1871)

- Living things are peculiar aggregates of ordinary matter and of ordinary force which in their separate states do not possess the aggregates of qualities known as life. (Bastian, 1872)

- If I had to define life in a single phrase … I should say: Life is creation. (Bernard, 1878)

- No physiology is held to be scientific if it does not consider death an essential factor of life … Life means dying. (Engels ca., 1880)

- The broadest and most complete definition of life will be the continuous adjustment of internal relations to external relations. (Spencer, 1884)

- It is the particular manner of composition of the materials and processes, their spatial and temporal organization which constitute what we call life. (Putter, 1923)

- A living organism is a system organized in hierarchical order … of a great number of different parts, in which a great number of processes are so disposed that by means of their mutual relations with wide limits with constant change of the materials and energies constituting the system and also in spite of disturbances conditioned by external influences, the system is generated or remains in the state characteristic of it, or these processes lead to the production of similar systems. (Von Bertalanffy, 1933)

- Life has the following characteristics: (1) character of animal or plant manifested to life by the metabolism, growth, reproduction, and internal powers of adaptation to the environment; (2) vital force distinguished from inorganic matter; (3) experience of animal from birth to death; (4) conscious existence; (5) of being alive; (6) duration of life; (7) individual experience; (8) manner of living; (9) life of the company; (10) the spirit; and (11) a duration of similarity. (*Webster's International Dictionary*, 1934)

- Life is replication plus metabolism. Replication is explained by the quantum-mechanical stability of molecular structures, while metabolism is explained by the ability of a living cell to extract negative entropy from its surroundings in accordance to the laws of thermodynamics. (Reformulated by Dyson (1997) from Schrödinger (1944))

- The essential criteria of life are twofold: (1) the ability to direct chemical change by catalysis; (2) the ability to reproduce by autocatalysis. The ability to undergo heritable catalysis changes is general, and is essential where there is competition between different types of living things, as … in the evolution of plants and animals. (Alexander, 1948)

- Life is not one thing but two, metabolism and replication, …, that are logically separable. (Von Neumann, 1948)

- Life is potentially self-perpetuating open system of linked organic reactions, catalyzed stepwise and almost isothermally by complex and specific organic catalysts which are themselves produced by the system. (Perrett, 1952)

- Life is the repetitive production of ordered heterogeneity. (Hotchkiss, 1956)

- The three properties "mutability, self-duplication and heterocatalysis" comprise a necessary and sufficient definition of living matter. (Horowitz, 1959)

- Any system capable of replication and mutation is alive. (Oparin, 1961)

- Life is "a hierarchical organization of open systems." (Von Bertalanffy, 1968)

- Life is "structural hierarchy of functioning units that has acquired through evolution the ability to store and process the information necessary for its own reproduction." (Gatlin, 1972)

- Life is a metabolic network within a boundary (Maturana and Varela, 1973; reformulated by Luisi 1993). All that is living must be based on autopoiesis, and if a system is discovered to be autopoietic, that system is defined as living; i.e. it must correspond to the definition of minimal life. (Maturana and Varela, 1973)

- Living organisms are distinguished by their specified complexity. (Orgel, 1973)

- Criteria of living systems: "metabolism, self-reproduction and spatial proliferation. Their more complicated kinds have also the ability of mutation and evolution too." (Gánti, 1974)

- We regard as alive any population of entities which has the properties of multiplication, heredity and variation. (Maynard-Smith, 1975)

- Life is the property of plants and animals which makes it possible for them to take in food, get energy from it, grow, adapt themselves to their surrounding and reproduce their kind. It is the quality that distinguishes an animal or plant from inorganic matter. Life is the state of possessing this property. (Webster, 1979)

- Life is that property of matter that results in the coupled cycling of bio-elements in aqueous solution, ultimately driven by radiant energy to attain maximum complexity. (Folsome, 1979)

- Living units are viewed as objects built up of organic compounds, as dissipative structures or at least dynamic low entropy systems significantly displaced from thermodynamic equilibrium. (Prigogine, 1980; Prigogine and Stengers, 1984; reformulated by Korzeniewski 2001)

- The sole distinguishing feature, and therefore the defining characteristic, of a living organism is that it is the transient material support of an organization with the property of survival. (Mercer, 1981)

- A living organism is defined as an open system which is able to fulfill the condition: it is able to maintain itself as an automaton … the long-term functioning of automata is possible only if there exists an organization building new automata. (Haukioja, 1982)

- Replication—a copying process achieved by a special network of inter-relatedness of components and component-producing processes that produces the same network as that which produces them—characterizes the living organism. (Csanyi and Kampis, 1985)

- Life is synonymous with the possession of genetic properties. Any system with the capacity to mutate freely and to reproduce its mutation must almost inevitably evolve in directions that will ensure its preservation. Given sufficient time, the system will acquire the complexity, variety and purposefulness that we recognize as alive. (Horowitz, 1986)

- Life is characterized by maximally-complex determinate patterns … life is matter that learned to recreate faithfully what are in all other respects random patterns. (Katz, 1986)

- Living system is an open system that is self-replicating, self-regulating, and feeds on energy from environment. (Sattler, 1986)

- Animate, and only animate matter can be said to be organized, meaning that it is a system made of elements, each one having a function to fulfill as a necessary contribution to the functioning of the system as a whole. (Lifson, 1987)

- The characteristics that distinguish most living things from nonliving things include a precise kind of organization, a variety of chemical reactions we term metabolism, the ability to maintain an appropriate internal environment even when the external environment changes (a process referred to as homeostasis), movement, responsiveness, growth, reproduction and adaptation to environmental change. (Vilee et al., 1989)

- Life is the ability to communicate. (de Loof, 1993)

- Life is an expected, collectively self-organized property of catalytic polymers. (Kauffman, 1993)

- Life is a self-sustained chemical system capable of undergoing Darwinian evolution. (NASA's working definition of life; Joyce (1994, 2002))

- Life may be described as "a flow of energy, matter and information." (Baltscheffsky, 1997)

- The existence of the dynamically ordered region of water realizing a boson condensation of evanescent photons inside and outside the cell can be regarded as the definition of life. (Jibu et al., 1997)

- Living organisms are systems characterized by being highly integrated through the process of organization driven by molecular (and higher levels of) complementarity. (Root-Bernstein and Dillon, 1997)

- A living entity is defined as "a system which owing to its internal process of component production and coupled to the medium via adaptative changes which persists during the time history of the system." (Luisi, 1998)

- Life is ... a recursive (self-producing and self-reproducing) organization where dynamic and informational levels are mutually dependent. (Bergareche and Ruiz-Mirazo, 1999)

- Life is defined as "a material system that can acquire store, process, and use information to organize its activities." (Dyson, 2000)

- Life is defined as a system of nucleic acid and protein polymerases with a constant supply of monomers, energy and protection. (Kunin, 2000)

- Life is (1) composed of particular individuals, that (2) reproduce (which involves transferring their identity to progeny) and (3) evolve (their identity can change from generation to generation). A living individual is defined as a network of inferior negative feedbacks (regulatory mechanisms) sub-ordinated to (being at service of) a superior positive feedback (potential of expansion of life). (Korzeniewski, 2001)

- A living system occupies a finite domain, has structure, performs according to an unknown purpose, and reproduces itself. (Sertorio and Tinetti, 2001)

- The characteristics of artificial life are emergence and dynamic interaction with the environment. (Yang et al., 2001)

- Life is the "symphony" of dynamic and highly integrated algorithmic processes which yields homeostatic metabolism, development, growth, and reproduction. (Abel, 2002)

- Life is the process of existence of open non-equilibrium complete systems, which are composed of carbon-based polymers and are able to self-reproduce and evolve on basis of template synthesis of their polymer components. (Altstein, 2002)

- Any living system must comprise four distinct functions: (1) increase of complexity; (2) directing the trends of increased complexity; (3) preserving complexity; and (4) recruiting and extracting the free energy needed to drive the three preceding motions. (Anbar, 2002)

- Life as a system capable of (1) self-organization; (2) self-replication; (3) evolution through mutation; (4) metabolism; and (5) concentrative encapsulation. (Arrhenius, 2002)

- Life is … a self-sustained molecular system transforming energy and matter, thus realizing its capacity of replication with mutations and anastrophic evolution. (Baltcheffsky, 2002)

- Life is … a set of symbiotically-linked molecular engines, permanently operating out of equilibrium, in an open flow of energy and matter, although recycling a great deal of their own chemical components, through cyclic chemistry. (Boiteau, 2002)

- Life is a chemical system capable of transferring its molecular information independently (self-reproduction) and also capable of making some accidental errors to allow the system to evolve (evolution). (Brack, 2002)

- Life is what the scientific establishment (probably after some healthy disagreement) will accept as life. (Friedman, 2002, paraphrasing Theodosius Dobzhansky)

- Life is matter that makes choices, binds time and breaks gradients. (Guerrero, 2002)

- Living beings are complex functional systems. Life is an abstract concept describing properties of cells, concrete objects. Life is the process manifested by individualized evolutionary metabolic systems. The functions, which are called life, are: metabolism, growth, and reproduction with stability through generations. (Guimarães, 2002)

- Life is energetic-dependent chemical cyclic process which results in increasing of functional and structural complexity of living systems and their inhabited environment. (Gusev, 2002)

- Life is simply a particular state of organized instability. (Hennet, 2002)

- Life is synonymous with the possession of genetic properties, i.e., the capacities for self-replication and mutation. (Horowitz, 2002)

- Life is a system which has subjectivity. (Kawamura, 2002)

- Life is metabolism and proliferation. (Keszthelyi, 2002)

- Life is a new quality brought upon an organic chemical system by a dialectic change resulting from an increase in quantity of complexity of the system. This new quality is characterized by the ability of temporal self-maintenance and self-preservation. (Kolb, 2002)

- Life is a high-organized form of the intensified resistance to spontaneous processes of destruction developing by means of the expedient interaction with the environment and regular self-renovation. (Kompanichenko, 2002)

- Any system that creates, maintains and/or modifies dissymmetry is alive. (Krumbein, 2002)

- Life is the form of existence of the substance capable of self-reproduction and maintenance of permanent metabolism with the environment. (Kulaev, 2002)

- A living entity is an ensemble of molecules which exhibit spatial organization and molecular-informational feedback loops in utilization of materials and energy from the environment for its growth, reproduction and evolution. (Lahav and Nir, 2002)

- It's alive if it can die. (Lauterbur, 2002)

- From a chemical point of view, life is a complex autocatalytic process. This means that the end products of the chemical reactions in a living cell (nucleic acids, polypeptides and proteins, oligo- and polysaccharides) catalyze their own formation. From a thermodynamic point of view, life is a mechanism which uses complex processes to decrease entropy. (Markó, 2002)

- Life is an attribute of living systems. It is continuous assimilation, transformation and rearrangement of molecules as per an in-built program in the living system so as to perpetuate the system. (Nair, 2002)

- Any definition of life that is useful must be measurable. We must define life in terms that can be turned into measurables, and then turn these into a strategy that can be used to search for life. So what are these? (a) structures; (b) chemistry; (c) replication with fidelity and (d) evolution. (Nealson, 2002)

- Life is "a system which can reproduce itself using genetic mechanisms." (Noda, 2002)

- Life as a structurally stable negentropy current supported by the self-correction for the biological hereditary genetic code ... providing an energy inflow. (Polishchuck, 2002)

- Life is instantiated by the objects that resist decay by means of constructive assimilation. (Rizzotti, 2002)

- Living systems as those that are: (1) composed of bounded micro-environments in thermodynamic equilibrium with their surroundings; (2) capable of transforming energy to maintain their low-entropy states; and (3) able to replicate structurally distinct copies of themselves from an instructional code perpetuated indefinitely through time despite the demise of the individual carrier through which it is transmitted. (Schulze-Makuch et al., 2002)

- Life is a form of matter organization that is energetically and informationally self-supported, with a good capacity of self-instruction and creation. (Scorei, 2002)

- Life is a historical process of anagenetic organizational relays. (Valenzuela, 2002)

- Life: A population of functionally connected, local, non-linear, informationally-controlled chemical systems that are able to self-reproduce, to adapt, and to coevolve to higher levels of global functional complexity. (Von Kiedrowski, 2002)

- A living system is one capable of reproduction and evolution, with a fundamental logic that demands an incessant search for performance with respect to its building blocks and arrangement of these building blocks. The search will end only when perfection or near perfection is reached. Without this build-in search living systems could not have achieved the level of complexity and excellence to deserve the designation of life. (Wong, 2002)

- The existence of a genome and the genetic code divides living organisms from non-living matter. (Yockey, 2002)

15.3 PHILOSOPHY OF A LIFE DEFINITION

The equivalent of a race presently exists for intellectual credit from producing "the ultimate definition of life" (Koh and Ling, 2013). One out of the box view is that since we cannot settle on any definition, we should assume that we do not need one and behave accordingly. Apart from being a sophism, this opinion does not help fields such as astrobiology, artificial life, and the legal system, where technical, quantitative, and unambiguous definitions are needed. Furthermore, we cannot choose what should be defined based on how difficult the subject is. Aiming at shortcutting years of disputes about life, Trifonov (2011) has proposed collecting definitions, identifying most frequently cited features, distinguishing between causal and derived properties, and assembling them in a hopefully "most commonly accepted definition" (Popa, 2012). Yet, new definitions are constantly produced and modified for various purposes and audiences. Furthermore, personal opinions about life are quite inflexible and the palette of levels of understanding of life is broad. Therefore, no single definition (including one established through a popularity contest) will ever satisfy everybody. The weight each author gives to various aspects of life also plays an important role in accepting, or not, emphases from definitions of others. Last but not least, descriptions of life evolve with each person's understanding of it, which has led to a historical collection of life definitions that is larger than the number of people producing them.

Philosophical avenues about how life should be explained include two key directions:

1. Importance given to collective properties of a system (i.e., holism) relative to properties of its individual parts (i.e., minimalism) in the functioning of living systems

2. Emphasis on causality relative to probability in the production of life's organization and in shaping the origin and evolution of life (Figure 15.1)

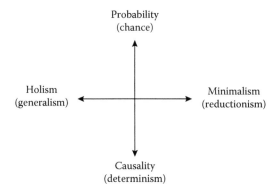

FIGURE 15.1   Opposite views about explaining the living state.

Other theoretical and philosophical avenues about life (such as the cause of life, mechanisms of life organization, the basic functional mechanism of all living systems, and the role of life in nature) are summarized in Section 15.6.

15.3.1 Holism (or Generalism)

Holism (or generalism) considers that life is a collective property, absent from its individual parts. Suggestive for this view is the aphorism *nothing is alive in a cell except the whole of it* (Olomucki, 1993). Prigogine (1980) proposed that living systems have a derived property called *coherence*; that is, even though interactions between individual parts from living cells (i.e., atoms and molecules) occur at the angstrom level, the system behaves like it was permanently informed about what happens at small scale and uses it for a grander purpose. Benefits of microscale transformations may not be obvious to the analytic observer, and *dissection* of all individual part properties (irrespective of how thorough) cannot explain life.

15.3.2 Minimalism (or Reductionism)

Minimalism (or reductionism) is an attempt to *distil* life to simple mechanisms or forces. The extreme reductionist will believe that what you see is what you get and that a system can be fully explained from properties of its individual parts or their direct interactions. According to the *onion heuristic* approach, by *peeling off* layer after layer the most derived and least important features of living systems, we should be capable to reveal the most basic mechanism of life and describe the functioning of the most primitive life forms.

Very few authors fully support either holism or minimalism at the total exclusion of the other. Yet, discussing these extremes helps one form a balanced opinion about life.

15.3.3 Causality (or Determinism)

In this view, things (life included) always happen due to underlying cause(s), which are more important than chance. This view may take many forms. For example, *vitalism* is the medieval belief that life is caused by a transcendental principle of organization or an esoteric force called *vital force* or *life-giving force*. Some philosophies have called this force *entelechy*. In *spiritualism*, life is produced, and possibly maintained, by *supernatural forces* or *divine will* beyond our power of observation and comprehension. Most scientists now consider *vitalism* an obsolete view (akin to astrology) and agree that the functioning of living systems has nothing to do with *spirituality*. Modern science has not fully replaced vitalism, but rephrased it in a rationalistic form. For example, scientists now propose that causality (in various forms such as *physical driving forces, cause–effect deterministic relationship*, and *derived properties*) was important for the origin of life.

15.3.4 Probability (or Chance)

In this view, life is the outcome of random chanceful circumstances, and no hidden causes, driving forces, or underlying ordering principles are needed to explain it. The origin of life was the outcome of an (admittedly) rare, yet plausible, set of environmental conditions.

In this case, the evolutionary history of life was controlled by chance, changes in the environment, and selection of the fittest.

Neither causality nor probability is 100% right or wrong. It is likely that both played a role in the origin of life, albeit relative contributions may have shifted during early evolution.

15.4 TECHNICAL FORM OF A LIFE DEFINITION

When produced, life definitions take one of few technical forms:

- *Parametric* (i.e., a collection of general and particular properties)

- *Cybernetic* (i.e., a numerical and/or quantitative analysis of life properties)

- *Cellularist/genetic* (i.e., a list of most important features of earthly biological life)

- *Non-Earth-centric* (i.e., features believed to be present in all forms of life in the universe)

15.5 LANGUAGE OF A LIFE DEFINITION

The terminology used by a life definition is often hotly disputed. Specialists seek technical definitions with precise scientific terms. Philosophers favor context-providing, causality- and purpose-addressing definitions and favor universal properties described in broad terminology. The general public likes concise definitions, a colloquial language, and expect for a definition to match the description of what they already believe living systems are.

15.6 LIFE'S CAUSALITY AND PATH TO ORGANIZATION

Next, we summarize hypotheses on mechanisms helping life originate, exist, and evolve.

15.6.1 Life's Drivers and Role in Nature

Living entities are energy dissipative systems, a class of dynamic systems otherwise common in the abiotic nature. Relative to other dynamic systems, living systems show notable upgrades regarding complexity, energy use, regulation, information, and evolution. Yet, much can be gained by making comparisons between living systems and abiotic energy dissipative systems, especially with regard to driving forces, the role of life in nature, basic functional mechanisms, and the transition from nonlife to life.

Two avenues help find life's cause, that is, the driving force (or physical attractor):

1. We can inquire if certain collective properties of living systems interact with laws of nature in ways that individual parts do not and lead to increased organization.

2. We can deduce from known physical drivers of abiotic energy dissipative systems.

Life driver(s) may be internal or external. Ideas fitting the description of internal drivers are as follows:

- Vital force, life-giving force, or entelechy (many medieval sources)

- Internal search for perfection or near perfection (Wong, 2002)

- Intrinsic tendency toward self-organization (Hazen, 2002)

- Internal tendency opposite in sense to entropization (Valenzuela, 2002)

- Selfish gene, a molecule created by chance and self-replicated (Dawkins, 1989)

- Self-controlled order, a non-Earth-centric extension of the selfish gene (Popa, 2004)

With regard to external drivers, analogies are often made with systems such as fire, growing crystals, convection cells, tornadoes, energy dissipative storms, molecular self-assembly, and evolution of ecosystems. These dynamic systems become organized because gradients of free energy build up in the environment and cannot be dissipated by turbulence or simpler forms of organization. For example, hurricanes (a type of energy dissipative storm) form when more heat accumulates at the ocean' surface than it can be dissipated by atmospheric convection cells. Hurricanes are more complex than convection cells yet more efficient at dissipating heat. Another example is the evolution of ecosystems toward *climax* (a system state of high adaptation). As they change toward climax, ecosystems also show increase in energy flow density (i.e., larger energy flux) and thus become better at dissipating free energy. Since both hurricanes and ecosystems also show lifelike properties (e.g., self-organization), it was proposed that their physical drivers also play a role in the origin and evolution of cellular life.

It is known that the organization of energy dissipative systems is linked with the natural tendency of an energy gradient to be dissipated and the availability of a practically infinite sink for the by-product of this transformation (i.e., heat). Thus, the reason energy gradients form, the reason for the direction of the second law of thermodynamics, and the cause of the infinite size of the heat sink may also *cause* life. Chaisson (2003) supports that energy dissipative systems (which includes life) are caused, among others factors, by the expansion of the universe. The increase in volume is exponential, and the production of free energy (a secondary outcome of matter being degraded in stars) and of heat (produced by various systems) cannot keep pace with the overall cooling. This results in an infinite heat sink capable of absorbing all the heat energy dissipative systems may produce. In this view, the universal product of life is heat and disorder (i.e., entropy), and life cannot originate in an environment lacking free energy, in a closed environment, or in an environment where the heat produced cannot be dissipated efficiently (Cimpoiasu and Popa, 2012). Aiming at summarizing the origin of life's order in a general principle, Peer Bak has proposed that a fourth law of thermodynamics should exist: *The flow of energy from a source to a sink through an intermediate system tends to order the intermediate system for the purpose of facilitating the energy flow* (modified after Bak 1996).

15.6.2 Conditions for Life's Organization

Once a physical driver exists, certain conditions are required for a system to become self-organized. In a generalization, Bonabeau and his collaborators (1999) have identified three such key conditions: (1) strong dynamic nonlinearity, (2) a balance of exploitation and exploration, and (3) multiple interactions. In the special case of the origin of life, other important requirements are sufficient environmental diversity to allow producing a complex system, a system composition that is adapted to the most easily available building blocks from the environment, fair statistical odds for a structure or mechanism to originate, reasonably slow and steady pace of environmental change to give a system sufficient time to evolve, a modular design to create a practically infinite number of construction variants, and functional continuum between simple and complex organization. Last but not least, the energy extracted by the system has to be large enough to compensate the system's maintenance costs. According to disequilibrium thermodynamics, energy dissipative order can only exist if the energy cost of maintaining internal order is less than the overall free energy degraded. The internal loss in entropy should be lower than the overall production of entropy by the system. From a broad perspective, systems such as stars, hurricanes, fire, and life are but temporary mechanisms of dissipating free energy and producing heat and entropy. When competition between two or more forms of organization exists in the same space and free energy builds up, the energy dissipative system that is the most efficient at seizing energy, matter, and space and at using and dissipating free energy will also have a competitive edge over the others and its probability of existence will increase.

15.6.3 Mechanisms of Life Organization

With regard to the primordial mechanism for life's organization, we have to cite the following:

- The *self-organization principle* formulated by Ashby (1947)

- The *origin of order from noise* by Von Foerster (1960)

- The concepts of *order through fluctuations* (Nicolis and Prigogine, 1977) and *order out of chaos* (Prigogine and Stengers, 1984)

In many respects, these principles are similar. In layman terms, the universe "abhors" order, free energy, and gradients of energy across space and time. Energy dissipative mechanisms (life included) are efficient at degrading free energy, increasing entropy, and *smoothing out* the spread of entropy across space and time. The organization of a biological system is a physical response to the need to eliminate energy gradients across space. The need to eliminate energy gradients across time causes competition between energy dissipative systems and adaptive evolution and adjusts energy dissipation to energy availability or evolution toward climax.

Because in energy dissipative systems the energy flow is coupled with a *self-controlled* (*self-rewarding*) *order*, a positive feedback is created in which the level of order increases with the energy flux, while in turn more order leads to larger energy flux. The system's growth stops when the energy cost of maintaining order becomes too high relative to the availability of free energy. For this reason, it is proposed that *the key mechanism in the functioning of all ordered energy dissipative systems (including life) is symbiosis between an energy flux through a narrow path and a self-rewarding order capable of assisting the energy flow*. As the overall energy flow increases, the system becomes more efficient at harvesting free energy from the environment, and as order increases, the system seizes more space and materials than its competitors. Such systems of organization become favored during competition for space and for building materials relative to other systems of organization that are less efficient at dissipating energy.

Based on the earlier text, *living systems are self-organized and adaptive mechanisms selected to exist because they are better at dissipating free energy, and the level of complexity life forms achieve is controlled (among other factors) by the availability of free energy.*

15.6.4 Fuzzy Boundary between Nonlife and Life

Describing the threshold between lifeless systems and living systems helps produce technical definitions. The main issues here are as follows:

1. Do we believe it or not that living systems originate in nonliving systems?

2. Did a sufficiently sharp threshold existed during this transition?

3. If many major changes have occurred during this evolution, which threshold should we select as the most important, because various authors have very different ideas about what living system mean?

Since we did not witness the origin of life and no artificial life has ever been created, we cannot be 100% sure that living systems can evolve from nonliving systems. Yet, a large body of evidence now makes scientists confident that (given appropriate conditions and sufficient time) lifeless networks can become alive. If the origin of life was a *phenomenological continuum* (Bedau, 1998), finger-pointing a very specific nonlife/life boundary may be difficult. This is called *the dilemma of endless gradualism* (Popa, 2010) and is akin to trying to distinguish black from white in a long and smooth series of grays. A potential solution to this dilemma is to seek for revolutionary jumps or phase transitions in a system's state, with regard to the level of organization, function, or consequences of change. It was often proposed that life has originated as a smooth evolution with sudden jumps. This view about the origin of life is called *punctuated gradualism* (Popa, 2010). Opinions about *the key achievement* during prebiotic evolution that should be acknowledged as a *decisive nonlife to life threshold* vary. Potential candidates include the first symbiosis between a catalyst and a gene, the first selfish information, the first autocatalytic network, the adaptive evolution, or the transition from automata to informata (Section 15.8.5).

15.7 WHAT IS TO BE DEFINED?

When definitions are produced, great confusion may come from insufficient clarity about what is to be defined. Life-related definitions vary in both essence and form whether we describe *life* (in general), *living entities* (in particular), *types of life* (or forms of life), or *life-relevant system states* (such as being *alive, dormant, dead,* or *not alive*). For example, a *living entity* should be described as a concrete system with specific internal components and properties, while *life* (in general) is a theoretical concept, a specific property of living systems, or the presence, or sum, of all living systems from a specified environment.

Living entities are often described based on their composition, dimensions, structure, function, and information capacity, while *life* (in general) is composition independent and dimensionless and has the capacity for adaptive evolution, which some living entities do not. When describing the composition of the earthly biological life forms, we use features such as water, carbon, proteins, and DNA, but when describing *life* in general, we use non-Earth-centric parameters such as energy, information, order, and complexity, and we may need to include considerations about origin, evolutionary trend, and the role played by life in nature.

15.8 UNIVERSAL FEATURES OF LIFE

Definitions are generally assembled from general to particular. They first give the broad category the object or subject to be defined belongs to, and then they provide particular properties that are necessary and sufficient to singularize it. In this section, we discuss universal features of life that can be used in parametric and non-Earth-centric definitions. Some of the most frequently cited features of life concern energy, organization, regulation and repair, information, and evolution.

15.8.1 Energy

Free energy is the potential to do work. Living entities maintain organization, which means they do work, by using part of the free energy they dissipate. With regard to how energy is used, the four key features of living systems are *catalysis, feedbacking free energy, energy currencies*, and *energy reserves*. Catalysis (in the broad sense) is the ability to speed up transformations and it is often used as a means to divert energy from a source on a desired path. All living systems that we know show catalysis, either because they have to compete for energy with other systems and abiotic processes or because their functioning requires high energy intensity that can be achieved by channeling energy through a narrow, low-resistance, path. Catalysis is not exclusive to life; in some form or another, catalysis (in the broad sense described earlier) or autocatalysis also exists in energy dissipative systems that are not alive, for example, in fire, iron rusting, the growth of crystals, and molecular self-assembly. Using catalysis alone to define life is confusing because most viruses do not contain catalysts, yet they are closer to what we call life than fire or growing crystals. *Feedbacking free energy*, also called *reflexive activity* (Popa, 2004), is the diversion of part of an energy flow for the purpose of maintaining organization or any type of work needed by the system. This feature is widespread in cell-based living systems, absent in viruses and common in many lifeless energy dissipative systems such as convection cells, tornadoes, and energy dissipative storms. As far as we know it, no energy dissipative order may exist without this property; thus, all living systems must have it except dormant life and energy parasites. *Energy currencies* are energy-rich carriers made in a process called *energy transduction*, which is the transformation of one form of energy in another. Energy currencies are means of temporary energy storage and are used for supporting the system's need for work, for compatibility between external energy sources and the energy types needed by internal mechanisms, and for standardizing the type of energy used by various parts in a system. Apparently, no abiotic system has energy currencies. *Energy reserves* are long-term stocks of free energy, helping the system withstand long-term fluctuations of energy availability. In theory, if the environment is sufficiently homogeneous across space and time, a living system may exist without energy reserves. Perishability distinguishes what carriers are best as an *energy currency* (i.e., short half-life) and what can be used as an *energy reserve* (i.e., long half-life). Examples of energy currencies are ATP in biological cells, electricity in human society, and money in financial systems. Examples of energy reserves are sugars and fats in living cells, stockpiles of food and combustible materials in human society, and gold bullion, real estate, and other long-term assets in financial systems. None of these features (i.e., *catalysis, feedbacking free energy, energy currencies*, and *energy reserves*) is in itself diagnostic for life and some nonliving systems have one or more of these properties. Yet, cellular living systems and socioeconomical systems are the only energy dissipative systems showing all four of these features.

15.8.2 Organization

As a system parameter, organization has two aspects: *order* and *complexity*. *Order* measures how much of a system is nonrandom with regard to composition, spatial arrangement, relationship between parts, and behavior. The *order parameter* (Anderson, 1997) has limited applicability, but it is a useful tool for learning the meaning of order. *Complexity* measures the intricacy of organization and is the *meaningful information capacity* of a system or *virtual information* of the ordered part of a system (Cimpoiasu and Popa, 2012). No easy way exists to measure complexity, and results are often observer and method dependent (Edmonds, 1999). Great confusion happens when *complexity* is confused with *diversity*. *Diversity* is a property of all parts of a system (i.e., random and nonrandom), while complexity only describes the nonrandom (or ordered) part of a system. By definition, the random part of a system has zero complexity. Networks with edges (or flows) and nodes (or stocks) are compared using a parameter called *cyclomatic complexity* (McCabe, 1976). *Kolmogorov complexity*, or computational effort needed to describe a system's organization (Abel, 2002), has more general application. Combinations of various compression algorithms are sometimes used to detect order and evaluate the level of complexity (Corsetti and Storrie-Lombardi, 2003). The idea behind *Kolmogorov complexity* and compression algorithms is that complex information is difficult to compress. Two problems arising here are that

- No software can identify all rules of organization and hidden messages from a system.
- Random data may appear complex because it is hard to compress, while in fact, random data contain no order and should have zero complexity.

Kolmogorov analysis and compression algorithms will overestimate complexity in disorganized and compositionally diverse systems. Living systems should have increased order relative to their environment, but the level of order and complexity needed for a system to be alive is unclear. Also unclear is how perishable, or stable, organization should be.

15.8.3 Regulation and Repair

The capacity to regulate the functioning of a system or to repair it is a key requirement of life. Viruses have no regulation and repair; their information content is produced and maintained by the living cells they parasitize. The ability to control all functional parameters of a living system is called homeostasis. It is unclear how much regulation a system must have to be alive. Kauffman (1993) has proposed that only complex networks can be truly self-maintained, which implies that living systems can only exist if they are very complex. This leads to a dilemma regarding prebiotic evolution. It makes it difficult to explain how prelife networks evolved toward ever-increasing complexity and eventually life, if they were simpler than what a self-controlled living state requires. Most prebiotic networks should have become increasingly disorganized with time, rather than evolve toward more organization and eventually life. Yet, self-organization of abiotic energy dissipative systems proves that exceptions from this logic do exist.

In a definition of life, we can state that self-regulation has to be present at a level sufficient to counteract environmental challenges. This leads to the philosophical dilemma that what we define as alive may be context dependent; that is, a system is alive in one environment and not alive in another. Alternatively, the measure of being alive may be linked with the resilience of a system over time. It is obvious that regulation and self-repair alone are insufficient to define life. For life to exist, we have to assume that regulation and repair have to reach a level sufficient to keep the system state from drifting uncontrollably. It is proposed that nonlife to life transition occurred when the capacity for self-control became extensive enough to make the evolution of a prebiotic system toward increased organization irreversible in a given environment.

15.8.4 Evolution

Darwinian evolution is the property most frequently proposed as diagnostic for life. In general, this term is reserved to describe the evolution of biological life forms and artificial life simulations. In non-Earth-centric definitions, the term *adaptive evolution* is recommended. A frequent topic of discussion in origin of life classes is that adaptive evolution represents change of a system with time, but not all changes mean adaptive evolution. For example, *quasi states* (described in qualitative dynamics as successions of system states occurring in a specified sequence) are an example of change and of system evolution, but are not adaptive. The capacity of a system to assume various states in response to environmental stress is also not an example of adaptive evolution, but regulation, irrespective of how many virtual states has the system memorized or how many states can the system summon from its genetic bank. Spurious changes, which randomly appear and disappear, drift in a system's state and changes that are not memorized by the system are also not examples of adaptive evolution.

In life definitions, explaining the mechanism of *Darwinian* (or *adaptive*) *evolution* is less important than the history of evolution itself. For such analysis, we can use the term *evolution of evolution* (a seemingly pleonasm). It is fair to assume that during the origin of earthly biological life, more primitive forms of evolution (than Darwinian evolution) have existed. Examples are *molecular Darwinism* and *Lamarckian evolution*. *Molecular Darwinism* is the selection of molecules (on features such as dimension, configuration, or sequence) due to their fitness (i.e., activity or performance) during competition with other molecules or in response to environmental stress. *Molecular Darwinism* is thought to have existed in molecular networks that have preexisted life. Thus, albeit important for understanding the origin of life, *molecular Darwinism* is not important for defining life. Furthermore, *molecular Darwinism* only applies to chemical networks, while life may exist in other (nonchemical) forms as well. *Lamarckian evolution* means inheritance of characters that are acquired during the lifetime of an individual. Lamarckism is unsuitable for explaining the evolution of modern biological life, where information only flows from genotype to phenotype (i.e., from DNA to proteins). It is unclear whether this direction rule is a requirement of life in general or it is a feature specific to the biological life on Earth. The evolution of complex automata and cells from the hypothetical RNA world (where single molecules, presumably RNA, have served simultaneously as genes and enzymes) resembles *Lamarckian evolution*. It is interesting that in an RNA world, the boundaries between *Darwinism* and *Lamarckism* are difficult to distinguish. *Lamarckism* is also useful for explaining the evolution of collective life forms such as ecosystems, socioeconomical systems, and GAIA. *Note that GAIA is the hypothesis that our entire planet is alive.* Another, albeit fictional, case of *Lamarckian evolution* may have occurred in the living ocean Solaris (Lem, 1961).

The term evolution has to be used with caution in definitions because some forms of life show adaptive evolution but only at the collective level and in a time frame beyond the life expectancy of individuals. Most eukaryotic living entities on Earth do not have *Darwinian evolution*; their genetic succession or genetic information has it. *In a definition, it is recommended to attribute the property of adaptive evolution to life in general or to state that adaptive evolution may be present at the level of individuals, collective or line of descent.*

15.8.5 Information

The word information is used for various properties of living systems and often confusing to the nonspecialist. Dissimilar parameters containing this word include information content, information capacity, meaningful information, shared information, and residual information. Information capacity is a log of a total number of microstates and measures the number of choices needed to single out a system state from a virtual collection of microstates. To distinguish life from nonlife, we have to identify types of information and their place in prebiotic evolution. Based on the carrier of information, systems may contain *contextual information*, *nominative information*, and *symbolic information*:

1. *Contextual information* means that no dedicated component in a system is in *charge* of storing and sharing specific information or controls the system. The system changes in response to the environment and changes may be seen in later states but are not memorized.

2. *Nominative information* is stored by a specific carrier. Here, we distinguish between the following:

   a. *Explicit information* (also called analogical information), which is used as is without translation. Examples include RNA molecules from the hypothetical RNA world, which have functioned as enzymes and their sequence information replicated; mineral arrays that can be used as templates for crystal growth; and simple regulators that are used to control the state of a system. Systems that only have explicit information have one or a very limited number of potential states.

   b. *Cryptic information* is hidden in a code (or language) and requires translation to be used. Examples include the mRNA sequence information (which has to be translated by ribosomes) and written law (which requires literacy in order to be interpreted as intended). Relative to explicit information, cryptic information has the advantage of compressibility and can exist hidden as a backup virtual reality without interfering with the actual state of a system and used if needed. Systems using cryptic information may contain many potential system states in reserve.

3. *Symbolic information* is not recorded in the genetic makeup of an individual, may be evolved in only a subgroup of individuals from a genetic community, and may be meaningless to individuals from other groups of the same genetic community. Examples include social behavior, body language, spoken language, writing, and various forms of culture and religion.

Here, we inquire about the type of information marking the transition between lifeless systems and living systems. It is proposed that a system only using *contextual information* is too simple to be alive, while a system using *symbolic information* is alive already. Hence, the nonlife to life transition must have occurred during the evolution of *nominative information*. Based on the type of nominative information used, systems can be differentiated in *automata* and *informata*. In the field of the origin of life, *automata* are systems only using *analogic information* for functioning and inheritance, while *informata* use *encrypted information* or a mixture of analogic and encrypted information (Popa, 2004; Popa and Cimpoiasu, 2012). Biological life forms are informata that use explicit information as well.

It makes sense that during prebiotic evolution automata has preceded informata. The evolution of modern biological informata toward increased complexity uses Darwinian selection, but principles controlling the evolution of automata are different. Informata keep records (within physical limits) of historical changes in organization that were successful in the past and thus store (either as individuals or as a collective) a memory (or cache) of *virtual realities*. These *virtual realities* are states the system may assume, or responses the systems may have, in specific conditions. They are backup system states and, potentially, most appropriate responses to the environment (according to past experiences). In contrast, automata have no specialized means of remembering the past and thus very limited (if any) recollection of former states.

This makes informata better adapted at withstanding complex and variable environments (albeit informata are complicated and expensive to build and maintain). Automata are less complex than informata and their fitness increases in homogeneous environments. Automata contain no hidden information—that is, what you see is what you get—and are believed to show a variant of Lamarckian evolution. They can only survive stringent competition if they are open to innovation and if they continuously *reinvent the wheel*. This may explain why during prebiotic evolution, automata (i.e., molecular networks with no genes) have been replaced by informata, and all living systems on Earth are presently informata.

With regard to defining life, it is proposed that the first living systems were informata and that the nonlife to life transition has occurred when regulation in a system became more cryptic than explicit and when encrypted information was used for adaptive evolution.

15.9 RECOMMENDED STEPS FOR PRODUCING A LIFE DEFINITION

1. Settle on a philosophy of what life means to you and take in consideration what the target audience would accept as being alive. Most definitions fit in the virtual space from Figure 15.1.

2. Decide what has to be defined (life in general, living systems, being alive, type of life, etc.).

3. Commit to a technical form of the definition based on scope. Definitions may be parametric, cybernetic, cellularist/genetic, non-Earth-centric, or a combination.

4. Construct the definition from general to particular; for example, "A living entity is an energy dissipative system capable of catalysis, self-regulation, adaptive evolution, etc."

5. Reword using terminology that is accurate yet commensurate with the target audience.

15.10 CONCLUSIONS

In this chapter, we have analyzed the diversity of life definitions and theoretical and technical aspects when producing a life definition. No absolute or universally agreed upon definition of life exists. This is due to the complexity of life and the diversity of living systems but also to the fact that a life definition is a summary of each person' understanding of life. Life definitions are constantly produced and tailored to various professional purposes and target audiences. This chapter emphasizes that we do not define life, we describe it; we do not accept a definition of life, we understand it; and we are not told what life is, we are taught about it. Many scientific definitions of life exist that are sufficiently rigorous and comprehensive, for example, the 1994 NASA definition (see Section 15.3). Yet, a definition of life that one will eventually accept remains a reflection or how much understanding he or she has about the essence of life and its role in nature. The definitions that I, presently, prefer are as follows:

Living entities are energy dissipative systems using cryptic information for regulation and inheritance and capable of adaptive evolution at the level of individual, collective or line of descent.

Being alive is the state of expressing the features of living entities.

Life is a state of organization of energy dissipative systems, can manifest at the individual or collective level, and indicates that the capacity to express their attributes is virtual or present.

A type of life (or life form) represents all living entities with similar means of construction, function, and information storage.

GLOSSARY

Anagenesis: Evolution of species involving an entire population, rather than a branching event.

Anastrophe: A turning back or about. A change in the order of parts from a message or process.

Autopoiesis: The capacity of a system to reproduce or maintain itself. The literal translation (i.e., self-creation) is misleading.

Boson condensation (or Bose–Einstein condensate): State of matter of a dilute gas of bosons near 0 K, where bosons occupy the lowest quantum state and quantum effects become apparent at macroscopic scale (i.e., macroscopic quantum phenomena).

Compression algorithms: Algorithms used to reduce data to fewer bits.

Convection cells: A system of vertical circulation of fluids leading to heat dissipation. Earth's atmosphere contains a total of six convection cell systems at various latitudes.

Cybernetics: Interdisciplinary way of studying regulatory systems, their structures, constraints, and possibilities. In cybernetics, all systems are abstractions analyzed for adaptation, control, learning, cognition, emergence, communication, and other general features.

Disequilibrium (or nonequilibrium) thermodynamics: A branch of thermodynamics that deals with thermodynamic systems that are not in thermodynamic equilibrium.

Dissipative system: Thermodynamically open system operating out of, and often far from, thermodynamic equilibrium in an environment with which it exchanges energy and matter.

Energy dissipative storms: Large systems of air circulation, such as hurricanes and cyclones, organized in a way that dissipates energy from the Earth's surface (in most cases from the ocean surface) upward into the atmosphere.

Entropy: Often interpreted as a measure of disorder or progression toward thermodynamic equilibrium. In thermodynamics, entropy is the amount of heat energy absorbed by the disordered part of a system when temperature increases by 1 K.

Evanescent: Exponential decay.

GAIA hypothesis: Hypothesis proposing that all organisms interacting with their inorganic surroundings on Earth form a self-regulating, complex system that contributes to maintaining the conditions for life on the planet. GAIA theory proposes that the entire planet is a living system.

Heterocatalysis: A catalytic process resulting in a product different from the starting material.

Kolmogorov complexity: A measure of the computational effort needed to describe a subject (object, phenomenon, or data set).

Negentropy: Entropy exported by a system to keep its internal entropy low or change in a system toward increased order and lower entropy.

Parametric: The use of specified features or parameters to describe a subject.

Physical driver: Natural force or combination of forces driving a transformation or system change toward a specified direction.

Pleonasm: The use of more words than it is necessary for clear expression: examples are *black darkness* or *burning fire*. A form of redundancy.

Self-rewarding order: The capacity of a system of organization, dynamic system, or object to instruct its own reproduction or organization. Examples include selfish genes, autocatalytic RNA molecules, fire reinitiating fire, and autocatalysis of crystal growth.

Solaris: A 1961 Polish science fiction novel by Stanisław Lem. In this novel, the entire ocean of a planet was alive and capable of interaction and analysis. On Earth, we are accustomed with discrete living entities interacting and evolving in an abiotic environment. In Solaris, the entire environment was alive, without the need for discrete and independently evolving entities.

Sophism: A seemingly well-reasoned, plausible, or true argument (yet hiding a fallacy) used to deceive somebody.

REVIEW QUESTIONS

1. Identify errors in the following sentences:

   Living system can accumulate entropy inside because they import free energy from the outside.

   Automata are systems only using analogic information for regulation and evolution, while informata systems only use cryptic information.

   Living entities are entropy dissipative systems.

   Living entities are systems specialized on producing free energy from heat.

   Sugar and fats are examples of energy currencies.

2. Fill in the blanks in the following sentences:

   Living systems degrade free energy and dissipate _____ in the environment.

   In Holism, life is seen as a _____ property, absent from its individual parts.

   Energy is defined as the potential to _____.

   Living entities are _____ dissipative systems.

   Organization has two quantitative aspects: order and _____.

   Quasi state means a succession of various system _____ in a given sequence.

3. Is the following sentence accurate? *Complexity is the same with diversity.* Explain your choice.

4. Catalysis is one of the features of living systems. Abiotic energy dissipative systems such as crystal growth show catalysis, but most viruses do not. Yet viruses are considered closer to life than growing crystals. Explain this contradiction and propose a solution when producing a life definition.

5. Although viruses cannot repair themselves, some scientists believe they are alive. If viruses are alive, then are viroids (i.e., viruses without a protein shell), prions, and computer viruses alive as well?

6. When we describe living systems we sometimes assume that the level of self-regulation has to be sufficient to adjust the system to environmental challenges. This leads to the dilemma that a system may be considered alive in one environment and not alive in another. Is life an absolute state or a relative state?

7. Give your own definition of life or living system and explain the choices of general and particular features that you have selected.

8. Make a list of all (general and particular properties) that you believe are necessary to distinguish living systems from nonliving systems. In making this list, you can only use universal properties of life; that is, you cannot use any properties that may be particular to biological life on Earth (such as water, carbon, proteins, sugars, lipid membranes, DNA, RNA, ATP, chromosomes, and ribosomes). Use the list that you have produced to analyze whether the following are alive: socioeconomical systems, the Internet, culture, and religion.

9. Assuming that the socioeconomical system is alive, at which point in history would you consider was the most important to mark the transition from nonliving state to living state? You can use examples such as the invention of fire or the wheel, transition from stone age to bronze age, the advent of agriculture, the formation of first villages or city states, the invention of writing, the creation of the first legal system, the origin of Greek democracy, the organization of market economy and financial systems, and the invention of Internet.

10. According to Abel (2002), complexity may be measured through the computational effort required to describe the uncertainty reflected by the object or state. Is this correct? Explain.

REFERENCES

Books

Abel, D.L. Is life reducible to complexity? Chapter 1.2, pp. 57–71. In: *Fundamentals of Life*, Pályi, G., Zucchi, C., and Caglioti, L. (eds.). Paris, France: Elsevier, 2002.

Anderson, P.W. *Basic Notions of Condensed Matter Physics*. Reading, MA: Addison-Wesley, 1997.

Bak, P. *How Nature Works: The Science of Self-Organized Criticality*. New York: Copernicus, 1996.

Bonabeau, E., Dorigo, M., and Theraulaz, G. *Swarm Intelligence: From Natural to Artificial Systems*, pp. 9–11. New York: Oxford University Press, 1999.

Dawkins, R. *The Selfish Gene*. New York: Oxford University Press, 1989.

Edmonds, B. What is complexity?—The philosophy of complexity per se with application to some examples in evolution. In: *The Evolution of Complexity*, Heylighen, F. and Aerts, D. (eds.). Dordrecht, the Netherlands: Kluwer, 1999, pp. 1–8.

Hazen, R.M. Emergence and the origin of life, Chapter II.5, pp. 277–286. In: *Fundamentals of Life*, Pályi, G., Zucchi, C., and Caglioti, L. (eds.). Paris, France: Elsevier, 2002.

Kauffman, S.A. *The Origins of Order*. New York: Oxford University Press, 1993.

Lem, S. *Solaris*. Google eBook, 1961.

Nicolis, G. and Prigogine, I. *Self-Organization in Nonequilibrium Systems: From Dissipative Structures to Order through Fluctuations*. New York: Wiley, 1977.

Olomucki, M. *La Chimie du Vivant*. Paris, France: Hachette, 1993.

Pályi, G., Zucchi, C., and Caglioti, L. (eds.). *Fundamentals of Life*. Paris, France: Elsevier, 2002.

Popa, R. *Between Necessity and Probability: Searching for the Definition and Origin of Life*. Heidelberg, Germany: Springer-Verlag, 2004.

Prigogine, I. *From Being to Becoming*. San Francisco, CA: W.H. Freeman, 1980.

Prigogine, I. and Stengers, I. *Order Out of Chaos*. London, U.K.: Haynemann, 1984.

Valenzuela, C.Y. A biotic Big-Bang, pp.197–202. In: *Fundamentals of Life*, Pályi, G., Zucchi, C., and Caglioti, L. (eds.). Paris, France: Elsevier, 2002.

Von Foerster, H. On self-organizing systems and their environments, pp. 31–50. In: *Self-Organizing Systems: Proceedings of an Interdisciplinary Conference*, Yovits, M.C. and Scott, C. (eds.). New York: Oxford, 1960.

Wong, J.T.F. Short definitions of life, p. 53. In: *Fundamentals of Life*, Pályi, G., Zucchi, C., and Caglioti, L. (eds.). Paris, France: Elsevier, 2002.

Journal Articles

Ashby, W.R. 1947. Principles of the self-organizing dynamic system. *Journal of General Psychology* 37: 125–128.

Bedau, M.A. 1998. Four puzzles about life. *Artificial Life* 4: 125–140.

Chaisson, E. 2003. A unifying concept for astrobiology. *International Journal of Astrobiology* 2: 91–101.

Cimpoiasu, V.M. and Popa, R. 2012. Biotic Abstract Dual Automata (BiADA): A novel tool for studying the evolution of prebiotic order (and the origin of life). *Astrobiology* 12: 1123–1134.

Corsetti, F.A. and Storrie-Lombardi, M.C. 2003. Lossless compression of stromatolite images: A biogenicity index? *Astrobiology* 3: 649–655.

Koh, Y.Z. and Ling, M.H. 2013. On the liveliness of artificial life. Human-level intelligence. *Online Research in Philosophy* 3: 1–17.

McCabe, T.J. 1976. A complexity measure. *IEEE Transactions on Software Engineering* 2: 308–320.

Popa, R. 2010. Necessity, futility and the possibility of defining life are all embedded in its origin as a punctuated-gradualism. *Origins of Life and Evolution of Biospheres* 40: 183–190.

Popa, R. 2012. Merits and caveats of using a vocabulary approach to define life. Comments on Trifonov. *Journal of Biomolecular Structure and Dynamics* 29: 607–608.

Trifonov, E.N. 2011. Vocabulary of definitions of life suggests a definition. *Journal of Biomolecular Structure and Dynamics* 29: 259–266.

CHAPTER 16

Language and Communication as Universal Requirements for Life

Guenther Witzany

CONTENTS

16.1 INTRODUCTION AND OBJECTIVES

Nowadays, we know that it is an empirical fact that if cells, tissues, organs, and organisms coordinate their behavior, this needs signals. Biotic signaling serves as a primary tool to coordinate groups of individual living agents such as cells and organisms, that is, the whole process we term communication. Current knowledge indicates communication as a basic interaction within and between organisms in all domains of life. Communicative interactions are necessary within organisms—intraorganismic—to coordinate cell–cell interactions, similar to tissue–tissue and organ–organ coordinations especially in complex bodies. This includes also the interpretation of abiotic environmental indices such as light, temperature, gravity, water, and nutrient availability as sensing, monitoring, and feedback control against stored background memories. We find interorganismic communication in all signal-mediated interactions between same and related species. If species communicate with nonmembers, we term this transorganismic communication. Throughout all kingdoms of life, we do not find any coordination and organization that does not depend on communication. In this chapter, I will shortly summarize these communicative levels in various domains of life.

Additionally, all that protein-based life we investigate in cells, tissues, organs, and organisms throughout all domains of life depends on stored information about structure, processing, development, and regulation. The genetic code has four kinds of nucleic acid components that serve as characters of an alphabet (guanine, adenine, thymine, and cytosine) that continuously is translated into amino acid language with 20 amino acids that form the whole variety of protein bodies and, additionally, an abundance of ribonucleic acids (RNAs) (with one varying character in the alphabet; uracil instead of thymine) that regulate all fine-tuned processes of cell replication, transcription, translation, and repair. The nucleotide sequences of the genetic code are base pairing according to the Chargaff rules (adapted by Watson and Crick in their double helix) in which they build complementary sequence structures and serve as information storage medium. We will look at some agents that are competent to edit the genetic code such as viruses and subviral RNAs.

First of all, let us have a look at the current background knowledge on natural languages/codes and on the basic biological features of communication processes. Because of its empirical significance, we can easily adapt this biocommunication approach to nonhuman organisms as well as the natural genome-editing competences of viruses and subviral agents to the evolution and content order of the genetic code.

16.2 SUMMARY OF CURRENT KNOWLEDGE ABOUT NATURAL LANGUAGES/CODES AND COMMUNICATION

Current knowledge in linguistics and communication theory identified three basic features that are essential characteristics to all natural languages/codes:

1. No natural language speaks itself as no natural code codes itself. In natural languages or codes, *living agents* that are competent to generate sign sequences are the ultimate prerequisite for the existence and occurrence of natural languages and codes.

2. The emergence of natural languages and codes depends on populations of living agents. This means that natural languages/codes in communication processes are primarily *social interactions.* Concrete social interactions are the essential experience for socializing children to learn the connections between linguistic utterances and their meaning. (Utterances are sentences with which we do something: convince someone, explain something, implement something, and similar. We do not think and then formulate sentences: we think in language and sentences. Also, every nonverbal expression is an utterance.)

3. Living agents that use natural languages/codes to initiate social interactions must be competent to follow *three levels of rules* that are obligatory and are inherent in any natural language or code: (a) competence to correctly combine signs to sequences (syntactic rules), (b) competence to correctly initiate communicative interactions according to the context specificity (pragmatic rules), and (c) competence to correctly designate objects by appropriate signs (semantic rules). If one level of rules is missing, one cannot seriously speak about a real natural language or code.

16.2.1 Language Is a Natural Language Only If Living Agents Use It

Language use depends on communities, a historically grown group of members that share these three levels of rules. Language use is a social action and is a priori intersubjective. If the time window of childhood learning and training in language words and sentences in social interactions is disturbed, linguistic and communicative competences can be deformed and even be lifelong. From this perspective, we can avoid monological concepts of language, all of which share an essential problem: how to make the move from a state of private consciousness to a state of mutual agreement and cooperation. Monological concepts include metaphysical, philosophy of mind, or other solipsistic approaches such as sender–receiver or coding–decoding narratives.

16.2.2 Mathematical Theories of Language vs. Pragmatics of Languages

The mathematical theory of language and its derivatives (systems theory, information theory, game theory, bioinformatics, synthetic biology, biolinguistics) tried the other way around: there is a logic of relations within material reality that is inherent also in biological matter. In the case of human evolution, this logic determines finally the architecture of neuronal brain construction. If the brain uses a language that depicts this logic, it must be possible to depict material reality with this language. The only language that is able to depict material reality is a formalizable, algorithm-based language, that is, mathematics. Therefore, the human brain must use formalized mathematical language so that it can scientifically depict and explain material reality.

The crucial deficit in this method is that one cannot explain everyday language with it. Everyday language is the ultimate metalanguage. This means that there cannot be any language that could go beyond everyday language. Since it serves as a primary tool for everyday life communication and socialization of humans, it is the source of the original meaning of words in sentences. Everyday language cannot be formalized: we cannot explain deep grammars and illocutionary acts we use primarily to transport a variety of meanings with identical superficial syntactic structures.

For the benefit of the readers, I would like to explain the terms locutionary, illocutionary, and perlocutionary, which are used in this chapter. Locutionary speech act is represented by the superficial syntax of the sentence. In contrast, illocutionary speech acts transport context-dependent meaning. For example, "I will come tomorrow"; its illocutionary force could be a promise, a threat, a secret code, etc., depending on the circumstances and intentions. With perlocutionary speech acts, someone fulfills a complete action. If someone asks you if you are willing to take XY as your wife and you answer "yes," then the yes is a perlocutionary speech act.

The same holds true for information theory: in a recently published article, Sydney Brenner states that biology is, in his opinion, physics with computation. The fundamental concept that integrates biological *information* with matter and energy is the universal Turing machine and von Neumann's self-reproducing machines. However, no single self-reproducing machine had ever been observed within the last 80 years since they presented their concept. There are good reasons for this, because machines cannot create new programs without algorithms. In contrast to the artificial machines that cannot reproduce themselves, the living cells and organisms can reproduce themselves and additionally generate an abundance of behavioral motifs for which no algorithm can be constructed, such as de novo generation of coherent nucleotide sequences.

16.2.3 How to Generate Correct Scientific Sentences: Results of the Philosophy of Science Debate in the Twentieth Century

To get things methodologically straight, we have to remind ourselves of the discussion between 1920 and 1980 in the history of philosophy of science and the transition of metaphysics to the *linguistic turn* and afterward to the foundation and justification of scientific sentences in the *pragmatic turn*.

1. The *linguistic turn* was the result of an attempt to delimit the logic of science from philosophy and other *nonscientific* methods. The term *delimit* generally means to define the conditions that must be fulfilled as validity claims of exact sciences. Specifically, the logic of science tried to find a language whose sentences are strictly scientific in contrast to the sentences from poetry, theology, astrology, and other nonscientific fields. Based on good reasons, the linguistic turn states that we do not understand per se objects, relations, structures, intelligence, mind, consciousness, cognition, construction, matter, energy, information, system, and natural laws, but only linguistic sentences in utterances. (Utterances are sentences with which we do something: try to convince someone, explain something to someone, call someone to do something, and similar.) Until linguistic turn, it was assumed in philosophy and sciences in general that sentences somehow depict reality.

   Only protocol propositions of observations that are reproducible in experimental setups are capable of depicting reality on a 1:1 basis, thus a direct correspondence. This is also valid for propositions of a language of theory that would have to be brought into agreement with these protocol propositions. What is required for a language is that it can be formalized, as in logical calculations, algorithms. This language would represent a universal syntax that would be universally valid (a) in the things of the external world, (b) in the physical laws, and (c) in the material reality of the brain of humans speaking in formalizable propositions. But after several unsuccessful attempts, logical empiricism had to abandon its efforts to achieve the ultimate validity claim of a universal scientific language.

2. As a result of this, the *pragmatic turn* refers to the communicative everyday interactions of historically evolved groups and communities that are the basis for learning and training linguistic and communicative competences. Historically grown communicative practice of linguistic communities is the prerequisite for organization and coordination of social interactions, and later of linguistic abstractions, such as scientific languages in communities of specialized disciplines (see Figure 16.1).

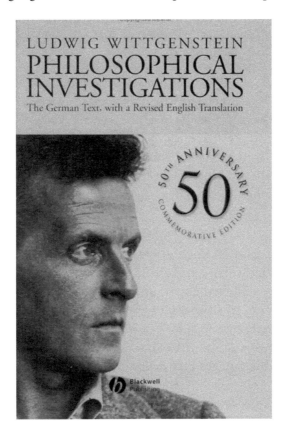

FIGURE 16.1 In his early thinking, which was outlined in the *Tractatus Logico-Philosophicus*, Ludwig Wittgenstein led the foundations for the project *exact scientific language*, that is, formalizable sentences that depict reality in a 1:1 manner. This served as common foundation of the *linguistic turn*. In his late *Philosophical Investigations*, Wittgenstein refuted his early thoughts completely. The basic features of natural languages are the real-life worlds of social groups that use sign systems in everyday use. The context of use determines the meaning of sign sequences. Natural languages serve as an essential tool in social interactions. This was the start of the *pragmatic turn*. (From Wittgenstein, L., *Philosophical Investigations: The German Text with a Revised English Translation*, Blackwell Publishing, Malden, 2001. With permission.)

The pragmatic turn founded and justified the *intersubjective–communicative character of thought, experience, and research.*

Everyday language analysis shows speech acts or *how to do things with words.* As we are both, in parallel, subjects and objects of our utterances, we are in a privileged position to take into account, at the same time, our historically evolved everyday language. We understand utterances as participants in the communicative, representative, imperative, and constitutive speech acts, rather than acting as lonely (*solus ipse*) isolated observers. This enables us to explain central marks of speech acts as there are

- Simultaneous understanding of identical meanings in two interacting partners, as expressed in successfully coordinated activity

- Differentiation between deep and superficial grammar of a statement along with differentiation between locutionary, illocutionary, and perlocutionary speech acts with which the statements are made

The pragmatic turn replaces all *monological* subjects of knowledge by the primacy of linguistic communities. This is the end for a methodological ideal that lasted nearly 2000 years and that maintained as a principle that one subject alone could—monologically—get knowledge and construct a language/code in a process that Thomas McCarthy has described as follows: "The monological approach preordained certain ways of posing the basic problems of thought and action: subject vs. object, reason vs. sense, reason vs. desire, mind vs. body, self vs. other, and so on."

16.2.4 Meaning of Messages Depends on Contextual Use (Pragmatics), Not Syntax

It is a deep grammar, or as outlined in great detail by John Austin and John Searle, it is the illocutionary act we undertake with what we say. So, besides the locutionary aspect, which is represented by the superficial syntax of the sentence, there is a variety of (hidden) possibilities of what we want to do (intend) with this sentence, namely, the illocutionary act. In extreme cases, we can intend contradictory goals with the same syntactic structure. This is the reality of everyday language that is impossible to capture in the formalized languages that cannot represent both locutionary and illocutionary aspects.

An important consequence of this is that pure language analyses that want to extract deep grammar/illocutionary action out of available syntactic sequences must necessarily fail. This is because the analyses of syntactical rules cannot explain pragmatic interactional contexts that finally determine the meaning of syntactic structures. This has serious consequences:

1. The variety of words combined to form sentences in everyday languages and dialects are not the result of copying errors or damage of preexisting sentences or available sentences. According to Gödel, natural language users are principally capable to produce new sequences that have never been generated before (see Figure 16.2).

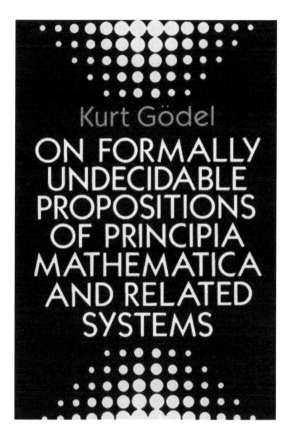

FIGURE 16.2 Gödel discovered the crucial deficit of the linguistic turn: in his investigation, he convincingly proved the fact that in open systems, there is the possibility to create new sentences that have never been created before and cannot be predicted out of preexisting ones in principle. This was the end of Hilbert's program of a self-consistent axiomatic system (Hilbert space) as the ultimate foundation and justification of mathematics. (From Gödel, K., *On Formally Undecidable Propositions of Principia Mathematica and Related Systems*, Dover Publications, New York, 1992. With permission.)

It is an inherent feature of all natural languages that living agents that use them can produce new ones. They may be unpredictable and thus incalculable. As such, they cannot be deduced out of former ones or available ones.

2. In natural languages/codes, there is no syntax based on a universal grammar that transports identical meanings (e.g., with a unique position in a formalizable sequence space as suggested by Manfred Eigen).

3. To learn a natural language means to communicate basic everyday needs with community members. This is how we learn what a word means ("The meaning of a word is its use"—Ludwig Wittgenstein). We can understand words and their sequences, because we have learned a practice of interaction, which includes learning from the community which words are combined with which interactional patterns; we then memorize those.

16.3 KEY LEVELS OF BIOCOMMUNICATION IN PROKARYOTES AND EUKARYOTES

Communication processes within and between organisms are rather complex sign-mediated interactions that significantly differ in prokaryotes (unicellular organisms without a true nucleus) and eukaryotes (uni- and multicellular organisms with true nucleus). Prokaryotes represent a success story in evolution and exist since nearly 3.8 billion years. However, the transition to eukaryotes was a real revolution because it incorporates a variety of former free-living prokaryotes within a double-layered cell and its information-bearing nucleus. The latter was most probably derived from a large double-stranded DNA virus. This means that the basic components of eukaryotic cells are both cellular and viral symbionts that are genetically conserved into a social organism.

16.3.1 Prokaryotes

The bacterial world is a social and communicative one. The production and the exchange of messenger molecules enable unicellular organisms to coordinate their behavior like a multicellular organism. The biocommunication in bacteria communities is not restricted to species-specific levels but represents a clear multilevel communication that enables hundreds of different bacteria groups to co-occupy one and the same ecological niche.

It has been proven that bacteria groups use quorum sensing to determine their strength and to react by coordinating their behavior such as in biofilm formation.

> Studies of quorum sensing systems demonstrate that bacteria have evolved multiple languages for communicating within and between species. Intra- and interspecies cell–cell communication allows bacteria to coordinate various biological activities in order to behave like multicellular organisms.
>
> SCHAUDER AND BASSLER (2001)

Figure 16.3 demonstrates this as a snapshot from a video in which a group of bacteria sense available nutrients and coordinate redirected group movements in the correct direction. Recently, all key levels of biocommunication of soil bacteria have been categorized.

16.3.2 Eukaryotes

16.3.2.1 Fungi

Fungi also communicate and therefore are able to organize and coordinate their behavior. Coordination and organization processes in fungi are seen at the intraorganismic level, for example, during the formation of fruiting bodies, between species of the same kind (interorganismic), and between fungal and nonfungal organisms (transorganismic).

The semiochemicals (from the Greek word *semeion*, meaning sign) used are of biotic origin, in contrast to abiotic indicators that trigger the fungal organism to react in a specific manner. The roles of some of these signaling molecules are as follows: (1) Mitogen-activated protein kinase signaling (MAPK) is involved in cell integrity, cell wall construction,

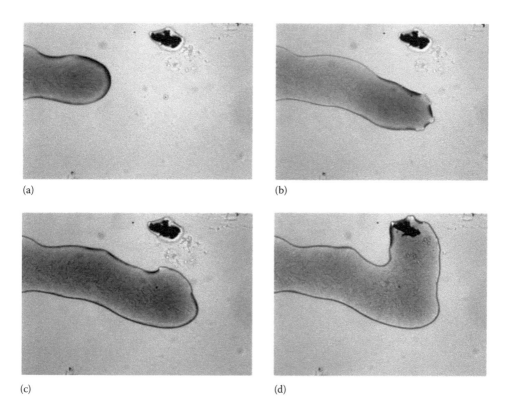

FIGURE 16.3 Swarming intelligence in bacteria: some extracellular food was positioned near a bacteria colony (a). After sensing of the food (b), the colony swarm moved toward the resource (c and d). Intra- and intercellular signaling is necessary to sense, interpret, and coordinate common motile behavior. (From Ben-Jacob, E., *Ann. N.Y. Acad. Sci.*, 1178, 78, 2009. With permission.)

pheromones/mating, and osmoregulation; (2) the cyclic adenosine monophosphate/protein kinase A (cAMP/PKA) system is involved in fungal development and virulence; (3) the RAS (protein family members that belong to a class of small GTPase) protein is involved in the cross talk between signaling cascades; (4) calcium, calmodulin, and calcineurin are involved in cell survival under oxidative stress, high temperature, and membrane/cell wall perturbation; (5) rapamycin is involved in the control of cell growth and proliferation; (6) aromatic alcohols tryptophol and phenylethylalcohol are used as quorum-sensing molecules; and also (7) a variety of volatile (alcohols, esters, ketones, acids, lipids) and nonvolatile inhibitory compounds (farnesol, H_2O_2).

To date, 400 different secondary metabolites have been documented. Development and growth of fungal organisms depend upon successful communication processes within, and between, cells of fungal organisms.

In order to generate an appropriate behavioral response, fungal organisms additionally must be able to sense, interpret, and memorize important indices from the abiotic environment and adapt to them appropriately (see Figure 16.4). Interestingly, certain rules of fungal communication are very similar to those of animals, while others more closely resemble those of plants.

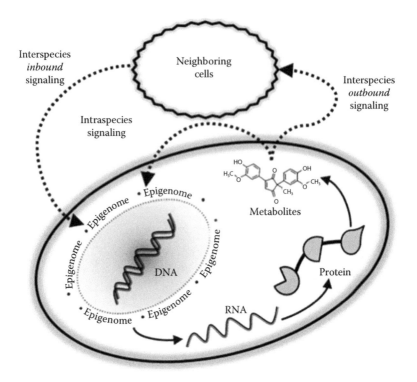

FIGURE 16.4 Depending on the real-life context of fungal organisms, epigenetic regulation can suppress or amplify incoming or transmitted secondary metabolites, an important signal resource of fungal organisms. Therefore, not for every message, a novel sequence has to be produced (multiple meanings of identical syntax structures). (From Cichewicz, R., *Epigenetic Regulation of Secondary Metabolite Biosynthetic Genes in Fungi*, Springer Science+Business Media, Dordrecht, the Netherlands, 2012. With permission.)

16.3.2.2 Animals

Animals also depend primarily on volatile substances such as pheromones to identify group identity of self and nonself. However, in addition, they depend on a variety of signs that convey meaning via vocal sounds and visual gestures. This opens up a variety of combinatorial possibilities and broadens the communicative competencies. Such complexity increases exponentially in comparison to biocommunication of bacteria, fungi, and plants. The signaling molecules, vocal and tactile signs, gestures, and their combinations differ throughout all species according to their evolutionary origins and variety of adaptation processes. However, certain levels of biocommunication can be found in all animal species:

1. Abiotic environmental indices such as temperature, light, water, and gravity that affect the local ecosphere of an organism are sensed and interpreted (against stored background memory). Then they are being used for organization of behavioral response to adapt accordingly (taking into account also optimal energy cost).

2. Trans-specific communication with nonrelated organisms as found in attack, defense, and symbiotic (even endosymbiotic) sign-mediated interactions.

3. Interorganismic communication between same or related species.

4. Intraorganismic communication, that is, sign-mediated coordination within the body of the organism. This means two sublevels, such as cell–cell communication and intracellular signaling between cellular parts.

16.3.2.3 Plants

Plants are sessile organisms that actively compete for environmental resources both above and below the ground. They assess their surroundings, estimate how much energy they need for particular goals, and then realize the optimum variant. They take measures to control certain environmental resources. They perceive themselves and can distinguish between *self* and *nonself*. They process and evaluate information and then modify their behavior accordingly (see Figure 16.5). Plant communication centers are the stem and the rhizosphere (the entire area of interactions within the root zone). The rhizosphere of plants is a realm of overlapping communicative interactions and a dynamic environment featuring dense microbiological life, high growth rates and metabolic activities, as well as rapidly changing physical conditions. The communication processes between tissues and cells in plants are incredibly complex and encompass nucleic acids, oligonucleotides, proteins and peptides, minerals, oxidative signals, gases, mechanical signals, electrical signals, fatty acids, and oligosaccharides, growth factors, several amino acids, various secondary metabolite products, and simple sugars.

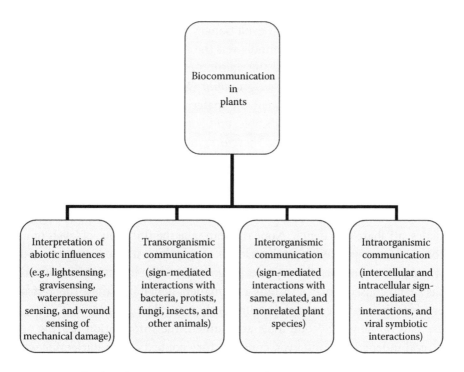

FIGURE 16.5 **Key levels of biocommunication in plants.** (From Witzany, G., *Key Levels of Biocommunication in Plants*, Springer-Verlag, Berlin, Germany, 2012. With permission.)

As in every biocommunication process of real lifeworld (pragmatic) situations, context determines semantic meaning of signals. Auxin, for example, is an ancient signaling molecule in plants. It functions in different hormonal, morphogen, and transmitter signaling pathways. Thus, it is very difficult to decipher the actual semantics of auxin, since it depends on whether it is used as hormonal or morphogen or transmitter signal. The same signal can take on different meanings and trigger different effects, depending on whether it reaches the whole plant, a tissue, or a cell. Because of this, the respective developmental status of the organism serves as a memory for each individual plant.

16.4 NUCLEIC ACID LANGUAGE AS NATURAL CODE

Concepts such as *genetic code, nucleic acid language, recognition sequences, translation process, amino acid language, immune responses,* and *cell–cell communication* represent irreplaceable core concepts in molecular biology. These concepts were not introduced into biochemistry and molecular biology by linguists, communication experts, or language philosophers. Rather, they were independently coined by molecular biologists to explain observed phenomena and were clearly invoked due to the strong analogy to processes of human communication. Francis Crick termed the genetic code a *code without commas*. However, Manfred Eigen investigated the genetic code as real language and not just as a metaphor, as revealed by the following citations: "The relative arrangement of the individual genes, the gene map, as well as the syntax and semantics of this molecular language are (...) largely known today" (Eigen and Winkler, 1983, p. 207). "All the words of the molecular language are combined to a meaningful text, which can be broken down into sentences" (p. 305). "At any rate one can say that the prerequisite for both great evolutionary processes of nature—the origin of all forms of life and the evolution of the mind—was the existence of a language" (p. 314).

As we know today, Eigen followed the opinion of his time, that language follows the structure of a *universal* and *context-free grammar* (Noam Chomsky) that underlies strict natural laws as it represents the logic of the material reality. The core functions of languages are limited, formalizable, predicable, and computable. The only real language that depicts material reality is mathematics. Therefore, the molecular genetic code can be investigated and described sufficiently by physics and chemistry. Eigen adapted the opinion of the linguistic turn as described earlier.

Several other derivatives of the linguistic turn and its mathematical theories of language, such as systems theory, cybernetic information theory, synthetic biology, and even biolinguistics, all share this deficit. None of them take the full range of levels of rules into their theoretical assumptions but let these levels of rules be restricted to syntax and semantics. But the demission of the primary role, pragmatics, to the investigations of natural languages has a fatal consequence. It installs a permanent deficit into these theoretical realms that determines their failure to explain sufficiently natural languages.

Because natural language/code tools are limited, the information-bearing sequences denote several independent and even contradictory contexts. One nucleotide word, such as a pseudoknot (a type of the nucleic acid secondary structure), may have several different meanings. Because living agents cannot invent new signs for every new situation or designation (energy costs), this evidently makes sense. Similar or equal combinations of signs, characters, and words that result in sentences can be used as informational tools to transport different meanings about a whole genome. Examples include overlapping epigenetic marking (the genetic sequence is marked through environmental influences that determine the context-relevant meaning/expression pattern) and silencing of transposons (DNA sequences that move), which induce repression of maternal cytotype (having different chromosomal factors) in animals, among others.

From human communication, we know that different gestures or spelling may indicate different meanings of the same words. Without contextual explanation, the phrase *the shooting of the hunters* cannot be understood unequivocably. The identical sequence may transport contradicting messages (see Figure 16.6). The marking of syntactic sequences by marking tools is common use in natural languages/codes and determines semantic content according to the needs of the pragmatic interacting agents.

To investigate syntactic sequences without knowing something about the real-life context of code using agents is senseless because syntactic structures do not represent unequivocable semantic meaning. Quantifiable analyses of signs, words, or sequences cannot extract context-dependent meaning. In a restricted sense, this is possible through sequence comparison, for example, if we know which sequences determine certain functions. But all these features are absent in nonanimate nature. If water freezes to ice, no living agents nor semiotic rules or signs are necessary.

FIGURE 16.6 *The shooting of the hunters*: in natural languages/codes, the meaning of syntactical identical sequences depends on the real-life world context in which competent sign users are interwoven. The use of identical syntax structures to transport different (and even contradictory) meanings saves energy costs. Algorithm-based machines (computers) that must extract the meaning of given syntax structures cannot decide between superficial grammar and deep grammar (illocutionary acts) intended by sign users. (Reprinted by permission from Macmillan Publishers Ltd. *EMBO Reports*, Witzany, G. and Baluška, F., Life's code script does not code itself. The machine metaphor for living organisms is outdated, 13, 1054–1056, Copyright 2012. Graphics by Uta Mackensen.)

16.5 AGENTS OF NATURAL GENOME EDITING

Within the last decade, views on natural genetic engineering and natural genome editing have changed dramatically. In particular, research in virology has opened perspectives on early evolution of life, as well as on viruses as essential agents within the roots and stem of the tree of life. From the early RNA world perspective, the whole diversity of processes within and between evolutionarily later-derived cellular life depends on various RNAs. The precellular RNA world must have been dominated by quasi-species consortia-based evolution, as are current RNA viruses.

Viruses can parasitize almost any replication system—even prebiotic ones. RNA viruses store crucial and dynamic information. Based on this and the results of phylogenetic analyses and comparative genomics, it is possible to establish viral lines of ancestral origin. These lines of origin can also be nonlinear because different parts of viruses contain different evolutionary histories. Since viruses with RNA genomes are the only living beings that use RNA as a storage medium, they are considered to be witnesses of an earlier RNA world. Current negatively stranded RNA viruses have genome structures and replication patterns that are dissimilar to all known cell types.

No similarity between RNA-viral replicases and those of any known cell types has been identified. DNA viruses, too, do not give any reference to a cellular origin. DNA-repair proteins of DNA viruses do not have any counterparts in cells. One milliliter of seawater contains one million bacteria and 10 times more viral sequences. 10^{31} bacteriophages infect 10^{24} bacteria each second. The enormous viral genetic diversity in the ocean has established pathways for the integration of complete and complex genetic data sets into host genomes, for example, acquisition of complex new phenotypes. A prophage can provide the acquisition of >100 new genes in a single genome-editing event. Today, it is assumed that the gene word order in bacterial genomes is determined by viral settlers of bacterial host genomes. Not only bacterial life is determined by nonlytic viral settlements, but also the evolution of eukaryotes has strongly depended on viral properties. In contrast to the mitochondria and other eukaryotic parts of bacterial descent, the eukaryotic nucleus was formerly a large double-stranded DNA virus. All properties of the eukaryotic nucleus are lacking in bacterial life forms but are typical features of DNA viruses. Even lethally irradiated viruses can often repair themselves. They are competent to recombine combinations of defective viral genomes in order to assemble intact viruses. Therefore, viruses are the only living agents capable of meaningfully recombining text fragments of a damaged genome into a fully functional viral genome that is capable of self-replication.

Lytic diseases that are caused by viral infections are the exception in viral life strategies, although they might have epidemic and pandemic and therefore catastrophic consequences for infected populations. The most dominant viral life strategy is the nonlytic but persistent viral settlement of cytoplasm of cellular hosts and even more of cellular host genomes. Addiction modules are the result of integration of former competing viral infections. As symbiotic neutralization and counterpart regulation, they represent new host phenotypic features. One feature is regulated exactly by the antagonist according to developmental stages in the cell cycle, replication, and tissue growth. Should this suppressor function become unbalanced, then the normally downregulated part might become lytic again. We can identify virus-derived addiction

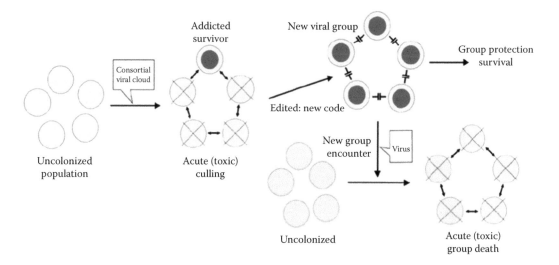

FIGURE 16.7   Basic interactional motif of infection-derived group identities: the addiction module as a result of counterbalanced, infection-derived, and persistent genetic parasites that initiate evolutionary inventions by natural genetic engineering of host genetic identities, some we can find as toxin/antitoxin, restriction/modification, or insertion/deletion modules. (From Villarreal, L.P. Viruses and host evolution: Virus-mediated self identity. In López-Larrea, C. (ed.), *Self and Nonself*, Springer Science+Business Media, Austin, LandesBioscience, New York, pp. 185–217, 2012b. With permission.)

modules in every toxin/antitoxin, restriction/modification, or insertion/deletion modules in which former competing viral clouds are now immunologically balanced (Figure 16.7). If a balanced status is reached, this means a changed genetic identity of the host organism and, in consequence, a changing genetic identity of the viral settler. Current knowledge indicates that most evolutionarily novel derived species are the result of changed and expanded genomic identities caused by persistent viral colonization. Research results in virology have led to the assumption that, besides communicative competences of cellular organisms, which are involved in coordinating behavior, there are *linguistic* competences of viruses and virus-derived viral parts (e.g., env, gag, pol), which not only regulate all cellular processes but edit the genetic content of living organisms. This viral genetic text-editing competence depends on living organisms that are different from each other, and it therefore needs a biotic matrix to expand this competence. Without living and interacting organisms and cells, genomic creativity would only be a possibility that is restricted to mere RNA combinatorial events (in an early precellular RNA world), which has no relevance to the generation of a biosphere.

16.5.1 Biocommunication in the RNA World: RNA Sociology

The ancient RNA world hypothesis is currently updated with RNA world facts and increasing knowledge about the abundance of different but compatible RNAs. In this world of life processes actively dominated by RNA, DNA is increasingly cast in the role of the *habitat* of genetic information storage, whereas the interacting RNAs seem to be the *inhabitants*

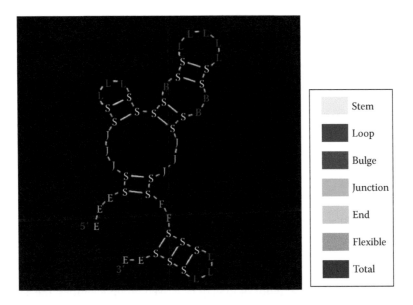

FIGURE 16.8 The RNA stem loops have several distinct parts/subunits: stems consisting of base-paired nucleic acids and loops/bulges/junctions consisting of unpaired regions limited by stems. Important: any RNA is part of such stem loops. (From Smit, S. et al., *RNA*, 12, 1, 2006.)

within this habitat. If we look at these network-like cooperating inhabitants, we can see a secondary structure that is shared by all these RNA nucleic acid sequences: it is the stem-loop structure or, in the case of more complex agents such as tRNA or ribosomal subunits, the ligated consortia of such stem loops (see Figure 16.8).

The rather astonishing result of investigations is that randomly associated RNAs that have no evolutionary history show the same structure-dependent compositional bias as natural derived (ribosomal) RNAs. This means that the differences do not depend on selection processes but on the overall composition of the RNA consortium.

16.5.2 Biocommunication at the Level of the RNA Group Membership

RNA group membership can never be completely specified, since it can always be further parasitized by as yet to be encountered members or parasites. This essential and most important feature renders the ability to absolutely specify membership (absolute immunity) as basically indefinable. Thus, an RNA group can never be fully secure from as yet undefined parasite agents. But a crucial consequence from this *insecurity* is that it provides the inherent capacity for novelty, that is, the precondition for evolutionary innovation such as greater complexity.

To introduce sociological terms, we now have to ask: How do agents emerge from chemicals to form identity and then form groups that learn membership? Single RNA stem-loop generation occurs by physical chemical properties solely as demonstrated by natural and randomized RNA experiments. If stem-loop consortia build complex consortia, they initiate social interactions not present in a pure chemical world, that is, biological selection emerges. This designates the crucial step from inanimate world to life.

Now we have molecular structures coherent with physical laws that store genetic information. In contrast to inanimate nature, they actively generate behavioral motifs and patterns of interaction, that is, coordinate common behavior according to rules that lead to consortia of self and nonself groups. This resembles some kinds of social group behavior with shared features:

- De novo initiation of behavior that cannot be deduced from former behavioral patterns

- Highly adaptive processes

- Lacking central or fittest type control

- Retaining a contextual history

- Smart (optimal energy costs)

- Solves problems beyond the capacity of its individual members

- Fast-changing reactions against nonmembers

Together these features are clearly and exclusively at the foundation of all living nature. If we were to eliminate these complementary competences out of the life processes, would there remain a living organism, or would it now simply be a chemical state? It seems not; thus, social RNAgents are essential.

16.5.3 Cooperation Outcompetes Selfishness

If we look at some interactional motifs of RNAgents to form consortial biotic structures that follow biological selection processes and not mere physical chemical reaction patterns, we must look at the group building of RNA stem-loop structures.

Recently, it has been found that single stem loops interact in a pure physical chemical mode without selective forces, independently whether they are derived randomly or are constructed under in vitro conditions. In contrast to this, if these single RNA stem loops build groups, they transcend pure physical chemical interaction pattern and emerge biological selection forces, biological identities of self/nonself identification and preclusion, immune functions, and dynamically changing (adapting) membership roles. A single alteration in a base-pairing RNA stem that leads to a new bulge may dynamically alter not solely this single stem loop but may change the whole group identity of which this stem loop is part of.

Simple self-ligating RNA stem loops can build much larger groups of RNA stem loops that serve for increase in complexity (Figure 16.9).

Significantly, RNA fragments that self-ligate into self-replicating ribozymes spontaneously form cooperative networks. For example, three-membered networks showed highly cooperative growth dynamics. When such cooperative networks compete directly against selfish autocatalytic cycles, the former grow faster, indicating competence of RNA populations to evolve greater complexity through cooperation. In this respect, cooperation clearly outcompetes selfishness.

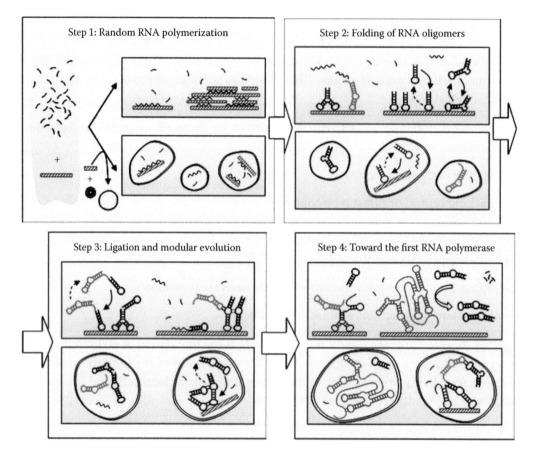

FIGURE 16.9 Schematic representation of the stepwise process toward a template-dependent RNA polymerase. In every step, we depict two possible and compatible scenarios: evolution on mineral surfaces (shown as brown rectangles) in bulk solution and evolution inside vesicles that could also encapsulate mineral particles. Functional hairpin structures (with ligase activity) are shown in red. Solid and dotted arrows stand for the surface-bound to in-solution equilibria. The RNA polymerase emerging from this process is depicted in blue. (From Briones, C. et al., *RNA*, 15, 743, 2009.)

16.6 CONCLUSION

What is the benefit of this concept of language and communication as universal requirements of life in contrast to mechanistic, holistic, objectivistic, mathematically derived formalizable concepts such as system theoretical, bioinformatics, or synthetic biology approaches in molecular biology, genetics, epigenetics, and microbiology? The biocommunication and natural genome-editing approach on processual reality of living agents brings some advantages to traditional scholarly conviction:

- Clear distinction between life and nonlife.

- Empirical nonmechanistic and nonreductionistic description method of biotic interactional patterns throughout all organismic kingdoms.

- Any observed coordination within and between organisms can be deciphered by research that identifies signaling molecules and syntactic, pragmatic, and semantic rules underlying the mode of signal use.

- Biological research must not concentrate any longer on experimental setups and theoretical approaches that want to elucidate language and communication in nonhuman living nature by mathematical (algorithm-based) modeling.

GLOSSARY

Biochemically related terms: Please consult the following recommended readings.

Biofilm: Bacterial group building by signaling interactions.

Communication: Interactions via signals between living organisms according to syntactic, pragmatic, and semantic rules.

Evolutionary biology: Genotype/phenotype novelty as a result of variation (mutation = replication error) and selection.

Language: Any repertoire of signs that is used according to syntactic, pragmatic, and semantic rules.

Linguistic turn: We do not understand the outer and the inner world, but sentences about it.

Natural genome editing: Evolutionarily relevant variation results out of de novo sequence generation and recombination by competent RNA consortia, not of replication error.

Noncoding RNAs: RNAs that shortly after transcription out of DNA are not translated into proteins but serve as gene regulatory tools, increasing the main part of genetic information in eukaryotes.

Pragmatics: Level of rules that determine how to correctly combine words/sentences with real-life context.

Pragmatic turn: Language use is a kind of social interaction, that is, needs user groups.

Semantics: Level of rules that determine the correct designation of objects with words/sentences, that is, meaning.

Syntax: Level of rules that determine how parts of an alphabet can be correctly combined.

REVIEW QUESTIONS

1. If we would find extraterrestrial life, how do you think we would identify communication within it? Suggest how the material on language and communication you have learned from this chapter may or may not be helpful for answering this question.

2. Write a short essay on one way of biocommunication in life, such as bacteria and fungi, based on the material in this chapter. Research further a subtopic that you like, for example, the way plants communicate.

3. Why is context dependency crucial for identifying *meaning* of words and sentences? Try to formulate examples in which illocutionary acts determine the meaning of syntax of sentences (e.g., *the shooting of the hunters*).

4. Who was John Searle and how is his philosophy of language relevant to sufficiently explain communication? Explore this topic; exemplify the difference of locutionary, illocutionary, and perlocutionary speech acts; and write a short essay to present to the class.

5. What are the basic differences of natural languages/codes and mathematical concepts of languages?

6. What are the essential differences between single RNA stem loops and (self-)ligated groups of RNAs? Is there any evolutionary important divergence?

7. What is meant by *RNA sociology*? Nominate the differences to a strict physical/chemical vocabulary.

8. For decades, it was assumed that errors in replication (mutations) are key for evolution, that is, variations that then are subject to biological selection. Now it is recognized that genetic novelty is better explained by natural genetic engineering and natural genome editing initiated by competent RNA agents. Why do you think was the error-based narrative insufficient to explain genetic novelty?

RECOMMENDED READINGS

Books
Atkins, J.F., Gesteland, R.F., and Cech, T. (eds.). *RNA Worlds: From Life's Origins to Diversity in Gene Regulation*. New York: Cold Spring Harbor Laboratory Press, 2010.

Austin, J.L. *How to Do Things with Words*. London, U.K.: Harvard University Press, 1975.

Carter, J.B. and Saunders, V.A. *Virology, Principles and Applications*, 2nd edn. Chichester, U.K.: Wiley, 2013.

Cichewicz, R. Epigenetic regulation of secondary metabolite biosynthetic genes in fungi. In Witzany, G. (ed.) *Biocommunication of Fungi*. Dordrecht, the Netherlands: Springer Science+Business Media, 2012.

Eigen, M., and Winkler, R. *The Laws of the Game: How the Principles of Nature Govern Chance*. Princeton, NJ: Princeton University Press, 1983.

Frisch, K. *Bees: Their Vision, Chemical Senses and Language*, revised edn. Ithaca, NY: Cornell University Press, 1971.

McCarthy, T. Translator's introduction. In *The Theory of Communicative Action*, Vol. 1. Boston, MA: Beacon Press, p. ix, 1984.

Searle, J. *Speech Acts: An Essay in the Philosophy of Language*. Cambridge, U.K.: Cambridge University Press, 1976.

Shapiro, J.A. *Evolution: A View from the 21st Century*. Washington, DC: FT Press, 2011.

Tomasello, M. *Origins of Human Communication*. Cambridge, MA: MIT Press, 2008.

Villarreal, L.P. *Viruses and the Evolution of Life*. Washington, DC: ASM Press, 2005.

Villarreal, L.P. The addiction module as social force. In Witzany, G. (ed.), *Viruses: Essential Agents of Life*. Dordrecht, the Netherlands: Springer, pp. 107–145, 2012a.

Villarreal, L.P. Viruses and host evolution: Virus-mediated self identity. In López-Larrea, C. (ed.), *Self and Nonself*. New York: Springer Science+Business Media, Austin, Landes Bioscience, pp. 185–217, 2012b.

Voet, D. and Voet, J.D. *Biochemistry*, 4th edn. Hoboken, NJ: Wiley, 2011.

Witzany, G. (ed.). *Natural Genetic Engineering and Natural Genome Editing*. New York: Annals of the New York Academy of Sciences, 2009.

Witzany, G. (ed.). *Biocommunication in Soil Microorganisms*. Heidelberg, Germany: Springer, 2011.

Witzany, G. (ed.). *Biocommunication of Fungi*. Dordrecht, the Netherlands: Springer, 2012.
Witzany, G. (ed.). *Viruses: Essential Agents of Life*. Dordrecht, the Netherlands: Springer, 2012a.
Witzany, G. *Key Levels of Biocommunication in Plants*. Berlin, Germany: Springer-Verlag, 2012b.
Witzany, G. (ed.). *Biocommunication of Animals*. Dordrecht, the Netherlands: Springer, 2013.
Witzany, G. and Baluška, F. (eds.). *Biocommunication of Plants*. Heidelberg, Germany: Springer, 2012.

Journals

Ben-Jacob, E. (2009) Learning from bacteria about natural information processing. *Annals of the New York Academy of Sciences* 1178: 78–90.
Briones, C., Stich, M., and Manrubia, S.C. (2009) The dawn of the RNA world: Toward functional complexity through ligation of random RNA oligomers. *RNA* 15: 743–749.
Gwiazda, S., Salomon, K., Appel, B., and Müller, S. (2012) RNA self ligation: From oligonucleotides to full length ribozymes. *Biochimie* 94: 1457–1463.
Mercer, T.R. and Mattick, J. (2013) Structure and function of long noncoding RNAs in epigenetic regulation. *Nature Structural & Molecular Biology* 20: 300–307.
Schauder, S. and Bassler, B.L. The languages of bacteria. *Genes & Development* 15 (2001): 1468–1480.
Smit, S., Yarus, M., and Knight, R. (2006) Natural selection is not required to explain universal compositional patterns in rRNA secondary structure categories. *RNA* 12: 1–14.
Vaidya, N., Manapat, M.L., Chen, I.A., Xulvi-Brunet, R., Hayden, E.J., and Lehman, N. (2012) Spontaneous network formation among cooperative RNA replicators. *Nature* 49: 72–77.
Villarreal, L.P. and Witzany, G. (2010) Viruses are essential agents within the roots and stem of the tree of life. *Journal of Theoretical Biology* 262: 698–710.
Villarreal, L.P. and Witzany, G. (2013) The DNA habitat and its RNA inhabitants: At the dawn of RNA sociology. *Genomics Insights* 6: 1–12.
Villarreal, L.P. and Witzany, G. (2013) Rethinking quasispecies theory: From fittest type to cooperative consortia. *World Journal of Biological Chemistry* 4: 79–90.
Witzany, G. (1995) From the "logic of the molecular syntax" to molecular pragmatism. Explanatory deficits in Manfred Eigen's concept of language and communication. *Evolution and Cognition* 1: 148–168.
Witzany, G. (2011) The agents of natural genome editing. *Journal of Molecular Cell Biology* 3: 181–189.
Witzany, G. and Baluška, F. (2012) Life's code script does not code itself. The machine metaphor for living organisms is outdated. *EMBO Reports* 13: 1054–1056.

CHAPTER 17

Transition from Abiotic to Biotic

Is There an Algorithm for It?

Sara Imari Walker

CONTENTS

17.1 INTRODUCTION

Our daily lives are dictated by the ebb and flow of information. The invention of computers and the Internet in the last century heralded an *information revolution* and led to the current *Information Age* where traditional industry has yielded to an economy based on digital technologies. Nowadays, information is everywhere: from computers and smartphones to the World Wide Web and social networks, we are embedded in a sea of information exchange. It therefore seems fitting that information-based concepts be applied to our understanding of the natural world. Indeed, information theory plays an important role in much of modern science. This is particularly true of the relatively young scientific discipline of complex systems research, which explores everything from the structure of social networks and ecological food webs to the birth and death of cities. There are many reasons to suspect that information plays a pervasive role throughout the physical world in general.

Often, in the history of science, we find that scientific thought closely parallels contemporary technology. The history of our prevailing views about the universe is an example. A popular viewpoint in Isaac Newton's day was that the universe behaves as a giant clock, describable as a perfectly predictable machine, with gears governed by the laws of physics. This *mechanistic* picture of the universe is reflective of the state-of-the-art technology of the day, where the regularity of the rules underlying the behavior of mechanical machines was being elucidated for the first time by newly discovered laws of classical physics (such as Newton's laws of motion). The viewpoint of a clockwork universe eventually yielded to a *thermodynamic* picture, where the universe was viewed as a giant engine. The notion of an engine universe was in vogue during the industrial revolution, when steam engines transformed both the economic landscape through industrialization and the scientific landscape with the discovery of the first, second, and third laws of thermodynamics. These newly discovered laws described how energy can be used to produce useful work at the expense of dissipation of heat, which had both practical applications and deep implications for our understanding of a possible *heat death* for the universe.*

In the current Information Age, it is popular to view the universe as a giant computer, where the entire universe is to be understood as either the output of a computer program or at the very least mathematical describable in such terms. This *informational* picture is reflective of the ubiquitous influence of computers on our daily lives and reflects a long-standing tradition of scientific thought colored by a technological lens. Likewise, one might equally well take a mechanistic, thermodynamic, or informational view of living systems. Whereas in previous centuries the mechanistic and thermodynamic pictures predominated scientific thought regarding the nature of life, it is currently fashionable to study biology using analogies from information and computer science.

The connection between information and the operation of living systems seems to be deeper than just a passing fad—in many ways, life and information seem inextricably enmeshed. Many discoveries in biology have been a direct result of the application of informational concepts. For instance, the *cracking of the genetic code* (where the mapping between the sequence of nucleotides in DNA and the composition of the resultant translated protein was first solved) was the result of using an information-based analogy to coding in communication theory. John Maynard Smith, in his short treatise *The Concept of Information in Biology*, boldly suggests that the code would never have been solved if informational concepts were not applied, that is, if translation had been treated in terms of the chemistry of protein–RNA interactions alone (Maynard Smith, 2000).

---

* The heat death of the universe is a direct consequence of the second law of thermodynamics, which states that the entropy of a closed system cannot decrease with time. Applied to the entire universe, this implies that all useful energy (work) will eventually be converted to heat (thus no stars, no life, no computation, etc.).

Applying informational concepts is clearly useful and relevant in many areas of biology (e.g., the use of coding, translation, and transcription to describe molecular processes). The challenge is that we do not yet understand the implications. Information-based concepts are not required to understand many aspects of the physical world (i.e., though they may be cast in such terms, it is not necessary to take an information-based description in order to understand Newton's laws). However, information appears to be an essential concept in biology, raising the question, "are informational concepts necessary to understand the phenomenon of life?" In particular, we do not know if biology is fully reducible to the underlying rules of chemistry and physics or if new information-based laws that only come into effect in complex biological systems are necessary to describe life. An important question then is information just a useful metaphor to describe life, which has been taken from our current phase of digitally based technologies, or is the application of informational concepts to biology hinting at something more fundamental?

Biologists have gotten along just fine without needing to answer this question; you do not need to define what life is in order to study it. However, there is one area of research that would significantly benefit from an answer to this question, that is, the emergence of life, where chemistry transitions to biology. It is not at all clear if chemistry alone will provide a sufficient account or if it is necessary to invoke information-based concepts to explain life's origin. This is perhaps one of the deepest questions facing modern science. To understand why the dichotomy between chemistry and information poses such a challenge, we must first consider the nature of information in biology and the applications of information-theoretic concepts to life and its origins. Therefore, the first important question we must address is, "What is information?"

17.2 WHAT IS INFORMATION?

Information is a concept that everyone implicitly seems to understand but is nonetheless difficult to explicitly define.* Any kind of information must always be defined in the context of a source and a receiver. A signal carries information about a source if one can predict the state of the source from the signal. Information is represented and conveyed by characters or letters, but it can also convey meaningful messages, the interpretation of which depends on the context. It is the division between the structure (string of characters or letters) and meaning of a message that presents the greatest conceptual hurdle to rigorously define what we mean when we say *information*.

A common way to organize the conceptual difficulty posed by defining information is to make a distinction between two senses of information: syntactic and semantic. The former readily lends itself to mathematical formalization, while the latter is much more problematic. In his book *Information and the Origin of Life*, Bernd-Olaf Küppers identifies what he calls the *syntactic* aspect of information as comprising the relationships between individual characters used to construct a signal (Küppers, 1990). Characters carrying

---

* Defining information therefore faces many of the same pitfalls as defining the related concepts of complexity, emergence, and even life. Often, people will take the position "I know it when I see it," a concept that does not readily lend itself to formalization.

meaning are *symbols*. Recognition of a symbol by a receiver requires a prearranged agreement between sender and receiver on what a particular symbol represents. The *semantic* aspect of information, on the other hand, comprises relationships between individual characters and what they stand for. In other words, it deals with the *meaning* of messages. An important point is that syntactic information is meaningless unless the recipient of the message already possesses the semantic information necessary to interpret the message.

Claude Shannon formalized the syntactic sense of information in his seminal paper "A mathematical theory of communication" published in 1948 (Shannon, 1948). In information theory as presented by Shannon, anything can be a source of information if it has a range of possible states, where one variable carries information about a second variable such that the state of the second is physically correlated with the first (see Section 17.5). For example, Newtonian mechanics provides the algorithm that maps the state of the solar system today onto its state tomorrow by specifying a trajectory through position and momentum space. One may describe this input–output mapping using Shannon's definition of information since the initial and final states are physically correlated. In general, a signal carries more information about a source if its state is a better predictor of the source, and a signal carries less information if its state is a worse predictor.

In the sense presented by Shannon, any physical system can be described in informational terms. However, biology seems to make use of a richer and much more challenging concept than that described by Shannon, where the expression of information, the execution of programs, and the interpretation of codes play an important role. Thus, biological information appears to also encompass an active quality associated with the interpretation of coded messages and the execution of programmed tasks that is associated with the meaning of molecular messages.

Meaningful information is referred to as *semantic*, *intentional*, or *functional* information. An example to illustrate the distinction between the semantic and syntactic aspect of information can be drawn from language. Consider, for instance, the two arrangements of letters: tlbfyture and butterfly. They both share the same syntactic (or Shannon) information content (e.g., they share the same characters) but carry very different semantic information content. Speakers of the English language will recognize the first arrangement of letters as nonsense and the second as a symbolic representation of a flying insect well known for its beautiful colors and the patterning of its wings. The word *butterfly* therefore carries semantic information, whereas *tlbfyture* does not. The prearranged agreement between sender (me) and receiver (you) for you to recognize the word *butterfly* is the assignment of meaning to words in the English language. In some areas of biology, such as perception, cognition, and language processing, the validity of similar concepts of information and representation seems obvious (i.e., as in human language). However, in other areas of biology, such as biochemistry and molecular biology, the role of semantic information content is much less clear-cut. Thus, we are left to puzzle, "At what level does *language* and symbolic representation emerge from chemical interactions?" In molecular biology, information clearly must act through chemistry but the connection between chemistry and semantic information is far from obvious.

17.3 CHEMISTRY AND INFORMATION

Biology is replete with the colloquial use of informational terms such as transcription, translation, proofreading, messenger, redundancy, editing, and programs, inter alia. While such terms are routinely applied in the biological realm, they do not readily carry over to the underlying chemistry. In fact, many features of biological information appear to be in some sense independent of chemistry.

Maynard Smith points out two particularly illustrative examples (Maynard Smith, 2000). The first is the genetic code, which supplies the mapping between the information encoded in DNA and translated protein (via RNA intermediates). The RNA codon table, detailing the assignment of the 20 coded amino acids to triplet codons, is shown in Table 17.1.

One of the most important observations about the genetic code is that the correspondence between a particular triplet codon and the amino acid it codes for is somewhat arbitrary. In fact, a codon does not even need to be a triplet (see Q3 under Review Question, for example, life could be based on a doublet or quadruplet code [Chen and Schindlinger, 2010]). While decoding necessarily depends on the rules of chemistry, the decoding machinery can be modified to alter the triplet codon assignments. In fact, synthetic biologists have already successfully altered the code by reassigning codons to specify unnatural amino acids (defined as amino acids outside of the standard biological set of the canonical 20 coded amino acids). Thus far, codon reassignment has been achieved by engineering aminoacyl transfer RNA (tRNA) synthetase enzymes to carry unnatural amino acids.*

TABLE 17.1 Codon Assignments for the Canonical Genetic Code

	U		C		A		G	
U	UUU	Phe	UCU	Ser	UAU	Tyr	UGU	Cys
	UUC		UCC		UAC		UGC	
	UUA		UCA		UAA	*Stop*	UGA	*Stop*
	UUG	Leu	UCG		UAG	*Stop*	UGG	Trp
C	CUU	Leu	CCU	Pro	CAU	His	CGU	Arg
	CUC		CCC		CAC		CGC	
	CUA		CCA		CAA	Gln	CGA	
	CUG		CCG		CAG		CGG	
A	AUU	Ile	ACU	Thr	AAU	Asn	AGU	Ser
	AUC		ACC		AAC		AGC	
	AUA		ACA		AAA	Lys	AGA	Arg
	AUG	**Met**	ACG		AAG		AGG	
G	GUU	Val	GCU	Ala	GAU	Asp	GGU	Gly
	GUC		GCC		GAC		GGC	
	GUA		GCA		GAA	Glu	GGA	
	GUG		GCG		GAG		GGG	

Note: AUG is the start codon (indicated in bold) and UAA, UAG, and UGA are stop codons (indicated in italics).

---

* During translation, tRNAs carry amino acids to the ribosome. Aminoacyl tRNA synthetases attach the appropriate amino acid onto the corresponding tRNA.

The result is a codon with an altered meaning. In this sense, the code is symbolic—it does not wholly depend on the specific attributes of the medium (chemistry). Of course, all reactions must obey the laws of chemistry, but there is no necessity as to which codons specify which amino acids (within a set of chemical constraints), implying that codons act as biochemical symbols.

Nobel laureate Jacques Monod called this symbolic aspect of biochemistry *gratuité* (roughly translated to gratuity) in his famous book *Chance and Necessity: Essay on the Natural Philosophy of Modern Biology* (Monod, 1971). In collaboration with François Jacob, Monod demonstrated an early mechanism for genetic regulation, an achievement that earned Monod and Jacob the Nobel Prize.* Jacob and Monod showed how expression of a gene could be switched off by a repressor protein that is produced by a second, regulatory gene. The same gene can be switched back on by an inducer, which is typically a small molecule that binds to the repressor protein and alters its shape (e.g., in the lac operon, the inducer molecule is lactose itself). The important point Monod made is that the inducer has no direct contact with the gene; thus, at least in principle, any inducer molecule could act as a switch for any gene. The connection between inducers and the genes they control is therefore arbitrary in the same sense that codon assignments are arbitrary, thus providing a second example of symbolic representation in biochemistry. Both inducers and codon assignments rely on molecular symbolism, with no fundamental underlying connection between their chemical composition and the resultant meaning or interpretation.

The role of symbolic representation in biology suggests a vast logical divide between the realm of chemistry and physics, where phenomena are described in terms of matter, energy, and forces, and that of biology, which is described in terms of signals and codes. In contrast to the semantic aspect of information, the Shannon aspect of information is easy to quantify, including in biological and chemical systems. We will therefore first look at the application of Shannon information theory to biology. We will then tackle the much more challenging problem of understanding the role of semantic information in life and its origin(s).

17.4 SHANNON INFORMATION CONTENT OF BIOPOLYMERS

Shannon originally developed information theory for describing reliable transmission of messages from a source to a receiver over a noisy communication channel (Shannon, 1948). In subsequent decades, Shannon's theory of information has been applied much more broadly, including in disciplines as diverse as linguistics, computer science, electrical engineering, physics, and biology to name a few. Here, we are most interested in the application of Shannon information theory to biology.

Let us consider the information content of genes composed of the four-letter alphabet of nucleobases A, G, C, and T. *Shannon information*, or Shannon entropy (*S*), provides a

---

* The specific system Jacob and Monod studied was the *Escherichia coli* lac operon, which encodes proteins for the transport and breakdown of the sugar lactose.

measure of the diversity of nucleobases in a given gene, which takes into account the relative representation of the different bases. The Shannon entropy is defined as

$$S = -\sum_i p_i \log p_i \qquad (17.1)$$

where
    i represents a letter of the alphabet
    p_i is the frequency of i in the sequence

If applied to the sequence of a gene, i corresponds to one of the four bases A, G, C, and T. A natural measure of information content is the bit—defined by a system that has two equally likely alternatives (e.g., an unbiased coin toss), or $\log_2 2 = 1$ bit of information, as shown in Figure 17.1 (for $p = 0.5$). Thus, the information content of a single genetic base, if all four bases are equally likely, is 2 bits.

In general, Shannon entropy is highest for sequences with the highest statistical diversity, that is, those representing an equal distribution of all possible base alphabet letters. In reality, the bases are not equally likely due to chemical effects such as correlations between neighboring bases. So, there is some reduction in the quantity of information, but 2 bits per base pair (bp) provides a good estimate. As such, Shannon entropy is highest for genetic sequences containing an evenly distributed number of each of the four bases. In contrast, for homogeneous, or uniform, sequences, such as polyguanine (poly-G, a repeating string of Gs), knowledge of the identity of a specific base does not provide any information, since a guess of G for any position in the sequence would always be correct and therefore relays 0 bits of information.

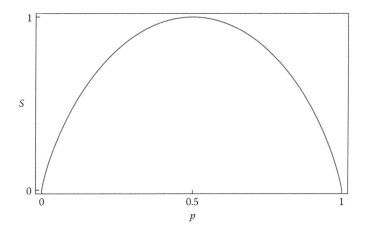

FIGURE 17.1 Shannon information content for a system with two alternatives. Shannon entropy is maximized at 1 bit per trial when the two possible alternatives are equally probable (e.g., an unbiased coin toss). Information content is 0 if the outcome relays no information.

The Shannon information content also depends on the choice of alphabet. So far, we considered the information content of DNA sequences in terms of the composition of bases. However, one might equally well choose a different base alphabet and calculate the information content in terms of the triplet codons. Taking the 64 codons as the base alphabet yields a maximal possible Shannon information content of 6 bits per triplet, for the case where all codons are equally likely. The coding regions of DNA are read out and translated to make proteins, which are based on yet another alphabet of 20 (or in some cases more) coded amino acids. The alphabet composed of 64 codons permits higher maximal Shannon information content than the alphabet composed of the 20 amino acids (see Q3 under Review Question). A coding region and the protein it specifies can therefore have very different Shannon information content, even if the role of coding regions is precisely to specify the composition of proteins. This is a direct result of the degeneracy of the genetic code (where multiple codons encode the same amino acid) and suggests that not all aspects of biological information content are accurately captured by the Shannon information measure alone.

Shannon entropy considers the sequence of a biopolymer as if it was a digital string of characters and contains no reference to biological function. Shannon entropy is therefore related to the more abstract measure of *algorithmic information*, used in computer science to study the complexity of digital strings (Chaitin, 1987). Algorithmic information measures the complexity of a given sequence by calculating its incompressibility by a computer algorithm and is related to the work of Alan Turing on the foundations of computing (see Section 17.8). In the words of Gregory Chaitin, a pioneer in algorithmic information theory, algorithmic information is "the result of putting Shannon's information theory and Turing's computability theory into a cocktail shaker and shaking vigorously" (Chaitin, 1987).

An example from the short and colorful article "The complexity of songs" by computer scientist Donald Knuth provides an informative illustration of algorithmic information (Knuth, 1984). Knuth notes that a song like "99 Bottles of Beer on the Wall" has very low information content due to the high redundancy in the lyrics. This should be in contrast with the information conveyed by a more complex song like Queen's "Bohemian Rhapsody," which relays a larger amount of information for a fixed number of letters. Nearly all of the lyrics of "99 Bottles of Beer on the Wall" are compressed to a single line of information contained in the name of the song. This is not true for "Bohemian Rhapsody." The compressibility of the lyrics of "99 Bottles of Beer on the Wall" makes it similar to a crystal in its Shannon or algorithmic information content. (A crystal has very low algorithmic information content because it may be specified by a compact instruction set of the form, "Add another atom at distance x and repeat N times." Likewise for "99 Bottles of Beer on the Wall.")

Structures that contain patterns contain redundancy and can therefore be specified by algorithms with low information content. Random structures, however, are algorithmically incompressible: they cannot be specified as the output of algorithms much simpler than themselves. Biopolymer sequences are typically random in this respect, that is, they do not contain much patterning or redundancy (there are of course exceptions to this; see the example of leucine zippers in Section 17.6). Therefore, biological information appears to be algorithmically incompressible (and thus more like "Bohemian Rhapsody" than "99 Bottles of Beer on the Wall"). In the very early days of molecular biology, Erwin Schrödinger, the renowned quantum physicist, recognized this when he postulated that the genetic material of life must be *an aperiodic crystal*, in his now famous book entitled *What Is Life?** His reasoning was that the genetic material must be incompressible, such that it could contain an instruction set of complexity comparable to the system it describes (i.e., a living cell).

Shannon information, and the related measure of algorithmic information, provides the formalism for quantifying the syntactic information content of the genome and the information acquired through the process of evolution (which we turn to in the next section). It also has implications for our understanding of (Shannon or algorithmic) information transfer between genetic polymers in the early evolution of life.

As briefly mentioned earlier, there is a mismatch between the Shannon information content of proteins and that of the DNA that encodes them: the Shannon entropy of DNA is much higher than that of proteins due to the redundancy of the genetic code (e.g., see Table 17.1). Prior to the evolution of translation, which mediates information transfer from DNA to proteins via coded RNA intermediates (or RNA messengers), early biopolymers would have "talked" directly with no coded intermediates. This would have occurred in much the same way as DNA "talks" to RNA. This is possible in modern life because RNA and DNA are compatible nucleic acids, allowing the transfer of sequential Shannon information between the two polymeric species. In part because of this compatibility, RNA is believed to have preceded DNA as the genetic material. This also opens the possibility that other primitive genetic polymers may have preceded RNA in this role (Hud et al., 2013).

While there are no known relics of this putative *pre-RNA* chemistry remaining in modern life, the hypothesis has some constraints and is not based purely on wild speculation. Not all related nucleic acids can transfer information with one another, even if they can interact directly with a common intermediate. For example, both threose nucleic acid (TNA) and glycol nucleic acid (GNA) have been suggested to be precursors to RNA. However, while TNA and GNA can both exchange sequential information with RNA, they cannot exchange information directly with one another (Yang et al., 2007). As such, TNA and GNA could not have been consecutive polymers in the same evolutionary pathway to RNA because Shannon information cannot be transferred directly between the two polymer species.

Mapping the landscape of genetic polymers that can exchange Shannon information with one another might allow us to sequentially step further back in time from RNA-based systems to prebiotic chemistry populated by very different pre-RNA genetic system(s). It also allows the possibility of stepping outward from extant life in new directions, which have not been explored by life on Earth. Much of this landscape of nucleic acid architectures has yet to be mapped (Eschenmoser, 1999). In particular, how far in the structural space of nucleic acids we can depart from RNA or DNA through transfer of sequential information and still maintain biological functionality remains an open question.

---

* Schrödinger's book *What Is Life?* is credited with stimulating much of the mid-twentieth century revolution in molecular biology, including inspiring both Watson and Crick to uncover the structure of DNA.

17.5 INFORMATIONAL LIMITS OF EVOLUTION

The Shannon entropy measure permits us to quantify the information content of genetic polymers, such as DNA and RNA (as well as other putative prebiotic genetic polymers such as TNA and GNA), thus allowing quantification of the information content of genomes. Shannon information therefore provides insights into the process of the biological acquisition of information through Darwinian evolution.

At its core, Darwinian evolution describes how entities with traits that maximize reproductive success will increase in frequency relative to their competitors through the process of natural selection. Novelty that produces differences in the reproductive viability of organisms can be introduced via mutation. Natural selection acts on this differential reproductive success, where reproductive success is typically quantified in terms of *fitness*. Thus, Darwinian evolution describes the process of *survival of the fittest*.

As a preamble, we describe the information-theoretic content of the replicator equation (RE), which will allow us to introduce some of the mathematical concepts before addressing the complications introduced by mutation. This treatment of the RE will provide us with a foundation for understanding the dynamics of replication and selection as embodied by *survival of the fittest*. We will then add mutation to the dynamics of selection and replication with the quasispecies equation. Introduction of copying errors via mutation allows us to derive a fundamental limit on the amount of information that can be reliably propagated from generation to generation, the so-called error threshold, which—as we will discuss—has important implications for the origin and early evolution of life.

17.5.1 Replicator Equation

Consider a population of N competing species—these could be RNA molecules in a test tube or lions and hyenas competing for prey in the African savannah. Selection acts to retain only the *fittest* variants, that is, those best adapted to survive and reproduce in the current environment. In a test tube, this is the most efficient RNA replicator, whereas on the savannah, it is the best hunter. The *survival of the fittest* dynamics of replication and selection are captured by the RE (Nowak, 2006).

The RE is the most fundamental equation of evolution. It describes selection among N different genotypes with fixed total population size and is written as

$$\dot{x}_i = x_i(f_i - \phi) \quad i = 1, \ldots, N$$

$$\phi = \sum_i f_i x_i \tag{17.2}$$

The structure of the population is given by the vector $\vec{x} = (x_1, x_2, \ldots, x_N)$ where $\sum_{i=1}^{N} x_i = 1$, such that each x_i represents the relative abundances, or frequencies, of a species i within the population. The equation is written in differential form, such that $\dot{x}_i$ represents the time rate of change of the abundance of species x_i. The dynamics are dependent on the fitness, f_i, of species i and the average fitness of the population given by ϕ.

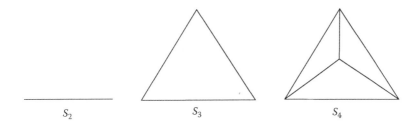

S_2 S_3 S_4

FIGURE 17.2 For fixed total population size and $i = 1, 2, 3, \ldots, N$ different genotypes, selection dynamics may be written in terms of frequencies or relative abundances, labeled x_i, where the sum over all x_i is one. The population structure is thus captured by the set of points $\vec{x} = (x_1, x_2, \ldots, x_N)$ where $\sum_{i=1}^{N} x_i = 1$. This describes a simplex, S_N. A simplex is the set of all points with nonnegative coordinates that add up to one. For N genotypes, the selection dynamics of the RE occur on the simplex S_N. Shown are the simplexes S_2, S_3, and S_4. Each simplex S_N is an $n - 1$ dimensional structure embedded in an N-dimensional Euclidean space.

The frequency of species i within the population will increase for $f_i > \phi$ (x_i is positive) and decrease for $f_i < \phi$ (x_i is negative). In words, species with fitness higher than the average survive and those with fitness lower than the average do not. Thus, we see that the RE captures the dynamics expected of survival of the fittest where the species with the highest relative fitness (largest value of f_i) is selected and all other variants become extinct.*

The set of points with $\sum_{i=1}^{N} x_i = 1$ is called the simplex S_N. S_N is an N-dimensional Euclidean space; examples for S_2, S_3, and S_4 are shown in Figure 17.2. Solutions to the RE live at a corner of a simplex where $x_i = 1$ for the fittest variant ($i = \max(f_i)$), and all other species $j \neq i$ are extinct. An illustrative trajectory showing the selection dynamics and convergence on a fittest variant is shown in Figure 17.3.

The mathematical structure of the RE readily allows characterizing evolution in terms of Shannon information. Following Krakauer (2011), the total information gathered by a population about its environment may be written in terms of the Shannon information content as

$$IG(t) = \sum_i x_i(t) \log x_i(t) - \log\left(\frac{1}{N}\right) \qquad (17.3)$$

such that the information accumulated by the process of selection is the difference between the observed Shannon information of the population (first term) and the maximum possible certainty over the choice of genomes (second term). For large times, $IG(t \to \infty) = \log(N)$. Approximately, N variants must be lost for a population to gain $\log(N)$ bits of information about its environment. Selection therefore acts to channel all resources into a

---

* Selection eventually leads to the survival of only the fittest species because ϕ is a dynamic variable and increases with time as the less fit species die out and only fit variants are left in the population. Eventually, only the species with the highest fitness survives this culling process.

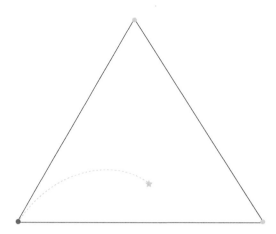

FIGURE 17.3   Selection dynamics of the RE on the simplex S_3. Each point in the simplex refers to a particular structure of the population. The dashed line traces a trajectory from the initial population (star) to a corner of the simplex (red circle) where only the fittest variant in the population remains.

single genotype. Information gained about the environment through the process of selection is therefore directly proportional to the life that is lost. Thus, the information gathered through evolution has a maximal value that is directly related to how many organismal variants did not survive (Krakauer, 2011).

17.5.2 Molecular Quasispecies and the Quasispecies Equation

Earlier, we focused on the information gained through the process of selection at the population level. We next consider the amount of information an individual genome carries within a population and introduce the possibility of mutation. Replication of genetic material, such as DNA and RNA, can result in errors (or mutations) in the copied sequences. Introduction of copying errors to the selection dynamics of the RE yields the quasispecies equation. Molecular quasispecies were first introduced by Manfred Eigen and Peter Shuster to describe an ensemble of related genomic sequences that are rapidly mutating, where offspring typically contain one or more mutations relative to their parent (Eigen, 1971; Eigen and Schuster, 1977). Quasispecies are to be contrasted with the notion of *species* in chemistry, which refers to an ensemble of identical molecules (i.e., the species of all water molecules).* For example, a quasispecies population of RNA molecules in a test tube, or a viral population undergoing rapid mutation between genotypes, does not contain an ensemble of strictly identical sequences and thus represents a *quasispecies* of molecular variants. Due to rapid mutation among related sequences, quasispecies are often described as a diffuse *cloud* of related genotypes, linked by mutation, that collectively contribute to the characteristics of the population, as shown in Figure 17.4.

---

* Here is an example where the same word can be interpreted differently in different disciplines, sometimes leading to confusion. There is a popular misconception that species in quasispecies refers to the notion of a biological species, but the original intention was with reference to the use of the word in chemistry, which has a very different interpretation.

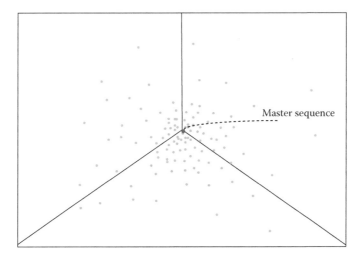

FIGURE 17.4 Quasispecies is a population of closely related reproducing genetic molecules. Quasispecies represent a diffuse *cloud* of related sequences in genotypic space, where the master sequence is the fittest variant.

Introducing mutation rates to the replication and selection dynamics of the RE captures the effects of errors in the copying process. Mutation rates are described in matrix notation and denoted as q_{ij}, where each q_{ij} describes the probability that replication of genome i results in genome j. Including these mutation rates in the RE yields the quasispecies equation

$$\dot{x}_i = \sum_{j=1}^{N} x_j f_j q_{ij} - x_i \phi \quad i = 1, \ldots, N \tag{17.4}$$

This equation is also sometimes referred to as the replicator–mutator equation. Here, as with the RE, each x_i represents the relative abundances of a genotype i, f_i represents the fitness of species i, and ϕ is the average fitness of the population. Quasispecies evolve until equilibrium between mutation and selection is achieved. Therefore, unlike solutions of the RE (which converges to a single *fittest* genotype), the solutions of the quasispecies equation contain a large diversity of genotypes. It is therefore the fitness of the cloud of related genotypes, rather than the fitness of a single genotype, that evolution is optimizing for the case of the quasispecies.

It is important to note that any genomic information (or any other kind of information) can be encoded in binary sequences. Thus, when *modeling* the dynamics of informational sequences, such as in the quasispecies model, binary sequences are often used for convenience even though biology implements larger alphabets (e.g., the set of four nucleotides, or 20 amino acids). To understand why, consider the total number of sequences of length L composed of a binary alphabet, compared to an alphabet composed of four letters. For a binary alphabet of 0 and 1, there are 2^L possible sequences. Compare to the case for genetic polymers composed of four bases with 4^L possible sequences. This represents a vast jump in complexity over a binary alphabet and therefore a vast jump in the complexity of a model required to describe it. As such, in theoretical modeling, binary representations are often used due to their relative simplicity, often without loss of generality. In the quasispecies model, the variants x_i each encode a unique binary sequence of 0s and 1s with $i = 0, 1, ..., N$ where $N = 2^L - 1$. In example, the most natural enumeration of sequences is obtained if the sequences encode the binary representation of the index i, that is, for a population of sequences of length $L = 5$, $i = 0$ corresponds to the sequence 00000, $i = 1$ corresponds to 00010, and so on.

As with the RE, quasispecies dynamics are defined on the simplex S_N. Sequence i is generated by replicating any sequence j at rate f_j times the probability that sequence j generates sequence i, which is given by p_{ij}. Sequences are removed at a rate ϕ such that the total population size remains constant with $\sum_{i=1}^{N} x_i = 1$. In the limiting case of no mutation, replication produces perfect copies with probability $q_{ii} = 1$, with all other $q_{ij} = 0$ for $i \neq j$, and we recover the RE.

For the set of all sequences of length L, the form of the mutation rates q_{ij} can be written out explicitly for the case of point mutations as

$$q_{ij} = u^{h_{ij}} (1-u)^{L-h_{ij}} \tag{17.5}$$

Here, u is the mutation rate for an individual position along a sequence. Thus, the probability that an individual position is copied correctly is given by $(1 - u)$. The h_{ij} are referred to as Hamming distances and track the number of positions that differ between sequences i and j. The *distance* is one in genotypic space. A Hamming space for $L = 3$ is shown in Figure 17.5.

For two sequences sharing all positions in common (i.e., identical sequences), $h_{ii} = 0$. For two sequences sharing no positions in common (i.e., separated by the maximum possible

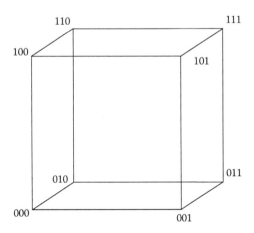

FIGURE 17.5 Genomes can be described in an abstract sequence space called a Hamming space. The length of the genome defines the number of dimensions of the Hamming space. Short viruses, such as the RNA Qβ virus, occupy a genomic space of several thousand dimensions; for humans, the space is roughly 3 billion dimensions; shown is the Hamming space for genomes of length $L = 3$.

distance L), $h_{ii} = L$ (i.e., for the space shown in Figure 17.5, three edges of the cube are traversed to change all three positions in a sequence). Thus, $L - h_{ij}$ is the number of sites shared in common by sequences i and j.

17.5.3 Error Threshold and the RNA World

The quasispecies model has important implications for our understanding of the origin and early evolution life. A popular view is the *RNA world* hypothesis, a term coined by Nobel laureate Walter Gilbert, which outlines how the catalytic properties of RNA support the idea of a prominent role for RNA in ancient life (Gilbert, 1986). In the RNA world, RNA played the role of both catalyst and genetic information carrier prior to the advent of the translation machinery, when the respective roles of protein catalysts as biochemical workhorse and DNA as genetic information storage repository were established.

There are weak and strong versions of the RNA world hypothesis. An extreme viewpoint is that life originated with self-replicating RNA molecules, which were initially synthesized abiotically and then gradually accrued additional functions through the process of Darwinian evolution. This strong RNA world scenario purports that RNA-based organisms (with RNA as the only biopolymer) were the very first form of life on Earth. A more moderate perspective is a weak RNA world scenario, where an early period of life on Earth utilized RNA as the only genetically encoded component of biological catalysis. This is an important distinction: under this second view, the RNA world may have been preceded by even more primitive living systems, which have left few relics in modern organisms. It is therefore not necessary that RNA is the very first chemistry of life. Regardless of one's preference for the strong or weak scenario, there is strong evidence to support RNA playing a prominent role in early evolution.

The RNA world hypothesis has fostered much interest in the evolution of RNA populations due to the implications for our understanding of early RNA-based systems. An important consideration is that any rRNA organisms (RNA-based organisms) populating the putative *RNA world* (strong or weak) would have been subject to high copying error rates, as they would have lacked the sophisticated and highly evolved error-correcting (protein-based) enzymes of modern life. Such highly mutating populations may be described by the quasispecies equation.

As noted earlier, quasispecies evolve until equilibrium between mutation and selection is achieved. Calculating the properties of this equilibrium distribution for complex fitness landscapes is difficult, but tremendous insight can be gained by considering an example with a simplified fitness landscape. We consider all binary sequences of length L and identify the all 0 sequence $x_0 = 000...0$ as the fittest sequence, dubbed the *master sequence* (the fittest sequence is also often referred to as the wild type). The fitness of the master sequence is given by $f_0 > 1$. All other sequences, or *mutants*, are given a fitness of 1 and are thus less fit than the master sequence. Using Equation 17.5, we can calculate the probability that the master sequence produces a perfect copy of itself ($h_{00} = 0$), which is given by $q = (1-u)^L$. Since perfect copies are produced with probability q, mutant sequences are generated by the master sequence with a probability given by $1-q$. For the purposes of discussion here, we neglect back mutation from the mutants to the master sequence (here, we are using the tools of the theorist by using carefully chosen approximations to make our equations more tractable to solve). This simplification allows us to reduce the system of equations in Equation 17.4 to two equations:

$$\dot{x}_0 = x_0(f_0 q - \phi)$$
$$\dot{x}_1 = x_0 f_0 (1 - q) + (1 - \phi)x_1$$

(17.6)

The first equation describes the dynamics of the master sequence x_0: perfect copies are produced at a rate $f_0 q$ and are lost at a rate ϕ. The second equation sums over the dynamics of all mutant sequences, where x_1 is the sum over the frequencies of all mutants. The first term describes production of mutations via error in copying of the master sequence, with new mutants produced at a rate $f_0(1 - q)$. The second term describes the selection dynamics of mutants, as would be written in the absence of mutation (recall $f = 1$ for all mutants). The average fitness is given by $\phi = f_0 x_0 + x_1$.

We can simplify this set of two equations even further by noting that the total population size is given by $x_0 + x_1 = 1$. Substituting this into our expression for ϕ, and substituting the result into Equation 17.6, yields

$$\dot{x}_0 = x_0(f_0 q - 1 - x_0(f_0 - 1))$$

(17.7)

This one equation describes the full dynamics of our population. The approximations made earlier are precisely what allow us to write the dynamics of the entire cloud of related genotypes in one relatively simple equation. Equation 17.7 has two steady-state (equilibrium) solutions found by solving the earlier equation when $\dot{x}_0 = 0$. Steady state is achieved if $x_0^* = 0$ or if

$$x_0^* = \frac{f_0 q - 1}{f_0 - 1}$$

(17.8)

(here, * is used to denote steady-state values). For $f_0 q < 1$, this second solution yields an unphysical value of a negative population size for the master sequence, that is, $x_0^* < 0$. The second solution is therefore unstable if $f_0 q < 1$, and the system converges to the first steady-state solution $x_0^* = 0$. The condition $f_0 q < 1$ therefore provides an *error threshold* whereby the fittest sequence cannot be maintained in the population. Using the expression $q = (1 - u)^L$, this inequality can be rewritten as $\log f_0 > -L \log(1 - u)$. However, this expression does not provide much insight in its current form. We can further simplify it by assuming small mutation rates $u \ll 1$, such that $\log(1 - u) \approx -u$, and by assuming that the fitness advantage of the master sequence is not too large nor too small such that $\log f_0 \approx 1$. With these assumptions, the error threshold condition reduces to the succinct expression

$$u < \frac{1}{L}$$

(17.9)

This is the central result we were after. This error threshold places an upper limit on the maximum mutation rate compatible with adaptive selection. Adaptation requires that on average, at least one perfect copy must be made per replication event. Therefore, for a sequence of L symbols, the probability of producing an error when replicating a symbol must be less than $1/L$. Stated another way, Equation 17.9 states that the maximum mutation rate must be less than the inverse of the genome length. If the genome size or the mutation rate rises above this limit, the result is an accumulation of mutants in the population at the expense of the master sequence. Systems that cross the error threshold experience an error catastrophe (an extinction event due to excessive mutation), where selection cannot retain the fittest variants within the population.

Information content scales with the length of a sequence; therefore, the error threshold places stringent limits on the amount of genomic information that may be reliably propagated from generation to generation for fixed mutation rate. Extant organisms can have large genomes due to the low mutation rates enabled by the presence of efficient error-correcting enzymes. These enzymes provide mechanisms of proofreading and mismatch repair during the copying of genetic material. Mutation rates vary among species and even among locations within the genome of the same organism. Many organisms and most viruses have mutation rates that place them at the border of the error threshold, such that their mutation rates are as large as possible while still permitting an adaptive response. Mutation rates vary from as low as 1 error per 100 million (10^{-8}) to 1 billion (10^{-9}) bases, primarily in bacteria, to as high as 1 mistake per 100 (10^{-2}) to 1000 (10^{-3}) bases in viruses. If DNA repair were perfect with no mutations ever accumulating, there would be no genetic variation and thus no genetic diversity to fuel evolution. Thus, the mutation rate of an organism represents an evolutionary trade-off, or compromise, between the need to reduce the frequency of harmful mutations and the need to sustain sufficient genetic diversity support evolution. It remains an open question as to the extent to which the mutation rates of extant organisms (and viruses) are optimized for evolvability (see, e.g., Q5 under Review Question).

The error threshold introduces a paradox in our understanding of the origin of life. In the absence of error-correcting enzymes, the error rate of nucleic acid replication is at best approximately 1 mistake per 100 (10^{-2}) bases. Thus, according to Equation 17.9, the maximum genome size, in the absence of error correction, is *at most* a few hundred digits.* A paradox arises because the enzymes required for error correction cannot be encoded in such short molecules. This dilemma, known as *Eigen's paradox*, may succinctly be stated as "no error-correcting enzymes without a large genome, and no genome without error-correcting enzymes" (Maynard Smith, 1983). More colloquially, Maynard Smith has referred this to as the *catch-22 of molecular evolution*.

The Eigen paradox presents a significant hurdle any viable model for the first replicating biopolymers must cross, but there are several possible resolutions. Eigen's own solution

---

* Viruses, as noted earlier, can get by with high error rates of order 1 mistake per 100. However, we must note that viruses co-opt the translation machinery of cells to reproduce. Information specifying the translation machinery is replicated with an error rate commensurate with that of the host organism; thus, even viruses rely on efficient error correction, albeit in an indirect manner.

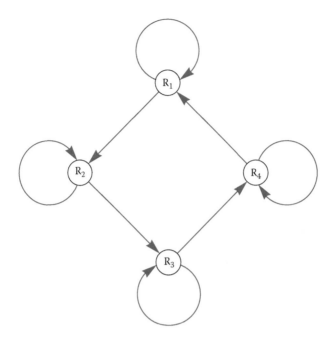

FIGURE 17.6 Hypercycle is a set of mutually catalytic molecules, each capable of self-replication and catalyzing the replication of another member of the set. Shown is a hypercycle with four cooperative catalysts; arrows indicated catalytic interaction.

was the introduction of the hypercycle, a set of mutually catalytic molecules, each capable of self-replication and catalyzing the replication of another member of the set as shown in Figure 17.6.

Eigen reasoned that by sustaining stable coexistence of multiple quasispecies, each with short lengths (below the error threshold), diversity could be maintained within a population while averting an error catastrophe. However, hypercycles are unstable systems and undergo oscillatory behavior in the population size of replicators for sets of four or more cooperative catalysts. This instability leads hypercycles to be vulnerable to perturbations, including those introduced via mutation of members of the hypercycle and through parasitism by other replicators, both of which lead to collapse of the hypercycle.

More recently, Kun et al. have looked for resolutions to the Eigen paradox by more closely studying the physical properties of RNA fitness landscapes (Kun et al., 2005). In the above, we calculated the error threshold for the limiting case where all bases contribute equally to the fitness of the genome. However, in reality, mutation of different bases will differentially impact the function of an RNA molecule, and the error threshold will therefore depend on the shape of the fitness landscape. In particular, the information that must be preserved through the replication process is not the sequence information per se, but the functionality of the molecule. In RNA fitness landscapes, *neutral* mutations can lead to changes in the genotype that do not lead to a corresponding change in the functionality of the molecule (its phenotype). That is to say that many related sequences will yield similar folded structures and thus similar functionality. Thus, the mapping of genotype to phenotype is not 1:1, as we assumed in our overly simplistic derivation. Taking the phenotype of a molecule into account, the error threshold for real ribozymes (catalytic RNA molecules) has been shown to rise to lengths as much as several thousand nucleotides (Kun et al., 2005). This is sufficient for a putative riboorganism to persist with a sizeable genomic inventory of a few thousand bases, even in the absence of modern protein-based error-correcting enzymes. Further studies are necessary to understand the implications of a relaxed error threshold for other plausible primitive genetic polymers, whose fitness landscapes are yet to be mapped.

The relaxation of the error threshold is permissible because what really counts is not strictly the sequence information, but instead the function of a biopolymer. For much of the foregoing discussion, we have implicitly assumed that all of the relevant information for the operation of biological systems is quantified by the number of bits in biopolymer sequences as quantified by their Shannon information content. However, Shannon information content quantifies the number of bits necessary to specify the sequence of a biopolymer, without reference to its biological function. By looking at the properties of real fitness landscapes and not blindly accepting the limits imposed on informational heredity by considering only sequential (Shannon) information content, we have inadvertently stumbled into the territory discussed in Section 17.1. That is, we have come to the point where we must address the functional and semantic aspects of biochemistry and the nature of biological information.

17.6 WHAT IS BIOLOGICAL INFORMATION?

In addressing the nature of biological information, it is much easier to state what it is not, rather than what it is. In particular, while it is true that the Shannon definition of information is legitimate and useful in many areas of biology, we have already seen hints that it does not fully capture the essence of biological information. We will use one more example to bring this point home, by assessing if Shannon information correlates with biological function.

Chen and Schindlinger have provided a nice analogy from music to illustrate how selection for function could correlate with selection on Shannon information content (see Chen and Schindlinger, 2010). Music relaying maximum Shannon information content is exemplified by the 12-tone technique (aka dodecaphony), a method of composition developed by Arnold Schoenberg in the early twentieth century, in which all 12 notes are sounded equally with no emphasis on any one particular note. While those in the art world may appreciate this style of music, most listeners prefer a tonal aesthetic where a particular key is chosen (emphasizing particular notes above others). Popular songs are usually in a key and thus have lower Shannon information content than compositions implementing the 12-tone technique. However, the opposite extreme of minimal Shannon information is also typically viewed as not being aesthetically pleasing. Selection in this case is for the function of listener enjoyment. While there are clear trends, the correlation between the function of listener enjoyment and Shannon information is tenuous and not readily defined. Chen and Schindlinger provided the counter example of the repeating note in Antonio Carlos Jobim's "One Note Samba," which serves the function of listener enjoyment despite its low Shannon information content.

Like with these examples from music, on first inspection, it appears that functionality is associated with biopolymer sequences that have relatively high, but submaximal, information content. Proteins, for example, are composed of an alphabet of 20 possible amino acids (with few exceptions), but are rarely composed of nearly equal numbers of all 20 amino acids. (For example, start [TUG] and stop [TAA, TAG, and TGA] codons appear only once in a gene sequence.) Many proteins contain repeat elements that are necessary for their function, but reduce their Shannon information content relative to the maximal probable value. For instance, leucine zippers are a common 3D protein structural motif, typically involved in regulating gene expression, which consist of multiple leucine spaced at approximately seven-unit intervals. The repeating leucines yield a hydrophobic region along one side of the protein, where it can *zip up* with a complimentary protein that also has a leucine-rich region. Due to the predominance of leucines over other species of amino acid in the sequence of a leucine zipper, proteins containing such a zipper have submaximal Shannon information content. Like selection for listener enjoyment, selection for function in biopolymers does not necessarily select sequences with maximum Shannon entropy. In fact, more generally, biopolymer function does not appear to readily map to Shannon information content. It is therefore at present unknown if (and if so, how) Shannon information content is correlated with biological function.

Due to the limitations of the Shannon measure when it comes to quantifying *function* in biology, various other measures of biological functional information have been proposed. One such measure, readily applicable to chemical systems, is the functional information measure of Hazen et al. (2007) who define functional information content as

$$I(E_x) = -\log[F(E_x)] \qquad (17.10)$$

This is calculated with reference to a specific function (i.e., catalytic activity of an RNA ribozyme), labeled as x, where E_x is a specific degree of that function (i.e., efficiency of a ribozyme catalyst). This measure is closely related to the Shannon measure of information; however, an important distinction is in the definition of the probability distribution in the logarithm. In the case of the Shannon measure we studied in Equation 17.1, the probability distribution p_i was over all possible states, for example, all letters of the defined alphabet. Here, the probability distribution $F(E_x)$ is instead over the specified degree of function. In words, $F(E_x)$ is a probabilistic measure of the small fraction of all possible configurations of a system that achieve the specified degree of function given by E_x. In this case, functional information is thus defined by the probability that an arbitrary configuration of a system will achieve a specific function (x) to a specific degree (E_x). Applying this measure to artificial genomes and biopolymers has demonstrated distinctive step features in plots of information versus function (Hazen et al., 2007). These islands of function represent distinct solutions to functional optimization with different maximum degrees of function.

While this seems promising as a framework for defining biological functional information, this definition carries with it an important caveat: calculating the functional information content requires a precise definition for E_x in a given system. One must therefore know a priori both the function and the specified degree of functionality before calculating information content. Therefore, while the measure may be useful for in vitro and in silico experiments where function and degree are easily defined, it cannot be the ultimate definition we seek to characterize biological systems of unknown function.

We have now encountered three information measures—Shannon information, algorithmic information, and functional information—which each fail to capture the essence of biological information. What is it that we are missing? It is at this point that we must unpack our earlier discussion of biological information as manifest in the expression of information, the execution of programs, and the interpretation of codes and address its deeper consequences for the emergence of life.

17.7 PARADOX OF SELF-REFERENCE, SELF-REPLICATING MACHINES, AND LIFE

Consider again as an example the nature of genetic information encoded in DNA. A genome provides a (mostly) passive access on demand database, which contributes biologically meaningful information by being read out to produce functional noncoding RNAs and coded proteins. The biologically relevant information stored in DNA therefore has very little to do with the chemical structure of DNA (recall Monod's gratuité). The crucial point here is that no information is actively processed in the DNA alone. DNA provides a set of instructions—a genetic program—and the processing of those instructions are distributed throughout a myriad of molecular machinery in the cell. The expression of DNA is subject to control by some of this machinery (e.g., gene silencing via methylation). Thus, the current functional state of a living system—that is, the relative level of gene expression—depends on the composition of the proteome, environmental factors, etc., that regulate the expression of individual genes. It is those same genes that dictate which proteins are expressed and as such dictate the future state of the system. Neither subsystem—genotype (genome) or phenotype (proteome)—acts in isolation.

More colloquially, the dynamics presented by this overly simplified example is often referred to as a chicken-or-egg problem, where the cause and effect of genotype and phenotype cannot be determined since each is a cause for the other. Such dynamics is well known from the paradoxes of self-reference. Picture, for example, the artist M.C. Escher's *Drawing Hands* where each of a pair of hands is drawing the other with no possibility of separating the two: it is unclear which hand is the cause and which is the effect. Another example comes from music in the form of the fugue, a style of composition where a subject (e.g., a recognizable melody) is introduced at the start of a piece and subsequently is recapitulated at the end of the piece such that you end up where you started. Masters of the style, such as Bach, are so artful in their composition you might feel as though you are traveling along a Mobius strip to listen to their fugue and be unable to tell if you had just started or just completed your journey. The crux of the problem in defining the role of information in living systems is that life behaves much the same way as Escher's drawings and Bach's fugue and is replete with self-reference (Hofstadter, 1979).

The mathematician John von Neumann provided one of the earliest treatments of this puzzling aspect of life. von Neumann is famous for his work on the foundations of computing, game theory, and quantum mechanics among others. In his later years, one of the problems he most liked to ponder was how to build a machine to perform any physical task—including building itself. He thus sought to design a self-replicating automaton, machines now referred to as von Neumann automata. While designing a self-replicating machine might sound like a straightforward task, in practice, it is fraught with conceptual difficulties, not the least of which are the paradoxes associated self-reference.

von Neumann devised an abstract machine called a universal constructor (UC), a machine capable of taking materials from its host environment to build any possible physical structure (consistent with available resources and laws of physics) including itself. However, UCs are mindless robots and must be told very specifically exactly what to do in order to build the correct object(s).* The UC therefore requires a blueprint of itself in order to replicate. And herein lie the challenge; the blueprint must contain a blueprint of the UC blueprint to be copied (i.e., it must reference itself). But this would only allow the machine and blueprint to be copied once, so the blueprint must contain a blueprint of the blueprint, which contains a blueprint of the blueprint ad infinitum, for the copying to proceed over successive generations. To avoid the logical fallacy presented by *blueprints all the way down*, von Neumann recognized that the blueprint must provide a set of instructions—or a program—for the operation of the UC, rather than an exact description of the UC and blueprint. The blueprint could then be blindly copied without reference to the instructions it contains. This dual role is familiar from the role of DNA in modern life, where the genome acts both passively as a hereditary structure to be copied and actively as a source of instructions (the genetic program). Thus, von Neumann discovered a route to managing the paradox of self-reference that living organisms seem to manage quite well.

To a rough approximation, all known life functions as a von Neumann automaton, where DNA provides the blueprint and ribosomes act as the core of the UC. What is remarkable about von Neumann's self-replicating automata is how closely they parallel the operation of living systems even though von Neumann formulated his ideas before the discoveries of modern molecular biology. Instead, his only tool was logic. The logic von Neumann followed was inspired by the work of Turing on universal computation. von Neumann's quest for a self-replicating machine therefore provides a bridge between the realm of the living and the foundations of computing that could provide insights into the question we posed at the very beginning: are informational concepts necessary to understand the phenomenon of life?

17.8 LIFE, THE FOUNDATIONS OF COMPUTING, AND THE HARDWARE–SOFTWARE DICHOTOMY

Turing is credited with founding the fields of computer science and artificial intelligence, catalyzing the information revolution of the twentieth century. His most impactful contribution was the formalization of computation with his model of a general-purpose

---

* von Neumann was inspired by Turing's work on universal computers, that is, computers that, given sufficient time, can output the results of any computable function. He thus designed the UC in close analogy with a universal computer.

computer—the universal Turing machine (Turing, 1937). Turing showed that it is possible to build a simple computer, which given enough time could output any computable function. Conceptually, a Turing machine is a simple device, consisting of a tape containing symbols and a machine that can read and write symbols on the tape, one at a time. In the words of Turing, "At any moment there is one symbol in the machine; it is called the scanned symbol. The machine can alter the scanned symbol and its behavior is in part determined by that symbol..." (Turing, 1948). The important piece is this last part—that the behavior of the machine is in part determined by the scanned symbol.

Recall our discussion of self-reference in living systems in the previous section where we discussed how the time evolution of the genotype–phenotype dichotomy depends, much like a Turing machine, on both the current state (composition of the proteome, environmental factors, etc.) of the cell and the expressed genes. Biological systems appear to be reading in symbols that in part affect their behavior. This suggests that biology is doing more than just passively using information; a living cell is actively executing programs. If the analogy with computation is apt, living cells must have a program or *software* stored in the underlying biochemical hardware. This software is not stored in any individual molecule—it is not in the DNA alone—but distributed throughout the cell. Thus, unlike with a computer where the software may be confined to a tape, it is difficult to disentangle software from hardware in a biological system: all biochemical circuitry encodes some part of the software in biology. This brings us to an important debate on the origin of life—did the software or hardware come first (Walker and Davies, 2013)?

Before addressing this debate, let us first take a detour to discuss yet another dichotomy in the origin-of-life research—the debate between *genetics-first* and *metabolism-first* scenarios. Traditionally, origin-of-life theorists have tended to split into two camps: those who stress the origin of genetics and those who stress the origin of metabolism. In informational language, genetics and metabolism may be unified under a common conceptual framework by regarding metabolism as a form of analog information processing, to be contrasted to the digital information of genetics. *Digital systems* are distinguished by their use of discrete, discontinuous representations of information, as is the characteristic of the information coded in genes (with a discrete nucleotide alphabet). In contrast, *analog systems* are characterized by continuous representations of information, as exemplified by the expression level of proteins and the concentration of metabolites in a cell. The debate between *metabolism first* and *genetics first* may be ill posed; in much the same way that genotype and phenotype cannot be disentangled, metabolism and genetics are not separable in known life. While these two pictures may differ in the representation of information, they both concentrate on the hardware of life, that is, they are concerned with the chemistry.

In contrast, a software-based view of the emergence of life is consistent with the idea of *information transfer* whereby primitive living systems based on rudimentary chemical hardware may have undergone a hardware upgrade (or a succession of upgrades) to arrive at the biochemistry we observe today. We saw an example of this with information transfer between genetic polymers. An even more drastic transition could involve information transfer between radically different chemical systems and not just nucleic acids.

Graham Cairns-Smith, for example, has proposed that genetic information was originally encoded in inorganic clays, where growth of clay crystals enabled replication and mutation—and thus evolution (Cairns-Smith, 1985). Cairns-Smith goes on to describe *genetic takeover* of these primitive life-forms whereby genetic information encoded in clays eventually transitioned to being stored in more familiar organic genetic polymers leading to the evolution of life we would recognize today. Thus, although the software of life must be instantiated in a physical media (i.e., chemistry) in this picture, the software preceded the hardware of life as we now know it.

However, life does not simply accumulate and store information in chemical hardware: it processes it. Thus, software-based pictures should be concerned with more than just the sequential transfer of (Shannon) information. Some form of chemical hardware must have initiated the processes leading to life's emergence; however, the *lifelike* attributes of this hardware were a result of information taking on an active role. In such a system, it is the software that will be preserved over geological timescales and not the exact hardware.*

Under an informational software-based view, the essential transition in driving the emergence of life from nonliving matter is the decoupling of software from hardware, such that informational language (e.g., programs, symbols, messages) begins to apply and provides appropriate descriptors of physical phenomena without needing to explicitly reference the underlying chemistry. In effect, asking when software emerges is another way of stating our earlier question: at what level does *language*, programs, and symbolic representation emerge from chemical interactions? The emergence of these attributes was likely critical transitions in the early evolution of life (e.g., the evolution of the genetic code). Explicitly taking into account the software, or program, as an active contributor to the functioning of life may therefore be an important factor in pinpointing life's origins that goes beyond than the substrate-based chemical narrative in identifying the essential transition(s) on the pathway from nonliving to living matter. Thus, just as the genetic code may never have been deciphered if the coding analogy had not been applied, identifying the key steps in the pathway to life may require application of information-based concepts. The utility of this approach is that it will apply to life-forms based on any chemistry (not just the chemistry we observe in extant life) as the informational principles underlying biological organization could be universal to any life we encounter, even if based on a radically different chemistry.

17.9 CONCLUSIONS

As with many of the most compelling scientific endeavors, we have left more questions open than answered. New approaches to understanding life's origins are needed to finally resolve the debate about whether the answer to how life emerged from nonlife will ultimately come from chemistry or information-based formalisms or, more likely, some combination of both. Perhaps, the answer to this question will be inspired by future technologies, as our current information-based understanding of the operation of biological systems has been

---

* For example, the chemistry of life is continually regenerated—we inherit genetic information from our parents copied in newly formed DNA; it is not the original DNA molecule that is inherited down a line of descent but the copied information.

by current technology. Or perhaps, the answer will come from uncovering genuinely new principles underlying how living matter first emerged that might in turn inspire the next technological revolution.

GLOSSARY

Algorithmic information: A measure of the complexity of a given sequence, calculated as the compressibility of a computer program whose output is the sequence of interest.

Error threshold: Mutation rate whereby accumulation of errors becomes so great that the fittest sequence cannot be maintained in a quasispecies population.

Functional information: A measure of information capturing the probability that an arbitrary configuration of a system will achieve a specific function to a specific degree.

Genetic code: The mapping between information stored in DNA and the composition of the translated protein.

Hypercycle: A set of mutually catalytic molecules, each capable of self-replication and catalyzing the replication of another member of the set.

Master sequence: The fittest sequence in a quasispecies population, representing the most efficient replicator.

Mutation: A change in the nucleotide sequence (Shannon information content) of a genome.

Quasispecies: An ensemble of related genomic sequences that are rapidly mutating.

RNA world hypothesis: A hypothesis suggesting that RNA played the role of genetic information carrier and the only genetically encoded catalyst in early life.

Self-reference: When a sentence, an idea, or an object refers to itself in natural or formal language.

Semantic information: The set of relationships between individual characters used to construct a message and what they stand for, that is, the meaning of a message.

Shannon information: A syntactic information measure that quantifies the average number of bits needed to store or communicate a character in a message.

Syntactic information: The relationship between individual characters used to construct a signal or message.

Wild-type sequence: See *master sequence.*

REVIEW QUESTIONS

Q1: In the text, we discussed mechanistic, thermodynamic, and information-based pictures of the universe and noted that life could also be described in each of these frameworks. Describe for each what aspect of living systems a mechanistic, thermodynamic, and information-based picture might focus on.

A1: A mechanistic view would focus on the substrate, or chemistry of life, and define the interactions among chemical components like gears in a machine. A thermodynamic picture would focus on energy flows through biological systems and how useful energy is converted to work. An information-based picture would focus on the expression and translation of biological programs, as discussed throughout this chapter.

Q2: The chemist Steve Benner has proposed that life could be based on an expanded genetic alphabet with a larger number of possible nucleotides than observed in known life (Geyer et al., 2003). The chemical rules of combination dictate that for Watson–Crick base pairing (the chemical mechanism for pairing DNA bases G:C and A:T in biological systems), the total possible number of bases is at most 12. If an alien or engineered life-form were to use all 12 possible bases, how much more information per base could its genome contain than known life with just 4 bases?

A2: The Shannon information content would be $\log_2 12 = 3.58$ bits, representing an increase of 1.58 bits per base over the standard genetic alphabet.

Q3: What is the information content of an individual amino acid within a protein if all 20 amino acids are equally likely? Compare this to the case calculated in the text for DNA, which carries 6 bits of information per triplet codon. Why is there a discrepancy between the potential information content of DNA and proteins? What advantage does a degenerate code confer? (*Hint*: Look at the amino acid codon assignments in Table 17.1.)

A3: The information content of one amino acid, if all are equally likely, is $\log_2 20 = 4.32$ bits of information, representing a drop in Shannon information content of 1.58 bits below that of the triplet codon alphabet. The discrepancy arises because of the degeneracy of the genetic code. A degenerate code is beneficial because it reduces the impact of mutational error: if two codons are closely related and code for the same amino acid, a mutation substituting the second codon for the first will not affect the sequence, and hence the functionality, of the expressed protein.

Q4: There are many hypotheses about how the genetic code may have evolved. One such hypothesis posits that the triplet genetic code may have been preceded by a doublet code, where each codon was composed of only two bases. The maximal number of possible codons for a doublet code is $4^2 = 16$. Thus, for the case where no doublet codons are degenerate (i.e., no amino acids are encoded by more than one doublet codon), only 16 coded amino acids are possible. A doublet code would therefore not work for modern life, which uses a set of 20 coded amino acids. However, there is no fundamental limit in the opposite direction that would dictate a maximum codon size. Many researchers have therefore suggested that life could utilize a quadruplet code rather than a triplet code. What is the maximum number of amino acids that could be encoded in a quadruplet code? What is the maximum Shannon information content per quadruplet codon?

A4: The number of quadruplet codons is 256; thus, the maximum number of amino acids that could be encoded in a quadruplet code is 256. The Shannon information per quadruplet codon is $\log_2 256 = 8$ bits.

Q5: The mutation rate per nucleotide per generation for the RNA virus Qβ is estimated to be approximately 1.5×10^{-3}. Estimate the maximum genome length permitted by the error threshold for Qβ, and compare it to the genome length of 4.2×10^3 bases. How does the actual length of Qβ compare to that permitted by the error threshold? Why might the length of the Qβ virus not agree with that predicted by error threshold theory?

A5: The maximum size for a Qβ viral genome permitted by the error threshold is roughly 667 bases. This is far less than the observed genome length, indicating that Qβ survives beyond the error threshold! Possible explanations include a complex fitness landscape (i.e., in the mapping of genotype to phenotype as discussed in the text) or a high rate of adaptability. However, the high rates of Qβ mutation that push it beyond the error threshold are currently unexplained.

REFERENCES

Books

Cairns-Smith, A.G. *Seven Clues to the Origin of Life: A Scientific Detective Story.* Cambridge, U.K.: Cambridge University Press, 1985.
Chaitin, G.J. *Algorithmic Information Theory.* Cambridge, U.K.: Cambridge University Press, 1987.
Hofstadter, D. *Gödel, Escher, Bach.* New York: Basic Books, 1979.
Küppers, B.-O. *Information and the Origin of Life.* Cambridge, MA: The MIT Press, 1990.
Monod, J. *Chance and Necessity: An Essay on the Natural Philosophy of Modern Biology.* New York: Knof, 1971.
Nowak, M. *Evolutionary Dynamics.* Cambridge, MA: Belknap Press, 2006.
Turing, A. 1948. Intelligent machinery. Reprinted in *Cybernetics: Key Papers.* C.R. Evans and A.D.J. Robertson, eds. Baltimore, MD: University Park Press, 1968.

Journal Articles

Chen, I.A. and Schindlinger, M. 2010. Quadruplet codons: One small step for the ribosome, one giant leap for proteins. *Bioessays* 32(8): 650–654.
Eigen, M. 1971. Self-organization of matter and evolution of biological macromolecules. *Die Naturwissenchaften* 58: 465–523.
Eigen, M. and Schuster, P. 1977. The hypercycle: A principle of natural self-organization. *Die Naturwissenchaften* 64: 541–565.
Eschenmoser, A. 1999. Chemical etiology of nucleic acid structure. *Science* 284: 2118–2124.
Geyer, R., Battersby, T., and Benner, S. 2003. Nucleobase pairing in expanded Watson–Crick-like genetic information systems. *Structure* 11: 1485–1498.
Gilbert, W. 1986. The RNA world. *Nature* 319: 618.
Hazen, R., Griffin, P., Carothers, J., and Szostak, J. 2007. Functional information and the emergence of biocomplexity. *Proc. Natl. Acad. Sci. USA* 104: 8574–8581.
Hud, N., Cafferty, B.J., Krishnamurthy, R., and Williams, L.D. 2013. The origin of RNA and "My Grandfather's Axe". *Chem. Biol.* 20: 466–474.
Knuth, D. 1984. The complexity of songs. *Commun. ACM* 27: 344–346.
Krakauer, D.C. 2011. Darwinian demons, evolutionary complexity, and information maximization. *Chaos* 21: 037110.
Kun, A., Santos, M., and Szathmáry, E. 2005. Real ribozymes suggest a relaxed error threshold. *Nature* 37: 1008–1011.
Maynard Smith, J. 1983. Models of evolution. *Proc. Roy. Soc. London B* 219: 315–325.
Maynard Smith, J. 2000. The concept of information in biology. *Philos. Sci.* 67(2): 177–194.
Shannon, C.E. 1948. A mathematical theory of communication. *Bell Syst. Tech. J.* 27: 379–423.
Turing, A. 1937. On computable numbers, with an application to the Entscheidungs problem. *Proc. Lond. Math. Soc.* 42: 230–265.
Walker, S.I. and Davies, P.C.W. 2013. The algorithmic origins of life. *J. R. Soc. Interface* 6: 20120869.
Yang, Y.-W., Zhang, S., McCullum, E., and Chaput, J.C. 2007. Experimental evidence that GNA and TNA are not sequential polymers in the prebiotic evolution of RNA. *J. Mol. Evol.* 65: 289–295.

CHAPTER 18

Extraterrestrial Life

What Are We Looking For?

Dirk Schulze-Makuch, Louis N. Irwin,
and Alberto G. Fairén

CONTENTS

18.1 PROBLEM OF FINDING EXTRATERRESTRIAL LIFE

At the present time, our entire science of biology is based on the life we know on Earth, which came into existence (or was introduced) under conditions peculiar to the early environment of our planet. We cannot be sure as to which of the existing components—the catalysts, membranes, genetic system, or metabolism—are unique solutions to problems inherent in the very nature of life and which are local solutions dictated by the geochemical circumstances of the early Earth (Shapiro and Schulze-Makuch, 2009). If the latter situation is the case, which seems to be supported by laboratory work and simulations, the discovery of another form of life with an alternative biochemistry is distinctly possible and would be a revolutionary event in the history of science. A discovery of this magnitude, however, will be less likely as long as we concentrate only on the details displayed by life on Earth. With this in mind, our search for extraterrestrial life must take into account the alternative planetary histories that could lead to forms of life unknown, as well as those known, to us.

18.2 EXTRATERRESTRIAL LIFE AS A FUNCTION OF ITS ENVIRONMENT

Life and its environment cannot be separated. This fact is clearly shown by the three terrestrial planets, Earth, Venus, and Mars, in our solar system. Earth, Mars, and also potentially Venus had liquid water on their surfaces early in their geological histories, indicating habitable conditions and the possible presence of life at that time. However, drastic environmental changes took the three planets in very different directions (Schulze-Makuch et al., 2013). Venus became a runaway greenhouse planet; any liquid water on its surface evaporated, with only traces of water remaining in the atmosphere. Earth underwent various rapid and drastic environmental changes, ranging from *snowball Earth* events to *hothouse* conditions, with liquid water and life today being abundant in its subsurface, on its surface, and, at least transiently, in its atmosphere. Mars lost its presumably thicker atmosphere and developed into a very cold and hyperarid *icehouse* world, with permanent liquid water today only possibly present in the subsurface. These drastic and diverse environmental changes affected the habitability of Venus, Earth, and Mars in markedly different ways.

18.2.1 Example 1: The Venusian Atmosphere

The Venusian climate today is controlled by a carbon dioxide–water greenhouse effect and the radiative properties of its global cloud cover (Huang et al., 2013; Figure 18.1). Ambient surface temperatures exceed 460°C, resulting in an extremely desiccated crust (Marcq et al., 2013). Most of the planet is covered by lava flows and volcanic terrains, although

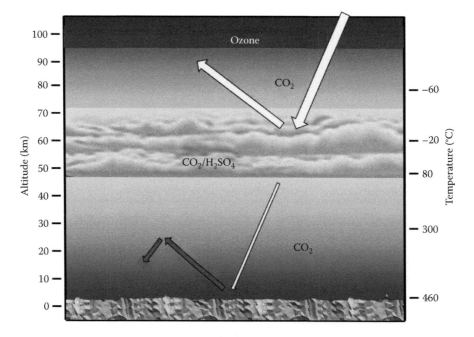

FIGURE 18.1   Atmospheric structure of Venus with massive inventory of CO_2 and sulfuric acid clouds that reflect about 80% of the Sun's radiation. The 20% that gets through the clouds is absorbed along with the emitted infrared energy from the surface by the dense CO_2 at lower levels. Only from the lower cloud deck up (at about 48 km) is the temperature cool enough for the stability of organic compounds and possible microbial life.

some of the channels dissecting the volcanic landscape have been suggested to be water related (Walker et al., 1981). The actual timing of the climatic shift from a habitable early Venus to the baked and desiccated surface of today is unclear (Schulze-Makuch and Irwin, 2002), with estimates ranging from a few hundred million to 2 billion years (Van Thienen et al., 2004). The atmospheric change is thought to have occurred gradually due to a positive feedback loop that amplified the greenhouse effect (Bézard et al., 2009). The loss in surface water would have prevented the formation of carbonates, with the result that volcanic outgassing increased CO_2 partial pressures to the point of producing the runaway greenhouse (Svedhem et al., 2007). If plate tectonics were active on early Venus (Limaye et al., 2009), they likely ceased to exist as lubricants boiled away. The evaporated water then dissociated in the atmosphere, with hydrogen escaping to space and oxygen forming sulfur oxides in the atmosphere and various other oxides with surface rocks (Marcq et al., 2013). Water vapor concentrations in the lower atmosphere today are at a scarce 44 ± 9 ppm (Bézard et al., 2009).

Today, Venus has a thick carbon dioxide–nitrogen atmosphere with little water, and the planet has no water in any form on the surface. Venus is gravitationally locked into near synchrony with the Sun, with one rotation of its axis taking longer than one orbit around the Sun. Nevertheless, surface temperatures on the planet's daytime and nighttime are nearly identical, because of the high heat conductance of an atmosphere many times denser than Earth's. The upper atmosphere is super rotating, circling the planet in 4 days with wind speeds up to 100 m/s (Svedhem et al., 2007). Circulating vortices deep within the planet's atmosphere, with dynamical and morphological similarities to terrestrial hurricanes, appear to maintain the superrotation (Limaye et al., 2009). H_2S and SO_2 have been found together in the Venusian atmosphere. Since these two compounds react with each other, there must be a yet unidentified mechanism producing them. Venus Express results also confirm latitudinal variation of another sulfur compound, COS, at concentrations of about 2–4 ppm in the troposphere that are anticorrelated to CO concentrations (Marcq et al., 2013). One surprising recent result was the detection of an ozone layer at an altitude of about 100 km (Montmessin et al., 2011), hundreds of times less dense than on Earth, but possibly indicating that some of the same key chemical reactions occurring in Earth's stratosphere may also operate on Venus. The most earthlike conditions that can be found today on Venus are in the lower cloud layer of the atmosphere. If water exists in the subsurface, it has to be in a supercritical state (Schulze-Makuch and Irwin, 2002) incompatible with the structural stability required of macromolecules in living systems. In the lower cloud layer, however, where temperatures range from about 300 to 350 K, the pH is approximately 0, and the pressure is about 1 bar; water vapor concentrations of up to several hundred ppm are found (Montmessin et al., 2011). Due to the thickness and superrotation of the Venusian atmosphere, particles of micrometer dimensions have much longer residence times than in Earth's atmosphere, on the order of months compared to days (Schulze-Makuch et al., 2004). If microbial life on Venus ever gained a foothold, either by independent origin or by panspermia from Earth, thermoacidophiles could have evolved as the surface waters turned warmer and more acidic and then retreated to the large liquid droplets of the lower cloud layer, where they might still float as microbial extremophiles today (Schulze-Makuch and Irwin, 2008), provided that the changes occurred slowly enough for microbial life to adapt to an airborne state. It is important to understand the climate history of Venus to elucidate what the original water endowment of Venus was, as this has a profound implication on how benign the environmental conditions of Venus were in the past and whether they were consistent with the requirements of life (Taylor and Grinspoon, 2009).

18.2.2 Example 2: The Martian Near-Surface Environment

From its origin like Earth as a rocky planet endowed with standing reservoirs of liquid water, Mars evolved into the very different, cold, and hyperarid planet it is today (e.g., Fairén, 2010). Too small and too far from the Sun to hold a significant atmosphere and be kept consistently warm, the oceans of water evaporated and were largely lost to space, but partly became sequestered beneath the surface. A very thin atmosphere of mostly carbon dioxide registers a mean pressure of only 6 mb today. Due to the thin atmosphere, surface temperatures vacillate considerably, from about −100°C at night to up to +25°C during the summer days near the equator. Surface liquid water may appear in highly restricted locations under special circumstances (Haberle et al., 2001), but has not yet been detected since doing so has become technically possible.

Mars is thought to have become a very cold desert about 3.7–3.8 Ga ago, after the endogenic activity steadily decreased, the magnetosphere collapsed, and the atmosphere and hydrosphere were mostly lost to space (Fairén et al., 2010). Also, around that time, whatever plate tectonic activity had been occurring probably ceased (Fairén and Dohm, 2004). However, the long-persistent hyperarid state was punctuated by short-duration episodes (~10^4–10^5 years) of considerably wetter conditions (Baker et al., 2005; Fairén, 2010). These episodes appear to have been induced by magmatic-driven activity at Tharsis and Elysium (Fairén et al., 2003) triggering cataclysmic outbursts of huge floods that carved the Martian outflow channels and led to temporary water bodies ranging from lakes to oceans (Baker et al., 1991; Fairén et al., 2003; Fairén, 2010). Further, cycles of exaggerated tilting of the Martian axis (Jakosky et al., 1995) resulted in the redeposition of water–ice accumulations from the pole to the equator and vice versa (Levrard et al., 2004). Major impacts on the surface of Mars, like those that formed the Hellas, Argyre, and Isidis basins, surely melted frozen reservoirs and released subterranean water on a global scale. These periods must have been accompanied by a thicker atmosphere, leading to a transiently warmer planet with conditions more likely to be habitable. The scenario that thus emerges for Mars is that of a planet that has been cold and dry for most of its history, with the possible exception of the first few hundred million years. During these earlier times, Mars was already cold, albeit warmer than today, and certainly wet (Fairén, 2010; Figure 18.2). The overall trajectory toward a colder and drier planet, however, has been punctuated by (1) sporadic global flooding triggered by episodic volcanisms and asteroid/cometary bombardment (Segura et al., 2002; Toon et al., 2010) and (2) localized flow from snowmelt or groundwater eruptions (Mangold et al., 2004). Many surface features on Mars are consistent with this picture, both for catastrophic flooding (Toon et al., 2010) and local sapping or seepage of groundwater (Grant, 2000).

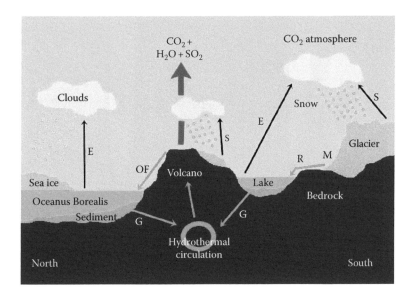

FIGURE 18.2 Schematic representation of land–ocean–atmosphere interactions associated with the presence of a long-term hydrological cycle on Mars under a very cold climate. S, sublimation; E, evaporation; M, meltwater; R, runoff; OF, outburst flooding; G, groundwater flow systems recharge. The cycle would have been active with an atmospheric pressure over ~1 bar and associated greenhouse effect raising temperatures over ~250 K during the first few hundred million years. (Compiled from Fairén, A.G. et al., *Nature*, 459, 401, 2009; Fairén, A.G., *Icarus*, 208, 165, 2010; Fairén, A.G. et al., *Astrobiology*, 10, 821, 2010.)

Even today, the existence of near-surface liquid water is indicated by distinctive features, such as numerous small gullies likely generated by surface runoff down hilly slopes or eruptions near the crest of canyon walls (Mellon and Phillips, 2001). Organics are probably present on the surface on Mars only in oxidized form (Benner et al., 2000), but locally limited habitable areas might exist not far beneath the surface. Life, whether it originated on Mars or was brought there by panspermia from Earth (Fairén and Schulze-Makuch, 2013), would have followed the liquid water and could still be present in subsurface and near-subsurface niches today.

Given that life on Earth originated at least 3.5 billion years ago (Schidlowski, 1988; Mojzsis et al., 1996), and that Earth and Mars shared similar environmental conditions early in their histories, it seems reasonable to speculate that life on Mars originated during its earlier warmer and wetter Noachian Eon. If so, as the planetary conditions became drier and colder (Fairén, 2010), the biosphere would have had to adapt accordingly and retreat to the subsurface, where locally habitable conditions would have still prevailed (Figure 18.3). Microorganisms may have retreated to the deep subsurface, beneath the ice sheets and permafrost, pursuing a psychrophilic lifestyle and scavenging organic nutrients still accessible there. Or, microbes, and possibly larger commensal or independent organisms feeding on them (Irwin and Schulze-Makuch, 2011), may have retreated to remaining favorable niches on or near the surface, such as hydrothermally active areas (Schulze-Makuch et al., 2007) and caves (Léveillé and Datta, 2010).

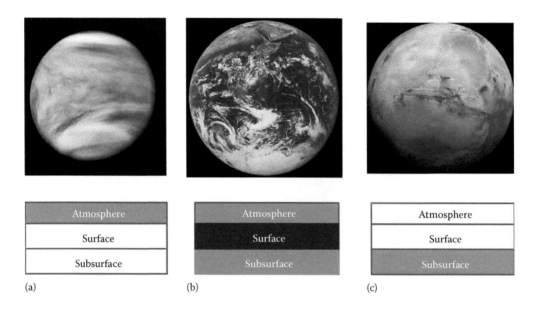

| Atmosphere |
| Surface |
| Subsurface |

(a)

| Atmosphere |
| Surface |
| Subsurface |

(b)

| Atmosphere |
| Surface |
| Subsurface |

(c)

FIGURE 18.3   (a) Today, liquid water is present on Venus in the atmosphere only; (b) on Earth, on the surface, in the subsurface, and at least transiently in the atmosphere; and (c) on Mars, in the subsurface.

Life on Mars could have developed adaptations similar to those known to exist among Earth's extremophiles, perhaps to an even greater degree. They may have evolved dormant stages for persisting through long periods of extreme drought and severe cold (Jakosky et al., 2003; Schulze-Makuch et al., 2005). When conditions grew warmer and wetter, because of transient volcanic activity or cyclic tilting of the Martian axis, the latent forms could have undergone a sporadic burst of growth and reproduction. Spores, cysts, and other dormant stages are common features of organisms on Earth, which enable survival for very long periods of time when environmental conditions cannot support their proliferative stages. In theory, spores can stay viable for an indefinite length of time, expending little if any energy (Schulze-Makuch and Irwin, 2008). Thus, organisms on Mars could have adapted to survive in this fashion for cold and arid spells lasting tens or hundreds of millions of years—punctuated by occasional periods of a few thousand years during which liquid water and warmth were available on the Martian surface. The drastically changing environmental conditions on Mars would certainly have exerted selective pressure to form such dormant stages of life.

Life on Mars could also have evolved a uniquely Martian adaptation to its cold and arid conditions, by using oxidants such as hydrogen peroxide and perchlorates, which became abundant after Mars turned cold and dry (Hecht et al., 2009). Incorporation of these compounds would have provided Martian organisms with the ability to lower the freezing point of their intracellular fluids, support oxidative metabolism, and absorb water hygroscopically from the atmosphere and substrate (Houtkooper and Schulze-Makuch, 2007). Hygroscopicity would be especially advantageous, similar to the use of hygroscopic salt crystals for attracting water by organisms in the Atacama desert (Davila et al., 2010). Another advantage of these highly oxidative compounds for putative near-surface microbes on Mars would be an enhanced ability to counter radiation damage, as the biochemical protective mechanisms against both oxidants and the radiation that generates them are complementary.

Thus, the extremophilic adaptations known on Earth, plus some other strategies perhaps unique to Martian biota, suggest the distinct possibility that life could exist today on Mars—though most likely in sequestered and microbial or miniaturized form.

18.2.3 Example 3: Titan, An Exotic World

Titan is the only satellite in our solar system with a substantial atmosphere mainly composed of nitrogen. It has many similarities to Earth in respect to atmospheric composition, surface pressure, and organic chemistry (Sohl et al., 2014). It is clearly a dynamical planet based on recent albedo changes in the Sotra Patera region and extensive surface changes spanning more than 500,000 km^2 in the wake of a seasonal storm (Turtle et al., 2011). Environmental conditions are generally thought to be conducive for life if it can be shown that (1) polymeric chemistry, (2) an energy source, and (3) a liquid solvent are present in appreciable quantities (Irwin and Schulze-Makuch, 2001), which is the case for Titan. Titan cannot only be considered a habitable planet today, but the presence of liquid methane presumably improves the chances for the origin of life, because the assemblage of organic macromolecules that could give rise to life appears to be more straightforward in a hydrocarbon environment (Schulze-Makuch and Grinspoon, 2005). The biochemistry of an organism in a hydrophobic solvent would be quite different though. For example, the miniaturization of cellular life in water on Earth may be a misleading model for life in a nonaqueous environment (Schulze-Makuch and Irwin, 2008). In an extremely cold, hydrophobic (but liquid) environment, surface/volume ratio considerations may be less constraining than at higher temperatures in polar solvents, because of the lower viscosity of the solvent and the slower diffusion rates permitted by the greatly reduced rate of metabolism. Thus, life on Titan could involve huge (by Earth standards) and very slowly metabolizing cells (Schulze-Makuch and Irwin, 2008).

Life also could utilize acetylene to transfer high-altitude solar UV energy to surface chemical reactions. Acetylene is produced high in the stratosphere from solar UV radiation and then, for the most part, condenses and falls to the surface (Lorenz et al., 2000). Spectroscopic evidence suggests that one by-product of this reaction, methane, is found to be isotopically lighter than would be expected from theories of Titan's formation (Lunine et al., 1999) and thus may hint toward microbial fractionation. Thus, a reasonable energy-yielding reaction for a metabolizing microbe would be the catalytic hydrogenation of photochemically produced acetylene (Schulze-Makuch and Grinspoon, 2005).

Or, free radical reactions could be used as a basis for metabolism. Raulin (1998) suggested that Titan's stratosphere is an active site of complex carbon and nitrogen radical chemistry. For example, a chemoautotrophic organism could use the following reactions:

$$CH_2 \text{ radical} + N_2 \text{ radical} \rightarrow CN_2H_2 \tag{18.1}$$

or

$$2CH \text{ radical} + N_2 \text{ radical} \rightarrow 2HCN \tag{18.2}$$

All reactants have been detected in Titan's environment based on data from Voyager 1 (Kunde et al., 1981; Smith et al., 1982), and these reactions would produce a sufficiently high energy yield in Titan's cold environment. The radical reactions (18.1) and (18.2) would also produce the biologically important compounds cyanamide and hydrocyanic acid, respectively. Life on Titan would thus most likely be quite different from any forms of life known on Earth. A variety of organisms thriving in both hydrophobic liquids, and in hydrophilic vacuoles or sequestered pockets of ammonia–water mixtures, have been envisioned (Irwin and Schulze-Makuch, 2011). Cellular units could be expected to be much larger, and metabolic rates much lower, than those known on Earth.

18.3 NEW FRONTIER OF EXOPLANETS AND POSSIBLE HINTS OF LIFE

As of this writing, over 1000 exoplanets have been detected, many in solar systems with multiple planets (for an update, see http://exoplanet.eu/). This includes now many planets with minimum masses smaller than 10 Earth masses, meaning that a large fraction of them could potentially be terrestrial. Given that we have already a great diversity of planetary bodies within our own solar system (from Venus to Triton and Earth to Titan), one can only speculate how different environmental conditions might be on worlds we have just begun to explore outside of our solar system. Technologies already operational or under development are focusing on the discovery of the smaller terrestrial planets, which are good candidates for possibly hosting life (Schulze-Makuch et al., 2011). There are different approaches to assess whether an exoplanet may be a good candidate for hosting life, which can be based on geoindicators of life or biosignatures of life.

18.3.1 Geoindicators of Life

Geoindicators are markers that are consistent with life, but do not prove that life is present on an analyzed planetary body. Schulze-Makuch et al. (2011) discussed the presence of geoindicators and proposed a measure, the planetary habitability index (PHI), to assess the habitability of exoplanets and prioritize the investigation of planets and moons that are likely habitable. The PHI is based on planetary parameters such as the presence of a substrate, the availability of energy, a suitable chemical environment, and the availability of liquids.

A substrate is important because life tends to thrive at discrete interfaces (Norde, 2003), which occur most notably on rocky and icy planetary bodies. Habitability is further favored on substrates that are protected from too intense shortwave radiation by an atmosphere.

The availability of energy is critical to life and for planetary bodies in reasonable proximity to a Sun, where energy harvested through light and cyclic oxidation–reduction chemistry is possible. While in principle any energy source has to be considered to be of possible biological utility, sunlight and chemistry are the most effective sources of free energy for driving biological processes (Schulze-Makuch and Irwin, 2008; Schulze-Makuch and Grinspoon, 2005).

Polymeric chemistry appears to be an essential requirement for life (Westall et al., 2000). For a variety of reasons, carbon has the most desirable chemical properties and molecular bonding characteristic for the formation of biopolymers (Schulze-Makuch and Irwin, 2008); thus carbon, especially in organic form, is a good geoindicator for the possible presence of life. In addition, the presence of nitrogen, sulfur, and phosphorus further enhances the probability of organic molecules, which makes the presence of complex chemistry more likely.

The presence of liquids is favorable for living processes because macromolecules and nutrients can be concentrated within a bounded internal environment without immobilizing interaction constituents (Schulze-Makuch and Irwin, 2008). The need for liquids is usually taken as a requirement for an aqueous medium, though this is not necessarily the case. Liquids in the atmosphere, on the surface, or beneath the surface are a function of chemistry, pressure, and temperature and can exist on very cold planetary bodies, such as the methane/ethane mixture of Saturn's moon Titan and other subsurface liquids seen to erupt periodically on worlds like Saturn's moon, Enceladus, and Neptune's moon, Triton.

18.3.2 Biosignatures

Biosignatures are markers that are a direct consequence of biological activity rather than being only consistent with the presence of life. Biosignatures can range from the presence of specific molecules such as chlorophyll to fossilized remnants of life.

Chlorophyll is the prime example and can be identified by radiance spectra in the visible region (Gordon et al., 1980; Hovis et al., 1980) and by advanced very high–resolution radiometer (AVHRR) measurements (Gervin et al., 1985). In addition, methylhopanoids have also been suggested as biomarkers and have the additional advantage of distinguishing between cyanobacteria (2-methyl) and methanotrophic (3-methyl) bacteria (Farrimond et al., 2004). Other examples of macromolecules that are linked to life include proteins, polypeptides, and phospholipids. In general, any macromolecule of a size larger than 500 Da (protein size) can be considered a biosignature (Schulze-Makuch and Irwin, 2008), particularly if those molecules would exhibit a preferred handedness. Chirality, or nonracemic handedness, is a fundamental property of terrestrial biogenic molecules and thus may be used as an indicator of possible extraterrestrial life detectable by remote sensing in the near future (Schulze-Makuch and Irwin, 2008). Large macromolecules are not symmetrical and thus inevitably exhibit chirality.

Metabolic by-products and end products are well known for organisms on Earth and may thus also serve as biosignatures on other planets and moons. They include various biochemical compounds such as ATP and lipids and also electron donor and acceptor pairs such as Fe^{3+}/Fe^{2+}, NH_3/N_2, and H_2S/S enriched in lighter isotopes. This isotope enrichment or fractionation occurs as part of the metabolic reactions for organisms on Earth and may also occur for life elsewhere (Schulze-Makuch and Irwin, 2008).

A related biosignature may be the metabolic multistep pathways that run close to equilibrium for some internal steps, but are coupled to a last step, which is energetically downhill, thus pulling the whole reaction to completion (Voet and Voet, 2004). Baross et al. (2007) considered this feature as a possible universal biosignature as it exploits most economically a surrounding chemical disequilibrium.

18.4 CONCLUSIONS

Based on the natural history of life on Earth, life and its environment are intrinsically interwoven. This is to be expected on other planets and moons as well if biology plays a role. Life, if it ever existed on Mars, Venus, Titan, and other worlds, would differ in form and function from the examples we observe on Earth. It will be an even greater challenge to find evidence for life on exoplanets as we know even less about the existing environmental conditions on these faraway places. Yet, some commonalities can be expected, which are usable for future searches. These include geosignatures consistent with dynamic worlds that have liquids on their surface and possess protecting atmospheres and biosignatures that include selected macromolecules and specific isotopic fractionation ratios.

GLOSSARY

Albedo: A measure (reflection coefficient) of how strongly an object reflects light from a light source such as the Sun.

Argyre: The second largest impact basin on Mars, up to about 5000 m deep, located in the southern highlands of Mars.

Chemoautotrophic: Organisms that obtain energy through metabolizing inorganic substrates and using carbon dioxide as a carbon source.

Commensal organism: Organism that lives with another organism in a type of symbiotic relationship in which the other organism is neither helped nor harmed.

Elysium: The second largest volcanic province on Mars after Tharsis, south of Elysium Planitia, a broad plain that straddles the equator of Mars.

Exoplanet: A planet outside our solar system.

Extremophile: An organism that thrives in environmental conditions not tolerated by most forms of life, mostly referring to pH levels, temperatures, solute concentrations, and pressures.

Hellas: The largest impact basin on Mars, up to about 7000 m deep, located in the southern hemisphere of Mars.

Hothouse: Refers to climate conditions on Venus that are extremely hot as a result of a runaway greenhouse effect (see also runaway greenhouse effect).

Isotope fractionation: Processes that affect the relative abundance of isotopes (elements that differ by the number of neutrons). In regard to biology, a typically seen effect is that organisms prefer the lighter isotopes of an element resulting in a specific isotope fractionation.

Noachian eon: Oldest time period of Mars, from the formation of the planet to about 3.7 billion years ago, at a time when surface water is thought to have been abundant and the climate more benign.

Panspermia: Hypothesis that some organisms can survive the effects of space by becoming trapped in rocks or dust that is ejected into space after collisions between solar system bodies that harbor life.

PHI: Planetary habitability index, measure used to assess the habitability on various planets and moons, inside and outside of our solar system.

Runaway greenhouse: Process by which the coupling of the atmosphere and the surface temperatures of a planet increases the greenhouse effect to such a degree that the planet's oceans boil away. This is usually thought to have happened in Venus' history.

Snowball Earth: An event in Earth's history when the surface of the Earth was largely or entirely frozen over. The last time this event occurred was about 650–700 million years ago.

Sotra Patera: A roughly circular volcano on Titan measuring about 65 km across with two peaks and multiple craters.

Tharsis: A volcanic plateau located near the equator in the western hemisphere of Mars.

Thermoacidophile: Microbe that thrives in highly acidic solutions at high temperatures.

REVIEW QUESTIONS

1. How does the environment affect the form and function of an organism?

2. How did the natural history of the terrestrial planets influence the evolution of life on Earth and possibly on neighboring Mars and Venus?

3. Why would we consider possible life on Titan as exotic?

4. Which biosignatures and geoindicators would indicate specific types of organisms and which would be general indicators for life?

5. Prioritize which signatures and indicators of life would be the most important to focus on.

6. Which were the main triggers for the environmental change on Venus? And on Mars?

7. How much has life itself influenced the habitability of Earth?

8. What kind of different adaptations can we expect in organisms inhabiting liquid water microenvironments in the Venus atmosphere versus the Mars subsurface? And how are those different relative to the adaptations to the liquid environments on Titan?

9. Which would be the limiting factors for microbial growth on Venus, Mars, and Titan? Are any of them coincident? Why?

10. Is biological transfer (*panspermia*) possible between Venus, Mars, and Titan? Specifically, could Martians colonize Venus? Now or at any moment in the past?

REFERENCES

Baker, V. R., Dohm, J. M., Fairén, A. G. et al. (2005) Extraterrestrial hydrogeology. [Invited]. *Hydrogeology Journal* 13: 51–68.

Baker, V. R., Strom, R. G., Gulick, V. C., Kargel, J. S., Komatsu, G., and Kale, V. S. (1991) Ancient oceans, ice sheets and the hydrological cycle on Mars. *Nature* 352: 589–594.

Baross, J. A., Benner, S. A., Cody, G. D. et al. (2007). *The Limits of Organic Life in Planetary Systems*. National Academies Press, Washington, DC.

Benner, S. A., Devine, K. G., Matveeva, L. N., and Powell, D. H. (2000) The missing organic molecules on Mars. *Proceedings of the National Academy of Sciences* 97: 2425–2430.

Bézard, B., Fulchignoni, M., and Lazzarin, M. (2009) Water vapor abundance near the surface of Venus from Venus Express/VIRTIS observations. *Journal of Geophysical Research* 114: E00B40. doi:10.1029/2008JE003251.

Davila, A. F., Duport, L. G., Melchiorri, R. et al. (2010) Hygroscopic salts and the potential for life on Mars. *Astrobiology* 10: 617–628.

Fairén, A. G. (2010) A cold and wet Mars. *Icarus* 208: 165–175.

Fairén, A. G., Davila, A. F., Gago-Duport, L., Amils, R., and McKay, C. P. (2009) Stability against freezing of aqueous solutions on early Mars. *Nature* 459: 401–404.

Fairén, A. G., Davila, A. F., Lim, D. et al. (2010) Astrobiology through the ages of Mars: The study of terrestrial analogues to understand the habitability of Mars. *Astrobiology* 10: 821–843.

Fairén, A. G. and Dohm, J. M. (2004) Age and origin of the lowlands of Mars. *Icarus* 168: 277–284.

Fairén, A. G., Dohm, J. M., Baker, V. R. et al. (2003) Episodic flood inundations of the northern plains of Mars. *Icarus* 165(1): 53–67.

Fairén, A. G. and Schulze-Makuch, D. (2013) The overprotection of Mars. *Nature Geoscience* 6: 510–511.

Farrimond, P., Talbot, H. M., Watson, D. F., Schulz, L. K., and Wilhelms, A. (2004) Methylhopanoids: Molecular indicators of ancient bacteria and a petroleum correlation tool. *Geochimica et Cosmochimica Acta* 68(19): 3873–3882.

Gervin, J. C., Kerber, A. G., Witt, R. G., Lu, Y. C., and Sekhon, R. (1985) Comparison of level I land cover accuracy for MSS and AVHRR data. *International Journal of Remote Sensing* 6: 47–57.

Gordon, H. R., Clark, D. K., Mueller, J. L., and Hovis, W. A. (1980) Phytoplankton pigments from the Nimbus-7 coastal zone color scanner—Comparison with surface measurements. *Science* 210: 63–66.

Grant, J. A. (2000) Valley formation in Margaritifer Sinus, Mars, by precipitation-recharged groundwater sapping. *Geology* 28(3): 223–226.

Haberle, R. M., McKay, C. P., Schaeffer, J. et al. (2001) On the possibility of liquid water on present-day Mars. *Journal of Geophysical Research: Planets (1991–2012)* 106(E10): 23317–23326.

Houtkooper, J. M. and Schulze-Makuch, D. (2007) A possible biogenic origin for hydrogen peroxide on Mars: The Viking results reinterpreted. *International Journal of Astrobiology* 6: 147–152.

Huang, J., Yang, A., and Zhong, S. (2013) Constraints of the topography, gravity, and volcanism on Venusian mantle dynamics and generation of plate tectonics. *Earth and Planetary Science Letters* 362: 207–214.

Hecht, M. H., Kounaves, S. P., Quinn, R. C. et al. (2009) Detection of perchlorate and the soluble chemistry of Martian soil at the Phoenix lander site. *Science* 325: 64–67.

Hovis, W. A., Clark, D. K., Anderson, F. et al. (1980) Nimbus-7 CZCS coastal zone color scanner—System description and early imagery. *Science* 210: 60–63.

Irwin, L. N. and Schulze-Makuch, D. (2001) Assessing the plausibility of life on other worlds. *Astrobiology* 1: 143–160.

Irwin, L. N. and Schulze-Makuch, D. (2011). *Cosmic Biology: How Life Could Evolve on Other Worlds.* Praxis Publishing, Berlin, Germany.

Jakosky, B. M., Henderson, B. G., and Mellon, M. T. (1995) Chaotic obliquity and the nature of the Martian climate. *Journal of Geophysical Research* 100(E1): 1579–1584.

Jakosky, B. M., Nealson, K. H., Bakermans, C., Ley, R. E., and Mellon, M. T. (2003) Subfreezing activity of microorganisms and the potential habitability of Mars' polar regions. *Astrobiology* 3(2): 343–350.

Kunde, V. G., Aikin, A. C., Hanel, R. A., Jennings, D. E., Maguire, W. C., and Samuelson, R. E. (1981) C_4H_2, HC_3N and C_2N_2 in Titan's atmosphere. *Nature* 292: 686–688.

Léveillé, R. J. and Datta, S. (2010) Lava tubes and basaltic caves as astrobiological targets on Earth and Mars: A review. *Planetary and Space Science* 58: 592–598.

Levrard, B., Forget, F., Montmessin, F., and Laskar, J. (2004) Recent ice-rich deposits formed at high latitudes on Mars by sublimation of unstable equatorial ice during low obliquity. *Nature* 431(7012): 1072–1075.

Limaye, S. S., Kossin, J. P., Rozoff, C. et al. (2009) Vortex circulation on Venus: Dynamical similarities with terrestrial hurricanes. *Geophysical Research Letters* 36: L04204, doi:10.1029/2008GL036093.

Lorenz, R. D., Lunine, J. I., and McKay, C. P. (2000) Geologic settings for aqueous organic synthesis on Titan revisited. *Enantiomer* 6: 83–96.

Lunine, J. I., Yung, Y. L., and Lorenz, R. D. (1999) On the volatile inventory of Titan from isotopic abundances in nitrogen and methane. *Planetary and Space Science* 47: 1291–1303.

Mangold, N., Quantin, C., Ansan, V., Delacourt, C., and Allemand, P. (2004) Evidence for precipitation on Mars from dendritic valleys in the Valles Marineris area. *Science* 305(5680): 78–81.

Marcq, E., Bertaux, J.-L., Montmessin, F., and Belyaev, D. (2013) Variations of sulphur dioxide at the cloud top of Venus' dynamic atmosphere. *Nature Geoscience* 6: 25–28.

Mellon, M. T. and Phillips, R. J. (2001) Recent gullies on Mars and the source of liquid water. *Journal of Geophysical Research* 106(E10): 23165–23180.

Mojzsis, S. J., Arrhenius, G., McKeegan, K. D., Harrison, T. M., Nutman, A. P., and Friend, C. R. L. (1996) Evidence for life on Earth before 3,800 million years ago. *Nature* 384(6604): 55–59.

Montmessin, F., Bertaux, J. L., Lefèvre, F. et al. (2011) A layer of ozone detected in the nightside upper atmosphere of Venus. *Icarus* 216: 82–85.

Norde, W. (2003). *Colloids and Interfaces in Life Sciences*. CRC Press, Boca Raton, FL.

Raulin, F. (1998) Titan. In *The Molecular Origins of Life*, A. Brack (ed.). Cambridge University Press, New York, pp. 365–385.

Schidlowski, M. (1988) A 3,800-million-year isotopic record of life from carbon in sedimentary rocks. *Nature* 333(6171): 313–318.

Schulze-Makuch, D., Dohm, J. M., Fan, C. et al. (2007) Exploration of hydrothermal targets on Mars. *Icarus* 189: 308–324.

Schulze-Makuch, D. and Grinspoon, D. H. (2005) Biologically enhanced energy and carbon cycling on Titan? *Astrobiology* 5(4): 560–567.

Schulze-Makuch, D., Grinspoon, D. H., Abbas, O., Irwin, L. N., and Bullock, M. A. (2004) A sulfur-based survival strategy for putative phototrophic life in the Venusian atmosphere. *Astrobiology* 4: 11–18.

Schulze-Makuch, D. and Irwin, L. N. (2002) Reassessing the possibility of life on Venus: Proposal for an astrobiology mission. *Astrobiology* 2: 197–202.

Schulze-Makuch, D., Irwin, L. N., Lipps, J. H. et al. (2005) Scenarios for the evolution of life on Mars. *Journal of Geophysical Research—Planets* 110: E12S23, doi:10.1029/2005JE002430.

Schulze-Makuch, D. and Irwin, L. N. (2008) *Life in the Universe: Expectations and Constraints*, 2nd edn. Springer, Berlin, Germany.

Schulze-Makuch, D., Irwin, L. N., and Fairén, A. G. (2013) Drastic environmental change and its effects on a planetary biosphere. *Icarus* 225(1): 775–780.

Schulze-Makuch, D., Méndez, A., Fairén, A. G. et al. (2011) A two-tiered approach to assessing the habitability of exoplanets. *Astrobiology* 11(10): 1041–1052.

Segura, T. L., Toon, O. B., Colaprete, A., and Zahnle, K. (2002) Environmental effects of large impacts on Mars. *Science* 298: 1977–1980.

Shapiro, R. S. and Schulze-Makuch, D. (2009) The search for alien life in our solar system: Strategies and priorities. *Astrobiology* 9(4): 335–343.

Smith, G. D., Strobel, A., Broadfoot, B., Sandel, D., Shemansky, J., and Holberg, J. (1982) Titan's upper atmosphere: Composition and temperature from the EUV solar occultation results. *Journal of Geophysical Research* 87: 1351–1360.

Sohl, F., Solomonidou, A., Wagner, F. W. Coustenis, A., Hussmann, H., and Schulze-Makuch, D. (2014) Structural and tidal models of Titan and inferences on cryovolcanism. *Journal of Geophysical Research*, in press.

Svedhem, H., Titov, D. V., Taylor, F. V., and Witasse, O. (2007) Venus as a more Earth-like planet. *Nature* 450: 629–632.

Taylor, F. and Grinspoon, D. (2009) Climate evolution of Venus. *Journal of Geophysical Research* 114: E00B40, doi:10.1029/2008JE003316.

Toon, O. B., Segura, T., and Zahnle, K. (2010) The formation of Martian river valleys by impacts. *Annual Review of Earth and Planetary Sciences* 38: 303–322.

Turtle, E. P., Del Genio A. D., Barbara, J. M. et al. (2011) Rapid and extensive surface changes near Titan's equator: Evidence of April showers. *Science* 331: 1414–1417.

Van Thienen, P., Vlaar, N. J., and Van den Berg, A. P. (2004) Plate tectonics on the terrestrial planets. *Physics of the Earth and Planetary Interiors* 142: 61–74.

Voet, D. and Voet, J. (2004) *Biochemistry*. Wiley & Sons, Hoboken, NJ.

Walker, J. C. S., Hays, P. B., and Kasting, J. F. (1981) A negative feedback mechanism for the long-term stabilization of Earth's surface temperature. *Journal of Geophysical Research* 86: 9776–9782.

Westall, F., Steele, A., Toporski, J. et al. (2000) Polymeric substances and biofilms as biomarkers in terrestrial materials: Implications for extraterrestrial samples. *Journal of Geophysical Research* 105: 24511–24527.

CHAPTER 19

Evolutionary Approach to Viruses in Astrobiology

Matti Jalasvuori and Jaana K.H. Bamford

CONTENTS

19.1 WHAT IS A VIRUS? ESSENTIAL BASICS FOR AN ASTROBIOLOGIST

In this chapter, we review viruses from an astrobiological perspective. We go through the structural and functional qualities of viruses and then briefly overview viruses in evolutionary astrobiological research.

Take a sample from almost any environment on Earth and observe carefully what you have in your vial. You will find two types of entities. There are cells of various sizes, specialized to different lifestyles. But, more importantly, there are small protein capsules. Each of these capsules entraps genetic information that originated billions of years ago, probably even before most cellular genes. Yet, despite their abundance, we often fail to see the small capsules when we become fascinated by the diversity of larger cellular entities. However, we cannot ignore them when we start considering life as a universal process (Jalasvuori, 2012). As the capsules are overwhelmingly abundant, they probably have a significant role in the evolution of life. Therefore, any model attempting to explain the presence of different cellular entities on our "pale blue dot" should also be able to explain the presence of capsules. As we are immersed in a sea of capsules, all living systems composed of cells, including the hypothetical extraterrestrial ones, will have their own viruses.

Viruses are parasites, which means that they exploit the resources of other living beings for their own reproduction. New virus particles are assembled from individual

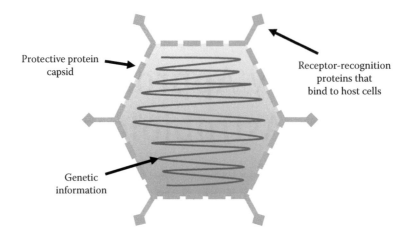

FIGURE 19.1 Schematic presentation of an infectious virus particle also known as a virion.

subunits within cells and thus they differ in this respect from cellular organisms. The sea of capsules is formed of these assemblages.

Cellular theory suggests that all life forms comprise cellular beings that reproduce by cellular division. You can take any cell in our biosphere, be it a cell in the skin of your finger or an archaeota at the bottom of the Pacific Ocean, and, by turning back time, you can follow the vertical history of the cell through billions of cell divisions to the origin of the very first cells on Earth. Yet, these very cells have been continuously parasitized by a large number of different viruses. Evidence suggests that even the first cells already had their parasites (Koonin et al., 2006; Ortmann et al., 2006). Viruses cast the shadow of the cellular tree of life—an ever-present partner—which cellular organisms are unable to escape. But are viruses just an immaterial, irrelevant component or are they genuine living entities that are of astrobiological importance?

Viruses are genetically reproducing entities: they can evolve and they can go extinct—feats that are generally associated with living beings. All of the 10^{31} virions (i.e., infectious virus particles; Hendrix et al., 1999) in this biosphere were assembled in a host cell in the past. Therefore, viruses are first and foremost genetic entities that parasitize cells by transforming cell vehicles into construction facilities that generate new virus particles. Virus is an information package that can modify cells into assembly lines generating new viruses that carry the information to do the same to other cells.

Figure 19.1 presents a schematic model of an infectious virus particle (virion).

In nature, there are many different variations of virions, and therefore, a simple presentation does not capture all the potential forms. Yet, all viruses share a few principles that are exemplified in Figure 19.1, which we go through one by one in the following sections.

19.2 BUILDING A FUNCTIONAL VIRUS: PROTEIN CAPSID AND THE LIFE OF A VIRUS

A virion has a protein capsid. This capsid provides the viral genetic information to a protective shell in the outside-of-host-cell environment. Naked genetic information is susceptible to many environmental hazards, and thus, protective capsid is essential for virus genes to survive the trip from one host cell to another.

There are many types of virus capsids (Ackermann, 2007; Akita et al., 2007): some are spherical while others are rigid or flexible rods. Archaeal viruses have many queer shells, which are, for example, lemon or bottle shaped (Prangishvili and Garrett, 2005). Nevertheless, all viruses construct a protein shell before they abandon the host and escape to the external environment in order to fulfill their ultimate mission: to find a new exploitable host.

The capsid itself is built out of just a few types of proteins. This allows viruses to carry only a couple of short genes that can generate thousands and thousands of capsid proteins during the replication cycle of a virus. Interestingly, capsid proteins of many viruses appear to be some of the most ancient genes in this biosphere (Koonin et al., 2006; Krupovic and Bamford, 2008a).

Viruses are completely inactive when they are enclosed within the capsid, and thus in many unfruitful discussions, viruses have been declared dead. Yet, the protective capsid is only a means for a virus to survive the trip from one host cell to another. It is the "living part" of a viral reproductive cycle that occurs within—not outside—the cell (Figure 19.2).

Mistaking viruses only for their inactive extracellular form would mean failing to recognize the true essence of what a virus is.

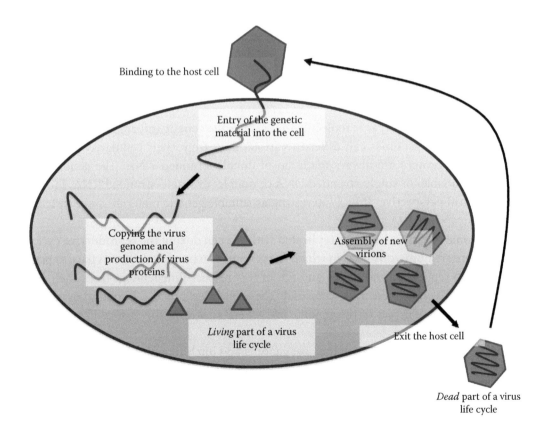

FIGURE 19.2   Virus life cycle.

19.3 BUILDING A FUNCTIONAL VIRUS: HOST RECOGNITION PROTEINS AND INFECTION

Viruses are not completely inactive even in the extracellular environment. As viruses are entities that rely on cells for their reproduction, it is essential for a successful virus to be able to distinguish potential host cells from all other features in the environment. Therefore, all virus particles have proteins that attach to structures on host cells. These host recognition proteins usually bind to natural and essential components on the cell surface, making it difficult for host cells to evolutionarily abandon them (Buckling and Rainey, 2002). The host recognition proteins are usually the most variable features of viruses when homologous viruses are compared. Similarly, closely related hosts may differ by their surface components as they have evolved to avoid virus infections. Indeed, we have previously suggested that the virus-induced selection on host surface components may have been a significant driver of the evolution of cellular interface with the external world (Jalasvuori and Bamford, 2008).

The recognition of a new host cell triggers a chain of events in either the virus or the host (Poranen et al., 2002). Some viruses forcefully inject their genetic material into the host cell. This brute approach requires the virus capsid to be constructed in such a manner that, while being stable and protective in the environment, it can upon a signal disassemble and fire the genome into the host. Cells may also have natural mechanisms to intake foreign particles—a property exploited by some viruses to facilitate their own entry. Many viruses have a membrane envelope that fuses with the host cell membrane in order to translocate the viral genome inside the cell.

19.4 BUILDING A FUNCTIONAL VIRUS: GENOME AND EVOLUTION OF A VIRUS

The virus genome is an organic molecule that carries the genetic information of the virus. Interestingly, viral genomes can differ in various ways from their cellular counterparts. Cellular chromosomes are always made up of double-stranded DNA. The genome of a virus can be double- or single-stranded DNA or double- or single-stranded RNA. From an astrobiological perspective, this is interesting as suitable genetic material is not restricted to double-stranded DNA alone.

The genome of a virus becomes replicated within a cell during the reproductive cycle of a virus. There are numerous ways in which the genome can be copied. Some viruses encode for their own replication molecules and are in this respect independent from their host. Other viruses use host molecules to mediate the replication process. Usually, tens or even tens of thousands of copies of the genome are generated within the host cell as a result of an infection.

Viruses can also integrate with the host genome and hitchhike there as an intragenomic parasite. Integration of viruses into the host chromosome is an evolutionarily important quality given that each integration of a virus can significantly modify the genome of the host organisms (Brussow et al., 2004). When these viruses are excised from the chromosome, they can take some of the host genes along with them and eventually transfer these genes to other host organisms (Canchaya et al., 2003; Choi and Kim, 2006). This shuffling of genes between entities allows organisms to evolve horizontally in addition to the traditional vertical evolutionary process. Moreover, a single integration event of a virus can provide the host with tens or hundreds of new genes. If the virus is inactivated for some reason, the host cell survives to live another day with a significantly altered genome.

19.5 VIRUSES AND THE ORIGIN OF LIFE

The origin of life is arguably one of the greatest challenges for science to understand. The self-organization of the inanimate universe into reproducing sentient vessels of genetic information is mind-boggling. One intriguing recent finding is that viruses appear to be ancient, probably predating contemporary cells.

There are primordial processes during which viruses may have played an important role. First of all, early life required some sort of cellular structures where selection operated already on multiple levels (Szathmary and Demeter, 1987) and possibly also required horizontal movement of information between cells in one way or another (Hogeweg and Takeuchi, 2003). Given that contemporary cells were already products of evolution (Jekely, 2006), the earliest proto-cells had to be of abiotic origin. These cells may have been inorganic structures incapable of induced division within, for example, alkaline hydrothermal vents (Koonin and Martin, 2005; Martin and Russell, 2007). In the absence of division, evolutionarily successful replicators would need to be transferred from one cell to another in order for their genetic information to spread within the system. It is also possible that the proto-cells were able to divide due to mechanical stress (Chen et al., 2004). In this case, replicators that get to spread to newly formed cells would intuitively be in evolutionary advantage as their success is not confined only to the direct descendants of their current proto-cells. Regardless of the exact nature of primordial proto-cells, the transfer of genetic information between cells must have been favorable sooner or later. Early community of life indeed appears to have been evolving horizontally rather than vertically (Woese, 1998, 2000, 2002).

Virions are genetically encoded structures that facilitate the transfer of genes from one cell to another (Jalasvuori and Bamford, 2008). Therefore, the emergence of virus-like entities may have been favorable even before the origin of contemporary cells, because these entities can provide a way for genetic information to move between proto-cells.

The movement of virus-like entities between various proto-cells may have given the early evolving system a mechanism to share innovations. Sharing of genetic novelty might have been critical for the survival of early life: it was possible for helpful genes to spread and become more common within the system (Vetsigian et al., 2006; Jalasvuori, 2011). Similar innovation sharing is commonplace even today (Ochman et al., 2000) and, therefore, even though we often mistake viruses to be harmful parasites, they may have been useful natural products of natural selection at the stem of life (Jalasvuori and Bamford, 2008). Perhaps life cannot survive long enough without the help of viruses (Jalasvuori et al., 2010). Thus, from an astrobiological perspective, it is important to realize that not only may viruses be present in other hypothetical, naturally emerging living systems, but that without viruses there may not be other living systems in the first place (Jalasvuori and Bamford, 2010; Villarreal and Witzany, 2010).

19.6 VIRUSES AS ANCIENT GENETIC ENTITIES

Viruses appear to be polyphyletic, indicating that they have multiple independent origins. Therefore, it was surprising to discover that viruses infecting eukaryal, archaeal, and bacterial organisms shared common features (Bamford et al., 2002; Khayat et al., 2005). Viruses that were previously considered unrelated shared common capsid architectures, major capsid proteins, and genome-packaging mechanisms (Bamford, 2003; Benson et al., 2004). Later on, several different families of viruses were grouped into deeply branching lineages (Krupovic and Bamford, 2008a; Jalasvuori et al., 2009). This has led to speculation that many of the contemporary viruses actually emerged before the separation of cellular life into the three domains (Krupovic and Bamford, 2008b). If viruses are truly ancient, then the study of viral genetic repertoire can provide us another way to investigate the genetic variety of emerging life on Earth (Jalasvuori and Bamford, 2009). Previously, the last universal common community from which all organisms eventually descended was investigated through bacterial, archaeal, and eukaryal cells. Now, viruses can help us discover novel features that may also have been there at the earliest stages of the evolution of life (Koonin, 2006).

Many widespread and ancient viruses have unique hallmark genes that are clearly primordial but that cannot be found from cellular chromosomes, or, if they can be found from chromosomes, they must have been acquired from a virus (Koonin et al., 2006). These hallmark viral genes include genes encoding certain major capsid proteins, various initiation and elongation proteins for DNA and RNA genome replication, and virus genome-packaging proteins. All these proteins have functions that may have contributed significantly to the evolution and ecology of primordial living systems. Therefore, genuine understanding of the origin of life may be better achieved if viruses are not forgotten (Villarreal and Witzany, 2010).

Viruses can help us answer astrobiological questions, and thus, they should not be treated only as diseases or mere indifferent parasites. Viruses are everywhere and they are a constitutive factor of life as we know it.

GLOSSARY

Capsid: Protective protein shell that covers the viral genome.
Virion: Infectious virus particle.

REVIEW QUESTIONS

1. What arguments would you use to argue that viruses are alive?

2. What arguments would you use to argue that viruses are dead?

3. Why is the capsid an important part of a virus?

4. In what ways do viral genomes differ from cellular chromosomes?

5. What essential functions may viruses have had in the early evolving community of life?

REFERENCES

Ackermann, H.W. 2007. 5500 Phages examined in the electron microscope. *Arch. Virol.* 152: 227–243.

Akita, F., Chong, K.T., Tanaka, H., Yamashita, E., Miyazaki, N., Nakaishi, Y., Suzuki, M., Namba, K., Ono, Y., Tsukihara, T., and Nakagawa, A. 2007. The crystal structure of a virus-like particle from the hyperthermophilic archaeon *Pyrococcus furiosus* provides insight into the evolution of viruses. *J. Mol. Biol.* 368: 1469–1483.

Bamford, D.H. 2003. Do viruses form lineages across different domains of life? *Res. Microbiol.* 154: 231–236.

Bamford, D.H., Burnett, R.M., and Stuart, D.I. 2002. Evolution of viral structure. *Theor. Popul. Biol.* 61: 461–470.

Benson, S.D., Bamford, J.K., Bamford, D.H., and Burnett, R.M. 2004. Does common architecture reveal a viral lineage spanning all three domains of life? *Mol. Cell* 16: 673–685.

Brussow, H., Canchaya, C., and Hardt, W.D. 2004. Phages and the evolution of bacterial pathogens: From genomic rearrangements to lysogenic conversion. *Microbiol. Mol. Biol. Rev.* 68: 560–602.

Buckling, A. and Rainey, P.B. 2002. Antagonistic coevolution between a bacterium and a bacteriophage. *Proc. Biol. Sci.* 269: 931–936.

Canchaya, C., Fournous, G., Chibani-Chennoufi, S., Dillmann, M.L., and Brussow, H. 2003. Phage as agents of lateral gene transfer. *Curr. Opin. Microbiol.* 6: 417–424.

Chen, I.A., Roberts, R.W., and Szostak, J.W. 2004. The emergence of competition between model protocells. *Science* 305: 1474–1476.

Choi, I. and Kim, A. 2006. Global extent of horizontal gene transfer. *Proc. Natl. Acad. Sci. USA* 104: 4489–4494.

Hendrix, R.W., Smith, M.C., Burns, R.N., Ford, M.E., and Hatfull, G.F. 1999. Evolutionary relationships among diverse bacteriophages and prophages: All the world's a phage. *Proc. Natl. Acad. Sci. USA* 96: 2192–2197.

Hogeweg, P. and Takeuchi, N. 2003. Multilevel selection in models of prebiotic evolution: Compartments and spatial self-organization. *Orig. Life Evol. Biosph.* 33: 375–403.

Jalasvuori, M. 2011. Horizontal movement of genetic information during early evolution of life and the timescale for the origin of the first gene. *J. Cosmol.* 16: 7017–7020.

Jalasvuori, M. 2012. Vehicles, replicators, and intercellular movement of genetic information: Evolutionary dissection of a bacterial cell. *Int. J. Evol. Biol.* 2012: 874153.

Jalasvuori, M. and Bamford, J.K.H. 2008. Structural co-evolution of viruses and cells in the primordial world. *Orig. Life Evol. Biosph.* 38: 165–181.

Jalasvuori, M. and Bamford, J.K.H. 2009. Did the ancient crenarchaeal viruses from the dawn of life survive exceptionally well the eons of meteorite bombardment? *Astrobiology* 9: 131–137.

Jalasvuori, M. and Bamford, J.K.H. 2010. Viruses and life: Can there be one without the other? *J. Cosmol.* 10: 3446–3454.

Jalasvuori, M., Jaatinen, S.T., Laurinavicius, S., Ahola-Iivarinen, E., Kalkkinen, N., Bamford, D.H., and Bamford, J.K.H. 2009. The closest relatives of icosahedral viruses of thermophilic bacteria are among halophilic archaea. *J. Virol.* 83: 9388–9397.

Jalasvuori, M., Jalasvuori, M.P., and Bamford, J.K.H. 2010. Dynamics of a laterally evolving community of ribozyme-like agents as studied with a rule-based computing system. *Orig. Life Evol. Biosph.* 40: 319–334.

Jekely, G. 2006. Did the last common ancestor have a biological membrane? *Biol. Direct* 1: 35.

Khayat, R., Tang, L., Larson, E.T., Lawrence, C.M., Young, M., and Johnson, J.E. 2005. Structure of an archaeal virus capsid protein reveals a common ancestry to eukaryotic and bacterial viruses. *Proc. Natl. Acad. Sci. USA* 102: 18944–18949.

Koonin, E.V. 2006. Temporal order of evolution of DNA replication systems inferred by comparison of cellular and viral DNA polymerases. *Biol. Direct* 1: 39.

Koonin, E.V. and Martin, W. 2005. On the origin of genomes and cells within inorganic compartments. *Trends Genet.* 21: 647–654.

Koonin, E.V., Senkevich, T.G., and Dolja, V.V. 2006. The ancient Virus World and evolution of cells. *Biol. Direct* 1: 29.

Krupovic, M. and Bamford, D.H. 2008a. Virus evolution: How far does the double beta-barrel viral lineage extend? *Nat. Rev. Microbiol.* 6: 941–948.

Krupovic, M. and Bamford, D.H. 2008b. Archaeal proviruses TKV4 and MVV extend the PRD1-adenovirus lineage to the phylum Euryarchaeota. *Virology* 375: 292–300.

Martin, W. and Russell, M.J. 2007. On the origin of biochemistry at an alkaline hydrothermal vent. *Philos. Trans. R. Soc. Lond. B: Biol. Sci.* 362: 1887–1925.

Ochman, H., Lawrence, J.G., and Groisman, E.A. 2000. Lateral gene transfer and the nature of bacterial innovation. *Nature* 405: 299–304.

Ortmann, A.C., Wiedenheft, B., Douglas, T., and Young, M. 2006. Hot crenarchaeal viruses reveal deep evolutionary connections. *Nat. Rev. Microbiol.* 4: 520–528.

Poranen, M.M., Daugelavicius, R., and Bamford, D.H. 2002. Common principles in viral entry. *Annu. Rev. Microbiol.* 56: 521–538.

Prangishvili, D. and Garrett, R.A. 2005. Viruses of hyperthermophilic Crenarchaea. *Trends Microbiol.* 13: 535–542.

Szathmary, E. and Demeter, L. 1987. Group selection of early replicators and the origin of life. *J. Theor. Biol.* 128: 463–486.

Vetsigian, K., Woese, C., and Goldenfeld, N. 2006. Collective evolution and the genetic code. *Proc. Natl. Acad. Sci. USA* 103: 10696–10701.

Villarreal, L.P. and Witzany, G. 2010. Viruses are essential agents within the roots and stem of the tree of life. *J. Theor. Biol.* 262: 698–710.

Woese, C.R. 1998. The universal ancestor. *Proc. Natl. Acad. Sci. USA* 95: 6854–6859.

Woese, C.R. 2000. Interpreting the universal phylogenetic tree. *Proc. Natl. Acad. Sci. USA* 97: 8392–8396.

Woese, C.R. 2002. On the evolution of cells. *Proc. Natl. Acad. Sci. USA* 99: 8742–8747.

CHAPTER 20

Virolution Can Help Us Understand the Origin of Life

Luis P. Villarreal

CONTENTS

20.1 OBJECTIVE: A VIRUS FIRST OVERVIEW

A *virus first* perspective for understanding human evolution will likely seem counterintuitive or even preposterous to many readers (Villarreal, 2004). Surely, these most selfish and destructive agents (virus) cannot be proposed to have contributed substantially to the many complex features that make us human. Yet, developing and supporting this assertion is the objective of this chapter. Viruses are genetic parasites, often capable of transmission and dependent on their host for replication and/or maintenance. They are thus fundamentally able to interact with and contribute to host genetic (and epigenetic) content. It is this capacity that allows virus to be editors of host genetic content (Villarreal and Witzany, 2010; Witzany, 2006, 2009a). We know viruses to be agents of disease, often serious and even fatal. In what way can this capacity relate to the complexity needed to generate human capabilities? But viruses are also capable of colonizing and persisting in host genomes and becoming one with them. In so doing, they bring new and diffuse instruction sets to their host that can promote new regulatory networks with new capacities. This process has been called *virolution*, virus-mediated evolution (Ryan, 2009). And it is the persisting viruses that are highly host specific which, usually sexually transmitted, also have the ability to differentially affect host survival. The relationship of persisting viruses to its host population has been proposed to contribute significantly to host survival and affect the tree of life (Villarreal, 1999, 2006, 2007, 2008; Villarreal and Ryan, 2011). This process is shown schematically in Figure 20.1.

Such a process is fundamentally synbiogenic (Pereira et al., 2012). Indeed, it will be asserted subsequently that persisting human viruses were likely involved in the Homo sapiens—Neanderthal evolutionary outcome.

Why would viruses promote novelty via the formation of complex networks able to contribute to host phenotypes? The currently accepted view is that viruses are simply providing an extended source of errors (diversity) that can occasionally become *exapted* by their host for host purposes. An infected individual host variant will survive and somehow adapt virus information for its own survival. Networks are then created from this information in stepwise series of selection events. The real answer, however, lies much deeper than is likely to be appreciated. Indeed, it relates directly to the earliest events in the evolution of life reaching all the way back into the RNA world. This world is characterized by consortial, cooperative, multifunctional, and transmissive

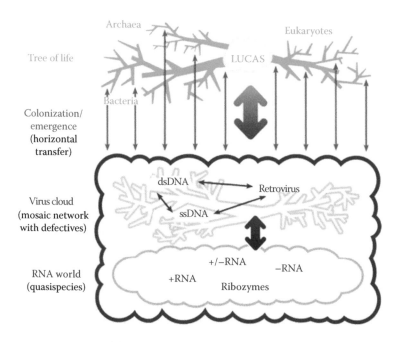

FIGURE 20.1 Overview of the viral cloud contribution to all the domains of the host evolutionary tree. Virus contribution to host is not occasional, but ongoing as shown. Also, such viral information is not due to *errors* as viruses are competent in host code and can be natural editors of host code. Viruses are thus an original and ongoing force for all life. (Reprinted from Villarreal, L.P. and Witzany, G., *J. Theor. Biol.*, 262, 698, 2010. With permission.)

RNA agents that operate in groups that can identify network membership and preclude nonmembers (immunity) (Villarreal and Witzany, 2013a,b).

We have long focused on the modest error-based genetic adaptations associated with neo-Darwinian selection and evolution. But whenever a host genome becomes colonized by nonancestral endogenous retroviruses (ERVs) and related elements that replicate via RNA, a quasispecies consortia (QS-C)-mediated process again applies to modify existing RNA societies that provided identity (control, immunity) superimposing new and often multiple uses of stem-loop RNAs that are now engaged in and providing new identity networks. This is a much more creative and punctuated process, able to promote complex regulatory shifts, but one that still essentially uses invasive stem-loop RNA agents.

20.2 SOME DEFINITION AND OTHER PROBLEMS

The term virus has a broad and almost instinctive meaning to many people with respect to disease. It is, however, worthwhile exploring a current definition of this term in order to employ it with greater consistency and precision. Since it is well known that many viruses can infect and exist within their host with no disease, clearly, disease cannot be a defining characteristic. Nor is uncontrolled replication a defining characteristic since many viruses have highly regulated replication cycles. Some do encode proteins involved in membrane synthesis. Some do not even encode their own capsid or membrane proteins, so this too cannot be a defining characteristic. But so far, no virus has been observed to code for a ribosome. Nor do they appear to encode many of the most fundamental metabolic proteins. Thus, viruses are fundamentally molecular entities that are parasitic to living systems (with ribosomes and energy production). But some viruses are parasitic to other viruses (thus parasitic to living systems plus virus), and most viruses can generate defective versions of themselves that are parasitic to the host system plus self-full virus. These situations can be very important for some specific viral lifestyles. Thus, our definition must be inclusive of all of these situations. I therefore propose the following characteristics for defining virus:

- A virus is a molecular genetic parasite.

- A virus must be competent in the instruction system of its host system.

- A virus must superimpose (edit) new instructions onto the host system (extending the code, bringing novelty, promoting symbiosis).

- Viral instructions must promote maintenance of the virus (i.e., self-identity compatible) which includes directed replication needed for either maintenance and/or transmission.

- Virus instructions can also simply include compelling the host to *maintain* the viral instruction set (persistence) and replicate it along with the host.

- Viral instructions must oppose (i.e., damage) competitive instruction sets (i.e., host immunity and/or virus competition).

- These viral instructions may subvert (colonize) opposing or competing instruction sets so as to maintain a coherent viral instruction system.

- The simplicity of RNA virus instructions requires that they be a coherent consortia of diverse RNA instructions (QS-C).

In addition to these defining characteristics, I would propose that the original *viral* instruction systems were simple stem-loop RNA replicators, as proposed for the RNA world (Briones et al., 2009). These parasitic replicators were able to transmit and occupy (ligate) their other RNA stem loops, including their own QS. Such self-invasion promotes the emergence of more complex functions (e.g., ribozymes and a consortial ribosome). RNA viruses still depend on these stem-loop instruction agents for basic identity and replication. The host (DNA) has become a habitat for these RNA societies.

There are other important problems involving definitions that should be mentioned. These include the terms networks and systems. Although popular use of these terms is consistent with the way I hope to apply them, the real problem, however, relates to attempts to mathematically define these terms so that calculation-based approaches can be applied to them. For example, a network stems from the concept of a net, with knots (nodes) connected to each other in binary links. This can be described mathematically. Similarly, formal complex systems posit a mathematical foundation for defining systems (von Bertalanffy and LaViolette, 1981). However, in the context of diverse but coherent RNA agents (QS-C), it is not possible to mathematically set either the potential interactions or the nature of these interactions for a single RNA entity as it will have conditional and context- and history-dependent activities (uses) within the population. This issue will be expanded further later. However, it does compel us to use the terms *networks* and *systems* in a less-defined (but popular) way. The concept of network in particular will be important for our discussions as it will relate directly to group identity that will require the specification of network membership characteristics. For an RNA agent, being a member of a network relates directly to its identity markers (often stem loops).

I will often consider the issues of group identity and group behavior as these are proposed to provide the foundations of social mechanisms. I will seek to define a network from the perspective of a consortium of RNA agents and apply the strategies of these diffuse transmissive agents to explain the creation of new networks and the editing of existing ones. However, it will be very difficult to think about and communicate these consortial or social issues. This is not because they are so inherently complex, but more because they are fundamentally interactive (social) phenomena that resist a linear explanation. For example, assigning a single function (or fitness) to an entity that is part of a consortium or network will inherently restrict out thinking about how the entity must function in the context of a consortium. A social system will have individual agents (such as RNA stem loops) that will fundamentally have multiple (often opposing) activities and uses. This is most apparent in the study of viral QS-C presented later. A system or network must always have this feature. We need to think socially, not serially.

I will present the case that immunity mechanisms are very the same as group identity mechanisms and that both operate using various strategies (such as addiction modules) that can destroy nonidentity and nongroup membership, while supporting group membership.

One more note of caution to keep in mind before further developing this (multi) line of reasoning. The types of RNA stem-loop-mediated changes we see in recent human evolution look hauntingly similar to those I proposed for the very early events in the evolution of life (see next chapter). It seems that from the very origins of life to recent genomic changes in human DNA, RNA societies have been acquired from infectious events and are mediating identity and social phenomena. It should therefore be considered that changes in RNA stem-loop composition may best define membership for all living systems and provide new identity systems in all domains of life, including virus.

20.3 CURRENT VIEW: INDIVIDUAL TYPE SELECTION AND EXAPTED VIRAL GENES

The development of neo-Darwinian thinking in the 1930s stems directly from the foundation that natural selection acting on variation (mostly from replication errors) in individuals selects for the survival of fittest type variant. Thus, the variation in offspring originates from the direct ancestor to the selected individual. However, when *nonancestor* virus-derived genes are seen to occur in host genomes, it is typically reasoned that such genes simply represent another form of variation (errors) that was also somehow associated with individual survival. The surviving host individual was then able to adapt (exapt) these genes for its own purpose and survival. This explanation still invokes a central role for individuals. What then results are various scenarios, such as kin selection, tit for tat, or arms race ideas involving a serial one-upmanship and linear process of selection. Any networks that emerge will then need to stem from this same serial process. The process is not prone to punctuated bursts in evolution nor is it particularly prone to rapid emergence of complexity or novelty. Also, any associative or group behavior that emerges, such as altruistic behavior, will similarly stem indirectly from individual survival, as described by various kin selection or game theory models. This view has been well accepted for numerous decades, and many current evolutionary biologists no longer question its basic tenets. Some even like to think of this as laws of evolution. But this is a view that emerged well before we understood the broad and ancient prevalence of virus. In the last several decades, analysis of comparative genomics and metagenomic sequencing of numerous habitats has shown us that virus-derived sequences dominate in all habitats so far evaluated (see Koonin, 2011). The term virosphere has been introduced to describe this vast cloud of genetic information. And within the genomes of organisms, virus-derived information is almost always the most dynamic component of host DNA for all domains of life. Much of this virus-derived host DNA, however, has long been seen to play no useful role; it was junk DNA that was the product of selfish replicators (Doolittle and Sapienza, 1980; Orgel and Crick, 1980). Yet, much of this *junk* was clearly viral derived, and often, its expression was associated with reproductive tissue (Ono et al., 1985). More recently, such *junk* has seemed much more important for the functioning of the organism (Volff, 2006).

But it is still basically seen as *exapted* stuff, put to some inadvertent good host use following individual selection. But the existence of a vast virosphere should compel us to think differently about virus-derived information in host genomes. All domains of life must survive in this ancient, unrelenting, and extremely adaptable virosphere. How can life thrive in this situation? In the following, we will look at the viruses themselves for an answer to this question. And from this question, we should rethink some of the tenets of evolutionary biology.

20.4 QS-C AND VIROLUTION

We should now ask: what is virolution and how does it affect host evolution? Can it provide some core, essential function needed for life? And more fundamentally, does virolution operate with additional principles, such as consortial group identities (QS-C) that can colonize and transform host, which fundamentally promote networks and complexity? If so, can these principles help us better understand the origin of life or provide insight into the origin of our human social capacities? Later, I attempt to summarize a large body of evidence from many domains of research that I feel help us better define virolution and the origin of host complexity. This will provide a core theme that will link all these diverse studies with the role of stem-loop RNA in viral and host identity. This role of QS-C in virolution is presented schematically in Figure 20.2.

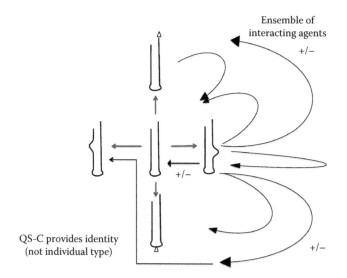

FIGURE 20.2 The crucial difference of QS-C with former QS concepts (fittest type—mutant spectra) is the basically consortial organization of functional RNA ensembles. Shown earlier are the possible consortial interactions (black arrows) of just one diversified RNA stem loop. These multiple activities (shown as +/−) preclude individual fitness definitions but require emergence and adaptation of group membership identities. Defectives with similar subviral RNA (stem-loop groups) remain relevant in both evolutionary and developmental processes. As a result of this basic evolutionary process of RNA stem-loop consortial building, we can look at the emergence of de novo identities. (Reprinted from Villarreal, L.P. and Witzany, G., *World J. Biol. Chem.*, 4, 71, 2013. With permission.)

As shown, such virolution is what promotes the creation of new *systems*, not serial selection from errors. But this looks like errors since most of the instructions are subviral. Virus, the ultimate and nearly invisible selfish agents, has finally taught us about the power of consortia. It is a big lesson and it applies to all levels of life. But why would a consortium of viral agents act to promote complexity? The answer I will develop later is that it is for the sake of superimposing group identity and group survival. The QS-C has to incorporate a new viral identity onto the host. This colonization will also clearly affect host survival in its extant virosphere. The virosphere matters for the success of all life. Such a colonizing event must promote the survival of this information and new viral identity/ecology that results. This is a very different perspective than that of selfish individual type selection. And this view suggests we adopt a new philosophy of biology in which collective behaviors are core for all life forms. And although virolution supports various forms of multilevel selection, it does not conflict with traditional individual type selection that emerged with DNA-based cells and virus. But whenever infectious sets of RNA-based replicating agents successfully colonize a host, they will again bring to bear the creative, cooperative, and distributed power of QS-C selection to their host. This is a most ancient process that still operates on DNA, using DNA as a stable habitat (Villarreal and Witzany, 2013a,b). I will now examine both the earliest events (RNA world) and most recent events (human-specific evolution) from this virolution perspective. One fundamental theme will apply throughout this examination that will be introduced both early and later in this chapter. This is the fundamental role of stem-loop RNA structures in the identity and function of infectious forms of RNA. When I focus on human-specific features, for example, such as our reproductive or social brain changes, it will be from the perspective of a role of infectious forms of stem-loop RNAs. The RNAs have multiple regulatory capacities that lead to a better understanding of RNA cascades and networks, which are the products of or promoted by serial colonization of virus (and often provide antiviral activity). These regulatory stem-loop RNAs will mostly occupy introns, 3′ untranslated region (UTR) and some 5′ promoter regions. We will also see that older identity/regulatory systems become subjected to manipulation (repurposed) or elimination following successful colonization.

20.5 ENCODE PROJECT: VIRAL *JUNK* AS *ESSENTIAL* RNA REGULATION

For many years, molecular biologist assumed that the complex RNA expression patterns observed by various techniques (such as hybridization kinetics) in the mammalian brain were due to the expression of many genes, which was expected for such a complex organ (see Chaudhari and Hahn, 1983). However, comparative genomics has made clear that gene transcription differs little between human and great apes (Khaitovich et al., 2004). Indeed, total gene numbers differ remarkably little between the simplest animals (*Caenorhabditis elegans*) and humans. But by far, the biggest differences between human and chimpanzee genomes were due to insertion and deletions (indels) (Mills et al., 2006; Watanabe et al., 2004; Wetterbom et al., 2006). The great majority of these indels are the result of retrotransposon activity of various types (ERVs, long terminal repeats [LTRs], long interspersed nuclear elements [LINEs], and alus being most numerous). Of these, the alu elements and transcripts are particularly active and affect RNA editing and intron splicing in the human genome (Sakate et al., 2007). In addition, they are frequently involved in epigenetic control and can emerge or expand rapidly in genomes (Zeh et al., 2009). Such a large-scale retroposon colonization would seem to pose a highly genotoxic situation for the human genome, an idea that seems supported by genomic analysis (Keightley et al., 2005, 2006). And yet this noncoding DNA is species specific (Toder et al., 2001), evolving quickly in humans (Bird et al., 2007), but also appears to be under very strong selective constraints (Bejerano et al., 2004; Bush and Lahn, 2005). This seems problematic in several ways: this is an inherently destructive event that should seldom result in novel or complex phenotype, plus it is both rapidly changing between species yet sometimes highly conserved. Indeed, this high rate of change was previously used to argue for the idea that it must be junk DNA. Yet, these are the changes that must be addressed and included to explain the emergence of the large and social human brain (Chapter 21). How then can we understand the origin of the most complex organ known (human brain) in the context of such massive introduction of errors? Clearly, we cannot. But perhaps the concept of *errors* is itself in error as implied earlier. Indeed, a major correction in our thinking has emerged from the ENCODE project. This project is a consortium of researchers that has sought to characterize all the RNA transcribed from the human genome, including RNA that is not cytoplasmic polyadenylated mRNA but is noncoding RNA (Mattick, 2005). It is now quite clear that most of this *junk* is transcribed and that 95% of the transcripts are from repeated sequences that were retrotransposed (Mattick, 2010,2011; Mattick et al., 2010). These transcripts include a previously poorly studied class of long noncoding RNA (lncRNA) (see Khalil et al., 2009). Furthermore, these noncoding transcripts appear particularly relevant to human brain and cognitive development and evolution (Barry and Mattick, 2012; Mattick and Mehler, 2008; Mehler and Mattick, 2007). Additionally, long-term memory also seems to use noncoding RNA (Mercer et al., 2008). These observations have led Mattick to propose that genetic programming in higher organisms (including human) has been misunderstood for 50 years (Mattick, 2001). Regulatory RNA derived from retrotransposons is key to eukaryotic complexity, compelling us to abandon the concept of selfish junk DNA. But in this realization, we also come to realize this regulatory RNA is operating mostly as stem-loop RNAs that have complex, multilevel, and even opposing functions. It is clearly operating and evolving as a network. But networks of stem-loop RNAs are also thought to have been crucial for the origin of RNA-based life. Could it be that the creative power of societies of stem-loop RNAs involved in the origin of life is still at work during recent human evolution? Let us further evaluate this idea.

20.6 FITTEST INDIVIDUAL TYPE RECONSIDERED

RNA viruses have long been recognized as distinct agents from their host cells in that they were the sole survivors of the RNA world that still used RNA as a genetic molecule. That they could replicate so readily and be characterized in the laboratory made them ideal systems to study variation in RNA replication. The variation was considered to result mostly from copy errors of a low-fidelity polymerase. And since viruses could be *cloned* (plaqued), they apparently adhered to the concept of individual fittest type selection. Since it was realized early on that RNA replications at the dawn of life in the RNA would also replicate with high error rates, this seemed to present a problem for the origin of life and the origin of the genetic code. It was from this perspective in the 1970s that Eigen developed the QS equations to explain the quantitative behavior of RNA populations that were generated via errors of the master fittest individual type template (Eigen, 1971). The basic assumptions were then that there was a master fittest type RNA template that would generate a cloud of RNA progeny due to copy errors but that this cloud would have certain overall behaviors (such as error threshold). Many more theoretical papers followed this early publication by Eigen, by his colleagues, and numerous others. And in the following decades, a large number of laboratory studies by RNA virologist sought to evaluate and measure various aspects of QS theory. It became very clear that the QS behavior of RNA viruses was very important for understanding clinical outcomes of human infections. And indeed, some of the insights of QS theory were observed, such as error threshold. The concepts of variation of the master fittest type became entrenched during this period as there seemed to be no conflict with more traditional neo-Darwinian selection. Thus, a consensus took hold for what QS is or means associated with master fittest type.

20.7 CORRECTIVE AND COLLECTIVE POWER OF VIRUS VIA MODERN QUASISPECIES: QS-C

In the ensuing several decades, many laboratory observations were made that indicated more complex collective behaviors for viral QS than were predicted by the QS equations. Two of the more active laboratories were those of John Holland and Esteban Domingo (see Domingo et al., 2008). The most recent compilation of these studies outlines many of the collective behaviors that have been made with QS (Domingo et al., 2012). In my opinion, the culmination study that most clearly reported that QS have more complex collective behaviors was the study from the Andino group of poliovirus pathogenesis in a mouse model in which diversity and cooperation were key to viral fitness (Vignuzzi et al., 2006). The QS collectives have distinct and measurable fitness. They can compete with and exclude related populations. They have minority populations that are crucial for overall fitness (Briones and Domingo, 2008; Briones et al., 2006). They can display heterogeneity important for fitness that is not observed in the consensus type (Borderia et al., 2012). They can suppress their own replication through lethal defection (Grande-Perez et al., 2005). They can be composed of members that can complement and interfere with replication of the collective, and many of these features can be observed in clinical infections such as humans with hepatitis C virus (Domingo and Gomez, 2007). Thus, QS are collectives that have positive and negative interacting members that are bound together for a combined fitness that depends on diversity (Arbiza et al., 2010; Lauring and Andino, 2010; Ojosnegros et al., 2011). It is thus ironic in that it is from the viruses, the most selfish of all genetic entities; we experimentally observe the characteristics of cooperative, collective behavior. And it was the *fittest type* assumptions of Eigen that generated QS equations and theory that stimulated the development of this modern collective QS view. But we are left with a conceptual contradiction. Modern QS observations do not depend on the master fittest type, and the consensus sequence may not predict to the fitness of the diverse collective. Diversity itself seems crucial. Such dynamic diversity allows a population of otherwise rather simple agents (such as HIV-1) to defeat a highly complex and evolved system of adaptive and innate immunity in their human host. If such infections were limited to the fittest type individuals, they would fail to overcome such a complex system. Not only can the power of QS defeat our human immune system, it has also largely defeated our combined human technology by frustrating the development of effective vaccines for 30 years. All this impressive biological competence comes from a small and *simple* virus! I therefore submit a modification to QS terminology to incorporate this collective and cooperative feature. The term QS-C will indicate a *collective* of *cooperative* character to the population. That way, the original term, QS, can still apply to fittest type models.

With this clarification, it should become apparent that all RNA replicators (especially simple ones) will have high rates of diversity generation (not error). In addition, all genetic entities that replicate via RNA will also be prone to QS-C (collective) behaviors. These behaviors will include both cooperative and competitive interactions, even within the same individual molecule. RNA, however, is not simply providing a syntax for genetic information. It is more than a code. It can also provide structure (stem loop), identity (stem loops, 5′, 3′ ends), and functional (ribozyme) activity. And it can be dynamic (e.g., pseudoknots) and responsive to the environment (riboswitches). Because of this much extended capacity relative to DNA, RNA can be considered as a more active entity, with behaviors that make it able to function as an *agent* to affect its own activity and survival (Witzany, 2009b, 2011). In that light, DNA can be considered as a habitat for various RNAs. It was from this perspective that Witzany and I proposed that DNA should be considered as a habitat for these active RNA agents (Villarreal and Witzany, 2013a,b). But this discussion of simple RNA replicators suggests that the concept of QS-C should also apply to the ideas and experiments concerning the *RNA world* hypothesis. Yet curiously, very little *RNA world* research has addressed any issues regarding QS (see Altman, 1989; Gesteland et al., 1999), let along the more modern QS-C idea. As many are starting to think that life originated in a cooperating situation (see Holmes, 2012), it is worth briefly considering if the QS-C concept will provide a different scenario for the origin of life.

20.8 RNA WORLD RECONSIDERED: INFECTIOUS STEM-LOOPS THAT OPERATE AS QS-C COLLECTIVES

To evaluate the QS-C and infectious perspective on the RNA world hypothesis, I will apply and explore the RNA agent concept introduced earlier to the role of stem-loop ribozymes in the origin of life. The main objective is to incorporate the historically absent QS-C and parasitic perspective (with its inherent feature for group fitness) into the process that creates RNA societies. I will not explore early chemical evolution that might have led to the emergence of RNA molecules but will instead assume that RNA has come into existence and follow its features from this perspective. One immediate consequence of this perspective is that we will be focused on collective features of RNA populations and will thus evaluate the chemical consequence of ribozyme QS societies, not individual replicators. This foundation immediately creates a situation in which *systems* of molecules with multiple behaviors will have the primary role in promoting the origin of life. It will also be important early on to consider how these systems maintain coherence (group identity, presented later), as this is an essential feature.

Indeed, a basic and continuing theme will be that a core function of stem-loop RNAs is to provide molecular identity through all of evolution, including recent human evolution. This identity theme will persist throughout this chapter and be frequently reintroduced. The idea is then that individual members of stem-loop RNA societies were collectively able to invade (ligate into) each other to form a more stable and capable (ribozyme active) consortia with emergent, transformative, and unpredictable abilities. These collectives would lead to the origin of various ribosome and RNA cellular societies (still linked to its stem-loop tRNA origin). Such a scenario also introduces the basic role of cooperation in the origin of life. It does not, however, eliminate competition, preclusion, and extinction that are also inherent features of QS-C behaviors. Furthermore, the identity and transmissive role for stem-loop RNAs sets the early (precellular) foundation for the origin of viruses whose emergence will further drive host evolution via colonization. The cooperative and parasitic features of QS-C will also promote the early participation of peptides in the identity and evolution of the ribonucleoprotein (RNP) complex society. The maintenance of these RNA societies as a coherent collective will generally be mediated by addiction modules (counteracting functions that recognize each other and harm nonrecognized partners) that underlie group identity and immunity in all living systems. With this foundation, the emergence of genes, DNA, cells, and individual fittest type selection can all be derived. But the emergence of DNA and cells and Darwinian evolution does not terminate the central role for transmissive RNA societies in the evolution of life. DNA becomes a habitat for these stem-loop *identity* RNAs, and it is from this perspective that I will subsequently examine recent events in human evolution. One issue should already be clear. This scenario posits that collective and cooperative behaviors were and remain essential for the emergence of living complexity. QS-C then provides a conceptual foundation for the study of cooperating chemical networks in which RNA mixtures of self-replicating ribozymes can form highly cooperative and dynamic autocatalytic cycles (Vaidya et al., 2012). Let us now put this into the perspective of virolution.

On the origin of the RNA world, short RNA oligomers formed by chemical processes needed to become longer RNAs able to perform template-based catalysis. It has been proposed that the initial chemical formation of hairpin-like RNAs (stem loops) could provide ribozyme activity following a ligation-based modular evolution that would yield ribozyme autocatalysis (Briones et al., 2009). Indeed, later, I present a series of studies that support this modular view. But according to the parameters of QS-C evolution, for a consortium of RNA stem-loop replicators to survive, they must form a coherent population. They must share their identity and survival. The recognition of the stem-loop sequence itself by catalytic agents could provide such common identity. Alternatively, chemical markers or initiators of catalysis could also mark the common population for priming or replication. Thus, it is very interesting that the smallest ribozyme so far reported consists of just five nucleotides able to catalyze aminoacylation of the 3′ end (Yarus, 2011). The addition of an amino acid to an RNA molecule has many interesting chemical implications. A ribozyme has rather limited chemical potential compared to proteins. This is mostly due to proton disassociation constant of various amino acid moieties that are not close to pH neutrality. Thus, amino acids are much more capable as chemical catalyst for this reason. Without the participation of amino acids, ribozymes must attain complex folds, often with some dynamic character (pseudoknots) to be effective catalyst allowing them to cleave and ligate RNA. Given this chemical advantage, we might expect that RNA evolution was greatly facilitated (but not coded) by peptides that contribute catalytically. In addition, such a modified RNA would likely also provide a chemical marker that could distinguish this RNA population. Indeed, this molecular identity idea is developed later as a way to better understand the origin of tRNA and its role in initiating replication of so many RNA viruses as well as how this chemical marker could promote the symbolic genetic code.

A good starting point for the accumulation of complexity seems to be hairpin ribozymes whose activity can be controlled by external effectors (Muller et al., 2012). Structural variation in these ribozymes allows progeny RNA to have different functions from their parental RNAs. The objective is to replicate RNA with RNA which hairpin ribozymes can perform via a sequence of ligation reactions that produce a longer ribozyme (Cheng and Unrau, 2010). Along these lines, two short hairpin RNAs can catalyze their own ligation to form larger RNA constructs (Gwiazda et al., 2012). Thus, we see interactions that promote more complex progeny. However, for a fully active ribozyme, complex RNA folding is needed. And such folding is cooperative (Behrouzi et al., 2012). Folded ribozymes can also interact with other small molecules promoting their function as riboswitches (Ferre-D'Amare, 2011). This includes amino acids that could promote either catalytic control or group identity marking. And the ribozyme folds can also be dynamic and context sensitive as seen in pseudoknots (Perreault et al., 2011). But ribozymes can also be invasive, including self-invasive (Kumar and Joyce, 2003). Thus, stem-loop RNAs have many behaviors that would allow them to function as a mixture of agents involved in their own recognition and synthesis. Of particular interest is their ability to self-ligate as this could promote the emergence of RNA societies with self-identity. We can also think of tRNA as stem-loop RNA with various functions and histories. Indeed, it appears that tRNAs evolved from two separate hairpins (Dick and Schamel, 1995), in which each of the stem loops interacts with a different ribosomal RNA subunit (presented later). This is a very interesting observation from an RNA society perspective. The invasive nature of intron ribozymes (endonuclease) also applies to tRNA from archaea, but here four distinct specificities are known (Fujishima et al., 2011). This very much resembles an identity system in which introns are marking central cellular (self) agents (tRNAs) for group identity but should destroy similar tRNAs (viral, other cellular, etc.) lacking the intron marking. It is thus also interesting that tRNAs with various linked amino acids themselves have been proposed to have originated before the translation system as genomic 3′ tags needed for RNA ribozyme replication (Rodin et al., 2011; Sun and Caetano-Anolles, 2008). This early function can also be explained as having served as a tag for group identity and could better explain the polyphyletic nature of the origin of tRNA (Di Giulio, 2013).

20.9 AGENTS JOIN GROUPS AND SOCIAL NETWORKS

The QS-C perspective allows us to consider the role of stem-loop RNAs in the origin of the RNA world in which the action of individual agents can cooperate and be combined into a more capable collective action of a population. Thus, the origin of spontaneous cooperating networks of stem-loop RNA replicators (Vaidya et al., 2012) can be understood from this perspective. However, I will use the term network to include some distinct features, specifically network membership. To designate this situation, I apply the term social network to distinguish networks that have no membership criteria. Networks can be either open or closed. Biological networks almost always have a closed feature to them. Basically, for a network to be coherent and able to act collectively, it must limit membership to promote coordination. Otherwise, it is simply a collection of uncoordinated agents, and there will be no selection for maintaining the network existence. If we are examining a network composed of stem-loop RNAs, it will be necessary for the individual RNAs to have some feature or behaviors that maintains membership (such as replication and recognition). This requires interaction. If only one type of RNA is supported (e.g., high-fidelity replication), there can be no complementation and complex function (i.e., ribozyme) for the collective. A diversity of behavior and type will be essential. Recall, however, that these RNAs act as agents in which various (multiple) behaviors will be possible even for the same sequence. This means there are diversity of interaction as well as diversity of type. Thus, overall interaction of an RNA agent with the collective must promote coherence and continued existence. What then are the features that promote continued existence (selection) for a network? This does not require that only positive (e.g., replication) interactions be supported. Negative interactions, including interference, will also be needed. For example, highly efficient runaway replicons would overtake a QS collective and yield only one RNA type. Thus, the QS would lose complementing functionality and would also consume all substrates if they were not regulated. This situation presents a problem in those habitats with limited substrates (likely a very common state). Therefore, some level of self-regulation (negation) in the collective would promote the survival of the collective, especially if these RNAs could interact with the substrate in a regulatory (riboswitch) manner. That efficient replicators become susceptible to parasitic replicators would provide an inherently spontaneous process of self-regulation. Yet, the collective will still need to promote replication when it is favored. Accordingly, it becomes important for members of the collective to be subjected to both positive and negative self-regulation via RNA–RNA interactions. However, here too, there must be some limits to self-regulation as the collective cannot tolerate overly active self-regulating members that will extinguish the collective. Thus, we see that being a successful member of a collective has many (and multiple) behaviors associated with it. On top of that, as a QS-C replicates, these features will drift with time in a dynamic manner. In this context, we can see that a random RNA stem loop or a stem-loop RNA from a different QS collective would likely not be coherent with the other members of a particular QS. A QS society is generally rather specific for its members. Group selection has already started. Indeed, many experiments with RNA viruses infecting humans and animals have shown that a particular QS will exclude other QS of the same virus. And such society membership is also time dependent in that the serial passage of the same viral QS will usually result in subsequent QS that precludes prior individual members of the QS. This behavior has often been called a *Red Queen* behavior, but such a classical neo-Darwinian view does not incorporate or acknowledge the issue of group membership. The membership view, on the other hand, allows us to understand the maintenance of minority types in the collective since these members can provide a needed but complementing catalytic control. Thus, a QS society is a network that will naturally promote the emergence of membership. And as noted, defective interfering agents can also contribute to membership control. As I have previously proposed (Villarreal, 2012), group membership can also be promoted by the combined action of toxic agents linked to antitoxic agents. A common version of a toxic agent is an endonuclease that will cleave sequences that are recognized (as foreign). The antitoxin in this case prevents the action of the endonuclease (e.g., via a bound protein or methylated base, dsRNA with another molecule, altered RNA fold). In this light, the endonuclease and ligation activities of stem-loop ribozymes are particularly interesting. A stem-loop ligase could provide a mechanism to recognize nonmember stem-loop RNAs and destroy them by ligation. Recall, however, that serial ligation can also be used to copy a stem-loop RNA. But such a situation has several very interesting implications. One of the problems with a society of stem-loop RNAs is that to attain their combined function, they need precise physical molecular placement relative to one another. This would normally require a high concentration dependence to counteract diffusion. By ligation, however, we could build a society of stem-loop RNAs that have covalently placed the various stem loops in the correct functional (or dynamic/regulatory) context and have lost their concentration dependence. It seems likely that such a process would involve an invasive self-colonizing stem-loop RNAs that result in one molecular entity with a common identity function. This would generate one entity that evolved from the ligation of a mixed set of stem-loop agents that now have a highly enhanced (collective) functional capacity. This collective would also have a highly enhanced capacity for persistence as it need not continually replicate individual stem-loop RNA agents to maintain its membership. The collective, however, would still need to oppose nonmember or other parasite participation. Additionally, a collective might attain a conditional (regulated) replication capacity if it incorporates stem-loop RNA riboswitches. It is by such a process that we can now consider the origin of the ribosome.

Membership is thus crucial for living networks (systems) to emerge. In examining the literature relevant to QS, the RNA world, and RNA network formation, we can indeed find some experimental evidence that supports QS and the spontaneous emergence of RNA networks. But almost completely lacking from such experiments is any evaluation of the membership issue. For example, QS-like behavior has been observed with in vitro RNA replicator studies (Arenas and Lehman, 2010). Nonenzymatic template (peptide)–directed autocatalytic systems can show network behavior (Dadon et al., 2012). And communities of RNA ribozyme replicator sets can also show lateral evolution (Hordijk and Steel, 2012). Also rule-based computing simulation has been applied to similar systems in an effort to understand the emergence of parasites and antiparasites (Jalasvuori et al., 2010). Along these lines, the hypercycle kinetic model was proposed to be a system of cross-catalyzing RNA replicators that depend on cooperation for growth, but this is not a collective autocatalytic system as proposed earlier (Szathmáry, 2013). But hypercycles as proposed are not able to tolerate parasites, let along depend on them for development. Yet, the biggest problem of all such studies is that there is no assumption regarding the basic importance of network or group membership. Without this network membership concept and its attending strategies and mechanisms, authentic collective action does not emerge. Systems do not develop. The dynamic nature of network membership and collective action poses many unsolved problems for existing theory.

For example, how is the multipotential of an individual RNA to be evaluated within the QS-C if we cannot specify all the other interactions and how they change with time? We cannot apply our current ideas of fitness to this individual RNA as the historical and population context is key. Network membership needs to be prominently considered if we are to understand the origin of the ribosome and the genetic code. As will be further outlined later, replicator identity marking via 3′ aminoacylated of a stem-loop RNAs appears most able to explain the origin of a tRNA-mediated genetic code. This is a big difference in our conceptual stance. For in contrast to Darwinian evolution, network members will generally have distinct ancestral histories. These members will mostly originate from separate parasitic lineages that were able to penetrate defenses and join the network (sometimes in mixtures). They do not need to descend from one individual or even be from the same type of agent (virus, transposon, intron, intene, etc.). From this perspective, we can understand why the two halves of tRNA have distinct evolutionary histories, yet tRNA is a core agent for the evolution of life. Thus, neither the amino acid-based (peptide) ancestors nor the RNA-based ancestors need a common origin to participate in a symbiogenic network. QS-C theory supports such a network process. And we will continue to apply the QS-C perspective for the rest of this chapter. In the following, network membership will provide the basis for examining noncoding RNA-based regulation needed for multicellular complexity (Lozada-Chavez et al., 2011).

20.10 CONCLUSIONS

The application of virolution overall to issues regarding the origin of life can provide us with a very distinct and new perspective on how living systems emerge from chemical replicators. The emergence of cooperative QS-C thinking from the more accepted QS equations of Manfred Eigen, based on individual type selection, has provided a conceptual foundation from which collective action of RNA agents can now be understood. As group membership becomes a basic criterion for the emergence of living systems, we also start to understand why the history and context of the RNA society become crucial for social survival and function. History and context dependence also lead to the emergence of symbolic code in living systems. Indeed, this QS-C thinking can also provide us with a transition point between the chemical world of RNA replicators and the living world of RNA agents that must belong to their respective society. The power of a consortium to solve complex, multilevel problems that can even use opposing and minority functions becomes evident. This power, which promoted the emergence of the RNA world, did not become extinct with the emergence of DNA-based life. As we will see subsequently, the consortial action of *parasitic* RNA stem-loop societies can also help us understand the emergence of our large social human brain.

GLOSSARY

ERVs: Endogenous retroviruses. Partial or complete sequences derived from retroviruses that have become part of the host genome.

Exapted: An evolutionary theory that proposes that DNA (genes) from other organisms, such as parasites, can become part of the host DNA following natural selection.

Genetic parasites: Genetic agents, such as viruses and transposable elements, that colonize and use host systems for their own maintenance.

Group identity: The capacity for a group of living and/or genetic agents to recognize other members of the same group.

LINEs: Long interspersed nuclear elements. Genetic agents, distinct from retroviruses, which derive from a type of retrotransposon, able to replicate from RNA via DNA.

LTRs: Long terminal repeats. The characteristic sequences found on both ends of a retrovirus genome.

Matagenomic: The study of populations of genomes as they occur in particular habitats.

Network membership: The capacity of a network to recognize allowed members from nonmembers.

Networks: A system of agents (or elements) that interact in a coherent fashion, usually associated with regulation.

Pseudoknots: The ability of a stem-loop-like RNA to fold itself into two distinct and dynamic conformations.

Quasispecies: A population of related genetic agents that are derived by variation during replication.

Riboswitches: A dynamic RNA fold structure that interacts with a regulator (such as a small molecule) to change its conformation and function.

Stem-loop RNA: Small regions of RNA that can fold back on themselves to form base pairs.

Synbiogenic: The generation of genetic novelty via the stable interaction with a symbiotic organism.

tRNA: Transfer RNA; the small clover-leaf-shaped RNA that binds to the ribosome and is responsible for providing the correct amino acid in the triplet genetic code.

UTR: Untranslated region. Sequences of DNA that are regulatory and do not code for protein synthesis.

Virolution: A term coined by Frank Ryan to describe host evolution mediated by the action (selection, protection, integration) of viruses.

Virosphere: The extended virus composition of a biological habitat.

REVIEW QUESTIONS

1. What is meant by the term *virosphere*?

2. What is meant by the term *virolution*? Explain how virolution promotes evolutionary novelty and why is this process more efficient than the serial selection from errors.

3. Explain *quasispecies* and QS-C. How do these two differ?

4. What are the individual agents in QS-C? How do they recognize each other? How is membership in the consortium determined and regulated?

5. Explain how the fitness of QS-C depends on diversity of its members.

6. Show the application of the concept of the ribozyme QS societies to the RNA world hypothesis. Why is it essential that one considers the QS societies rather than the individual replicators?

7. Spontaneous cooperating networks of stem-loop RNA replicators are at the origins of the RNA world. Discuss the concepts of networks versus *social* networks that have membership criteria. Why is membership important?

8. A QS society is specific for its members. Provide experimental data that illustrate this.

REFERENCES

Altman, S., 1999. The RNA world. *Scholar* 90812: 130421.

Arbiza, J., Mirazo, S., Fort, H., 2010. Viral quasispecies profiles as the result of the interplay of competition and cooperation. *BMC Evol. Biol.* 10, 137.

Arenas, C.D., Lehman, N., 2010. Quasispecies-like behavior observed in catalytic RNA populations evolving in a test tube. *BMC Evol. Biol.* 10, 80.

Barry, G., Mattick, J.S., 2012. The role of regulatory RNA in cognitive evolution. *Trends Cogn. Sci.* 16, 497–503.

Behrouzi, R., Roh, J.H., Kilburn, D., Briber, R.M., Woodson, S.A., 2012. Cooperative tertiary interaction network guides RNA folding. *Cell* 149, 348–357.

Bejerano, G., Pheasant, M., Makunin, I., Stephen, S., Kent, W.J., Mattick, J.S., Haussler, D., 2004. Ultraconserved elements in the human genome. *Science* 304, 1321–1325.

Bird, C.P., Stranger, B.E., Liu, M., Thomas, D.J., Ingle, C.E., Beazley, C., Miller, W., Hurles, M.E., Dermitzakis, E.T., 2007. Fast-evolving noncoding sequences in the human genome. *Genome Biol.* 8, R118.

Borderia, A.V., Lorenzo-Redondo, R., Pernas, M., Casado, C., Alvaro, T., Domingo, E., Lopez-Galindez, C., 2012. Initial fitness recovery of HIV-1 is associated with quasispecies heterogeneity and can occur without modifications in the consensus sequence. *PLoS ONE* 5, e10319.

Briones, C., de Vicente, A., Molina-Paris, C., Domingo, E., 2006. Minority memory genomes can influence the evolution of HIV-1 quasispecies in vivo. *Gene* 384, 129–138.

Briones, C., Domingo, E., 2008. Minority report: Hidden memory genomes in HIV-1 quasispecies and possible clinical implications. *AIDS Rev.* 10, 93–109.

Briones, C., Stich, M., Manrubia, S.C., 2009. The dawn of the RNA world: Toward functional complexity through ligation of random RNA oligomers. *RNA* 15, 743–749.

Bush, E.C., Lahn, B.T., 2005. Selective constraint on noncoding regions of hominid genomes. *PLoS Comput. Biol.* 1, e73.

Chaudhari, N., Hahn, W.E., 1983. Genetic expression in the developing brain. *Science* 220, 924–928.

Cheng, L.K.L., Unrau, P.J., 2010. Closing the circle: Replicating RNA with RNA. *Cold Spring Harb. Perspect. Biol.* 2, a002204.

Dadon, Z., Wagner, N., Cohen-Luria, R., Ashkenasy, G., 2012. Reaction networks, in: P.A. Gale and J.W. Steed (eds.), *Supramolecular Chemistry*. John Wiley & Sons, Ltd., New York.

Dick, T.P., Schamel, W.A., 1995. Molecular evolution of transfer RNA from two precursor hairpins: Implications for the origin of protein synthesis. *J. Mol. Evol.* 41, 1–9.

Di Giulio, M., 2013. A polyphyletic model for the origin of tRNAs has more support than a monophyletic model. *J. Theor. Biol.* 318, 124–128.

Domingo, E., Gomez, J., 2007. Quasispecies and its impact on viral hepatitis. *Virus Res.* 127, 131–150.

Domingo, E., Parrish, C.R., Holland, J.J., 2008. *Origin and Evolution of Viruses*, 2nd edn. Academic Press, San Diego, CA.

Domingo, E., Sheldon, J., Perales, C., 2012. Viral quasispecies evolution. *Microbiol. Mol. Biol. Rev.* 76, 159–216.

Doolittle, W.F., Sapienza, C., 1980. Selfish genes, the phenotype paradigm and genome evolution. *Nature* 284, 601–603.

Eigen, M., 1971. Selforganization of matter and the evolution of biological macromolecules. *Naturwissenschaften* 58, 465–522.

Ferre-D'Amare, A.R., 2011. Use of a coenzyme by the glmS ribozyme–riboswitch suggests primordial expansion of RNA chemistry by small molecules. *Philos. Trans. R. Soc. Lond.* 366, 2942–2948.

Fujishima, K., Sugahara, J., Miller, C.S., Baker, B.J., Di Giulio, M., Takesue, K., Sato, A., Tomita, M., Banfield, J.F., Kanai, A., 2011. A novel three-unit tRNA splicing endonuclease found in ultrasmall Archaea possesses broad substrate specificity. *Nucleic Acids Res.* 39, 9695–9704.

Gesteland, R.F., Cech, T., Atkins, J.F., 1999. *The RNA World: The Nature of Modern RNA Suggests a Prebiotic RNA, Cold Spring Harbor Monograph Series, 37*. Cold Spring Harbor Laboratory Press, Cold Spring Harbor, NY.

Grande-Perez, A., Lazaro, E., Lowenstein, P., Domingo, E., Manrubia, S.C., 2005. Suppression of viral infectivity through lethal defection. *Proc. Natl. Acad. Sci. USA* 102, 4448–4452.

Gwiazda, S., Salomon, K., Appel, B., Muller, S., 2012. RNA self-ligation: From oligonucleotides to full length ribozymes. *Biochimie* 94, 1457–1463.

Holmes, B., 2012. First life may have survived by cooperating. *New Scientist* 216, 10.

Hordijk, W., Steel, M., 2012. A formal model of autocatalytic sets emerging in an RNA replicator system. arXiv:1211.3473.

Jalasvuori, M., Jalasvuori, M.P., Bamford, J.K.H., 2010. Dynamics of a laterally evolving community of ribozyme-like agents as studied with a rule-based computing system. *Orig. Life Evol. Biosph.* 40, 319–334.

Keightley, P.D., Lercher, M.J., Eyre-Walker, A., 2005. Evidence for widespread degradation of gene control regions in hominid genomes. *PLoS Biol.* 3, e42.

Keightley, P.D., Lercher, M.J., Eyre-Walker, A., 2006. Understanding the degradation of hominid gene control. *PLoS Comput. Biol.* 2, e19 (author reply e26).

Khaitovich, P., Muetzel, B., She, X., Lachmann, M., Hellmann, I., Dietzsch, J., Steigele, S. et al., 2004. Regional patterns of gene expression in human and chimpanzee brains. *Genome Res.* 14, 1462–1473.

Khalil, A.M., Guttman, M., Huarte, M., Garber, M., Raj, A., Rivea Morales, D., Thomas, K. et al., 2009. Many human large intergenic noncoding RNAs associate with chromatin-modifying complexes and affect gene expression. *Proc. Natl. Acad. Sci. USA* 106, 11667–11672.

Koonin, E.V., 2011. *The Logic of Chance: The Nature and Origin of Biological Evolution*. FT Press, Upper Saddle River, NJ.

Kumar, R.M., Joyce, G.F., 2003. A modular, bifunctional RNA that integrates itself into a target RNA. *Proc. Natl. Acad. Sci. USA* 100, 9738–9743.

Lauring, A.S., Andino, R., 2010. Quasispecies theory and the behavior of RNA viruses. *PLoS Pathog.* 6, e1001005.

Lozada-Chavez, I., Stadler, P.F., Prohaska, S.J., 2011. "Hypothesis for the modern RNA world": A pervasive non-coding RNA-based genetic regulation is a prerequisite for the emergence of multicellular complexity. *Orig. Life Evol. Biosph.* 41, 587–607.

Mattick, J.S., 2001. Non-coding RNAs: The architects of eukaryotic complexity. *EMBO Rep* 2, 986–991.

Mattick, J.S., 2005. The functional genomics of noncoding RNA. *Science* 309, 1527–1528.

Mattick, J.S., 2010. The central role of RNA in the genetic programming of complex organisms. *Acad. Bras. Cienc.* 82, 933–939.

Mattick, J.S., 2011. The central role of RNA in human development and cognition. *FEBS Lett.* 585, 1600–1616.

Mattick, J.S., Mehler, M.F., 2008. RNA editing, DNA recoding and the evolution of human cognition. *Trends Neurosci.* 31, 227–233.

Mattick, J.S., Taft, R.J., Faulkner, G.J., 2010. A global view of genomic information—Moving beyond the gene and the master regulator. *Trends Genet.* 26, 21–28.

Mehler, M.F., Mattick, J.S., 2007. Noncoding RNAs and RNA editing in brain development, functional diversification, and neurological disease. *Physiol. Rev.* 87, 799–823.

Mercer, T.R., Dinger, M.E., Mariani, J., Kosik, K.S., Mehler, M.F., Mattick, J.S., 2008. Noncoding RNAs in long-term memory formation. *Neuroscientist* 14, 434–445.

Mills, R.E., Luttig, C.T., Larkins, C.E., Beauchamp, A., Tsui, C., Pittard, W.S., Devine, S.E., 2006. An initial map of insertion and deletion (INDEL) variation in the human genome. *Genome Res.* 16, 1182–1190.

Muller, S., Appel, B., Krellenberg, T., Petkovic, S., 2012. The many faces of the hairpin ribozyme: Structural and functional variants of a small catalytic RNA. *IUBMB Life* 64, 36–47.

Ojosnegros, S., Perales, C., Mas, A., Domingo, E., 2011. Quasispecies as a matter of fact: Viruses and beyond. *Virus Res.* 162, 203–215.

Ono, M., Toh, H., Miyata, T., Awaya, T., 1985. Nucleotide sequence of the Syrian hamster intracisternal A-particle gene: Close evolutionary relationship of type A particle gene to types B and D oncovirus genes. *J. Virol.* 55, 387–394.

Orgel, L.E., Crick, F.H., 1980. Selfish DNA: The ultimate parasite. *Nature* 284, 604–607.

Pereira, L., Rodrigues, T., Carrapiço, F., 2012. A symbiogenic way in the origin of life, in: J. Seckbach (ed.), *Genesis—In The Beginning, Cellular Origin, Life in Extreme Habitats and Astrobiology.* Springer, Dordrecht, the Netherlands, pp. 723–742.

Perreault, J., Weinberg, Z., Roth, A., Popescu, O., Chartrand, P., Ferbeyre, G., Breaker, R.R., 2011. Identification of hammerhead ribozymes in all domains of life reveals novel structural variations. *PLoS Comput. Biol.* 7, e1002031.

Rodin, A.S., Szathmary, E., Rodin, S.N., 2011. On origin of genetic code and tRNA before translation. *Biol. Direct* 6, 14.

Ryan, F., 2009. *Virolution.* Collins, London, U.K.

Sakate, R., Suto, Y., Imanishi, T., Tanoue, T., Hida, M., Hayasaka, I., Kusuda, J., Gojobori, T., Hashimoto, K., Hirai, M., 2007. Mapping of chimpanzee full-length cDNAs onto the human genome unveils large potential divergence of the transcriptome. *Gene* 399, 1–10.

Sun, F.J., Caetano-Anolles, G., 2008. The origin and evolution of tRNA inferred from phylogenetic analysis of structure. *J. Mol. Evol.* 66, 21–35.

Szathmáry, E., 2013. On the propagation of a conceptual error concerning hypercycles and cooperation. *J. Syst. Chem.* 4, 1–4.

Toder, R., Grutzner, F., Haaf, T., Bausch, E., 2001. Species-specific evolution of repeated DNA sequences in great apes. *Chromosome Res.* 9, 431–435.

Vaidya, N., Manapat, M.L., Chen, I.A., Xulvi-Brunet, R., Hayden, E.J., Lehman, N., 2012. Spontaneous network formation among cooperative RNA replicators. *Nature* 491, 72–77.

Vignuzzi, M., Stone, J.K., Arnold, J.J., Cameron, C.E., Andino, R., 2006. Quasispecies diversity determines pathogenesis through cooperative interactions in a viral population. *Nature* 439, 344–348.

Villarreal, L., Ryan, F., 2011. Viruses in host evolution: General principles and future extrapolations. *Curr. Top. Virol.* 9, 79–89.

Villarreal, L.P., 1999. DNA virus contribution to host evolution, in: E. Domingo, R.G. Webster, and J.J. Holland (eds.), *Origin and Evolution of Viruses.* Academic Press, San Diego, CA, pp. 391–420.

Villarreal, L.P., 2004. Can viruses make us human? *Proc. Am. Philos. Soc.* 148, 296–323.

Villarreal, L.P., 2006. How viruses shape the tree of life. *Future Virol.* 1, 587–595.

Villarreal, L.P., 2007. Virus–host symbiosis mediated by persistence. *Symbiosis* 44, 1–9.

Villarreal, L.P., 2008. The widespread evolutionary significance of viruses, in: E. Domingo, C.R. Parrish, and J.J. Holland (eds.), *Origin and Evolution of Viruses.* Academic Press, London, U.K.

Villarreal, L.P., 2012. The addiction module as a social force, in: G. Witzany (ed.), *Viruses: Essential Agents of Life.* Springer Science+Business Media, Dordrecht, the Netherlands.

Villarreal, L.P., Witzany, G., 2010. Viruses are essential agents within the roots and stem of the tree of life. *J. Theor. Biol.* 262, 698–710.

Villarreal, L.P., Witzany, G., 2013a. Rethinking quasispecies theory: From fittest type to cooperative consortia. *World J. Biol. Chem.* 4, 71–82.

Villarreal, L.P., Witzany, G., 2013b. The DNA habitat and its RNA inhabitants: At the dawn of RNA sociology. *Genom. Insights* 6, 1–12.

Volff, J.-N., 2006. Turning junk into gold: Domestication of transposable elements and the creation of new genes in eukaryotes. *BioEssays* 28, 913–922.

von Bertalanffy, L., LaViolette, P.A., 1981. *A Systems View of Man*. Westview Press, Boulder, CO.

Watanabe, H., Fujiyama, A., Hattori, M., Taylor, T.D., Toyoda, A., Kuroki, Y., Noguchi, H. et al., 2004. DNA sequence and comparative analysis of chimpanzee chromosome 22. *Nature* 429, 382–388.

Wetterbom, A., Sevov, M., Cavelier, L., Bergstrom, T.F., 2006. Comparative genomic analysis of human and chimpanzee indicates a key role for indels in primate evolution. *J. Mol. Evol.* 63, 682–690.

Witzany, G., 2006. Natural genome-editing competences of viruses. *Acta Biotheor.* 54, 235–253.

Witzany, G., 2009a. A perspective on natural genetic engineering and natural genome editing. Introduction. *Ann. N.Y. Acad. Sci.* 1178, 1–5.

Witzany, G., 2009b. Noncoding RNAs: Persistent viral agents as modular tools for cellular needs. *Ann. N.Y. Acad. Sci.* 1178, 244–267.

Witzany, G., 2011. The agents of natural genome editing. *J. Mol. Cell Biol.* 3, 181–189.

Yarus, M., 2011. The meaning of a minuscule ribozyme. *Philos. Trans. R. Soc. Lond.* 366, 2902–2909.

Zeh, D.W., Zeh, J.A., Ishida, Y., 2009. Transposable elements and an epigenetic basis for punctuated equilibria. *Bioessays* 31, 715–726.

CHAPTER 21

Can Virolution Help Us Understand Recent Human Evolution?

Luis P. Villarreal

CONTENTS

21.1 CHAPTER OBJECTIVES

In the previous chapter, I outlined how virus can contribute to the origin of life. Here, I present how a similar process applies to human evolution. The currently accepted view is that when virus-derived information is found in host, viruses are simply providing an extended source of errors (diversity) that can occasionally become *exapted* by their host for host purposes. An infected individual host variant will survive and somehow adapt virus information for its own survival. Any networks that are then created from this information would form in stepwise series of selection events. However, as presented in the previous chapter, virus-mediated evolution that involves cooperative quasispecies (populations of variants) relates directly to the earliest events in the evolution of life reaching all the way back into the RNA world (life before DNA emergence). This world is characterized by consortial, cooperative, multifunctional, and transmissive RNA agents that operate in groups and that can identify network membership and preclude nonmembers (immunity) (Witzany et al., 2013). In this chapter, I examine the action of similar RNA agents in human evolution as a way to understand the processes and selective forces that promoted the emergence of our large and very social human brain.

21.2 NOTE TO THE READER

For the readers who are not experts in virology or molecular biology, I provide two miniprimers, which explain more complex concepts. These primers are placed in the text as needed and can be recognized by the word *primer* in the subtitle. They are written in a more colloquial style and are referenced only sparingly. The readers who are experts in the field can skip the primers.

21.3 PRIMER 1: THE CONCEPT OF ADDICTION MODULES

The origin of the idea of an addiction module came from the study of persisting (transmissible) P1 phage in *E. coli* in the early 1990s (from M. Yarmolinsky and colleagues at NIH). This phage was episomally maintained, yet it was very stable (not needing chromosomal integration). It attains its stability by expressing sets of genes that are both toxic (and stable) and antitoxic (but unstable). If the phage genome is lost (cured, such as during sexual transfer), the *cured* cell still retains the toxic activity and will kill itself. Thus, the persisting phage insures that only persistently infected host will survive. This is survival of the persistently infected! However, the P1 phage will also oppose not only lytic infection by P1 but also lytic and lysogenic infection by other phage. Thus, the host has acquired a more generalized survival phenotype (especially in the virosphere). This concept can be generalized for any host-specific persisting virus (or its defectives) that can protect it but also harm noninfected host. This concept also provides a clear mechanism that promotes both symbiosis and group survival. In this, it provides a major insight into how cooperation (virus–host symbiosis) within groups can be generated and maintained without requiring kin selection or game theory. The concept can also be further generalized even more by proposing that group identity in general is attained by the presence of a protective (antitoxic) function that when absent elicits a harmful (toxic) response. Thus, addiction modules of various forms become essential for group identity (including extended social bonding). The concept of virus-mediated group selection also applies to the Spanish conquistador success in the New World. Thus, the addiction module represents a big generalized idea that is applicable to all levels of biology and drives symbiosis, identity, and complexity.

Why do addiction circuits exist in animal brains? Psychological pain has always presented a difficulty for evolutionary biology. In 1978, Panksepp hypothesized that social attachments functioned by piggybacking onto the physical pain system and did so through opioid-mediated processes. He noted the parallels between attachment processes and addiction.

But if this represents the toxic half of an addiction module that binds individuals together, it makes much more sense to be wired into brains. Strong social bonds must have been important for the origin of placental mother–infant relationship and care. How were they promoted? It is likely the placental regulatory network must have contributed to this. And when this bond is broken (e.g., by experiencing or learning infant death), very toxic (withdrawal-like) reactions are immediate. Similarly, the pleasure of love (infant and romantic) represents a beneficial (antitoxin) reaction that can also bind individuals. Thus, pair bonding and family bonding can be seen as an extension of the original maternal bond just as Panksepp proposed. But such thinking requires that group bonds be ingrained into the very origin and core of our biology (and are able to emerge from virus colonization). For this to be the case, we must reconsider some of the basic forces in evolutionary biology (including a virus role and group selection) and must now move beyond the long-held belief that individual fittest type selection can account for complex network formation and emergence. Clearly, such a challenging reevaluation is necessary.

21.4 SOCIETIES OF HUMANS FROM SOCIETIES OF NONCODING RNA

Let us start by outlining some human features we wish to understand from our perspective. One current view is that humans are a hypersocial species that exhibit extended social bonding and social intelligence as basis of their more general intelligence (Striedter, 2005). Many mammals show maternal bonding. A few mammals show mate bonding but very few show extended (nonfamilial) bonding. We can start by asking how evolution can promote the emergence of an extended large social brain from a maternal one, based on the assumption that complexity is added to preexisting systems. In addition, we can also acknowledge that these recent brain changes are mostly not mediated by genetic coding regions (information translated into protein) but by changes in networks of genetic regulatory regions (information that controls gene expression) and noncoding RNA, including epigenetic processing (environmental, not code-mediated regulation). This then informs us of our evolutionary objective: the addition of complex network regulation onto maternal bonding systems to promote a large brain that is competent for more extended social bonding and increased intelligence. A fundamental role of societies of RNA stem-loop structures in this brain evolution will be presented. And these regulatory RNAs are proposed to have derived mostly from the complex colonization by mixed virus and retroposon agents [e.g., ERVs (endogenous retroviruses)] and their defective elements (noninfectious ERV segments) (Witzany et al., 2013). These agents provide the raw material and initiating pressure to form new networks of regulatory programming for altering the identity of various stem cells.

But in order to understand the basis for this reasoning, we must first revisit some very dearly held views about the nature of evolution: the unvarying importance of individual fittest type selection. Ironically, it was from a premise of individual fittest type selection that Manfred Eigen first developed the ideas of quasispecies to explain the behaviors of diverse RNA virus populations (Eigen, 1971). But as presented previously, the experimental evaluation of quasispecies by virologist led us to conclude that they operate as cooperative consortia (collectives) that can recognize members and preclude nonmembers; see Domingo et al. (2012). Since RNA diversity per se is crucially important and individual fittest type selection becomes much less important, we employ the term quasispecies collective (QS-C) to describe the role of diverse RNA in evolution (Villarreal and Witzany, 2013). Previously, I presented a basal role of stem-loop RNA collectives in the emergence of the RNA world. A concept that emerges from the features of these collectives is that they are transmissive RNA societies (infectious QS) that play a central role for the origin of group identity (social membership) during evolution. Group identity resembles multilevel selection (Shelton and Michod, 2010). The concepts of identity, immunity, and the role of addiction modules (a linked set of toxic/antitoxic functions that must be kept together to prevent harm) all factor into understanding group identity. I will now trace how viruses (and other genetic parasites) have promoted the formation of new human identity networks. That pathway is initiated by virus (via RNA network) but runs through reproductive tissue, embryonic stem cells, and the placenta and onto the recent changes of our brain. Maternal behavior mechanisms are examined as the basis of human social adaptation and change. And finally, the genomes of *Homo sapiens* and Neanderthal will be considered from a virus first perspective. Here too, clear virus footprints are seen. This leaves us positioned to now seriously consider how viruses contributed to making us human.

21.5 PRIMER 2: THE CONCEPT OF QS-C AND SOCIAL COHERENCE

It can be very difficult to understand how a population of genetic entities can operate as a coherent whole. Many questions are raised.

For example, how can RNAs recognize each other within QS-C, and how can they prevent other viral RNAs from joining their consortium? What prevents two different viral QS-C from joining together? Would we not expect that the defective RNA agents would ultimately not compete with other RNAs in attacking/reproducing in the host? Why might they also be retained? There is too much relevant experimental evidence to summarize succinctly regarding such questions. However, the following metaphor might be helpful. Let us think of it in terms of language dialects that change but must retain competence for communication (via RNA and RNA–protein interactions). The defectives (and others) become efficient and better users of the regulatory dialect used by the current nondefectives. Eventually this makes them too efficient, which will crash the whole population. In addition, and often at this time, some new nondefectives generate dialect variations that are less susceptible to the past defectives. This leads to emergence of new populations of nondefectives that tolerate old dialects (although all the old dialects still stay around as minorities). Thus, any member of the QS must have evolved to be coherent with past and current regulatory dialects. They must be competent for the particular QS they are members of. This inherently promotes group identity and is the reason why different QSs will clash with each other.

21.6 SOME PROBLEMS WITH COMMUNICATION

As it is the aim of this chapter to address the emergence of complex group social behavior, it will be essential to define what networks (groups) are and how they form from the perspective of a consortium of RNA agents. However, it will be very difficult to think about and to communicate these issues. This is not because they are so inherently complex but more because they are fundamentally interactive (social) phenomena that resist linear explanations. The various relevant strategies involved in networks and group identity are linked. For example, the issue of group identity and individual immunity would seem to be separate topics. We can, for example, devise assessments of these features by separate measurements. Yet, as I will present in the following, immunity mechanisms are very the same as group identity mechanisms, and both operate using equivalent strategies (such as addiction modules), which can destroy nonidentity and nongroup membership. Apoptosis (programmed cell death) is a prime example of this: it is used for both pathogen immunity and self-identity (cell-type programming). I will describe how these very same strategies (addiction modules) can be applied to extended social bonds of humans in order to promote group identity. We cannot separate these issues into distinct linear ideas.

A social system will also have individual agents (such as RNA stem-loops) that will fundamentally have multiple (often opposing) activities and uses (see Figure 20.2). Thus, a system or network must always have internally opposing features. But this presents us with a problem. Humans are clearly capable of complex parallel processing of information, especially social information. But our language is necessarily a linear system that promotes linear communication (such as this text) and thinking. We tend to think about the linear functions of agents (such as stem-loop RNAs). Yet, as we will see, such RNAs (like Alu's) never resolve to one particular function. They are always multifunctional. But our tendency will be to think that we simply have not figured out how the original (master type) and most basal Alu RNAs function from which all the others must have subsequently evolved. We will likely go back to the bench to try to figure out this first putative master function. But try as we might, we always find different functions for different contexts. A network simply does not operate via linear processes. RNA acts like a word and the network like a language; a word will always have multiple meaning depending on context and use history. We can always put *not* in front of a word (or make it antisense for RNA) and completely transform its meaning and use. If we are alert to this problem in our thinking, we can avoid many logical pitfalls that will otherwise occur. We must start to think and communicate socially, not serially.

There is another dilemma posed by individual-type selection. This is the idea that *errors* are the main source of the variation needed for natural selection to promote the improved version of the genetic instruction in surviving individuals. Thus, we see indels (insertions and deletions of various transposons) as a large supply of errors for such recent selection in the human genome. And it is reasoned that from this seemingly catastrophically high level of transposon introgression (genome colonization), natural selection was able to promote the serial emergence of the human brain, the most complex of any entity we know. I suggest this is the error of errors! If we instead consider this situation from a virolution QS-C perspective, the formation of a coherent diverse set of viral agents becomes a key for success. What must then happen is not a rain of sequential errors by introgression but a coordinated *hack* of the complex instruction and identity network that controls the host nervous system. This is a high-level programming indeed, requiring intimate competence of the most basic operating system (the stem cells), in order to reprogram the development of the fetal and the adult brain. This will not be accomplished by the actions of a lone individual agent that just happened to find the Achilles heel of the host network. It will require an army of coordinated diffuse agents, and not necessarily of the same genetic lineage. Its competence will stem from the power of virus QS-C to adapt to many (even conflicting) situations in parallel by using RNA-based diversity and identity and operate via a diffuse (but still coherent) instruction set.

21.7 COMPLEX SOCIAL PHENOTYPES OF HUMANS: EXTENDED BONDING VIA REPURPOSED REGULATORY RNA NETWORK

The central importance of human social capacity has led some to propose the reintroduction of group selection in order to understand human evolution (Wilson and Sober, 1994). Let us examine this controversial suggestion from the perspective of QS-C virolution, which promotes group selection. The large human brain is metabolically costly and places major demands on the placenta to develop. But newborn humans are among the most helpless and incapacitated of all mammals as aside from defecation, they must learn essentially all the activities needed to survive, including eating. But following birth, social learning initiates and will develop into many social capacities (e.g., language) and complex social behaviors including extended bonds.

Strong arguments have been made that the enhanced human social brain is thus the crucial change to be understood; see Parr (2003). Biologically, we might therefore focus on the genomic changes that underlie the emergence of this large social brain, along with its capacity for language and other social abilities; see Konopka and Geschwind (2010). Our objective is to evaluate recent human-specific adaptations (such as Neanderthals) for evidence of virus (and QS-C) involvement. Along the way, we already have some strong clues to guide us. For example, we know that many social behaviors (monogamy, parental care, cooperative breeding) are strongly influenced or mediated by neuropeptides (vasopressin and oxytocin); see Konopka and Geschwind (2010) and Goodson (2013). So, this will provide some mechanistic focus for our evaluation. The distinctions between Neanderthal and *Homo sapiens* cognition will be of particular interest (Somel et al., 2013). As noted earlier (and further explored in the following), stem-loop RNAs mediate much of the identity of infectious RNAs. And they are mostly derived and transcribed from *junk* DNA (retroposon-derived DNA of no apparent function). Thus, the acquisition of noncoding and stem-loop RNAs in brain evolution (and immune/apoptosis response) will provide an initial focus; see Barry and Mattick (2012). Similarly, microRNAs (very small regulatory RNA) are involved in learning and social behavior disorders (Muiños-Gimeno et al., 2011) as well as RNA editing (changes to RNA code after copying) and can operate via stem-loop Alu RNA that will also be considered (Paz-Yaacov et al., 2010). Were the genetic regulatory networks underlying maternal care in placental mammals repurposed by RNA genetic colonization, and does this apply to more expanded social (group) bonds (Pedersen, 2004)? Considering the maternal bond, how the maternal brain is influenced by pregnancy (the placenta) (Kinsley and Amory-Meyer, 2011) and breast-feeding (Kim et al., 2011) is of interest. Rodents are particularly useful for this issue, and the role of oxytocin in inducing maternal behavior will be examined (Pedersen et al., 1982). The role of oxytocin in extended (group) human social bonds is also considered (Feldman, 2012). However, for extended human bonds, the fetal and male brain must become susceptible to neuropeptides that regulate bonding (paternal and mate), and it is likely that in both rodent and human males, vasopressin is more involved (Wynne-Edwards, 2001; Walum et al., 2008). The repurposing of such regulatory networks was mediated by parasitic DNA agents and stem-loop RNAs acting on their receptors (Hammock and Young, 2005; Donaldson et al., 2008). A major site for modifying brain development was the frontal lobe (Hoffmann, 2013) and also modifications to some dopamine (i.e., social) functions (Camperio Ciani et al., 2013). These modifications are linked to motivational and reward systems (Aron et al., 2005). Indeed, the great apes in particular have mostly extricated themselves from the use of olfaction for social learning as done by the rodents (Sanchez-Andrade and Kendrick, 2009). The role of the large olfactory bulb (brain region for odor processing) and vomeronasal organ (VNO) (a patch of sensory neurons in the nasal–oral cavity for pheromone detection) for social learning is mostly lost. In humans, social learning can be voice (language) and sight (facial) based with corresponding brain network modifications. These changes were also mediated by genetic colonization and network repurposing (see Figure 21.1 and Table 21.1).

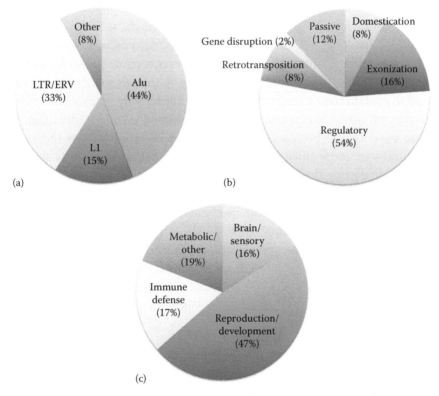

FIGURE 21.1 Human genome composition including various types of genetic parasites. (a) Transposable elements (TEs) implicated in the generation of primate-specific traits. (b) Types of events mediated by TEs underlying primate-specific traits. Passive events entail TE-mediated duplications, inversions, or deletions. (c) Aspects of primate phenotype affected by TEs. (Reproduced from Oliver, K.R. and Greene, W.K., *Mob. DNA*, 2, 8, 2011. With permission.)

TABLE 21.1   TE Activity in Human and Chimpanzee Lineages

Element	Chimpanzee[a]	Human[a]
Alu	2340 (0.7 Mb)	7082 (2.1 Mb)
LINE-1	1979 (>5 Mb)	1814 (5.0 Mb)
SVA	757 (>1 Mb)	970 (1.3 Mb)
ERV class 1	234 (>1 Mb)[b]	5 (8 kb)[c]
ERV class 2	45 (55 kb)[d]	77 (130 kb)[d]
(Micro)satellite	7054 (4.1 Mb)	11,101 (5.1 Mb)

Source: Modified from Chimpanzee Sequencing and Analysis Consortium, *Nature*, 437, 69, 2005. With permission.

[a] Number of lineage-specific insertions (with total size of inserted sequences indicated in brackets) in the aligned parts of the genomes.
[b] PtERV1 and PtERV2.
[c] HERV9.
[d] Mostly HERV-K.

That viruses might be important for human evolution is not a new idea; see Van Blerkom (2003). But all such prior proposals present the traditional *arms race* or *tit-for-tat* strategies (involving single-host adaptations followed by counter pathogen adaptations) that involve viral *plague sweeps* (killing of host by infections) in susceptible populations yielding occasional surviving individuals. As presented here, however, the very networks controlling cellular identities that led to complex social learning capacity were mediated by infectious RNA QS-C colonization events. These colonization events create networks, directly affect group survival, and alter host-specific viral ecology. Although social scientists have long felt the need to explain the extensive human social capacity by mechanisms that relate to group survival, no direct mechanism for group selection could be identified. Since the 1960s, group selection has been solidly rejected by neo-Darwinians as individual fittest type and natural selection seemed able to explain social behaviors (such as altruism) when kin selection and/or game theory is applied. But such theories emerged prior to the discovery of the massive and diverse virome (the virus composition of a habitat). Now, QS-C-mediated virolution can provide a clear mechanistic foundation for group identity and group survival (Villarreal, 2006, 2007, 2008, 2009a, 2012a). Thus, QS-C-mediated genetic phenomena inherently support group selection. Consider the trophectoderm (outer cell layer) that produces the placenta and mediates implantation in embryonic mammals. Was virolution involved in this novelty? Similarly, the placenta, the maternal brain, and our large human brain can also be examined for evidence of virolution.

21.8 ENCODE PROJECT: VIRAL *JUNK* AS *ESSENTIAL* FOR COMPLEX RNA REGULATION

How do we reregulate a complex network for even greater complexity, such as in our brain? Can the idea of QS-C/virolution apply to this? For many years, molecular biologist assumed that the complex RNA expression patterns observed by various techniques in the mammalian brain were due to the expression of many genes, as expected for such a complex organ; see Chaudhari and Hahn (1983). However, comparative genomics has made clear that gene transcription differs little between human and great apes

(Khaitovich et al., 2004). Indeed total gene numbers differ remarkably little between the simplest animals (*Caenorhabditis elegans*) and humans. But by far, the biggest differences between human and chimpanzee genomes were due to indels (insertion and deletions) (Watanabe et al., 2004; Mills et al., 2006; Wetterbom et al., 2006). The great majority of these indels are the result of retrotransposon activity of various types [ERVs, long terminal repeats (LTRs), long interspersed nuclear elements (LINES), and Alu's being most numerous]. This is shown in Figure 21.1 and Table 21.1. Of these changes, the Alu elements and their transcripts are particularly active and affect RNA editing and intron splicing (internal RNA removal from mRNA) in the human brain (Sakate et al., 2007).

In addition, they are frequently involved in epigenetic control and can emerge or expand rapidly in genomes (Zeh et al., 2009). Such a large-scale retroposon colonization would seem to pose a highly genotoxic situation for the human genome, an idea which seems supported by genomic analysis (Keightley et al., 2005, 2006). And yet this noncoding DNA is species specific (Toder et al., 2001), evolving quickly in humans (Bird et al., 2007), but also appears to be under very strong selective constraints (Bejerano et al., 2004; Bush and Lahn, 2005). This seems problematic in several ways: for one, such an inherently destructive event should only seldom result in novel or complex phenotype. Indeed, that this component of the genome shows a high rate of change was previously used to argue for the idea that it must be junk DNA. However, more recently, the ENCODE project has corrected out thinking. As mentioned in the last chapter, this project sought to characterize all the RNA transcribed from the human genome, including RNA that is not cytoplasmic polyadenylated mRNA (protein coding) but is noncoding RNA (Mattick, 2005). It has established that most of the *junk* DNA is transcribed and that 95% of the transcripts are from repeated sequences that were retrotransposed (Mattick, 2010; Mattick et al., 2010), including previously poorly studied class of long noncoding RNA (lncRNA); see Khalil et al. (2009). Furthermore, these noncoding transcripts appear particularly relevant to human brain and cognitive development (Mehler and Mattick, 2007; Mattick and Mehler, 2008; Barry and Mattick, 2012). Additionally, long-term memory also seems to use noncoding RNA (Mercer et al., 2008). These observations have led J. Mattick to propose that genetic programming in higher organisms has been misunderstood for 50 years (Mattick, 2011). We now realize that regulatory RNA derived from retrotransposons is key to eukaryotic complexity, compelling us to abandon the concept of selfish junk DNA. But in this realization, we also come to realize this regulatory RNA is operating mostly as stem-loop RNAs that have complex, multilevel, and even opposing functions. It is clearly operating and evolving as a network.

21.9 APPLYING QS-C/VIROLUTION TO NONCODING RNA

Human-specific ERVs (HERVs, especially HERV-K and HML2) do distinguish the various hominid species, such as humans and chimpanzee, from one another (Barbulescu et al., 2001; Polavarapu et al., 2006; Romano et al., 2006, 2007; Buzdin, 2010). Plus, there are a much larger number of solo LTRs derived from the HERVs, all of which appear to have evolved from HERV-K versions of virus (Buzdin et al., 2002). Some of these LTRs are clearly affecting RNA regulation (Buzdin et al., 2006; Gogvadze et al., 2009). Given the number of solo LTRs in the human genome (about 310,000) and the fact that they arise following excision of integrated full-length ERVs, it can be estimated that during its evolution, the human genome once hosted ERV DNA that would have equaled its full current DNA content (3×10^9 bp). Thus, retroviruses have indeed been very active editors of human DNA. In this light, it should also be added that of all chromosomes, Y chromosomes differ most dramatically between human and chimpanzee (visible via chromosome staining—heterochromatic C—bands). And the Y chromosome has been called a *graveyard for ERVs* since they harbor the largest numbers of these elements (Kjellman et al., 1995).

21.10 VIRUS, NONCODING RNA, AND REPROGRAMMING STEM CELLS

The reprogramming of neurons to promote the emergence of a large social brain must have included alteration to their predecessor stem cells. Currently, numerous studies have suggested that retroposons and transposable elements (TEs) have mediated many of the network changes that have been needed in stem cells; see von Sternberg and Shapiro (2005), Kunarso et al. (2010), and Alzohairy et al. (2013). These views propose the host-mediated *domestication* of these selfish elements (Presutti et al., 2006; Mehler and Mattick, 2007; Qureshi and Mehler, 2009; Mattick, 2010; Smalheiser et al., 2011). We know that such *rewiring* network events can be quick during evolution, quicker than changes to protein networks (Shou et al., 2011). What is missing from such proposals, however, is a mechanism that can rapidly initiate network reregulation. Often the vague idea of *genomic stress* has been proposed to contribute, but the reason why this initiates network reregulation is not obvious. Also missing from such ideas is any acknowledgment that the responsible retroagents all must reproduce via RNA and hence generate large diversity (via QS-C). Thus, current idea for retroposon involvement in networks is still basically one of providing *errors* that occasionally become useful. I do not see how such a serial error-based process creates networks. However, if the precipitating event is the genome colonization by what is often mixture of exogenous retroviruses and if such an event engages resident retroposons to respond in populations (networks), we begin to see how network reregulation can be initiated. Consistent with this idea, the HERVs have achieved high germ-line copy numbers mostly via reinfection with replication-competent versions of exogenous (transmissible) virus, which can also be complemented in trans by other viral agents (Belshaw et al., 2004, 2005). It is thus proposed that precisely such reinfection events will promote the modification of regulatory RNAs (such as stem cell–specific expression of lncRNA) (Khalil et al., 2009; Loewer et al., 2010; Pauli et al., 2011) and why this is mostly derived from retrotransposons (Kelley and Rinn, 2012).

Retroviral regulatory RNAs mostly operate as stem-loop RNAs that control ERV replication, transcription, and assembly typically by epigenetic mechanisms (Maksakova et al., 2008; Rowe and Trono, 2011). The retroviral regulatory regions (such as LTRs) will become distributed by integration into various compatible (mostly regulatory) genome sites. These newly acquired regulatory regions also have limited sets of transcription factor (proteins that regulate transcription) binding sites, and in this way, they can reprogram stem cell expression networks using only a limited set of stem cell–specific transcription factors (Kunarso et al., 2010; Loewer et al., 2010; Lynch et al., 2011; Gifford et al., 2013).

Indeed, it was precisely such a process that accounts for the global role of the HERV-K LTR in altering p53 binding changes and creating new regulatory networks in human evolution (Wang et al., 2007). Let us consider this same scenario in the context of mouse and human stem cells.

21.11 FROM MOUSE ZYGOTE TO EMBRYO: NETWORKS AND RETROPOSONS

Nowhere in the development of mammals is the implementation of complex gene control as crucial as it is early in development. Initially, the mammalian egg cell contains a high level of maternal RNA. Following fertilization, the paternal nucleus becomes transcriptionally active in the zygote, and maternal transcripts are decreased. Cell division then begins and soon thereafter, the cell programming begins that will soon produce the trophectoderm, the outer layer of cells that will become the placenta and allow implantation into the uterine wall. Indeed, all of these phases of early development are tightly linked to various aspects of expression and control by mostly ERV-derived RNA. We have increasingly come to realize that ERV and retroposon are not only tightly associated with these events but also involved in their regulation. As mentioned, the ENCODE project has also made clear that retroviruses and retrotransposons are actively transcribed into various RNAs that exert regulatory control over most events of gene expression (Harrow et al., 2012). Thus, retroposon-derived genomic DNA appears to more closely scale with the complexity of the eukaryotic organism that contains it, compared to gene content (Mattick, 2005; Mercer et al., 2009; Mattick et al., 2010). Such retroposon regulatory RNA is also much more involved in the origin and function of the human nervous system, and human cognition then was previously realized (see Sections 21.14, 21.15, 21.17, 21.20, and 21.21). Furthermore, such noncoding regulatory RNAs are generally multipurposed and are simultaneously or conditionally involved in several functions (Mattick and Gagen, 2001; Katayama et al., 2005; Mattick, 2011).

21.12 MOUSE OOCYTE AND EMBRYONIC STEM CELLS

It has been known for some years that early mouse embryos are highly active in transcribing from ERV LTRs (Peaston et al., 2004, 2007; Maksakova and Mager, 2005; Feschotte, 2008). Indeed, a considerable portion of a mouse oocyte RNA pool is derived from class III LTRs (Evsikov et al., 2004; Peaston et al., 2004, 2007). And this RNA appears to regulate embryo development; see "Systems biology of the 2-cell mouse embryo (Evsikov et al., 2004) and Rowe and Trono (2011). A systems analysis approach to evaluating total transcription patterns during oocyte, zygote, and early mouse embryo development has found that high levels of retrotransposon transcription occurs and changes during these crucial cell divisions. In oocytes, LTR class III ERVs and retrotransposons make an especially high contribution to maternal pools of mRNA (Evsikov et al., 2004; Peaston et al., 2004, 2007). And following fertilization, the paternal nucleus also transcribes an unexpectedly large amount of similar RNAs that can exert regulatory control over embryo development (Ohnishi et al., 2010, 2012). Not only are the ERVs providing small regulatory RNAs in development, but the expression of the encoded ERV-L gag protein (retroviral capsid) is also crucial for early mouse development (Kättström et al., 1989; Benit et al., 1997;

Kigami et al., 2003; Svoboda et al., 2004; Macfarlan et al., 2012). Human oocytes transcribe similar ERV sequences (Nilsson et al., 1992). Given that many of these ERVs are relatively young elements in evolution of host genomes, the question is raised of why they would be providing such a basic regulatory role for early development. In the case of the older MuERV-L gag (endogenous murine leukemia virus) expression, it is also clear that this viral capsid gene can provide a virus restriction function (able to inhibit similar virus, like the FV-1 locus does), thus affecting the host–virus relationship. Another crucial issue regarding regulatory RNAs relates to the epigenetic control of retrotransposons in early development. High-level ERV expression in the early embryo is followed by a generalized (but not total) suppression of ERV transcription by epigenetic repression (such as histone methylation, Huda et al., 2010). Yet, many of these ERVs also provide essential regulatory (and promoter) elements used during embryogenesis. This situation appears inconsistent with the generally held view that epigenetic ERV suppression was initially acquired as an immune system to protect the germ line against colonization by retrotransposons. Since the early embryo is most affected by and open to genetic modification by endogenizing retroviruses, its exclusion would have prevented these developments. And as the embryo develops further, generating the trophectoderm and inner cell mass, we also see ensuing high-level activation of specific but mixed and species-restricted ERVs, such as IAPs, HERV-K, and ERV-L, that were acquired during placental evolution (see Figure 21.2).

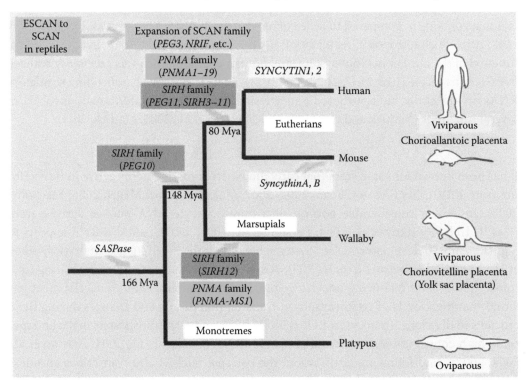

FIGURE 21.2 Acquisitions of ERVs during placental evolution. Patterns of ERV colonization linked to evolution of placental species. (Reproduced from Kaneko-Ishino, T. and Ishino, F., *Front. Microbiol.*, 3, 262, 2012. With permission.)

Why then is there so much activity by these parasitic agents at these crucial stages of development? According to the precepts of QS-C-mediated virolution, I would propose that these ERVs had colonized as QS consortia to superimpose new identity (and immunity) networks onto their host, which permanently alters the virus–host relationship and allows the maintenance of the new viral information. This new network is then available to modify host cellular identity and behavior. Accordingly, a particular focus of HERV-K-related ERVs that are found in our genomes is in order as these viruses are also specifically and precisely expressed in human embryonic stem cells (Santoni et al., 2012).

21.13 LINES AND ERVS

Retrotransposon activity in an early embryo is not limited only to ERVs. Following ERV activation in the oocyte and zygote, there is also a rapid induction of LINE-1 activity in the early embryo (van den Hurk et al., 2007; Kano et al., 2009; Vitullo et al., 2012). And this activity is also associated with the programming of pluripotent stem cells. Thus, we can see that reverse transcriptase activity appears to be essential for early mouse embryo development (Beraldi et al., 2006; Sciamanna et al., 2011). Given the competing and sometimes excluding patterns of ERV and LINE colonization in the genome, it also seems likely that there are various interactions between these two retroposon agents that are operating. Thus, it is interesting that most L1 elements are retrotransposition defective; nonetheless, there are approximately 100 full-length L1s potentially capable of retrotransposition in the diploid genome. The issue of possible interaction between ERVs and LINES will also be interesting in the context of the origin and evolution of Alu elements, which clearly required LINE-mediated 5′ (end) integration but were derived from 7S RNA (involved in transport and found in all domains of life). 7S RNA was itself initially discovered as specifically packaged into all retroviruses. As Alu introgression (genome colonization) into the human genome was a major event in our recent evolution, possible retroposon interactions (i.e., social RNA) are of crucial interest. Thus, it is noteworthy that the 5′ untranslated region (UTR) of full-length human L1 produces bidirectional transcripts that can also be processed to small interfering RNAs (siRNAs) that suppress retrotransposition by an RNA interference (RNAi) mechanism. Therefore, an RNAi triggered by antisense transcripts (complement of the mRNA sense) may modulate human L1 retrotransposition efficiently and economically (Yang and Kazazian, 2006; Sekita et al., 2008). Thus, retrotransposons clearly appear to be multifunctional and interact among themselves. A *society* of retroposon RNAs are interacting with and involved in host development as proposed (Witzany et al., 2013).

21.14 NONCODING RNA, NEURON IDENTITY, AND THE HUMAN BRAIN

The placenta can clearly affect the development of the maternal brain and learning the maternal bond. However, the maternal (social) bonding–learning processes had been previously well used in all other tetrapods. As noted previously, in all these vertebrates, olfaction-based social learning was of central importance throughout their evolution, but this was largely lost in human (and the other great ape) evolution, along with acquisition of trichromic (color) vision. Both the loss of this prior (olfactory) identity system and the gain of an extended new social identity system were mediated by the action of virus, retrotransposons, and regulatory RNA.

21.15 ERV-MEDIATED PLACENTA EMERGENCE

The placenta is considered a very significant and complex biological innovation, but explaining its emergence poses a serious test for any theory of evolution. Can QS-C-mediated virolution help us understand the origin of the placenta? The placenta is an astonishing organ. It feeds the embryo by invading and interfacing with the mother's blood supply, prevents rejection by the mother's immune system, and alters the mother's physiology and behavior from self-centered to offspring centered (Loke, 2013). This is shown in Figure 21.3.

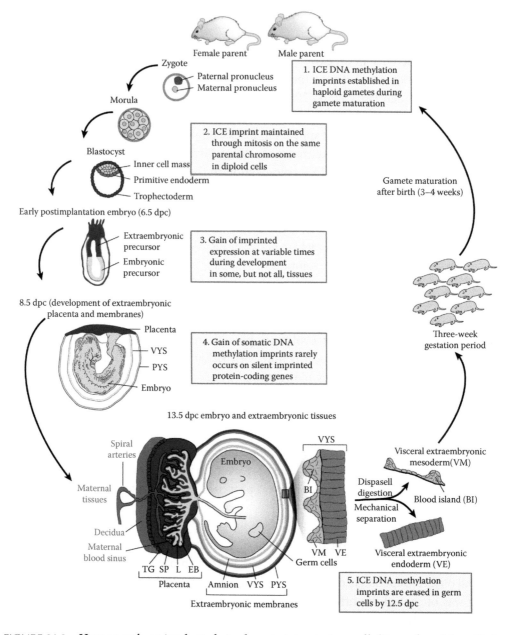

FIGURE 21.3   Human embryo implanted via placenta into uterine wall. (Reproduced from Barlow, D.P., *Annu. Rev. Genet.*, 45, 379, 2011. With permission.)

It is also very interesting that unlike other organs, the placenta is an organ that varies substantially between species, including clear differences in humans relative to chimpanzee placenta (Lynch et al., 2011). The control and coordination of such complex placental functions via gene expression rely on transcriptional and posttranscriptional regulatory networks. It is estimated that about 1500 genes had to acquire distinct regulatory control with the emergence of the placenta (Lynch et al., 2011, 2012; Chuong et al., 2013). The placental regulatory network is indeed highly affected by ERV (LTR) activity (Chuong et al., 2013). And placental phylogeny relates closely to ERV and retrotransposon composition (Murphy et al., 2007) (see Figure 21.2). These varied ERV changes clearly resemble a genome colonization event that involved many related viral regulatory elements and resulted in a new coherent network. Thus, this is also strong evidence that mixed ERV colonization was indeed the crucial initiating event in the origin of the trophectoderm and placenta (Feschotte, 2008; Feschotte and Gilbert, 2012; Lynch et al., 2012). And the resulting complex transcriptional network was adapted also to address complex host–fetus interface functions. Thus, at the origin of all the placental orders, there indeed does appear to have been major species-specific changes to the genomic ERV content as well as changes to other retroposons (Ohshima et al., 1993; Kim et al., 1999, 2000; Kim and Takenaka, 2001; Churakov et al., 2009; Nishihara et al., 2009; Suh, 2012).

21.16 CLASH OF ZYGOTE IDENTITY AND IMMUNITY NETWORKS

There is an immunological dilemma associated with the origin of viviparous (live, not via an egg) birth and the placenta. Since the implanted embryo is the product of sex, it will contain paternal genes not found in the maternal genome. Given the close blood contact between mother and embryo (including the exchange of antibodies), the mother's adaptive immune system should recognize and reject the embryo. But this usually does not happen. Previously, I have argued that the biological complexity posed by adaptive immunity and the problem of embryo implantation in placental species presented very complex problem that required a network-level solution. Virus population with their inherently invasive, endogenizing, and immune-suppressive functions provides a ready-made solution to solve such complex problems (Villarreal and Villareal, 1997). However, even this argument did not suggest a more fundamental role for virus in the emergence of the placenta (and host complexity) itself (or networks per se), as I am now suggesting. A role for QS-C-based virolution in network origination was previously absent. Indeed, virus-mediated evolution can even be applied to explain the origin of the adaptive immune system itself (Villarreal, 2007, 2009b, 2011, 2012b).

21.17 MIXED RETROVIRUSES AS ESSENTIAL REGULATORS

ERV mixtures were involved in human evolution. Historically, whenever it is observed that virus-derived information has become part of the host, we say that the host has *exapted* (taken) this viral information for host survival. This is always in the context of single viral *gene* acquisition, which would likely also be mediated by a surviving individual host. But network emergence is complex and requires a population-based mechanism and resists such simple serial explanations. Given the large role of endogenized retrovirus in placental networks, one might suspect that more complex colonization events must be applied to

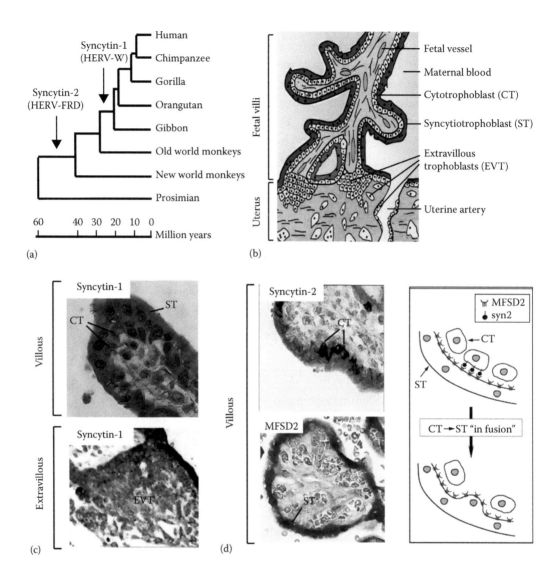

FIGURE 21.4 Expression of ERV envelope genes in a human placenta. The ERV-derived human genes are syncytin-1 and syncytin-2. Shown is the human placental villus with ERV expression in all types of trophoblast cells: syncytiotrophoblasts (ST) and cytotrophoblasts (CT). (a) Phylogenetic tree of primates. Arrows indicate the time of insertion into the genome of primates of the two retroviruses that have bequeathed the syncytin-1 and syncytin-2 genes. (b) Schematic representation of a human placental villus. (c) Immunohistochemical staining of human placental villous sections for syncytin-1 showing that it is expressed in all types of trophoblast cells. (d) Left panel: Immunohistochemical staining and in situ hybridization of human placental villous sections for syncytin-2 and MFSD2 expression, respectively. MFSD2 expression is restricted to the syncytiotrophoblast layer (ST), while syncytin-2 expression is restricted to underlying mononucleated cytotrophoblasts (CT). Right panel: "In fusion" model for syncytiotrophoblast formation, where interaction between syncytin-2 and MFSD2 drives polarized fusion of the cytotrophoblast into the syncytiotrophoblast. (Reproduced from Dupressoir, A. et al., *Placenta*, 33, 663, 2012. With permission.)

explain the role of virus in the emergence of the placenta. ERV-derived genes used by the placenta are uniformly proposed by researchers to have been exapted by the host. But these genes can clearly show complex mixed-virus control. This is best illustrated by the syncytin genes. *Syncytin* is an *env*-related DNA sequence found in the human and other mammalian genomes that was clearly derived from a retrovirus env gene as first discovered in humans (Blond et al., 2000; Mi et al., 2000). There are two syncytin genes that are derived from the different human-specific ERVs, HERV-W, and HERV-FRD, respectively. These became integrated into a primate lineage 25 and >40 million years ago (MYA). They provide essential fusogenic (membrane fusing) and immunoregulatory gene functions for the placenta and are required for placental development and embryo implantation. Related genes appear to exist in most orders of mammals, also derived from distinct ERVs, some of which appear to have recently integrated (Benit et al., 2003; Mangeney et al., 2007; Heidmann et al., 2009; Dupressoir et al., 2011; Vernochet et al., 2011). These syncytin genes themselves are controlled by a very complex mix of ERV regulatory domains derived from other distinct retroviruses (Pérot et al., 2012) (see Figure 21.4). Thus, proposing that such complex ERV regulation and expression results from an *exaptation* event seems clearly untenable.

In addition, other ERV env gene domains (such as a TM immunosuppressive domain) from the more recently acquired human-specific ERV (HERV K) are precisely expressed in the same crucial placental tissues (Prudhomme et al., 2005; Kämmerer et al., 2011; Haig, 2012). We do not yet understand the functional significance of this HERV-K env expression in the human placenta. However, we do know that HERV-K retains RNA stem-loop folding regions that have various activities on transcription (such as transport) of HERV-K transcripts (Yang et al., 2000).

21.18 PLACENTAL AND MATERNAL BEHAVIOR

The success of placental mammals was not only due to the major changes in reproductive structures and tissues, but it also required significant alterations to behavior, especially maternal behavior. Thus, it is of great interest to examine the placental role in the control of the maternal brain and the resulting maternal behavior. These behavioral adaptations principally required brain changes that expanded and social learning and bonding capacity. But why should any virus (let alone mixtures) be involved in this event? The biology of mammalian reproduction involves numerous cell types, including the oocyte, sperm, zygote, and the placenta. These tissues are also a rich habitat for viruses, many of which establish persistent lifelong infections and are maternally or sex transmitted. This species-specific mixture of persisting viromes can have a very profound and differential effect on the host reproductive success. Thus, sexual behavior becomes a big factor in virus–host ecology and host fitness. Can the mechanisms used for pair bonding and paternal bonding in rodents provide insight regarding human evolution?

21.19 BRUCE EFFECT AND EXTENDED SOCIAL BONDING

The Bruce effect is the prevention of implantation of a fertilized mouse oocyte following the olfactory exposure to a male that was not the father (Brennan, 2001, 2009; Keller et al., 2009). The olfactory signals involved are present in the male's urine. Exogenous progesterone can inhibit this implantation interference (Rajendren and Dominic, 2009). In addition, the more males exposed to the impregnated female, the greater the Bruce effect. The Bruce effect can also be found in other species, such as the domestic horse. Also, with the horse, a promiscuous mare can increase the pregnancy block (Bartoš et al., 2011). Clearly sexual and family behavior is important for the Bruce effect. It is not found in all rodent species, such as Syrian hamster. Although it can also be found in some wild primates, such as geladas (Roberts et al., 2012), it is not found in the African primates.

In rodents, the VNO is directly involved in the Bruce effect and can be prevented by cutting the VNO nerve (Halpern, 1987). The resolution of VNO urinary odor-type differentiation can be of very high and elicit aggression in response to different Y chromosomes of the same species (Monahan et al., 1993). VNOs are found in monkeys, but not great apes (Johns, 1986). And in rats, the Bruce effect can be perturbed by administering oxytocin (Narver, 2012). It is interesting to note that VNO-mediated mate learning also requires memory formation (Ichikawa, 2003). Mate and offspring identities are thus learned via VNO olfaction. Although a VNO is present in human fetus, it becomes vestigial (degraded) in adult humans (Brennan, 2001). Also, most human VNO receptors (V1R/V2R) are pseudogenes (nonfunctional gene segments) (Keverne, 1999; Kouros-Mehr et al., 2001; Zhang and Webb, 2003) and have thus been inactivated by the action or retroposons. One receptor gene does appear to have remained intact in humans and is also present in marmosets, a basal primate (Giorgi et al., 2000). These changes in olfaction appear linked to acquisition of trichromic (color) vision via the X chromosome in the African primates (Smallwood et al., 2002; Zhang and Webb, 2003; Gilad et al., 2004). Given the much more extended social groups in the African primates and the decreased dependence on olfaction for their social communication, it is interesting to note that the human odor receptor genes were made pseudogenes on large scale during recent evolution (Newman and Trask, 2003). Thus, overall we see a large-scale genetic invasion or colonization by retroposons of human DNA in which the VNO organ and their receptors were incapacitated (e.g., made pseudogenes). Yet the great apes (especially humans) underwent a much enhanced capacity for forming cohesive social groups. Also, primates acquired a much enhanced capacity for facial and emotional recognition associated with mirror neurons (neurons activated by seeing the actions of others), thought to be linked to language development (Parr et al., 2005; Cooper, 2006; Uddin et al., 2007). How then was this extended primate social capacity accomplished?

Clearly, human evolution has led to the emergence of a large social, emotionally perceptive, and empathic brain. This required many coordinated changes that included a much enhanced neocortex (outer surface neurons) and increased brain volume as well as enhanced visual processing along with a diminished olfactory lobe and odor receptors. The maternal brain of mammals undergoes modification and growth during pregnancy that also results in enhanced maternal bonding as well as an enhanced ability to understand the emotional state of an infant. This maternal process has been proposed to have likely provided the base mechanism used to expand human social capacity (Pedersen, 2004; Kinsley and Amory-Meyer, 2011; Swain, 2011). The dominant changes associated with the altered regulation of the human brain appear to have been most mediated via the action of various

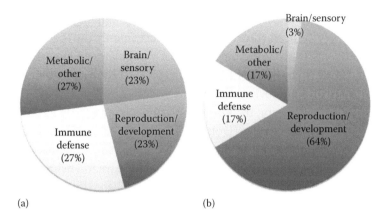

FIGURE 21.5 Distinct retroposon RNA expression activity in human reproductive tissue and brain tissue. Comparison of aspects of primate phenotype affected by (a) Alu elements and (b) LTR/ERVs. (Reproduced from Oliver, K.R. and Greene, W.K., *Mob. DNA*, 2, 8, 2011. With permission.)

small snapback RNAs (Mattick and Makunin, 2006; Mehler and Mattick, 2006; Oliver and Greene, 2011; Barry and Mattick, 2012) (see Figure 21.5).

Of these, the Alu elements appear to be the most numerous. Since Alu's are clearly related to 7S RNA, which were discovered in and are specifically packaged by retroviruses (Giles et al., 2004; Onafuwa-Nuga et al., 2005; Kriegs et al., 2007; Marz et al., 2009), a viral role in their human-specific expansion also seems likely.

This idea also appears consistent with recent changes in human sex chromosomes (X and Y), which overrepresent both recently acquired HERVs and the youngest and most brain-active versions of Alu's (Jurka, 2004; Jurka et al., 2004). Such prominent viral footprints in our recent evolution require better explanations than those that can be offered by serial selection that must operate via individual fittest type. Thus, virolution and its inherent capacity for consortia (QS-C)-based adaptations may provide a better perspective from which to understand human evolution.

21.20 PREGNANCY HORMONES: MATERNAL BONDING AND ERV NETWORKS

Pregnancy is associated with various significant shifts in hormonal control. One of the main pregnancy hormones is prolactin. Prolactin is made both in the brain (all mammals) and in reproductive tissue (pregnant endometrium) of human. This hormone plays a direct role in inducing maternal brain growth (in the lateral ventricle) associated with maternal behavior (Larsen and Grattan, 2012). As noted, maternal behavior is heavily mediated by olfaction in non-ape mammals (Brennan and Kendrick, 2006; Sanchez-Andrade and Kendrick, 2009). Interestingly, in most mammals, retrovirus-derived LTRs are also usually involved in prolactin regulation. However, in some species, such as human, a primate-specific MER39 has replaced older LTRs and now controls prolactin expression in reproductive tissue, unlike other mammal species (Emera and Wagner, 2012; Emera et al., 2012; Lynch et al., 2012). Thus, this is a distinct ERV-derived regulatory element not seen in tetrapod (e.g., rodents use MER77).

Estrogen is also a major regulatory hormone during pregnancy, and many pregnancy-related genes are under estrogen control. Thus, it is very interesting that the estrogen receptor has also been used to study the emergence of a regulatory network in the placenta. Here too, the LTRs of various ERVs (ERV-L and HERV-K) have contributed directly to the formation of this network (Testori et al., 2012).

With respect to social behavior, including maternal behavior, both vasopressin and oxytocin neuropeptides play major roles (Donaldson and Young, 2008). Arginine vasopressin (AVPR 1 a/b) receptors appear to be directly involved in social aggression and intruder attack (Wacker et al., 2010, 2011). They are also crucial for parental behavior and pair bonding and thus are linked to both associative and aggressive behaviors. The behavior changes that would be needed for maternal care to emerge are striking. The female must undergo a transition from self-centered to offspring-centered behavior. And it is clear that the maternal brain is altered during pregnancy (Kinsley and Lambert, 2006; Kinsley et al., 2006; Kinsley and Amory-Meyer, 2011). However, in order for adult human males to be similarly bonded in more extended family structures (mates, offspring, tribes), it seems obvious that male brains cannot be directly altered by pregnancy hormones. Instead, any significant male brain alterations would need to occur during development (in utero) as well as after birth (via learning). Additionally, bonding mechanisms would need to stem from *learned* social cues, not hormones or odorants affecting placental tissue. Human brain alterations thus needed to enhance social bonding by mechanisms that lie outside of the olfactory lobe. This is likely linked to the recently expanded human neocortex and changes to language and visual systems. There are mammalian rodent models (prairie vole) of extended social bonding that use vasopressin and oxytocin signaling for male bonding. From a virus perspective, it is also interesting that the vasopressin receptor is also involved in B-cell (immune cells that make antibody) receptor signaling (Hu et al., 2003). This is relevant since it has been asserted that altered immune regulation is involved in the regulation of conception and embryo implantation (Robertson, 2010). Along these lines, the deletion of the vasopressin 1b receptor results in a reduced Bruce effect (pregnancy abortion induced by non-male–male olfaction described as follows) (Stevenson and Caldwell, 2012). This observation is also intriguing in the context of needing to include nonmate males into extended social bonds since the pregnant females would need to eliminate the Bruce effect to allow nonmate males to be members of stable social or family groups. Indeed, the great apes all lack Bruce effects.

Oxytocin is also clearly involved in social memory, including maternal behavior and attachment (also via altered ventral septum of the mouse brain). It contributes to the recognition of conspecifics (belonging to the same species) and affiliative behavior (Lee et al., 2009). Brain oxytocin is used for social recognition in both male and female rats (Neumann, 2008). In females, levels are elevated during pregnancy but return to normal at postpartum. Oxytocin also interacts with dopamine and serotonin neurotransmitter systems. In monogamous prairie voles, oxytocin levels and receptor are regulated during pair bonding. Also, male–male aggression can be induced by oxytocin (Young, 2002; Hammock and Young, 2006; Smeltzer et al., 2006; Goodson, 2013). Thus, in monogamous rodents, oxytocin is involved in both mate bonding and paternal care. Variation in noncoding regions of these receptor genes appears to account for the biological variation.

The regulatory differences of the vasopressin receptor between monogamous and polygamous vole species are due to variations in noncoding repeated elements (Wang et al., 2000; Pitkow et al., 2001; Hammock and Young, 2002, 2005; Hammock et al., 2005). In rats, vasopressin and oxytocin receptors are linked by a LINE element (Schmitz et al., 1991). Although there is strong evidence that oxytocin is also involved in human social bonding (maternal and mate), clearly human oxytocin regulatory network would have needed to be edited (repurposed) to be used for human social learning.

21.21 SNAPBACK RNA NETWORKS, VIRUS, AND HUMAN BEHAVIOR

Gene regulatory changes are thus thought to be major factors driving mammalian evolution, with creation of new regulatory regions likely being instrumental in contributing to diversity among vertebrates (Baillie et al. 2011). Although the human genome contains over one million Alu repeat elements, their distribution is not uniform. These encode snapback regulatory RNAs on a massive scale and suggest massive network changes. Their expression is quite variable and may make most neurons virtually unique in their RNA expression profiles. Additional examples of small human-specific regulatory RNA, such as the gene, called miR-941 (a hairpin RNA in first intron of DNAJ, which is not present in chimp genome and varies in the human genome), are known. This RNA also appears to have played a crucial role in human brain development and could shed light on how we learned to use tools and language (Hu et al., 2012). It is expressed in the prefrontal cortex and cerebellum and mutations of this RNA affect language and speech. According to Martin Taylor, who led the study of this RNA at the Institute of Genetics and Molecular Medicine at the University of Edinburgh, "as a species, humans are wonderfully inventive - we are socially and technologically evolving all the time. But, according to scientists, this gene emerged fully functional out of noncoding genetic material, previously termed 'junk DNA,' in a brief interval of evolutionary time." This *junk* perspective is an example of reductive (nonnetwork) thinking. We must now start to think in terms of societies and networks being altered by virus societies. If we consider such RNAs to be part of a society (network) of snapback regulatory RNA, we might better understand how or why virus might have been involved in such networks.

This chapter has sought to emphasize the ancient, formative, and ongoing role for virus-derived transmissible small snapback RNAs in human evolution. The previous chapter similarly presented the role of small RNA societies in the origin of all domains' life and their viruses. Such RNAs have an inherently social (QS-C) character. This social theme has been applied to evaluate the recent changes in human evolution. Indeed, small RNAs (Alu's, HARS) underwent the most significant changes in recent human evolution. And these RNAs appear to have major regulatory roles in human brain function. We can now better evaluate how viruses were involved in the origins of such RNAs and what their relationship to virus might now be. Indeed, the recent sequencing of the Neanderthal and Denisovan genomes and comparison to anatomically modern human genome clearly shows differences not only in their corresponding ERVs composition (see Figure 21.6), but sexual exchange between humans and Neanderthals appears to have involved the core genes of the antiviral innate adaptive immune system (Agoni et al., 2012;Mendez et al., 2012; Contreras-Galindo et al., 2013) leaving little doubt that virolution applies to recent human evolution.

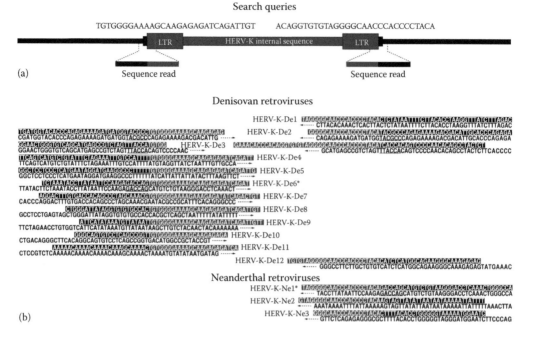

FIGURE 21.6 ERVs found in Neanderthals and Denisovans, absent from modern humans. (a) A provirus (red) inserted into a host genome (black). LTR sequences used as queries in searches are shown above the provirus. Positions of flanking duplications (black boxes) are shown as small black boxes adjacent to the LTRs. Sequence reads of virus–host junctions that were sought in the searches are shown below the provirus. (b) Individual Denisovan and Neanderthal sequence reads containing HERV-Ks. Archaic hominin sequences are shown in boxes with red background at the viral LTR sequences and black background around the flanking host genome DNA. (Reproduced from Lorenzo, A. et al., *Curr. Biol.*, 22, R437, 2012. With permission.)

GLOSSARY

Consult also glossary from the previous chapter.

ERVs: Endogenous retroviruses. Partial or complete sequences derived from retroviruses that have become part of the host genome.

Exapted: An evolutionary theory that proposes that DNA (genes) from other organisms, such as parasites, can become part of the host DNA following natural selection.

Genetic Parasites: Genetic agents, such as viruses and TEs, that colonize and use host systems for their own maintenance.

Group Identity: The capacity for a group of living and/or genetic agents to recognize other members of the same group.

LINES: Long interspersed nuclear elements. Genetic agents, distinct from retroviruses, that derive from a type of retrotransposon, able to replicate from RNA via DNA.

LTRs: Long terminal repeats. The characteristic sequences found on both ends of a retrovirus genome.

Metagenomic: The study of populations of genomes as they occur in particular habitats.

Network Membership: The capacity of a network to recognize allowed members from nonmembers.

Networks: A system of agents (or elements) that interact in a coherent fashion, usually associated with regulation.

Pseudoknots: The ability of a stem-loop-like RNA to fold itself into two distinct and dynamic conformations.

Quasispecies: A population of related genetic agents that are derived by variation during replication.

Riboswitches: A dynamic RNA fold structure that interacts with a regulator (such as a small molecule) to change its conformation and function.

Stem-loop RNA: Small regions of RNA that can fold back on themselves to form base pairs.

Synbiogenic: The generation of genetic novelty via the stable interaction with a symbiotic organism.

tRNA: Transfer RNA, the small clover leaf–shaped RNA that binds to the ribosome and is responsible for providing the correct amino acid in the triplet genetic code.

UTR: Untranslated region. Sequences of DNA that are regulatory and do not code for protein synthesis.

Virolution: A term coined by Frank Ryan to describe host evolution mediated by the action (selection, protection, integration) of viruses.

Virosphere: The extended virus composition of a biological habitat.

REVIEW QUESTIONS

1. Can the number of genes explain the fact that humans have larger and more social brains than simpler animals?

2. Viruses have promoted the formation of new human identity networks and the emergence of complex group social behavior. Explain why networks do not operate by nonlinear processes.

3. Provide examples how genetic regulatory networks underlying maternal care in placental mammals were repurposed by RNA genetic colonization.

4. Give examples how QS-C (quasispecies consortia) mediated virolution can provide a mechanistic foundation for group identity and group survival.

5. What is the ENCODE project? Did it support or not the concept of *junk* DNA?

6. How do ERVs contribute to the development of early mammalian embryos?

7. How do ERVs contribute to the origin and development of the placenta?

8. What are Alu RNAs and where are they expressed?

9. How might human social bonds be mediated if they do not use odor-based signals (pheromones)?

10. What is an addiction module and how can it mediate social bonds?

REFERENCES

Agoni, L., Golden, A., Guha, C., Lenz, J., 2012. Neanderthal and Denisovan retroviruses. *Curr. Biol.* 22, R437–R438.

Alzohairy, A.M., Gyulai, G., Jansen, R.K., Bahieldin, A., 2013. Transposable elements domesticated and neofunctionalized by eukaryotic genomes. *Plasmid* 69, 1–15.

Aron, A., Fisher, H., Mashek, D.J., Strong, G., Li, H., Brown, L.L., 2005. Reward, motivation, and emotion systems associated with early-stage intense romantic love. *J. Neurophysiol.* 94, 327–337.

Baillie, J.K., Barnett, M.W., Upton, K.R., Gerhardt, D.J., Richmond, T.A., De Sapio, F., Brennan, P.M. et al., 2011. Somatic retrotransposition alters the genetic landscape of the human brain. *Nature* 479, 534–537.

Barbulescu, M., Turner, G., Su, M., Kim, R., Jensen-Seaman, M.I., Deinard, A.S., Kidd, K.K., Lenz, J., 2001. A HERV-K provirus in chimpanzees, bonobos and gorillas, but not humans. *Curr. Biol.* 11, 779–783.

Barlow, D.P., 2011. Genomic imprinting: A mammalian epigenetic discovery model. *Annu. Rev. Genet.* 45, 379–403.

Barry, G., Mattick, J.S., 2012. The role of regulatory RNA in cognitive evolution. *Trends Cogn. Sci.* 16, 497–503.

Bartoš, L., Bartošová, J., Pluháček, J., Šindelářová, J., 2011. Promiscuous behaviour disrupts pregnancy block in domestic horse mares. *Behav. Ecol. Sociobiol.* 65, 1567–1572.

Bejerano, G., Pheasant, M., Makunin, I., Stephen, S., Kent, W.J., Mattick, J.S., Haussler, D., 2004. Ultraconserved elements in the human genome. *Science* 304, 1321–1325.

Belshaw, R., Katzourakis, A., Paces, J., Burt, A., Tristem, M., 2005. High copy number in human endogenous retrovirus (HERV) families is associated with copying mechanisms in addition to re-infection. *Mol. Biol. Evol.* 22, 814–817.

Belshaw, R., Pereira, V., Katzourakis, A., Talbot, G., Paces, J., Burt, A., Tristem, M., 2004. Long-term reinfection of the human genome by endogenous retroviruses. *Proc. Natl. Acad. Sci. USA* 101, 4894–4899.

Benit, L., Calteau, A., Heidmann, T., 2003. Characterization of the low-copy HERV-Fc family: Evidence for recent integrations in primates of elements with coding envelope genes. *Virology* 312, 159–168.

Benit, L., De Parseval, N., Casella, J.F., Callebaut, I., Cordonnier, A., Heidmann, T., 1997. Cloning of a new murine endogenous retrovirus, MuERV-L, with strong similarity to the human HERV-L element and with a gag coding sequence closely related to the Fv1 restriction gene. *J. Virol.* 71, 5652–5657.

Beraldi, R., Pittoggi, C., Sciamanna, I., Mattei, E., Spadafora, C., 2006. Expression of LINE-1 retroposons is essential for murine preimplantation development. *Mol. Reprod. Dev.* 73, 279–287.

Bird, C.P., Stranger, B.E., Liu, M., Thomas, D.J., Ingle, C.E., Beazley, C., Miller, W., Hurles, M.E., Dermitzakis, E.T., 2007. Fast-evolving noncoding sequences in the human genome. *Genome Biol.* 8, R118.

Blond, J.-L., Lavillette, D., Cheynet, V., Bouton, O., Oriol, G., Chapel-Fernandes, S., Mandrand, B., Mallet, F., Cosset, F.-L., 2000. An envelope glycoprotein of the human endogenous retrovirus HERV-W is expressed in the human placenta and fuses cells expressing the type D mammalian retrovirus receptor. *J. Virol.* 74, 3321–3329.

Brennan, P.A., 2001. The vomeronasal system. *Cell. Mol. Life Sci.* 58, 546–555.

Brennan, P.A., 2009. Outstanding issues surrounding vomeronasal mechanisms of pregnancy block and individual recognition in mice. *Behav. Brain Res.* 200, 287–294.

Brennan, P.A., Kendrick, K.M., 2006. Mammalian social odours: Attraction and individual recognition. *Philos. Trans. R. Soc. Lond. B Biol. Sci.* 361, 2061–2078.

Bush, E.C., Lahn, B.T., 2005. Selective constraint on noncoding regions of hominid genomes. *PLoS Comput. Biol.* 1, e73.

Buzdin, A.A., 2010. Functional analysis of retroviral endogenous inserts in the human genome evolution. *Bioorg. Khim.* 36, 38–46.

Buzdin, A., Khodosevich, K., Mamedov, I., Vinogradova, T., Lebedev, Y., Hunsmann, G., Sverdlov, E., 2002. A technique for genome-wide identification of differences in the interspersed repeats integrations between closely related genomes and its application to detection of human-specific integrations of HERV-K LTRs. *Genomics* 79, 413–422.

Buzdin, A., Kovalskaya-Alexandrova, E., Gogvadze, E., Sverdlov, E., 2006. At least 50% of human-specific HERV-K (HML-2) long terminal repeats serve in vivo as active promoters for host nonrepetitive DNA transcription. *J. Virol.* 80, 10752–10762.

Camperio Ciani, A.S., Edelman, S., Ebstein, R.P., 2013. The dopamine D4 receptor (DRD4) exon 3 VNTR contributes to adaptive personality differences in an Italian small island population. *Eur. J. Pers.* 27, 593–604.

Chaudhari, N., Hahn, W.E., 1983. Genetic expression in the developing brain. *Science* 220, 924–928.

Chimpanzee Sequencing and Analysis Consortium, 2005. Initial sequence of the chimpanzee genome and comparison with the human genome. *Nature* 437, 69–87.

Chuong, E.B., Rumi, M.A.K., Soares, M.J., Baker, J.C., 2013. Endogenous retroviruses function as species-specific enhancer elements in the placenta. *Nat. Genet.* 45, 325–329.

Churakov, G., Kriegs, J.O., Baertsch, R., Zemann, A., Brosius, J., Schmitz, J., 2009. Mosaic retroposon insertion patterns in placental mammals. *Genome Res.* 19, 868–875.

Contreras-Galindo, R., Kaplan, M.H., He, S., Contreras-Galindo, A.C., Gonzalez-Hernandez, M.J., Kappes, F., Dube, D. et al., 2013. HIV infection reveals wide-spread expansion of novel centromeric human endogenous retroviruses. *Genome Res.* 23, 1505–1513.

Cooper, D.L., 2006. Broca's arrow: Evolution, prediction, and language in the brain. *Anat. Rec. B New Anat.* 289, 9–24.

Domingo, E., Sheldon, J., Perales, C., 2012. Viral quasispecies evolution. *Microbiol. Mol. Biol. Rev.* 76, 159–216.

Donaldson, Z.R., Kondrashov, F.A., Putnam, A., Bai, Y., Stoinski, T.L., Hammock, E.A., Young, L.J., 2008. Evolution of a behavior-linked microsatellite-containing element in the 5′ flanking region of the primate AVPR1A gene. *BMC Evol. Biol.* 8, 180.

Donaldson, Z.R., Young, L.J., 2008. Oxytocin, vasopressin, and the neurogenetics of sociality. *Science* 322, 900–904.

Dupressoir, A., Lavialle, C., Heidmann, T., 2012. From ancestral infectious retroviruses to bona fide cellular genes: Role of the captured syncytins in placentation. *Placenta* 33, 663–671.

Dupressoir, A., Vernochet, C., Harper, F., Guégan, J., Dessen, P., Pierron, G., Heidmann, T., 2011. A pair of co-opted retroviral envelope syncytin genes is required for formation of the two-layered murine placental syncytiotrophoblast. *Proc. Natl. Acad. Sci. USA* 108, E1164–E1173.

Eigen, M., 1971. Self-organization of matter and the evolution of biological macromolecules. *Naturwissenschaften* 58, 465–522.

Emera, D., Casola, C., Lynch, V.J., Wildman, D.E., Agnew, D., Wagner, G.P., 2012. Convergent evolution of endometrial prolactin expression in primates, mice, and elephants through the independent recruitment of transposable elements. *Mol. Biol. Evol.* 29, 239–247.

Emera, D., Wagner, G.P., 2012. Transposable element recruitments in the mammalian placenta: Impacts and mechanisms. *Brief. Funct. Genomics* 11, 267–276.

Evsikov, A.V., de Vries, W.N., Peaston, A.E., Radford, E.E., Fancher, K.S., Chen, F.H., Blake, J.A. et al., 2004. Systems biology of the 2-cell mouse embryo. *Cytogenet. Genome Res.* 105, 240–250.

Feldman, R., 2012. Oxytocin and social affiliation in humans. *Horm. Behav.* 61, 380–391.

Feschotte, C., 2008. Transposable elements and the evolution of regulatory networks. *Nat. Rev. Genet.* 9, 397–405.

Feschotte, C., Gilbert, C., 2012. Endogenous viruses: Insights into viral evolution and impact on host biology. *Nat. Rev. Genet.* 13, 283.

Gifford, W.D., Pfaff, S.L., Macfarlan, T.S., 2013. Transposable elements as genetic regulatory substrates in early development. *Trends Cell Biol.* 23, 218–226.

Gilad, Y., Wiebe, V., Przeworski, M., Lancet, D., Pääbo, S., 2004. Loss of olfactory receptor genes coincides with the acquisition of full trichromatic vision in primates. *PLoS Biol.* 2, e5.

Giles, K.E., Caputi, M., Beemon, K.L., 2004. Packaging and reverse transcription of snRNAs by retroviruses may generate pseudogenes. *RNA* 10, 299–307.

Giorgi, D., Friedman, C., Trask, B.J., Rouquier, S., 2000. Characterization of nonfunctional V1R-like pheromone receptor sequences in human. *Genome Res.* 10, 1979–1985.

Gogvadze, E., Stukacheva, E., Buzdin, A., Sverdlov, E., 2009. Human-specific modulation of transcriptional activity provided by endogenous retroviral insertions. *J. Virol.* 83, 6098–6105.

Goodson, J.L., 2013. Deconstructing sociality, social evolution and relevant nonapeptide functions. *Psychoneuroendocrinology* 38, 465–478.

Haig, D., 2012. Retroviruses and the placenta. *Curr. Biol.* 22, R609–R613.

Halpern, M., 1987. The organization and function of the vomeronasal system. *Annu. Rev. Neurosci.* 10, 325–362.

Hammock, E.A., Lim, M.M., Nair, H.P., Young, L.J., 2005. Association of vasopressin 1a receptor levels with a regulatory microsatellite and behavior. *Genes Brain Behav.* 4, 289–301.

Hammock, E.A., Young, L.J., 2002. Variation in the vasopressin V1a receptor promoter and expression: Implications for inter- and intraspecific variation in social behaviour. *Eur. J. Neurosci.* 16, 399–402.

Hammock, E.A., Young, L.J., 2005. Microsatellite instability generates diversity in brain and sociobehavioral traits. *Science* 308, 1630–1634.

Hammock, E.A., Young, L.J., 2006. Oxytocin, vasopressin and pair bonding: Implications for autism. *Philos. Trans. R. Soc. Lond. B Biol Sci.* 361, 2187–2198.

Harrow, J., Frankish, A., Gonzalez, J.M., Tapanari, E., Diekhans, M., Kokocinski, F., Aken, B.L. et al., 2012. GENCODE: The reference human genome annotation for The ENCODE Project. *Genome Res.* 22, 1760–1774.

Heidmann, O., Vernochet, C., Dupressoir, A., Heidmann, T., 2009. Identification of an endogenous retroviral envelope gene with fusogenic activity and placenta-specific expression in the rabbit: A new "syncytin" in a third order of mammals. *Retrovirology* 6, 107.

Hoffmann, M., 2013. The human frontal lobes and frontal network systems: An evolutionary, clinical, and treatment perspective. *ISRN Neurol.* 2013, 1–34.

Hu, H.Y., He, L., Fominykh, K., Yan, Z., Guo, S., Zhang, X., Taylor, M.S. et al., 2012. Evolution of the human-specific microRNA miR-941. *Nat. Commun.* 3, 1145.

Hu, S.-B., Zhao, Z.-S., Yhap, C., Grinberg, A., Huang, S.-P., Westphal, H., Gold, P., 2003. Vasopressin receptor 1a-mediated negative regulation of B cell receptor signaling. *J. Neuroimmunol.* 135, 72–81.

Huda, A., Bowen, N.J., Conley, A.B., Jordan, I.K., 2011. Epigenetic regulation of transposable element derived human gene promoters. *Gene* 475(1), 39–48.

Huda, A., Marino-Ramirez, L., Jordan, I.K., 2010. Epigenetic histone modifications of human transposable elements: Genome defense versus exaptation. *Mob. DNA* 1, 2.

Ichikawa, M., 2003. Synaptic mechanisms underlying pheromonal memory in vomeronasal system. *Zoolog. Sci.* 20, 687–695.

Johns, M.A., 1986. The role of the vomeronasal organ in behavioral control of reproduction. *Ann. N. Y. Acad. Sci.* 474, 148–157.

Jurka, J., 2004. Evolutionary impact of human Alu repetitive elements. *Curr. Opin. Genet. Dev.* 14, 603–608.

Jurka, J., Kohany, O., Pavlicek, A., Kapitonov, V.V., Jurka, M.V., 2004. Duplication, coclustering, and selection of human Alu retrotransposons. *Proc. Natl. Acad. Sci. USA* 101, 1268–1272.

Kämmerer, U., Germeyer, A., Stengel, S., Kapp, M., Denner, J., 2011. Human endogenous retrovirus K (HERV-K) is expressed in villous and extravillous cytotrophoblast cells of the human placenta. *J. Reprod. Immunol.* 91, 1–8.

Kaneko-Ishino, T., Ishino, F., 2012. The role of genes domesticated from LTR retrotransposons and retroviruses in mammals. *Front. Microbiol.* 3, 262.

Kano, H., Godoy, I., Courtney, C., Vetter, M.R., Gerton, G.L., Ostertag, E.M., Kazazian, H.H., 2009. L1 retrotransposition occurs mainly in embryogenesis and creates somatic mosaicism. *Genes Dev.* 23, 1303–1312.

Katayama, S., Tomaru, Y., Kasukawa, T., Waki, K., Nakanishi, M., Nakamura, M., Nishida, H. et al., 2005. Antisense transcription in the mammalian transcriptome. *Science* 309, 1564–1566.

Kättström, P.-O., Bjerneroth, G., Nilsson, B.O., Holmdahl, R., Larsson, E., 1989. A retroviral gp70-related protein is expressed at specific stages during mouse oocyte maturation and in preimplantation embryos. *Cell Differ. Dev.* 28, 47–54.

Keightley, P.D., Lercher, M.J., Eyre-Walker, A., 2005. Evidence for widespread degradation of gene control regions in hominid genomes. *PLoS Biol.* 3, e42.

Keightley, P.D., Lercher, M.J., Eyre-Walker, A., 2006. Understanding the degradation of hominid gene control. *PLoS Comput. Biol.* 2, e19, author reply e26.

Keller, M., Baum, M.J., Brock, O., Brennan, P.A., Bakker, J., 2009. The main and the accessory olfactory systems interact in the control of mate recognition and sexual behavior. *Behav. Brain Res.* 200, 268–276.

Kelley, D., Rinn, J., 2012. Transposable elements reveal a stem cell-specific class of long noncoding RNAs. *Genome Biol.* 13, R107.

Keverne, E.B., 1999. The vomeronasal organ. *Science* 286, 716–720.

Khaitovich, P., Muetzel, B., She, X., Lachmann, M., Hellmann, I., Dietzsch, J., Steigele, S. et al., 2004. Regional patterns of gene expression in human and chimpanzee brains. *Genome Res.* 14, 1462–1473.

Khalil, A.M., Guttman, M., Huarte, M., Garber, M., Raj, A., Rivea Morales, D., Thomas, K. et al., 2009. Many human large intergenic noncoding RNAs associate with chromatin-modifying complexes and affect gene expression. *Proc. Natl. Acad. Sci. USA* 106, 11667–11672.

Kigami, D., Minami, N., Takayama, H., Imai, H., 2003. MuERV-L is one of the earliest transcribed genes in mouse one-cell embryos. *Biol. Reprod.* 68, 651–654.

Kim, H.S., Hyun, B.H., Choi, J.Y., Crow, T.J., 2000. Phylogenetic analysis of a retroposon family as represented on the human X chromosome. *Genes Genet. Syst.* 75, 197–202.

Kim, H.S., Takenaka, O., 2001. Phylogeny of SINE-R retroposons in Asian apes. *Mol. Cells* 12, 262–266.

Kim, H.S., Wadekar, R.V., Takenaka, O., Hyun, B.H., Crow, T.J., 1999. Phylogenetic analysis of a retroposon family in african great apes. *J. Mol. Evol.* 49, 699–702.

Kim, P., Feldman, R., Mayes, L.C., Eicher, V., Thompson, N., Leckman, J.F., Swain, J.E., 2011. Breastfeeding, brain activation to own infant cry, and maternal sensitivity. *J. Child Psychol. Psychiatry* 52, 907–915.

Kinsley, C.H., Amory-Meyer, E., 2011. Why the maternal brain? *J. Neuroendocrinol.* 23, 974–983.

Kinsley, C.H., Lambert, K.G., 2006. The maternal brain. *Sci. Am.* 294, 72–79.

Kinsley, C.H., Trainer, R., Stafisso-Sandoz, G., Quadros, P., Marcus, L.K., Hearon, C., Meyer, E.A.A. et al., 2006. Motherhood and the hormones of pregnancy modify concentrations of hippocampal neuronal dendritic spines. *Horm. Behav.* 49, 131–142.

Kjellman, C., Sjögren, H.-O., Widegren, B., 1995. The Y chromosome: A graveyard for endogenous retroviruses. *Gene* 161, 163–170.

Konopka, G., Geschwind, D.H., 2010. Human brain evolution: Harnessing the genomics (r)evolution to link genes, cognition, and behavior. *Neuron* 68, 231–244.

Kouros-Mehr, H., Pintchovski, S., Melnyk, J., Chen, Y.J., Friedman, C., Trask, B., Shizuya, H., 2001. Identification of non-functional human VNO receptor genes provides evidence for vestigiality of the human VNO. *Chem. Senses* 26, 1167–1174.

Kriegs, J.O., Churakov, G., Jurka, J., Brosius, J., Schmitz, J., 2007. Evolutionary history of 7SL RNA-derived SINEs in Supraprimates. *Trends Genet.* 23, 158–161.

Kunarso, G., Chia, N.-Y., Jeyakani, J., Hwang, C., Lu, X., Chan, Y.-S., Ng, H.-H., Bourque, G., 2010. Transposable elements have rewired the core regulatory network of human embryonic stem cells. *Nat. Genet.* 42, 631–634.

Larsen, C.M., Grattan, D.R., 2012. Prolactin, neurogenesis, and maternal behaviors. *Brain Behav. Immun.* 26, 201–209.

Lee, H.-J., Macbeth, A.H., Pagani, J.H., Scott Young 3rd, W., 2009. Oxytocin: The great facilitator of life. *Prog. Neurobiol.* 88, 127–151.

Loewer, S., Cabili, M.N., Guttman, M., Loh, Y.-H., Thomas, K., Park, I.H., Garber, M. et al., 2010. Large intergenic non-coding RNA-RoR modulates reprogramming of human induced pluripotent stem cells. *Nat. Genet.* 42, 1113–1117.

Loke, Y.W., 2013. *Life's Vital Link: The Astonishing Role of the Placenta.* Oxford University Press, Oxford, U.K.

Lorenzo, A., Golden, A., Guha, C., Lenz, J., 2012. Neandertal and Denisovan retroviruses. *Curr. Biol.* 22, R437–R438.

Lynch, V.J., Leclerc, R.D., May, G., Wagner, G.P., 2011. Transposon-mediated rewiring of gene regulatory networks contributed to the evolution of pregnancy in mammals. *Nat. Genet.* 43, 1154–1159.

Lynch, V.J., Nnamani, M., Brayer, K.J., Emera, D., Wertheim, J.O., Pond, S.L.K., Grützner, F. et al., 2012. Lineage-specific transposons drove massive gene expression recruitments during the evolution of pregnancy in mammals. *arXiv*:1208.4639.

Macfarlan, T.S., Gifford, W.D., Driscoll, S., Lettieri, K., Rowe, H.M., Bonanomi, D., Firth, A., Singer, O., Trono, D., Pfaff, S.L., 2012. Embryonic stem cell potency fluctuates with endogenous retrovirus activity. *Nature* 487, 57–63.

Maksakova, I.A., Mager, D.L., 2005. Transcriptional regulation of early transposon elements, an active family of mouse long terminal repeat retrotransposons. *J. Virol.* 79, 13865–13874.

Maksakova, I.A., Mager, D.L., Reiss, D., 2008. Keeping active endogenous retroviral-like elements in check: The epigenetic perspective. *Cell. Mol. Life Sci.* 65, 3329–3347.

Mangeney, M., Renard, M., Schlecht-Louf, G., Bouallaga, I., Heidmann, O., Letzelter, C., Richaud, A., Ducos, B., Heidmann, T., 2007. Placental syncytins: Genetic disjunction between the fusogenic and immunosuppressive activity of retroviral envelope proteins. *Proc. Natl. Acad. Sci. USA* 104, 20534–20539.

Marz, M., Donath, A., Verstraete, N., Nguyen, V.T., Stadler, P.F., Bensaude, O., 2009. Evolution of 7SK RNA and its protein partners in metazoa. *Mol. Biol. Evol.* 26, 2821–2830.

Mattick, J.S., 2005. The functional genomics of noncoding RNA. *Science* 309, 1527–1528.

Mattick, J.S., 2010. The central role of RNA in the genetic programming of complex organisms. *An. Acad. Bras. Cienc.* 82, 933–939.

Mattick, J.S., 2011. The central role of RNA in human development and cognition. *FEBS Lett.* 585, 1600–1616.

Mattick, J.S., Gagen, M.J., 2001. The evolution of controlled multitasked gene networks: The role of introns and other noncoding RNAs in the development of complex organisms. *Mol. Biol. Evol.* 18, 1611–1630.

Mattick, J.S., Makunin, I.V., 2006. Non-coding RNA. *Hum. Mol. Genet.* 15 Spec No. 1, R17–R29.

Mattick, J.S., Mehler, M.F., 2008. RNA editing, DNA recoding and the evolution of human cognition. *Trends Neurosci.* 31, 227–233.

Mattick, J.S., Taft, R.J., Faulkner, G.J., 2010. A global view of genomic information—Moving beyond the gene and the master regulator. *Trends Genet.* 26, 21–28.

Mehler, M.F., Mattick, J.S., 2006. Non-coding RNAs in the nervous system. *J. Physiol.* 575, 333–341.

Mehler, M.F., Mattick, J.S., 2007. Noncoding RNAs and RNA editing in brain development, functional diversification, and neurological disease. *Physiol. Rev.* 87, 799–823.

Mendez, F.L., Watkins, J.C., Hammer, M.F., 2012. A haplotype at STAT2 Introgressed from Neanderthals and serves as a candidate of positive selection in Papua New Guinea. *Am. J. Hum. Genet.* 91, 265–274.

Mercer, T.R., Dinger, M.E., Mariani, J., Kosik, K.S., Mehler, M.F., Mattick, J.S., 2008. Noncoding RNAs in long-term memory formation. *Neuroscientist* 14, 434–445.

Mercer, T.R., Dinger, M.E., Mattick, J.S., 2009. Long non-coding RNAs: Insights into functions. *Nat. Rev. Genet.* 10, 155–159.

Mi, S., Lee, X., Li, X., Veldman, G.M., Finnerty, H., Racie, L., LaVallie, E. et al., 2000. Syncytin is a captive retroviral envelope protein involved in human placental morphogenesis. *Nature* 403, 785–789.

Mills, R.E., Luttig, C.T., Larkins, C.E., Beauchamp, A., Tsui, C., Pittard, W.S., Devine, S.E., 2006. An initial map of insertion and deletion (INDEL) variation in the human genome. *Genome Res.* 16, 1182–1190.

Monahan, E., Yamazaki, K., Beauchamp, G.K., Maxson, S.C., 1993. Olfactory discrimination of urinary odortypes from congenic strains (DBA/1Bg and DBA1.C57BL10-YBg) of mice differing in their Y chromosomes. *Behav. Genet.* 23, 251–255.

Muiños-Gimeno, M., Espinosa-Parrilla, Y., Guidi, M., Kagerbauer, B., Sipilä, T., Maron, E., Pettai, K. et al., 2011. Human microRNAs miR-22, miR-138-2, miR-148a, and miR-488 are associated with panic disorder and regulate several anxiety candidate genes and related pathways. *Biol. Psychiatry* 69, 526–533.

Murphy, W.J., Pringle, T.H., Crider, T.A., Springer, M.S., Miller, W., 2007. Using genomic data to unravel the root of the placental mammal phylogeny. *Genome Res.* 17, 413–421.

Narver, H.L., 2012. Oxytocin in the treatment of dystocia in mice. *J. Am. Assoc. Lab. Anim. Sci.* 51, 10–17.

Neumann, I.D., 2008. Brain oxytocin: A key regulator of emotional and social behaviours in both females and males. *J. Neuroendocrinol.* 20, 858–865.

Newman, T., Trask, B.J., 2003. Complex evolution of 7E olfactory receptor genes in segmental duplications. *Genome Res.* 13, 781–793.

Nilsson, B.O., Kättström, P.-O., Sundström, P., Jaquemin, P., Larsson, E., 1992. Human oocytes express murine retroviral equivalents. *Virus Genes* 6, 221–227.

Nishihara, H., Maruyama, S., Okada, N., 2009. Retroposon analysis and recent geological data suggest near-simultaneous divergence of the three superorders of mammals. *Proc. Natl. Acad. Sci. USA* 106, 5235–5240.

Ohnishi, Y., Totoki, Y., Toyoda, A., Watanabe, T., Yamamoto, Y., Tokunaga, K., Sakaki, Y., Sasaki, H., Hohjoh, H., 2010. Small RNA class transition from siRNA/piRNA to miRNA during pre-implantation mouse development. *Nucleic Acids Res.* 38, 5141–5151.

Ohnishi, Y., Totoki, Y., Toyoda, A., Watanabe, T., Yamamoto, Y., Tokunaga, K., Sakaki, Y., Sasaki, H., Hohjoh, H., 2012. Active role of small non-coding RNAs derived from SINE/B1 retrotransposon during early mouse development. *Mol. Biol. Rep.* 39, 903–909.

Ohshima, K., Koishi, R., Matsuo, M., Okada, N., 1993. Several short interspersed repetitive elements (SINEs) in distant species may have originated from a common ancestral retrovirus: Characterization of a squid SINE and a possible mechanism for generation of tRNA-derived retroposons. *Proc. Natl. Acad. Sci. USA* 90, 6260–6264.

Oliver, K.R., Greene, W.K., 2011. Mobile DNA and the TE-Thrust hypothesis: Supporting evidence from the primates. *Mob. DNA* 2, 8.

Onafuwa-Nuga, A.A., King, S.R., Telesnitsky, A., 2005. Nonrandom packaging of host RNAs in moloney murine leukemia virus. *J. Virol.* 79, 13528–13537.

Parr, L.A., 2003. The discrimination of faces and their emotional content by chimpanzees (*Pan troglodytes*). *Ann. N. Y. Acad. Sci.* 1000, 56–78.

Parr, L.A., Waller, B.M., Fugate, J., 2005. Emotional communication in primates: Implications for neurobiology. *Curr. Opin. Neurobiol.* 15, 716–720.

Pauli, A., Rinn, J.L., Schier, A.F., 2011. Non-coding RNAs as regulators of embryogenesis. *Nat. Rev.* 12, 136–149.

Paz-Yaacov, N., Levanon, E.Y., Nevo, E., Kinar, Y., Harmelin, A., Jacob-Hirsch, J., Amariglio, N., Eisenberg, E., Rechavi, G., 2010. Adenosine-to-inosine RNA editing shapes transcriptome diversity in primates. *Proc. Natl. Acad. Sci. USA* 107, 12174–12179.

Peaston, A.E., Evsikov, A.V., Graber, J.H., de Vries, W.N., Holbrook, A.E., Solter, D., Knowles, B.B., 2004. Retrotransposons regulate host genes in mouse oocytes and preimplantation embryos. *Dev. Cell* 7, 597–606.

Peaston, A.E., Knowles, B.B., Hutchison, K.W., 2007. Genome plasticity in the mouse oocyte and early embryo. *Biochem. Soc. Trans.* 35, 618.

Pedersen, C.A., 2004. Biological aspects of social bonding and the roots of human violence. *Ann. N. Y. Acad. Sci.* 1036, 106–127.

Pedersen, C.A., Ascher, J.A., Monroe, Y.L., Prange, A.J., 1982. Oxytocin induces maternal behavior in virgin female rats. *Science* 216, 648–650.

Pérot, P., Bolze, P.-A., Mallet, F., 2012. From viruses to genes: Syncytins, in: Witzany, G. (Ed.), *Viruses: Essential Agents of Life.* Springer, Dordrecht, the Netherlands, pp. 325–361.

Pitkow, L.J., Sharer, C.A., Ren, X., Insel, T.R., Terwilliger, E.F., Young, L.J., 2001. Facilitation of affiliation and pair-bond formation by vasopressin receptor gene transfer into the ventral forebrain of a monogamous vole. *J. Neurosci.* 21, 7392–7396.

Polavarapu, N., Bowen, N.J., McDonald, J.F., 2006. Identification, characterization and comparative genomics of chimpanzee endogenous retroviruses. *Genome Biol.* 7, R51.

Presutti, C., Rosati, J., Vincenti, S., Nasi, S., 2006. Non coding RNA and brain. *BMC Neurosci.* 7, S5.

Prudhomme, S., Bonnaud, B., Mallet, F., 2005. Endogenous retroviruses and animal reproduction. *Cytogenet. Genome Res.* 110, 353–364.

Qureshi, I.A., Mehler, M.F., 2009. Regulation of non-coding RNA networks in the nervous system—What's the REST of the story? *Neurosci. Lett.* 466, 73–80.

Rajendren, G., Dominic, C.J., 2009. The male-induced implantation failure (the Bruce effect) in mice: Effect of exogenous progesterone on maintenance of pregnancy in male-exposed females. *Exp. Clin. Endocrinol.* 101, 356–359.

Roberts, E.K., Lu, A., Bergman, T.J., Beehner, J.C., 2012. A Bruce effect in wild geladas. *Science* 335, 1222–1225.

Robertson, S.A., 2010. Immune regulation of conception and embryo implantation—All about quality control? *J. Reprod. Immunol.* 85, 51–57.

Romano, C.M., de Melo, F.L., Corsini, M.A., Holmes, E.C., Zanotto, P.M., 2007. Demographic histories of ERV-K in humans, chimpanzees and rhesus monkeys. *PLoS One* 2, e1026.

Romano, C.M., Ramalho, R.F., Zanotto, P.M., 2006. Tempo and mode of ERV-K evolution in human and chimpanzee genomes. *Arch. Virol.* 151, 2215–2228.

Rowe, H.M., Trono, D., 2011. Dynamic control of endogenous retroviruses during development. *Virology* 411, 273–287.

Sakate, R., Suto, Y., Imanishi, T., Tanoue, T., Hida, M., Hayasaka, I., Kusuda, J., Gojobori, T., Hashimoto, K., Hirai, M., 2007. Mapping of chimpanzee full-length cDNAs onto the human genome unveils large potential divergence of the transcriptome. *Gene* 399, 1–10.

Sanchez-Andrade, G., Kendrick, K.M., 2009. The main olfactory system and social learning in mammals. *Behav. Brain Res.* 200, 323–335.

Santoni, F.A., Guerra, J., Luban, J., 2012. HERV-H RNA is abundant in human embryonic stem cells and a precise marker for pluripotency. *Retrovirology* 9, 111.

Schmitz, E., Mohr, E., Richter, D., 1991. Rat vasopressin and oxytocin genes are linked by a long interspersed repeated DNA element (LINE): Sequence and transcriptional analysis of LINE. *DNA Cell Biol.* 10, 81–91.

Sciamanna, I., Vitullo, P., Curatolo, A., Spadafora, C., 2011. A reverse transcriptase-dependent mechanism is essential for murine preimplantation development. *Genes* 2, 360–373.

Sekita, Y., Wagatsuma, H., Nakamura, K., Ono, R., Kagami, M., Wakisaka, N., Hino, T. et al., 2008. Role of retrotransposon-derived imprinted gene, Rtl1, in the feto-maternal interface of mouse placenta. *Nat. Genet.* 40, 243–248.

Shelton, D.E., Michod, R.E., 2010. Philosophical foundations for the hierarchy of life. *Biol. Philos.* 25, 391–403.

Shou, C., Bhardwaj, N., Lam, H.Y.K., Yan, K.-K., Kim, P.M., Snyder, M., Gerstein, M.B., 2011. Measuring the evolutionary rewiring of biological networks. *PLoS Comput. Biol.* 7, e1001050.

Smalheiser, N.R., Lugli, G., Thimmapuram, J., Cook, E.H., Larson, J., 2011. Endogenous siRNAs and noncoding RNA-derived small RNAs are expressed in adult mouse hippocampus and are up-regulated in olfactory discrimination training. *RNA* 17, 166–181.

Smallwood, P.M., Wang, Y., Nathans, J., 2002. Role of a locus control region in the mutually exclusive expression of human red and green cone pigment genes. *Proc. Natl. Acad. Sci. USA* 99, 1008–1011.

Smeltzer, M.D., Curtis, J.T., Aragona, B.J., Wang, Z., 2006. Dopamine, oxytocin, and vasopressin receptor binding in the medial prefrontal cortex of monogamous and promiscuous voles. *Neurosci. Lett.* 394, 146–151.

Somel, M., Liu, X., Khaitovich, P., 2013. Human brain evolution: Transcripts, metabolites and their regulators. *Nat. Rev. Neurosci.* 14, 112–127.

Stevenson, E.L., Caldwell, H.K., 2012. The vasopressin 1b receptor and the neural regulation of social behavior. *Horm. Behav.* 61, 277–282.

Striedter, G.F., 2005. *Principles of Brain Evolution*. Sinauer Associates, Sunderland, MA.

Suh, A., 2012. A retroposon-based view on the temporal differentiation of sex chromosomes. *Mob. Genet. Elements* 2, 158–162.

Svoboda, P., Stein, P., Anger, M., Bernstein, E., Hannon, G.J., Schultz, R.M., 2004. RNAi and expression of retrotransposons MuERV-L and IAP in preimplantation mouse embryos. *Dev. Biol.* 269, 276–285.

Swain, J.E., 2011. The human parental brain: In vivo neuroimaging. *Prog. Neuropsychopharmacol. Biol. Psychiatry* 35, 1242–1254.

Testori, A., Caizzi, L., Cutrupi, S., Friard, O., Bortoli, M.D., Cora', D., Caselle, M., 2012. The role of transposable elements in shaping the combinatorial interaction of transcription factors. *BMC Genomics* 13, 400.

Toder, R., Grutzner, F., Haaf, T., Bausch, E., 2001. Species-specific evolution of repeated DNA sequences in great apes. *Chromosome Res.* 9, 431–435.

Uddin, L.Q., Iacoboni, M., Lange, C., Keenan, J.P., 2007. The self and social cognition: The role of cortical midline structures and mirror neurons. *Trends Cogn. Sci.* 11, 153–157.

Van Blerkom, L.M., 2003. Role of viruses in human evolution. *Am. J. Phys. Anthropol.* 122, 14–46.

van den Hurk, J.A., Meij, I.C., Seleme, M.C., Kano, H., Nikopoulos, K., Hoefsloot, L.H., Sistermans, E.A. et al., 2007. L1 retrotransposition can occur early in human embryonic development. *Hum. Mol. Genet.* 16, 1587–1592.

Vernochet, C., Heidmann, O., Dupressoir, A., Cornelis, G., Dessen, P., Catzeflis, F., Heidmann, T., 2011. A syncytin-like endogenous retrovirus envelope gene of the guinea pig specifically expressed in the placenta junctional zone and conserved in Caviomorpha. *Placenta* 32, 885–892.

Villarreal, L.P., 2006. How viruses shape the tree of life. *Future Virol.* 1, 587–595.

Villarreal, L.P., 2007. Virus-host symbiosis mediated by persistence. *Symbiosis* 44, 1–9.

Villarreal, L.P., 2008. *Origin of Group Identity*. Springer, New York.

Villarreal, L.P., 2009a. Persistence pays: How viruses promote host group survival. *Curr. Opin. Microbiol.* 12, 467–472.

Villarreal, L.P., 2009b. The source of self: Genetic parasites and the origin of adaptive immunity. *Ann. N. Y. Acad. Sci.* 1178, 194–232.

Villarreal, L.P., 2011. Viral ancestors of antiviral systems. *Viruses* 3, 1933–1958.

Villarreal, L.P., 2012a. The addiction module as a social force, in: Witzany, G. (Ed.), *Viruses: Essential Agents of Life*. Springer Science + Business Media, Dordrecht, the Netherlands.

Villarreal, L.P., 2012b. Viruses and host evolution: Virus-mediated self identity, in: Lopez-Larrea, C. (Ed.), *Self and Nonself*. Landes Bioscience, Austin, TX; Springer + Business Media, New York, pp. 185–217.

Villarreal, L.P., Villareal, L.P., 1997. On viruses, sex, and motherhood. *J. Virol.* 71, 859–865.

Villarreal, L.P., Witzany, G., 2013. Rethinking quasispecies theory: From fittest type to cooperative consortia. *World J. Biol. Chem.* 4, 71–82.

Villarreal, L.P., Witzany, G., 2013. The DNA habitat and its RNA inhabitants: At the dawn of RNA sociology. *Genomics Insights* 6, 1.

Vitullo, P., Sciamanna, I., Baiocchi, M., Sinibaldi-Vallebona, P., Spadafora, C., 2012. LINE-1 retrotransposon copies are amplified during murine early embryo development. *Mol. Reprod. Dev.* 79, 118–127.

von Sternberg, R., Shapiro, J.A., 2005. How repeated retroelements format genome function. *Cytogenet. Genome Res.* 110, 108–116.

Wacker, D.W., Engelmann, M., Tobin, V.A., Meddle, S.L., Ludwig, M., 2011. Vasopressin and social odor processing in the olfactory bulb and anterior olfactory nucleus. *Ann. N. Y. Acad. Sci.* 1220, 106–116.

Wacker, D.W., Tobin, V.A., Noack, J., Bishop, V.R., Duszkiewicz, A.J., Engelmann, M., Meddle, S.L., Ludwig, M., 2010. Expression of early growth response protein 1 in vasopressin neurones of the rat anterior olfactory nucleus following social odour exposure. *J. Physiol.* 588, 4705–4717.

Walum, H., Westberg, L., Henningsson, S., Neiderhiser, J.M., Reiss, D., Igl, W., Ganiban, J.M. et al., 2008. Genetic variation in the vasopressin receptor 1a gene (AVPR1A) associates with pair-bonding behavior in humans. *Proc. Natl. Acad. Sci. USA* 105, 14153–14156.

Wang, T., Zeng, J., Lowe, C.B., Sellers, R.G., Salama, S.R., Yang, M., Burgess, S.M., Brachmann, R.K., Haussler, D., 2007. Species-specific endogenous retroviruses shape the transcriptional network of the human tumor suppressor protein p53. *Proc. Natl. Acad. Sci. USA* 104, 18613–18618.

Wang, Z.X., Liu, Y., Young, L.J., Insel, T.R., 2000. Hypothalamic vasopressin gene expression increases in both males and females postpartum in a biparental rodent. *J. Neuroendocrinol.* 12, 111–120.

Watanabe, H., Fujiyama, A., Hattori, M., Taylor, T.D., Toyoda, A., Kuroki, Y., Noguchi, H. et al., 2004. DNA sequence and comparative analysis of chimpanzee chromosome 22. *Nature* 429, 382–388.

Wetterbom, A., Sevov, M., Cavelier, L., Bergstrom, T.F., 2006. Comparative genomic analysis of human and chimpanzee indicates a key role for indels in primate evolution. *J. Mol. Evol.* 63, 682–690.

Wilson, D.S., Sober, E., 1994. Reintroducing group selection to the human behavioral-sciences. *Behav. Brain Sci.* 17, 585–608.

Wynne-Edwards, K.E., 2001. Hormonal changes in mammalian fathers. *Horm. Behav.* 40, 139–145.

Yang, J., Bogerd, H., Le, S.Y., Cullen, B.R., 2000. The human endogenous retrovirus K Rev response element coincides with a predicted RNA folding region. *RNA* 6, 1551–1564.

Yang, N., Kazazian, H.H., 2006. L1 retrotransposition is suppressed by endogenously encoded small interfering RNAs in human cultured cells. *Nat. Struct. Mol. Biol.* 13, 763–771.

Young, L.J., 2002. The neurobiology of social recognition, approach, and avoidance. *Biol. Psychiatry* 51, 18–26.

Zeh, D.W., Zeh, J.A., Ishida, Y., 2009. Transposable elements and an epigenetic basis for punctuated equilibria. *Bioessays* 31, 715–726.

Zhang, J., Webb, D.M., 2003. Evolutionary deterioration of the vomeronasal pheromone transduction pathway in catarrhine primates. *Proc. Natl. Acad. Sci. USA* 100, 8337–8341.

Index

For Product Safety Concerns and Information please contact our EU representative GPSR@taylorandfrancis.com Taylor & Francis Verlag GmbH, Kaufingerstraße 24, 80331 München, Germany

T - #0263 - 160425 - C504 - 254/178/22 - PB - 9781466584617 - Gloss Lamination